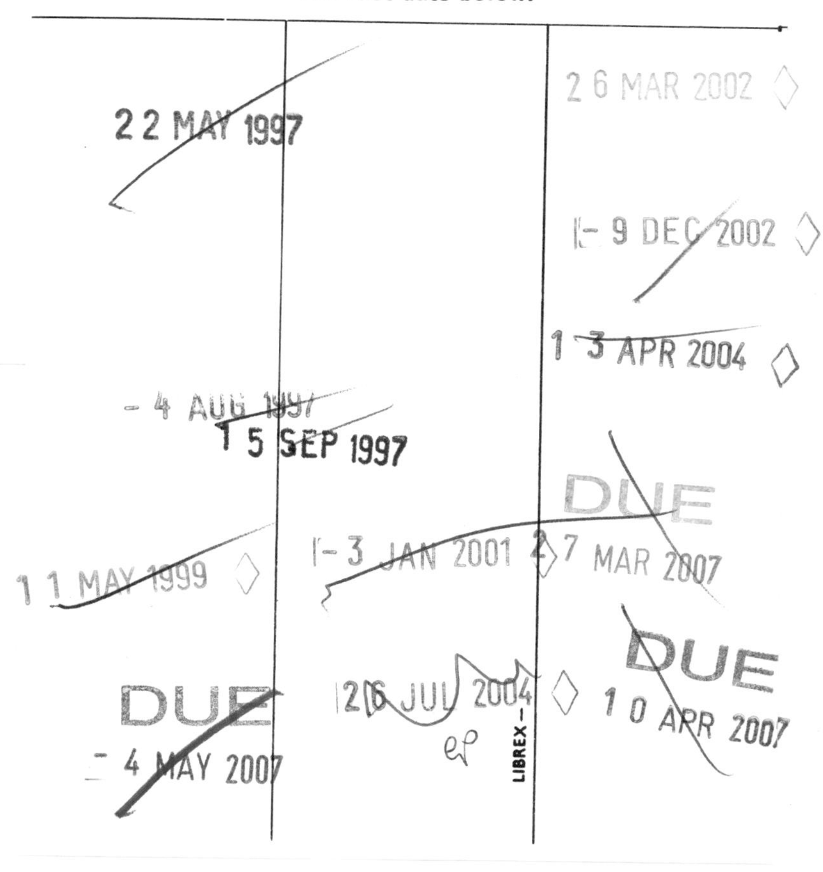

HANDBOOK OF NONPOINT POLLUTION

Sources and Management

HANDBOOK OF NONPOINT POLLUTION

Sources and Management

Vladimir Novotny, Ph.D., P.E.
Associate Professor of Civil Engineering
Marquette University, Milwaukee, Wisconsin

and

Gordon Chesters, Ph.D., D.Sc.
Professor of Soil Science and Director
of the Water Resources Center
University of Wisconsin, Madison, Wisconsin

Van Nostrand Reinhold Environmental Engineering Series

VNR **VAN NOSTRAND REINHOLD COMPANY**
NEW YORK CINCINNATI ATLANTA DALLAS SAN FRANCISCO
LONDON TORONTO MELBOURNE

Van Nostrand Reinhold Company Regional Offices:
New York Cincinnati Atlanta Dallas San Francisco

Van Nostrand Reinhold Company International Offices:
London Toronto Melbourne

Library of Congress Catalog Card Number: 81-1525
ISBN: 0-442-22563-6

Manufactured in the United States of America

Published by Van Nostrand Reinhold Company
135 West 50th Street, New York, N.Y. 10020

Published simultaneously in Canada by Van Nostrand Reinhold Ltd.

15 14 13 12 11 10 9 8 7 6 5 4 3 2 1

Library of Congress Cataloging in Publication Data

Novotny, Vladimir, 1938–
Handbook of nonpoint pollution.

(Van Nostrand Reinhold environmental engineering series)
Includes index.
1. Water–Pollution. I. Chesters, Gordon. II. Title. III. Series.
TD423.N69 363.7′394 81-1525
ISBN 0-442-22563-6 AACR2

Van Nostrand Reinhold Environmental Engineering Series

THE VAN NOSTRAND REINHOLD ENVIRONMENTAL ENGINEERING SERIES is dedicated to the presentation of current and vital information relative to the engineering aspects of controlling man's physical environment. Systems and subsystems available to exercise control of both the indoor and outdoor environment continue to become more sophisticated and to involve a number of engineering disciplines. The aim of the series is to provide books which, though often concerned with the life cycle—design, installation, and operation and maintenance—of a specific system or subsystem, are complementary when viewed in their relationship to the total environment.

The Van Nostrand Reinhold Environmental Engineering Series includes books concerned with the engineering of mechanical systems designed (1) to control the environmental within structures, including those in which manufacturing processes are carried out, and (2) to control the exterior environment through control of waste products expelled by inhabitants of structures and from manufacturing processes. The series includes books on heating, air conditioning and ventilation, control of air and water pollution, control of the acoustic environment, sanitary engineering and waste disposal, illumination, and piping systems for transporting media of all kinds.

Van Nostrand Reinhold Environmental Engineering Series

ADVANCED WASTEWATER TREATMENT, by Russell L. Culp and Gordon L. Culp

ARCHITECTURAL INTERIOR SYSTEMS—Lighting, Air Conditioning, Acoustics, John E. Flynn and Arthur W. Segil

SOLID WASTE MANAGEMENT, by D. Joseph Hagerty, Joseph L. Pavoni and John E. Heer, Jr.

THERMAL INSULATION, by John F. Malloy

AIR POLLUTION AND INDUSTRY, edited by Richard D. Ross

INDUSTRIAL WASTE DISPOSAL, edited by Richard D. Ross

MICROBIAL CONTAMINATION CONTROL FACILITIES, by Robert S. Rurkle and G. Briggs Phillips

SOUND, NOISE, AND VIBRATION CONTROL (Second Edition), by Lyle F. Yerges

NEW CONCEPTS IN WATER PURIFICATION, by Gordon L. Culp and Russell L. Culp

HANDBOOK OF SOLID WASTE DISPOSAL: MATERIALS AND ENERGY RECOVERY, by Joseph L. Pavoni, John E. Heer, Jr., and D. Joseph Hagerty

ENVIRONMENTAL ASSESSMENTS AND STATEMENTS, by John E. Heer, Jr. and D. Joseph Hagerty

ENVIRONMENTAL IMPACT ANALYSIS: A New Dimension in Decision Making by R. K. Jain, L. V. Urban and G. S. Stacey

CONTROL SYSTEMS FOR HEATING, VENTILATING, AND AIR CONDITIONING (Second Edition), by Roger W. Haines

WATER QUALITY MANAGEMENT PLANNING, edited by Joseph L. Pavoni

HANDBOOK OF ADVANCED WASTEWATER TREATMENT (Second Edition), by Russell L. Culp, George Mack Wesner and Gordon L. Culp

HANDBOOK OF NOISE ASSESSMENT, edited by Daryl N. May

NOISE CONTROL: HANDBOOK OF PRINCIPLES AND PRACTICES, edited by David M. Lipscomb and Arthur C. Taylor

AIR POLLUTION CONTROL TECHNOLOGY, by Robert M. Bethea

POWER PLANT SITING, by John V. Winter and David A. Conner

DISINFECTION OF WASTEWATER AND WATER FOR REUSE, by Geo. Clifford White

LAND USE PLANNING: Techniques of Implementation, by T. William Patterson

BIOLOGICAL PATHS TO SELF-RELIANCE, by Russell E. Anderson

HANDBOOK OF INDUSTRIAL WASTE DISPOSAL, by Richard A. Conway and Richard D. Ross

HANDBOOK OF ORGANIC WASTE CONVERSION, edited by Michael W. M. Bewick

LAND APPLICATIONS OF WASTE (Volume 1), by Raymond C. Loehr, William J. Jewell, Joseph D. Novak, William W. Clarkson and Gerald S. Friedman

LAND APPLICATIONS OF WASTE (Volume 2), by Raymond C. Loehr, William J. Jewell, Joseph D. Novak, William W. Clarkson and Gerald S. Friedman

STRUCTURAL DYNAMICS: Theory and Computation, by Mario Paz

HANDBOOK OF MUNICIPAL WASTE MANAGEMENT SYSTEMS: Planning and Practice, by Barbara J. Stevens

INDUSTRIAL POLLUTION CONTROL: Issues and Techniques, by Nancy J. Sell

WASTE RECYCLING AND POLLUTION CONTROL HANDBOOK, by A. V. Bridgwater and C. J. Mumford

WATER CLARIFICATION PROCESSES: Practical Design and Evaluation, by Herbert E. Hudson, Jr.

HANDBOOK OF NONPOINT POLLUTION: Sources and Management, by Vladimir Novotny and Gordon Chesters

ENVIRONMENTAL RISK ANALYSIS FOR CHEMICALS, edited by Richard A. Conway

NATURAL SYSTEMS FOR WATER POLLUTION CONTROL, by Ray Dinges

Preface

The title for this handbook conceivably could be "The Other Pollution." For almost a century engineers and scientists have been concerned with municipal and industrial pollution which was easy to locate and control. The other pollution, originating on farms and urban streets went unnoticed. The storm water runoff responsible for most nonpoint pollution has been considered sewage diluter and, therefore, does not require any treatment. Agriculturally related pollution—erosion, nutrient and pesticide losses to receiving waters—has been regarded as a necessary price for crop production.

The passage of the Water Pollution Control Act Amendments (PL 92-500) by the U.S. Congress in 1972 began an effort to identify and control the nonpoint sources of pollution. Shortly thereafter, the research on the magnitudes and effects of nonpoint pollution became prominent and since the late 1970's the nonpoint pollution problem has been an integral part of environmental science and engineering. Since then several established and new scientific journals have begun to publish articles on nonpoint pollution problems. By the end of the 1970's the number of scientific articles on nonpoint pollution surpassed more than 500 per year and no treatise or handbook could be written that could encompass even the most important articles that deal with the nonpoint pollution problem.

The authors have searched many articles and literature reviews and authored and coauthored several reviews on nonpoint pollution. The primary sources of information were found in many scientific journals among which the following should be acknowledged: *Journal of Environmental Quality*, *Journal of the Water Pollution Control Federation*, *Water Resources Bulletin*, *Water Resources Research*, *Water Research*, *ASCE Journals*, *Soil Science Society of American Journal*, *Soil Science*, *Environmental Health Perspectives*, *Environmental Science & Technology*, *Journal of Great Lakes Research*, *Atmospheric Environment*, *Journal of the Air Pollution Association*, *Science*, *Nature*, *Transactions of the American Society of Agricultural Engineers*, *Residue Reviews*, *Pesticide Monitor-*

ing Journal, and several others. A more complete list of sources is in the reference section for each chapter of this book.

A significant portion of information contained in this book was gathered by the authors during their participation on the research sponsored by the International Joint Commission (IJC) for the Great Lakes. Many U.S. and Canadian scientists shared their knowledge while trying to resolve the nonpoint pollution problem of the Great Lakes. The research activities under the auspices of PLUARG (Pollution from Land Use Activities-Reference Group) of the IJC have resulted in many pioneering works dealing with the magnitude and control of nonpoint pollution. Without this participation and international collaboration this treatise would not have been written. The U.S. Environmental Protection Agency provided most of the support for this research on the U.S. side of the Great Lakes.

This book is for all engineers, scientists, and governmental officials involved in pollution abatement programs. The nonpoint pollution problem is broad and the information contained in the handbook also will be of use to agricultural engineers, environmental scientisits, planners and developers, Soil Conservation Districts, and local government officials. Students—undergraduate or graduate—taking courses dealing with environmental pollution and its control also should find the book useful. The authors have made a substantial effort to present the materials in a "textbook" form, supported by numerous examples and without complicated mathematics.

The authors are grateful to their employing institutions—the University of Wisconsin and Marquette University—for providing support, sabbatical leave for one of the authors, and a general creative environment and understanding.

Finally, our wives and families must be thanked for their support and patience during the year long period of writing, rewriting, typing, and retyping. Sincere appreciation are extended to Lynn Novotny, Jennifer Diamantis, and Kari Sherman for typing, editing, and preparation of the manuscript. Dr. S. Walesh reviewed the outline and contributed valuable comments during the preparation and review of this handbook.

VLADIMIR NOVOTNY
GORDON CHESTERS

Contents

HANDBOOK OF NONPOINT POLLUTION

Sources and Management

1

Introduction

NONPOINT POLLUTION PROBLEM

Until recently, all pollution abatement efforts were aimed at controlling sewage and wastewater discharges into receiving waters. These discharges were very visible and if untreated caused obnoxious kinds of pollution—odorous anaerobic water, fish kills, and aesthetic impairment of the receiving waters during critical low-flow periods. Thus, since the end of the last century, man has tried to control municipal sewage and later industrial discharges first by employing simple suspended particle separation methods (screening and sedimentation) and later—practically in the middle of this century—by using more sophisticated secondary and tertiary treatment. With an increased level of treatment technology has come a marked increase in cost.

The cleaning up of municipal and major industrial sources has resulted in some cases in remarkable improvements in water quality. For example, the river Thames near London—which was severely polluted for more than a century and which became anaerobic almost every summer–is alive again, and fish have been caught there since the 1970s.

Nonpoint pollution–pollution from storm water and runoff–was not recognized generally until the late 1960s. In many areas of the world emphasizing industrial growth, for example the United States at the beginning of this century, most obvious signs of nonpoint pollution such as smoking industrial stacks, open mine pits, and construction are considered signs of progress rather than pollution. Engineering design practices under such conditions treat runoff and storm water as dilution of sewage that replaces sewage treatment. When the dilution of storm water and sewage in combined sewers reaches a certain "safe" ratio–ranging from 4:1 to 8:1–the mixture of storm water and sewage is allowed to overflow without any treatment into the receiving waters despite the fact that the overflow contains large amounts of untreated sewage and settled solids from sewers.

In nonurban areas, soil loss–one of the primary components of surface runoff pollution–has been recognized as harmful and often devastating to agricultural production since the dust storms of the 1920s and 1930s. Since that time, soil and water conservation programs have been implemented in the United States, resulting in significant reductions of soil loss by wind and water erosion. However, the water pollution impact of soil loss and agricultural activities in general was not realized. For example, a congressman from Indiana noted that in the late 1960s, groups representing agricultural interests testified with great feelings at hearings on the state's water quality standards that no such thing as agricultural pollution existed.[1]

Only recently has it been realized that the cleaning up of municipal and industrial discharges may not bring about dramatic improvements in water quality, especially in lakes, reservoirs, and estuaries, as well as many streams. Many water bodies receive significant pollution loadings from sources other than municipal and industrial discharges: generally, from sources related to use of the land by man and to natural processes occurring in the watershed. The term *nonpoint* (or *diffuse*) *sources* defines these sources of pollution. Lake Tahoe, California-Nevada (Fig. 1-1) is an example of a water body whose water quality and eventual degradation now is caused primarily by nonpoint pollution, since the major point sources have been eliminated.

Nonpoint sources of pollution account for more than 50% of the total water quality problem, and they are being recognized and investigated nationally and internationally. In many areas, nonpoint pollution, such as runoff from crop land, urban storm water, strip mining, and runoff from construction sites are becoming major water quality problems.[2]

The problem of nonpoint pollution does not only involve the "traditional" pollution parameters such as suspended sediment, biochemical oxygen demand (BOD), dissolved oxygen (DO), and nutrients (nitrogen and phosphorus). As a

Fig. 1-1. The rate of eutrophication and degradation of Lake Tahoe, California-Nevada, primarily depends on nonpoint pollution inputs from surrounding forests, urban and recreational centers.
(Photo authors)

matter of fact, some of the most serious nonpoint pollution problems do not have a parallel in the traditional point source oriented environmental pollution control area. These problems include PCB (polychlorinated biphenyl) contamination of Great Lakes fish, acid rain, and pesticide contamination of surface water bodies and aquatic biota.

It appears that engineers, scientists, and water quality officials of many countries are concerned with diffuse pollution. The problem is addressed in the Great Lakes basin under the U.S.–Canada Great Lakes Water Quality Agreement, and a monumental research effort has been undertaken on both sides of the Great Lakes to identify lands and land-use activities causing pollution of the lakes and to propose remedial measures.[3]

In other countries—notably Switzerland, the German Federal Republic, and the Scandinavian countries—extensive attention has been given to nonpoint pollution characterization and control. Switzerland, in particular, has launched a nonpoint abatement program that includes modification of roof drains, storage and chemical treatment of storm water, overland flow treatment, and several other techniques. Facilities of the Technical University of Denmark in Copenhagen incorporate nonpoint pollution abatement. All parking lots and open areas are designed to maximize infiltration by use of pervious pavement, lattice pavement, and vegetation. Several unique remedial schemes for storm-water control, including the partitioning of a lake for storm-water storage and subsequent treatment, have been investigated in Sweden.[4]

The U.S. Environmental Protection Agency (EPA) has sponsored numerous research projects that in past years have contributed to vastly increased knowledge concerning the extent of the nonpoint pollution problem and its control. Many watersheds have been studied and monitored for several years, and the amount of obtained data is such that it will continue to be analyzed for years to come. The major watershed studies on the U.S. side of the Great Lakes include the Black Creek Watershed in the Maumee River basin (a tributary to Lake Erie); the Menomonee River Watershed—an urbanizing watershed in Milwaukee, Wisconsin; the Genesee River Watershed in New York State; and other small basins. In addition, the EPA sponsors several large projects dealing with the urban stormwater and combined sewer overflow problems.

DEFINITIONS

Water Quality and Pollution

Any treatise dealing [illegible] Water quality in the minds [illegible] lution and, similarly, water [illegible] npoint sources—is equated w[illegible]

Water quality reflec[illegible]cesses and by man's cultur[illegible]es and related to intended w[illegible]ferent people, for example [illegible]n and propagation of fish [illegible]th the health and safety of [illegible] man is concerned with the [illegible]

The term *polluti*[illegible]ns "to soil" or "to defile." [illegible]*grada-tion* often are used s[illegible]waters. Various definitions [illegible]related terms.[5] Pollution h[illegible] physical, chemical, or bio[illegible]t may or will hostilely affect human life or that of other desirable species, or industrial process, living conditions, and cultural assets, or that may or will waste or deteriorate our natural resources".[2]

Pollution also can be defined as "the addition of something to water which changes its natural quality so that downstream riparian owners do not obtain the natural water of the stream transported to them."[6]

According to the State of California definitions,[7] pollution means adverse and unreasonable impairment of the beneficial use of water even though no actual health hazard is involved; contamination causes actual hazards to public health, and agencies are empowered to take immediate action; and nuisances involve damage caused by odors or unsightliness.

Most of the pollution sources can be controlled and/or managed. The term *pollution control* refers to regulation of pollutants from individual outfalls or areal (nonpoint) sources including urban developing lands, feedlots, agricultural areas, and other pollutions generating land uses. The purpose of all pollution controls should be: (1) to protect the capacity of surface waters to assimilate pollution without damage or impairment of their use; (2) to protect shellfish and wildlife; (3) to preserve or restore the aesthetic and recreational values of surface waters; and (4) to protect humans from adverse water quality conditions.[7]

Water quality management deals with all aspects of water quality problems for all beneficial uses of water or lands from which pollution originates, whereas pollution control is mostly understood as the safe disposal of wastewaters and their treatment.

Water quality and pollution are determined by comparing measured physical, chemical, biological, microbiological, and radiological quantities and parameters to a set of standards and criteria. The difference between *standards* and *criteria* should be explained.

A criterion is basically a scientific quantity upon which a judgment can be based. It is developed usually from scientific experiments. A water quality criterion can be based on morbidity or chronic toxicity of various substances to man or aquatic life, or it can be related to technical methods of removing the substances from water. A standard applies to any definite rule, principle, or measure established by an authority.

A water quality objective is a goal—usually established by an authority—to which water pollution control measures should be aimed and financial resources devoted. For planning purposes, the objectives must be translated into a set of standards to be maintained after the period of implementation. The objectives may express various levels of possible water uses and corresponding water quality.

The water quality standards used presently by water pollution engineers and scientists as well as pollution abatement authorities throughout the world are either stream standards or effluent standards. The effluent standards, which determine how much pollution can be discharged from municipal and industrial wastewater sources, are of lesser importance in nonpoint source pollution control and management. Performance standards, an equivalent of effluent standards to control pollution from lands, have been used by some local authorities to control pollution from subdivisions, construction areas, or mining. The stream standards can be related to the protection of aquatic habitat and biota and/or to intended downstream use of water.

Pollution Source

Sources of pollution can be divided basically into two groups: natural and cultural (those caused by man). The sources can be further classified as either point or diffuse (nonpoint) sources of pollution.

Point sources enter the pollution transport routes at discrete, identifiable locations and usually can be measured directly or otherwise quantified, and their impact can be evaluated directly. Major point sources include effluent from industrial and sewage treatment plants, and effluent from farm buildings or solid-waste disposal sites.

Pollution from diffuse sources can be related to weathering of minerals, erosion of virgin lands and forest including residues of natural vegetation, or artificial or semiartificial sources. The last can be related directly to human activities such as fertilizer application or use of agricultural chemicals controlling weeds or insects, erosion of soil materials from agricultural farming areas and animal feedlots, construction sites, transportation, cumulation of dust and litter on impervious urban surfaces, strip mining, and others.

Since pollution has been defined as the addition of substances to waters above their natural quality, pollution from nonpoint sources is caused mostly by man and his activities and should be distinguished from water quality contributions, sometimes called "background pollution," caused by the contact of water with rocks, undisturbed soils and geological formations, natural erosion and elutriation of chemical and biochemical components from forest litter, migration of salt water into estuaries, or other natural sources.

Besides the sewage and industrial wastewater discharges and erosion from agricultural lands or urban areas, many other activities can cause pollution. The cutting down of a forest results in a loss of protective cover and exposes soils, thus causing increased surface runoff and erosion; stream channelization and removal of trees and protective grass banks may have significant ecological effects, sometimes complete destruction of the original biological life in the stream.

There are several general characteristics that describe nonpoint source pollution:[8]

- Nonpoint source discharges enter surface waters in a diffuse manner and at intermittent intervals that are related mostly to the occurrence of meteorological events.
- Pollution arises over an extensive area of land and is in transit overland before it reaches surface waters.
- Nonpoint sources generally cannot be monitored at their point of origin, and their exact source is difficult or impossible to trace.

- Elimination or control of pollutants must be directed at specific sites.
- In general, the most effective and economical controls are land management techniques and conservation practices in rural zones and architectural control in urban zones.
- Compliance monitoring for nonpoint sources is carried out on land rather than in water.
- Nonpoint source pollutants cannot be measured in terms of effluent limitations.
- The extent of nonpoint pollution is related, at least in part, to certain uncontrollable climatic events, as well as geographic and geologic conditions, and may differ greatly from place to place and year to year.
- Nonpoint sources are derived from consecutive operations on extensive units of land, as opposed to industrial activities that typically use repetitive operations on intensive (small) units of land.

Rural (Nonurban) Nonpoint Sources. These sources are mostly related to agricultural activities. Agricultural pollutants have their origin in fertilizer use and pesticide applications, and generally, the primary causes are agricultural methods of disturbing soils by tillage (agricultural lands) or logging (silvicultural lands). Several other factors also affect pollution loading: soil type, climate, management practices, and topography. Land uses that produce the most pollution per unit area are animal feedlot operations and farming on steep slopes. Forested lands and pastures, on the other hand, produce the least amount of pollution, that is, approaching background levels. The impact of pollution on receiving waters depends on the distance of the source from the nearest concentrated flow—stream—and on the processes taking place during the overland flow phase of the pollutants' transport.

Urban Nonpoint Sources. Urbanization and related hydrologic modifications may cause increased pollution loadings that are significantly above the original or background levels. The source of urban nonpoint pollutants varies considerably, ranging from urban bird and pet populations, street litter accumulation, tire wear of vehicles, abrasion of road surfaces by traffic, street salting practices, and construction activities. Urban nonpoint pollution may contain many dangerous contaminants such as lead, zinc, asbestos, PCBs (polychlorinated biphenyls), oil, and grease.

Three basic urban land uses exist: residential, commercial, and industrial. Such categorization, however, is loose and can hardly be correlated with pollution generation. For example, residential areas can range from low-density, relatively "clean" residential suburban zones with 1 or 2 houses/ha, to high-density, congested urban centers with several hundred people residing on an area of 1 ha. Similarly, industrial zones can include "clean" light manufacturing as well as "dirty" heavy industries such as foundries, smelting operations, steel mills, and mining.

Least pollution originates usually from low-density residential zones and park and recreation areas, while highest pollution loadings—as one would expect—can be attributed to high-density downtown and industrial centers and, above all, construction sites.

Combined sewer overflows, although not often considered nonpoint sources of pollution, represent a special problem of urban storm-water management and control. These sewer systems, in addition to the pollution from urban lands carried by storm-water discharges into receiving waters, often contain untreated sewage, deposited sewage solids in sewers, and decaying contents of catch basins.

An interesting feature of nonpoint pollution is that—with few exceptions—the bulk of pollution is carried by surface runoff. Thus, the areas where surface runoff originates—called hydrologically active areas—also are sources of nonpoint pollution and should be subject to control and management. Hydrologically active areas include impervious urban and roadway surfaces, areas with a high groundwater table and/or tight soils, and frozen soils during spring rains. It has been found that during even large storms only a small part of a watershed contributes surface runoff and can be counted as a source of nonpoint pollution. The term hazardous land uses or areas has been used to describe hydrologically active areas that contribute the highest amount of pollution.

MAGNITUDE OF NONPOINT POLLUTION

Overview

The overall magnitudes of nonpoint pollution are given in Table 1-1. Associate problems can be summarized as:[2,9,10,11]

1. Over 4 billion tons of sediment are delivered annually into streams and rivers of the conterminous United States. Almost half of that amount originates from approximately 170 million ha of agricultural lands.
2. Strip mining, which affects approximately 150,000 ha of lands annually, results in the discharge of millions of tons of sediments and high amounts of acidity into receiving waters.
3. Nonpoint sources contribute roughly 80% of total nitrogen load and more than 50% phosphorus load into receiving waters.
4. Agricultural and urban runoff may contain large quantities of toxic metals, pesticides, and other organic chemicals.
5. Large amounts of decomposable organics originate as a part of soil loss from nonpoint sources that may form objectionable mud deposits in surface waters.

TABLE 1-1. Estimated Magnitudes of Pollutant Contributions to Surface Waters from Selected Nonpoint Sources in the Conterminous United States.[a]

Nonpoint Source Category	Sediment	BOD_5	Nitrogen	Phosphorus
	(million tonnes/yr)			
Cropland	1700	8.2	3.9	1.42
Pasture and range	1190	4.5	2.3	0.98
Forest	232	0.73	0.35	0.08
Construction	179			
Mining	54			
Urban runoff	18	0.45	0.13	0.017
Rural roadways	2	0.004	0.0005	0.001
Small feedlots	2	0.05	0.15	0.032
Land fills		0.27	0.024	
Subtotal	3377	14.2	6.9	2.5
Natural (background) loadings	1150	4.6	2.3	1.0
Total	4527	18.8	9.2	3.5

[a]From "Best Management Practices for Agriculture and Silviculture," Raymond C. Loehr, Douglas A. Haith, Michael F. Walter, and Colleen Martin; Ann Arbor Science Publishers, Inc., Ann Arbor, Michigan, 1979.

6. For fecal and total coliform counts, nonpoint sources account for over 98% of the total loadings.

Some of the quality contributions from nonpoint sources are unavoidable—natural erosion and nutrient loss from lands has shaped the earth's surface for billions of years (Fig. 1-2). Even some of man's activities resulting in nonpoint pollution, such as crop production, are necessary to sustain the economy and provide food. However, accelerated urbanization (about 2000 ha of U.S. rural land is converted daily to urban uses) has accelerated and increased nonpoint pollution loadings in the past 30 yr. Also, most profound water quality changes caused by nonpoint pollution have taken place during the same period. Next to agriculture, urbanization and related disturbing activities are now considered the primary contributors of pollution to surface waters.

In the Great Lakes region,[12,13] phosphorus contribution from nonpoint sources represents roughly 50% of the total loadings to Lakes Superior, Huron, and Erie, and exceeds municipal and industrial point sources (Table 1-2). The balance is due to atmospheric fallout and shoreline erosion. Phosphorus is the limiting nutrient controlling eutrophication of most of the Great Lakes. Over

Fig. 1-2. The beautiful landscape of the Grand Canyon National Park has for millions of years been formed by natural erosion of the Colorado River and its tributaries. (Photo R. M. Butterfield, NPS Grand Canyon, AZ.)

90% of lead loading to Lakes Superior, Michigan, Huron, and Ontario originate also from nonpoint sources. In addition, the International Joint Commission research[13] has found that diffuse sources including atmospheric inputs account for major loading of PCBs and organic chemicals to the lakes.

The magnitude of loadings from various areas varies considerably depending on the hydrological activity of the area. The loadings of sediment may vary from

TABLE 1-2. Loadings to the Great Lakes from the United States During 1975.[12]

Pollutant	Total from Diffuse Sources (%)	Diffuse Unit Area Load (kg/ha-yr)
Total phosphorus	71	0.40
Soluble orthophosphate	60	0.10
Suspended solids	98	310.00
Total nitrogen	85	6.4
Chlorides	66	74.00

zero or a few kilograms per hectare per year from forest lands and pastures on permeable soils or some low-density suburban residential areas and parks, to several hundred tons per hectare per year from construction sites, mining, or congested nonmaintained urban centers. Staggering amounts of nutrients and organic chemicals are lost to surface waters from animal feedlot operations. More on unit loadings from various land uses can be found in Chapter 10. The loadings exhibit seasonal variations; for example, the highest loadings from agricultural lands can be expected in spring after plowing, and the lowest, in fall and winter. Loadings are subject to meteorological factors; high-intensity storms generate the highest pollution.

Comparison of Point and Nonpoint Pollution

Pisano[15] reported that in cities where secondary treatment is provided for sewage (point source), storm-generated runoff accounts for 40 to 80% of the annual load of BOD. During intense storms, 94 to 95% of the BOD load is contributed directly to runoff.

Table 1-3 presents a comparison of concentrations typically found in sewage and in runoff from some nonpoint sources. It can be seen that urban stormwater runoff results in pollution levels of the same order of magnitude as treated or even untreated sewage. Suspended solids concentrations, however, far exceed those typical for sewage. This was confirmed by Soderlund and Lehtinen[18] for conditions in Sweden (Table 1-4).

Using a hypothetical example, Pitt and Field[19] showed that in an urban area of 100,000 people, COD contribution from urban runoff is approximately 50% of that from raw sewage. These results, together with comparisons for some other constituents, are shown in Table 1-5. It also was shown that if sanitary sewage is receiving adequate treatment (90 to 95% removal of solids and organic compounds), almost all—approximately 97%—of the total solids and about 70% of the BOD reaching the receiving waters arises from urban runoff.

Pollution loadings from so-called hazardous lands, such as animal feedlots, construction sites, and mining, may exceed sewage concentrations of pollutants or their loading per unit area by several orders of magnitude.

Pollution by Toxic Organic Chemicals

Some hazardous pollutants are almost uniquely generated from nonpoint sources and their contribution from point sources is negligible, for example, pesticides and PCBs. Pesticide contamination of surface waters originates from the use of agricultural chemicals to control weeds and pests, whereas PCB contamination can be traced to PCB-containing products primarily in use in urban areas.

The example of PCBs shows how a chemical component, which represents

TABLE 1-3. Comparison of Typical Magnitudes of Concentrations from Nonpoint Sources and Sewage.

Source	Suspended Solids (mg/liter)	BOD_5 (mg/liter)	COD (mg/liter)	Total N (mg/liter)	Total P (mg/liter)	Lead (mg/liter)	Total Coliform (no./100 ml)
Precipitation[a]	11–13	12–13	9–16	1.2–1.3	0.02–0.04		
Background levels[b]	5–1000	0.5–3		0.05–0.5	0.01–0.2	0.1	
Agricultural cropland[a]		7	80	9	0.02–1.7		
Animal feedlots[a]	30	1000–11,000	31,000–41,000	920–2100	290–380		
Urban storm water[b]	100–10,000 (630)[c]	10–250 (30)	20–600	3–10	0.6	0.35	10^3–10^8
Combined sewers[b]	100–2000 (410)	20–600 (115)	20–1000	9–10	1.9	0.37	10^5–10^8
Municipal sewage,[b] untreated	100–330 (200)	100–300 (200)	250–750	40	10		10^7–10^9
Municipal sewage, treated	10–30	15–30	25–80	30	5		10^2–10^4

[a]From data by Loehr.[9,14]
[b]From data by Lager et al.[16,17]
[c]( ) flow weighted averages.

TABLE 1-4. Comparison of Treated Sewage and Storm-Water Loadings During the Wet Period (600 Hr/Yr) from Swedish Residential Areas (kg/ha).

	Storm-Water-Treated Sewage (35 persons/ha)		Storm-Water-Treated Sewage (100 persons/ha)	
Suspended solids	173	7.0	275	19
BOD_5	14	7.0	43	19
Total N	1.4	9.0	3.5	25
Total P	0.04	0.4	0.2	1.0
Total coliforms[b]	10^{11}	10^{12}	10^{12}	10^{12}
Fecal coliforms[b]	10^{11}	10^{12}	10^{11}	10^{12}

[a]After Soderlund and Lehtinen.[18]
[b]No./ha.

only a minor point source problem, could contaminate the environment and become a most serious nonpoint pollution hazard.

Sources and Properties of PBCs. PCBs are in the class of chlorinated organic compounds. These substances have been manufactured in the United States since 1929 by the Monsanto Chemical Company. Other producing nations include Germany, France, Japan, Italy, the USSR, and Czechoslovakia.

PCBs are prepared individually by the reaction of biphenyl with anhydrous chlorine. PCBs are available as liquids, resins, or solids. The most important physcial properties of PCBs are low vapor pressure (high boiling point), low water

TABLE 1-5. Comparison of Areal Loadings of Pollutants from a Hypothetical American City of 100,000 People (Tonnes/Yr).[a]

Pollutant	Storm Water	Raw Sewage	Treated Sewage
Total solids	17,000	5,200	520
COD	2,400	4,800	480
BOD	1,200	4,400	440
Total P	50	200	10
Total Kjeldahl N	50	800	80
Lead	31		
Zinc	6		

[a]Reprinted from *Journal American Water Works Association*, Volume 69, Number 8 (August 1977), by permission. Copyright 1977, the American Water Works Association.

TABLE 1-6. Physical Properties of Some PCBs and DDT.[20]

	Aroclor				
	1242	1248	1254	1260	DDT
Molecular weight					
Range	154–358	222–358	290–392	324–460	352
Average	262	288	324	370	352
Chlorine (%)	42	48	54	60	50
Solubility in H_2O (μg/liter)	200	100	50	25	0.7
Vapor pressure at 20°C mm Hg	10^{-4}	3×10^{-5}	3.6×10^{-6}		1.5×10^{-7}

solubility, and high dielectric constant. They can be dissolved in many organic solvents, including some humic acids. The PCBs manufactured in the United States are sold under the trade name Aroclor. Table 1-6 shows some physical properties of PCBs as compared to another environmentally serious contaminant—DDT.

Since 1929 over 600,000 Tonnes of PCBs have been produced in the United States, and the annual production reached a maximum of 38,000 Tonnes in 1970. Figure 1-3 shows the approximate movement of PCBs and the amounts of PCBs that are said to contaminate the environment.

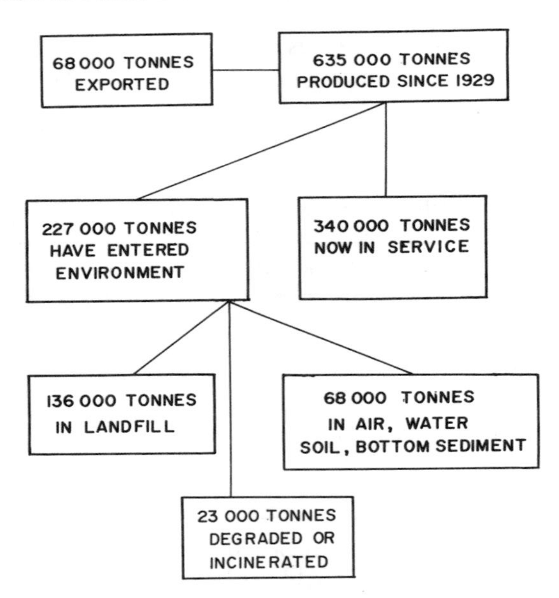

Fig. 1-3. Distribution of PCBs in the products and in the environment.[21]

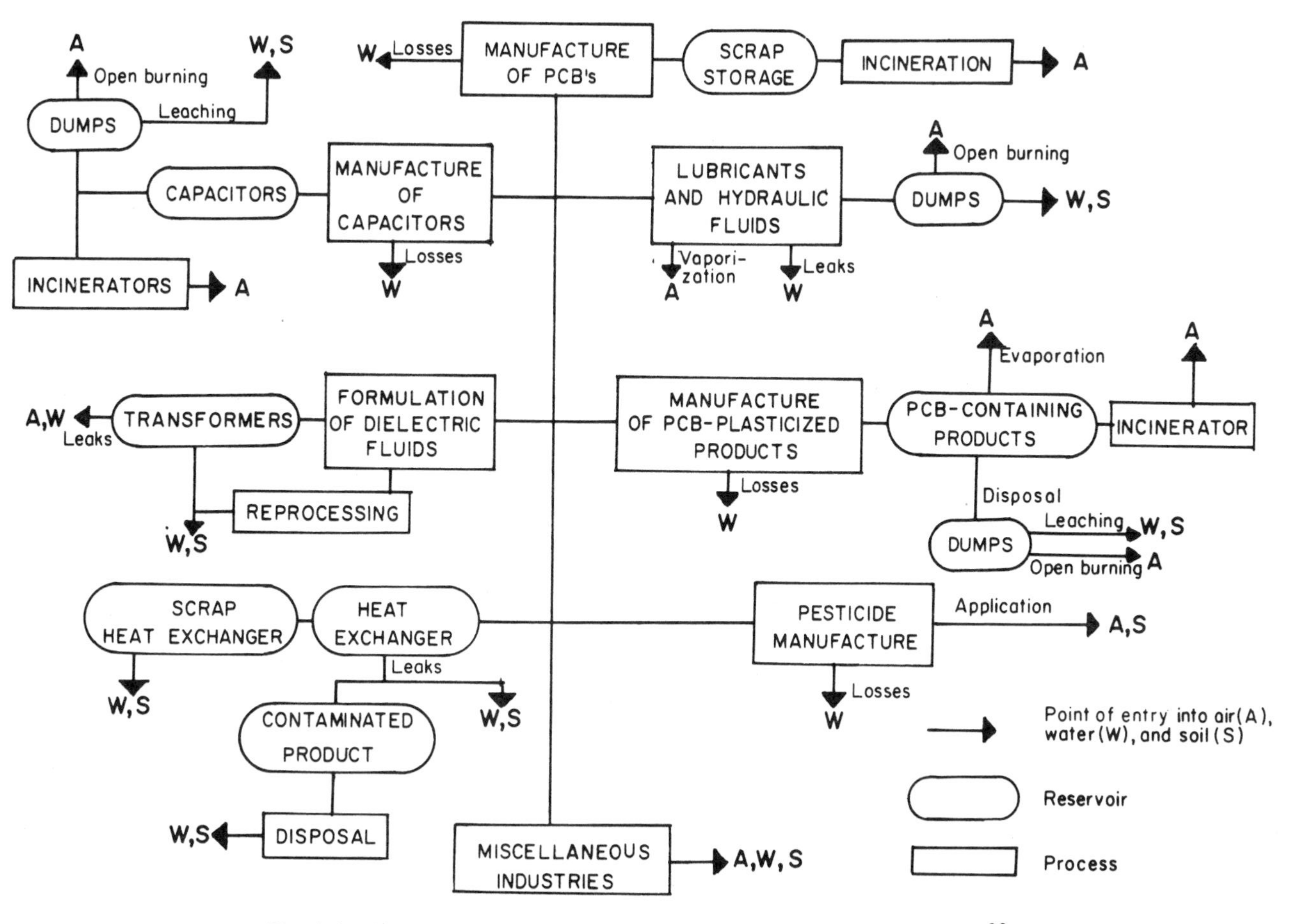

Fig. 1-4. Pathways of PCBs and points of entry into the environment.[20]

Sources of PCBs. From an environmental viewpoint, the commercial use of PCBs can be divided into three categories:

1. Controllable closed systems–PCBs used as dielectrics in transformers and large capacitors have a life equal to that of the equipment, and with proper design environmental contamination should not occur.
2. Uncontrollable closed systems–PCBs are used in heat transfer and hydraulic systems that permit leakage.
3. Dissipative use–PCBs in paint, paper, lubricants, plasticizers, etc. are in direct contact with the environment.

The major source of PCBs that have contaminated the environment are (Fig. 1-4):

1. Leaks from sealed transformers and heat exchangers;
2. leaks of PCB-containing fluids from hydraulic systems that are only partially sealed;
3. spills and losses in the manufacturing of either PCBs or PCB-containing fluids;
4. vaporization or leaching from PCB-containing formulations;
5. disposal of waste PCBs or PCB-containing materials.

The chemical property of extreme stability that made PCBs an ideal industrial and commercial compound also made them persistent and cumulative toxic components in the environment.

Because PCBs are soluble in lipid tissue, these components have been found to accumulate and concentrate in the fat of living organisms including man. Salmonoid fish are very succeptible to PCB accumulation because these species have high body fat.[21] Because of the toxic nature of PCBs, the U.S. Food and Drug Administration established a tolerance level of 2 mg/kg for PCBs in fish tissue. Fish testing has shown that most large Lake Michigan fish species (trout and salmon) contain over 5 mg/kg of PCBs.[21]

PCBs–in a fashion parallel to that for DDT in the 1950s and 1960s–have entered into global transport routes and can be detected in extremely remote areas. Their removal rate from the environmental ecosystems is very slow.

NONPOINT POLLUTION CONTROL LAWS

Until the U.S. Congress passed the Federal Water Pollution Control Act (PL 92-500) in 1972, there was no major legislation dealing with nonpoint pollution per se. As stated previously, the general population was not even aware of the problem until the late 1960s. However, legislation existed prior to 1972 which dealt with nonpoint pollution problems.

Institutionally, nonpoint pollution and related problems were a part of water resource protection and development and land management policy areas. In the water resources area, the federal government established its control and influence in areas of navigation, river development, dams and flood control, irrigation, and others. These programs and legislation began in the late 19th century. During the past decades, various federal agencies directly responsible for water resource development and pollution control were created—the U.S. Army Corps of Engineers, the Federal Power Commission (a part of the present Department of Energy), and the U.S. Environmental Protection Agency, which succeeded the Federal Water Pollution Administration in the Department of Interior.

While the federal government maintained a strong influence on water policy development and control, its effect on land management was minimal.[22] Policies of the federal government have been restricted to public lands under its proprietary control and to incentives for state and local governments, which developed land management ordinances, including zoning.

Local governments remain closest to the landowners, and despite federal and state influences, land use retains its local emphasis. This is evidenced by successful federal programs, namely soil conservation, where since the 1930s federal policy objectives have been carried out at the local level through state and local institutions—the Soil and Water Conservation Districts. Federal staff involved at this level provides technical assistance to local institutions, rather than administrative functions.[22]

Floodplain Management Laws. As indicated previously, nonpoint pollution is a closely related part of storm-water events and, consequently, floods. Minimizing and reducing floods is almost synonymous with reducing nonpoint pollution. Thus, the discussion of nonpoint pollution control laws should begin with an introduction to a series of "Flood Control Acts," the first of which was enacted in 1936 and amended in 1938, 1944, 1960, and 1970. This large body of laws enables the U.S. Army Corps of Engineers also to undertake projects and activities other than flood control, such as, drainage of wetlands.[23] Of particular interest is the Flood Control Act of 1960, Section 266 (PL 86-645), which authorizes the Corps of Engineers to provide state and local governments with information needed to regulate the use of floodplain lands. Most of the Corps of Engineers tasks under these acts are limited to large river basins and rivers. The Flood Control Act of 1936 also authorizes the U.S. Department of Agriculture to propose land treatment plans for reducing floods in major river basins.

U.S. Department of Agriculture Programs. The U.S. Department of Agriculture programs in the nonpoint pollution control area are carried out mainly by the U.S. Soil Conservation Service (USDA-SCS). The SCS was established in 1935 by the 74th Congress (P.L. 46). Its primary purpose is to carry out a national program to conserve and develop soil and water resources. More specifically, the SCS:

> "Develops and carries out a national soil and water conservation program through conservation districts (P.L. 46, 74th Congress, 1935).
>
> "Helps develop and carry out watershed protection and flood prevention projects in cooperation with other agencies (Flood Control Act, P.L. 534, 78th Congress, 1944).
>
> "Helps develop and carry out watershed protection and flood prevention projects and river basin investigations in cooperation with other agencies (P.L. 566, 83rd Congress, 1954)."

Through Public Law 46 programs, the SCS provides technical assistance to farmers and encourages them to adopt soil conservation practices such as contour cultivation, terracing, crop rotation, conversion of steep lands to pasture or woodland, and installation of gully control structures. This involves working with individual farmers on plans to implement erosion control measures. The SCS helps local organizations (usually Soil Conservation Districts) to install reservoirs, bank and channel protection measures, and grade stabilization structures through Public Law 566 programs.

The Rural Development Act of 1972 (PL 92-419) authorizes the Secretary of Agriculture to assist farmers and communities in rural areas to install pollution control measures. This should make erosion control as a multiple-purpose measure for reducing the rate at which nutrients are carried to receiving streams possible.[23]

Through the Rural Environmental Assistance Program, another agency of the Department of Agriculture–the Agricultural Stabilization and Conservation Service (ASCS)–assists farmers in installing soil conservation measures on a cost-sharing basis. The ASCS shifts a considerable amount of land into soil conserving uses and thus helps alleviate erosion, sedimentation, and water pollution.

The National Flood Insurance Program (PL 90-448) and Flood Disaster Protection Act of 1973 puts high insurance rates on properties located in floodplains, thus discouraging developments on floodplains. Indirectly, this reduces the nonpoint pollution impact from developing floodplains.

Environmental Pollution Laws. The National Environmental Policy Act–NEPA (PL 91-190)–was designed to protect natural and environmental resources. It requires elaboration of environmental impact statements wherever major federal actions significantly affecting the human environment are undertaken and requires those responsible for the proposed action to prove that the project is environmentally sound. The NEPA requires the utilization of expert advice from other agencies, accommodation of public views, and consideration of alternatives. It is designed (1) to make agencies more sensitive to environmental values in the early stages of the planning process, and (2) to develop useful information on all relevant factors for the ultimate decision-makers.

The Federal Water Pollution Control Act (PL 92-500) with its 1977 amendments (PL 95-217), known as "The Clean Water Act," is the first water pollution control law that recognizes that cleaning up the point sources will not solve the nation's water quality problems. The overall goals of the Clean Water Act are:

1. Eliminate discharges of pollutants into navigable water by 1985.
2. Make the nation's waters suitable for recreation and fish and wildlife propagation.
3. Eliminate discharges of toxic pollutants in toxic amounts.
4. Provide financial assistance to construct publicly owned treatment works.
5. Develop an area-wide waste treatment management planning process to assure adequate control of sources of pollutants in each state.
6. Provide incentives to major research and demonstration efforts to develop the technology necessary for eliminating discharge of pollutants into navigable waters, waters of the contiguous zone, and the oceans.

Specifically, Section 208 of the act calls for area-wide water pollution planning in areas designated by the governor of each state that would include both point and nonpoint sources and pollution abatement programs. The plans should include: (a) identification of the treatment works necessary to meet the anticipated municipal and industrial waste treatment needs of the area with associated construction priorities, time schedules, and the establishment of regulatory programs for such treatment works, including urban runoff and storm water; (b) identification of the sources of nonpoint pollution—agriculture (including runoff from irrigated fields), silviculture, runoff from land used for livestock and crop production or land that has had manure applied to it, mining, saltwater intrusion, waste disposal on lands, disposition of all residual waste generated in the designated area, and land and subsurface excavations; (c) setting forth of a procedure and methods (including land-use requirements) that feasibly will control such sources.

The area-wide water quality management planning process in each state is a continuous activity that should be updated periodically and reviewed annually by the governor of each state. The governor of each state, in consultation with the designated planning agency, shall also make one or more waste treatment management agencies responsible for carrying out the plan dealing with the sewage and wastewater disposal, including all of the urban nonpoint sources. The plan should include analysis over a 20-yr period of treatment works and alternative solutions, wastewater collection, and urban storm runoff system control and abatement.

Local governments (municipalities and counties) are the main regulators of nonpoint pollution and its sources resulting from urbanization processes, that is, suburban erosion, septic tanks, and storm-water runoff. Regulatory techniques include zoning, subdivision ordinances and performance standards, grading and fill requirements, solid-waste and litter disposal ordinances, and taxation. Local governments rarely regulate rural nonpoint sources (agriculture, silviculture, and mining). Some states even forbid the application of local zoning ordinances to agricultural lands.[24]

According to the area-wide Water Quality Management Plan, the Secretary of Agriculture acting through the SCS, is authorized to establish and administer a program incorporating so-called Best Management Practices to control nonpoint pollution from rural lands.

The Best Management Practices have been defined as:

Best Management Practice (BMP) means a practice or combination of practices that is determined by a state (or designated area-wide planning agency) after problem assessment, examination of alternative practices, and appropriate public participation to be the most effective practicable (including technological, economic, and institutional considerations) means of preventing or reducing the amount of pollution generated by nonpoint sources to a level compatible with water quality goals.

Rather than creating an enforcement agency for rural nonpoint sources, the SCS can enter into a contract with rural landowners and provide them with technical assistance and cost sharings for the conservation practices that will improve water quality.

State and federal environmental agencies, through the National Pollutant Discharge Elimination System (NPDES) permitting process, carry out most of the point source regulation.

The continuous planning process must ensure that the wastewater discharges, including those from present or future nonpoint sources, do not exceed the natural waste assimilation capacity of the receiving water bodies, as related to recreation and fish and wildlife protection and propagation standards. Once area-wide plans are approved, the U.S. Environmental Protection Agency will provide funds and construction grants only to those works and management practices included in the plan.

Mining pollution, including erosion and acid drainage, under PL 92-500 and PL 95-217, is controlled by the NPDES regulations and by Sec. 208. In addition, the Surface Mining Control and Reclamation Act (PL 95-87, August 1977) regulates surface mining operations and the acquisition and reclamation of abandoned mines. The primary purpose of this act is to establish a nationwide protection program from the adverse effects of surface coal mining operations, including erosion and acid mine drainage.

Additional nonpoint pollution control in estuarine and coastal areas is provided by the Coastal Zone Management Act of 1972 (PL 92-583). Under this act, states are to develop and administer management programs for their coastal zones subject to approval for the National Oceanic and Atmospheric Administration.

Laws Controlling Toxic Substances. Several laws control toxic substances. The majority of these are aimed primarily at consumer protection. The primary legislative document that controls toxic substances in the environment is the Toxic Substances Control Act (PL 94-469, October 1976). This law makes the U.S. Environmental Protection Agency responsible for establishing safe levels of various chemicals and toxic substances. The regulations may prohibit or limit the manufacturing, processing, distributing, and/or disposing of a chemical substance or mixture. Pesticides are excluded from the regulations by the act.

Emission of toxic substances into the atmospheric environment may be controlled by the 1977 Clean Air Act Amendments (PL 95-95, August 1977). This law regulates ambient air emissions of pollutants from stationary, fugitive, and mobile sources.

The ambient levels of toxic substances, including some pesticides and other organic chemicals, are also controlled by the Clean Water Act (PL 95-217, December 1977) and by the Safe Drinking Water Act (PL 93-523, December 1974).

Manufacture, use, and disposal of pesticides is covered by the Federal Insecticide, Fungicide, and Rodenticide Act (PL 95-516, October 1972). The primary purpose of this law is to control the entry of pesticides into the environment.

Legal Problems of Great Lakes Nonpoint Pollution. Castrilli and Dines[25] summarized and evaluated the possibilities of legislatively controlling nonpoint pollution inputs into the Great Lakes from the United States and Canada. In Ontario, development and planning legislation is the principal control instrument in urban zones. In the United States, the area-wide water quality management planning process (under Section 208 of PL 92-500) is the primary means that links planning and pollution control functions. Specifically, it has been found that:

1. Within the Great Lakes Basin, there is no direct Canadian or U.S. federal involvement controlling erosion from urban construction sites on nonfederal lands. Initiatives have been attempted on local levels. In Ontario, nonpoint pollution control from urban construction developments under the Planning Act is administered by the Ministry of Housing.

2. Permits or approvals for discharges respecting water quality from separate storm sewers are not required in the United States or Canada. The storm-water runoff disposal problem has been viewed at all levels of government more as a runoff disposal problem than as a water quality hazard. Few U.S. cities are under a court injunction to clean up their combined sewer overflows.

3. In the United States and Canada, pesticide regulation is premised upon the protection of ecological balances and the prevention of highly toxic or persistent pesticides to accumulate in the environment. At the federal level in both countries, the regulation of the agricultural use of pesticides emphasizes controlling their market availability.

4. Laws in the Great Lakes Basin regarding fertilizers are directed at health and consumer protection objectives. Fertilizer use and application rates that would be responsive to water quality control objectives are not controlled. Existing controls address manufacturing, registration, and distribution issues.

5. In both countries, feedlot operations and animal waste management practices are essentially unregulated and water quality protection relies on farmers to voluntarily comply with good farm practices and codes.

6. Soil erosion on both sides of the Great Lakes Basin can be controlled by voluntary farmers' programs assisted by state or federal authorities. Lack of funding programs appears to limit the wider development of these programs in Canada. In the United States, the most significant programs are those conducted by the SCS under the acts discussed previously.

7. Control of pollution from extractive mining in both nations is generally accomplished through discharge permit processes.

REFERENCES

1. Rousch, J. E. 1976. Section 208-A congressional view, p. 32–35. In: Proc. of the best management practices for nonpoint source pollution control seminar. EPA 905/76-005, U.S. Environmental Protection Agency, Chicago, Illinois.
2. Alexander, G. R. 1976. Best management practices for nonpoint source control, p. 3–5. In: Proc. of the best management practices for nonpoint source pollution control seminar. EPA 905/76-005, U.S. Environmental Protection Agency, Chicago, Illinois.
3. Anonymous. 1972. Great Lakes water quality agreement with annexes and texts and terms of references between the United States of America and Canada. International Joint Commission, Washington, D.C., Ottawa, Ontario.
4. Anonymous. 1978. Developments at ninth international conference on water pollution research. Stockholm, Sweden, USNC-IAWPR, U.S. Environmental Protection Agency.
5. Haney, P. D. 1966. What is pollution?—An engineer's viewpoint. *J. Sanitary Eng. Div., Proc. ASCE*, **92**:109–113.
6. Krenkel, P. A., and Novotny, V. 1980. "Water Quality Management." Academic Press, New York.
7. Patrick, R. 1975. Some thoughts concerning correct management of water quality. In: W. Whipple, Jr. (ed.), "Urbanization and Water Quality Control." AWRA, Minneapolis, Minnesota.

8. Adamkus, V. W. 1976. The future of nonpoint pollution control: A federal perspective. In: Proc. of the Conference on Nonpoint Sources of Water Pollution: Problems, Policies, and Prospects. Purdue University, West Lafayette, Indiana.
9. Loehr, R. C. 1972. Agricultural runoff–characterization and control. *J. Sanitary Eng. Div., Proc. ASCE*, **98**:909–923.
10. Midwest Research Institute. 1975. National assessment of water pollution from nonpoint sources. U.S. Environmental Protection Agency, Washington, D.C.
11. Barley, G. W., and Waddell, T. E. 1979. Best management practices for agriculture and silviculture: An integrated overview. In: R. C. Loehr et al. (eds.) "Best Management Practices for Agriculture and Silviculture." Ann Arbor Science Publishers, Inc., Ann Arbor, Michigan.
12. Sonzogni, W. C., Monteith, T. J., Bach, W. N., and Hughes, V. G. 1978. United States Great Lakes tributary loadings. International Joint Commission, Windsor, Ontario.
13. PLUARG. 1978. Environmental management strategy for the Great Lakes systems. International Joint Commission, Windsor, Ontario.
14. Loehr, R. C. 1974. Characteristics and comparative magnitude of nonpoint sources. *J. Water Pollut. Control Fed.* (WPCF), **46**:1849–1872.
15. Pisano, M. 1976. Nonpoint sources of pollution: A federal persepctive. *J. Environ. Eng. Div., ASCE*, **102**:555–565.
16. Lager, J. A., and Smith, W. G. 1974. Urban stormwater management and technology–An assessment. EPA-670/2-74-040, U.S. Environmental Protection Agency, Washington, D.C.
17. Lager, J. A., Smith, W. G., Lynard, W. G., Finn, R. M., and Finnemore, E. S. 1977. Urban stormwater management and technology: Update and users guide. EPA-600/8-77-015, U.S. Environmental Protection Agency, Washington, D.C.
18. Soderlund, G., and Lehtinen, H. 1972. Comparison of discharges from urban stormwater runoff, mixed storm overflow and treated sewage. "Proceedings of International Conference of the IAWPR, Jerusalem, Israel." Pergamon Press, Oxford.
19. Pitt, R., and Field, R. 1977. Water quality effects from urban runoff. *J. Amer. Water Works Assoc.*, **69**:432–436.
20. Nisbet, C. T., and Sarofim, A. F. 1972. Rates and routes of transport of PCBs in the environment. *Env. Health Perspectives*, **1**:21–38.
21. Sheffy, T. B. 1979. The management of PCBs in Wisconsin. 3rd Water Research Conference, Wisconsin Section AWRA, Oshkosh, Wisconsin.
22. Runge, C. 1976. Land management institutional design for water quality objectives. In: Proc. of the Best Management of Practices for Nonpoint Source Pollution Seminar. EPA-905/76-005, U.S. Environmental Protection Agency, Chicago, Illinois.
23. Anonymous. 1973. Water policies for the future. National Water Commission. Final Report to the President and to the Congress, Washington, D.C.
24. Lienesh, W. C. 1977. Legal and institutional approaches to water quality

management and implementation. Contract No. 68-01-3564, U.S. Environmental Protection Agency, Washington, D.C.

25. Castrilli, J. F., and Dines, A. J. 1978. Control of water pollution from land use activities in the Great Lakes Basin: Legislative and administrative programs in Canada and the United States. PLUARG, International Joint Commission, Windsor, Ontario.

2

Surface water problems

INTRODUCTION

As it is clearly demonstrated throughout this handbook, nonpoint pollution can affect, often deleteriously, all surface water bodies. Even some small headwater streams in protected national parks may have signs of nonpoint pollution caused by construction, roads, atmospheric inputs, and waste disposal by the visiting public. Water quality and contamination of the Great Lakes, especially lower Lakes Erie and Ontario, are the most serious problems caused by nonpoint pollution. Excessive algae growths have impaired water use of Lake Erie* and portions of other Great Lakes. Pesticides and PCBs have caused curtailment of commercial and recreational fishing. In the early 1970s, the International Joint Commission asked a large panel of experts from the United States and Canada to undertake the most comprehensive research, which would answer the following question:

---

*Due to curtailment of point and nonpoint discharges of phosphorus to Lake Erie, significant water quality improvements of the Lake have been reported in early 1980's.

> Are the boundary waters of the Great Lakes systems being polluted by land drainage (including ground and surface runoff and sediment) from agriculture, forestry, urban and industrial land development, recreational and parkland development, utility and transportation systems, and natural resources?[1]

After about five years of research and many millions of dollars spent, the answer was clearly affirmative. It was found that pollution from nonpoint sources by phosphorus, sediments, industrial and agricultural organics, and some heavy metals is detrimental to the Great Lakes. This pollution causes accelerated eutrophication (aging of the lakes), impairment of commercial and recreational fishing, local water quality problems, and aesthetic damage (Table 2-1).

Evaluation of the effects of nonpoint pollution on surface water quality may differ from the traditional approach designed mainly for point sources. The difference can be summarized as follows:

Point Sources	Nonpoint Sources
Fairly steady flow and quality. Variability ranges less than one order of magnitude.	Highly dynamic in random intermittent intervals. Variability ranges often more than several orders of magnitude.
The most severe impact during a low-flow summer period.	The most severe impact during or following a storm event.
Enter receiving water at identifiable points.	Point of entry often cannot be identified or defined.
Primary parameters of interest:	
BOD, dissolved oxygen, nutrients, suspended solids.	Sediment, nutrients, toxic substances, pH (acidity), dissolved oxygen.

Figure 2-1 is an example of pollution input from nonpoint sources to a lake.

A typical nonpoint pollution load to surface waters is a response of the drainage area to a storm event. It has a limited duration, lasting from a fraction of an hour to 2 days. However, some inputs from soil and groundwater zones last longer. The magnitude of the nonpoint pollution load depends on many uncontrollable factors, primarily the rain volume, intensity, and quality; duration of the preceding dry period; and others.

A plot of runoff flow or quality load versus time is called a histogram, or hydrograph in the case of flow. The term *pollutograph* is sometimes improperly used for the histogram of the quality loading. Figure 2-2 shows flow and quality load histograms.

The water flow and mass loading peaks for histograms may not coincide. The first portion of the storm-water runoff ("first flush") is usually more polluted and carries heavier loads of pollution than the rest of the discharge. This phenomenon is more profound in combined sewer systems, where it carries deposited solids and sewage content from sewers and catch basins.

WASTE ASSIMILATIVE CAPACITY AND STREAM STANDARDS

By definition, any pollution discharged into surface water is harmful to aquatic biota or to present or future beneficial uses of water. However, not all quality contributions from point and nonpoint sources should be called pollution. Receiving waters have a natural ability to accept and assimilate limited amounts of organic matter and other potential pollutants without visible harm and damage. Therefore, it may not be feasible, economical, or beneficial to require a complete removal of wastewater discharges, including those from nonpoint sources. This natural ability of surface water to accept potential pollutants without harmful effects is called the *waste assimilative capacity* and has been defined as:[2] that amount of potential pollutants that can be added to surface waters without impairing their beneficial uses.

Patrick[3] points out that the assimilative capacity of streams does not only mean the breakdown of waste materials by bacteria and fungi and some of the invertebrates; it also means the utilization of these waste products for the growth of algae and other forms of aquatic life.

While the waste assimilative capacity of surface water for biodegradable organic matter is fairly large, it is very low and sometimes almost nil for some toxic and carcinogenic cumulative organic chemical and toxic metals.

Research from the Academy of Natural Science has shown that often small amounts of trace metals such as manganese, vanadium, nickel, and selenium may affect the kinds of algae species that become prevalent and affect the whole structure and efficiency of functioning at the base of the food web.[3]

The most dramatic example of damage done to surface water by relatively small quantities of pollutants is the occurrence of PCBs in the Great Lakes. PCBs have been manufactured since the 1920s and have been recognized as pollutants since the late 1950s. Their manufacturing in the United States has been subsequently reduced. Meanwhile, a portion of existing PCBs, in spite of the production curtailment, have become part of the environment. PCBs are persistent and can travel via air and water particulate matter and cumulate in the fatty tissues of fish and aquatic wildlife species. Even though the levels of PCBs are rarely detectable in water due to their insolubility, PCB levels in sediments and fish tissue in the Great Lakes are quite high and have done almost irreparable damage to Great Lakes fishing.

Other cumulative toxicants with low assimilative capacity in surface water include most of the persistent pesticides such as the organochlorine compounds DDT, chlordane, aldrin, dieldrin, and heptachlor.

In engineering analysis and practice, the waste assimilative capacity of receiv-

TABLE 2-1. Great Lakes Water Quality Problems and Pollutants.[1]

I. Parameters for Which a Great Lakes Water Quality Problem Has Been Identified

	Problem		Sources				
			Diffuse				
Pollutant	Lakewide	Near-Shore or Localized	Land Runoff	Atmosphere	In-Lake Sediments	Point	Remarks
Phosphorus[1]	Yes	Yes	Yes	Yes	Yes	Yes	[a]Percentage unknown, not considered significant over annual cycle.
Sediment[b,1]	No	Yes	Yes[c]	Negligible	Under some conditions	Negligible	[b]May contribute to problems other than water quality (e.g., harbor dredging). [c]Including streambank erosion.
Bacteria of public health concern	No	Yes	Minor[d]	No	No	Yes	[d]Land runoff is a potential but minor source; combined sewer overflows generally more significant.
PCBs[1]	Yes	Yes	Yes	Yes	Yes	Yes	
Pesticides[1] (past)	Yes[e]	Yes[e]	Yes	Yes	Yes	No	[e]Some residual problems exist from past practices.
Industrial organics[1]	Yes	Yes	Yes	Yes	Yes	Yes	
Mercury[1]	Yes	Yes	Minor	Yes	Yes	Yes	
Lead[1]	Potential[f]	Potential[f]	Yes	Yes	Yes	Yes	[f]Possible methylation to toxic form.

II. Parameters for Which No Great Lakes Water Quality Problem Has Been Identified, but Which May Be a Problem in Inland Surface Waters or Groundwaters

Nitrogen	No	No[g]	Yes	Yes	Minor	Yes	[g]Some inland groundwater problems.
Chloride	No	No[h]	Yes	Negligible	No	Yes	[h]Some local problems exist in near-shore areas due to point sources.
Pesticides[1] (present)	No	No	Yes	No	No	Yes	[i]New pesticides have been found in the environment; continued monitoring is required.
Other heavy metals	Potential[f]	Potential[f]	Yes	Yes	Yes	Yes	
Asbestos[j]	No	Yes	No	?	Yes	Yes	
Viruses[j]	←	No Data Available			→	Yes	[j]Better detection methods needed.
Acid precipitation[k]	No	No[k]	No	Yes	No	No	[k]A potential problem for smaller, soft-water inland lakes.

[1]Sediment per se causes local problems; phosphorus and other sediment-associated contaminants have lakewide dispersion.

Fig. 2-1. Sediment discharge from a tributary into a lake. (Photo University of Wisconsin.)

ing water is determined from and based on stream standards. There are basically two types of stream standards used throughout the world.[2] Most of the European countries use a system of classifying streams into four "water quality classes." The origin of this system can be traced to the Kolkwitz-Marson taxonomic saprobic classification, which classifies water quality according to the tolerance of various benthic and aquatic organisms to different levels of pollution. The best quality is termed *oligosaprobic* and is usually considered water quality suitable for public water supply. The medium water quality is called α- or β-mesosaprobic, and the most polluted water quality is described by polysaprobic conditions. Later, chemical water quality parameters were correlated to the saprobic system giving thus the origin of a more comprehensive stream standard system. Table 2-2 gives an example of French water quality standards.

The present U.S. stream standards are based on the Water Pollution Control Act of 1972 (PL 92-500), which gives the authority to states to establish stream standards according to the Guidelines of the Environmental Protection Agency. The standards are related to intended water uses of the state surface water. The

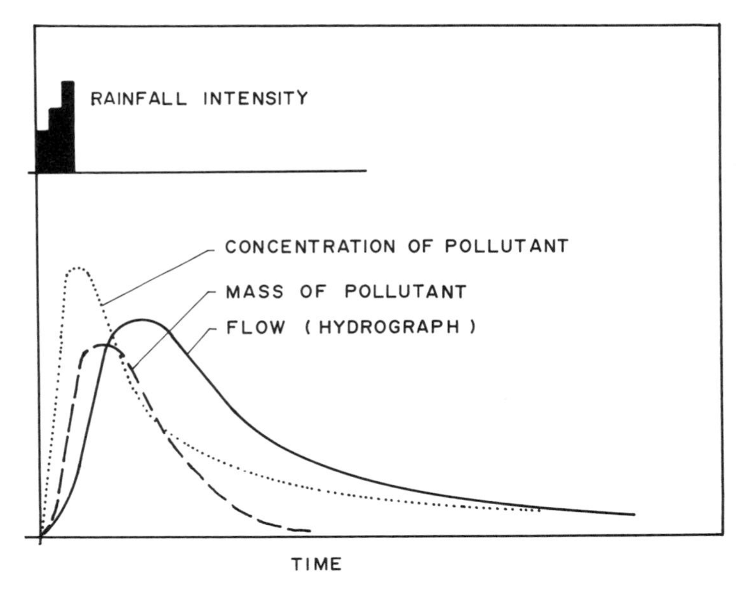

Fig. 2-2. Flow and quality histogram from nonpoint sources.

beneficial uses that determine which set of standards is to be used include the following:

a. Municipal, industrial, and domestic water supply.
b. Water contact recreation, i.e., swimming and water skiing.
c. Noncontact recreation including boating, fishing, and aesthetic.
d. Fish and wildlife propagation and protection.
e. Agricultural irrigation.

The Water Pollution Control Act (the Clean Water Act) requires that all navigable U.S. waters be suitable for contact recreation and fish and wildlife propagation and protection. Table 2-3 contains water quality criteria suggested by the EPA to the states for their incorporation in the state stream standards. In addition EPA listed criteria for a number of toxic substances.

Specific Water Quality Problems.

Water quality problems associated with nonpoint pollution can be divided into several categories:

1. Problems with sediment.
2. Dissolved oxygen and biodegradable organics.

TABLE 2-2. French Water Quality Standards.[4]

Quality	1A	1B	2	3	4
Temp.	≤20	20–22	22–25	25–30	>30
pH	6.5–8.5	6.5–8.5	6.5–8.5	5.5–9.5	<5.5 or >9.5
D.O. (mg/liter)	>7	5–7	3–5	<3	
% Sat.	>90	70–90	50–70	<50	
BOD_5 (mg/liter)	≤3	3–5	5–10	10–25	>25
COD (mg/liter)	≤20	20–25	25–40	40–80	>80
SO_4^{2-} (mg/liter)		<250		>250	
NH_4^+ (mg/liter)	≤0.1	0.1–0.5	0.5–2	2–8	>8
NO_3^- (mg/liter)		≤44		44–100	>100
Phenols (mg/liter)	≤0.001		0.001–0.05	0.05–0.5	>0.5
Orthophos. (mg/liter)	≤0.4	0.4–0.7		>0.7	
Deterg. anion (mg/liter)	≤0.2		0.2–0.5	>0.5	
CN^- (mg/liter)		≤0.05		>0.05	
Cr (mg/liter)		≤0.05		>0.05	
F (mg/liter)	≤0.7	0.7–1.7		>1.7	
Pb (mg/liter)		≤0.05		>0.05	
Se (mg/liter)		≤0.01		>0.01	
Cu (mg/liter)	≤0.05		0.05–1	>1	
Zn (mg/liter)	≤3	3–5		>5	
As (mg/liter)	≤0.05		0.05–0.1	>0.1	
Fe (mg/liter)	≤0.5	0.5–1	1–1.5	>1.5	
Mn (mg/liter)	≤0.1	0.1–0.25	0.25–0.5	>0.5	
Cd (mg/liter)		≤0.005		>0.005	
Subs. extrac. (mg/liter)	≤0.2	0.2–0.5	0.5–1	>1	
E. Coli N/100 ml	≤2000		>2000		
Streptoc. N/100 ml	≤20	20–1000	1000–10,000	>10,000	
Salinity	**0**	**1**	**2**	**3**	**4**
Conduct. μmhos/cm	≤400	400–750	750–1500	1500–3000	>3000
Cl^- mg/liter	≤100	100–200	200–400	400–1000	>1000

TABLE 2-3. Selected Water Quality Criteria for Surface Waters.[5]

Parameter	Primary (Contact) Recreation	Fish and Wildlife Protection and Propagation		
		Warm Water	Cold Water	Marine
pH	6.5–8.3	6–9	6–9	6.7–8.5[a]
Total alkalinity (mg $CaCO_3$/liter)	–	>20	>20	–
Turbidity				
Secchi disk (m)	>1.2	–	–	–
Jackson Units	–	<50	<10	–
Max. temperature (°C)	<30	–	<20	–
Max. temperature increase (°C)	–	<2.8[b]	<2.8[b]	<0.8[c] <2.2[d]
Change in salinity	–	–	–	<10% of natural variations
Dissolved oxygen (mg/liter)	–	>5	>6	
Total phosphorus (μg/liter)	–	<100[e]	<100[e]	
Coliform median MPN (No./100 ml)	–	–	–	<70[f]
Fecal coliforms (No./100 ml)	<200[g]			

[a]Normal pH range should not be altered by more than 0.1 pH units.
[b]For lakes <1.6°C.
[c]Period June–August.
[d]Period September–May.
[e]For stream entering lakes <50 μg/liter.
[f]For waters used for shellfish cultivation and harvesting. Maximum of 10% of samples not to exceed MPN of 230/100 ml for five tbe tests.
[g]Maximum of 10% of samples not to exceed 400/100 ml during any 30-day period.

3. Nutrient contribution and eutrophication.
4. Toxic chemicals and metals.
5. pH and acidity.

Other problems such as bacteriological and radiological nonpoint pollution impact on receiving waters are of lesser importance or are local. Most of the radiological nonpoint pollution originating from atmospheric testing of nuclear weapons is presently below dangerous levels due to the ban of the atmospheric testing by major nations with nuclear capability.

Fig. 2-3. Destruction of a suburban stream by sediment loads from construction activities. (Photo authors.)

Sediment Problem. Sediment from nonpoint sources is the most widespread pollutant of surface water. Sediment, especially its finer fractions, cuts down light penetration and, thus, greatly reduces algal production. The turbidity caused by the sediment also has deleterious effects on benthic biota and fish, and impairs most of the major beneficial uses of water. Figure 2-3 shows a suburban stream destroyed by sediment deposits caused by excavation and nearby construction.

Large amounts of sediment settle out annually in slow-moving sections of surface water, particularly in reservoirs, harbors, estuaries, and impounded streams.

In streams, sediment either moves in suspension or is shifted on the bottom. The suspended portion is called *washload*, while *bedload* is the portion that moves at or near the bottom in an erratic movement along the stream bed. The bedload portion of the sediment transport, with its instability, eliminates suitable habitats for aquatic life.

The concentrations of suspended sediment in streams are highly variable and are influenced by several factors including rainfall duration and intensity, soil condition, topography, geology, vegetation cover, and disturbing activities taking place in the watershed. Sediment concentrations may vary from a few miligrams to more than ten thousand miligrams per liter. Most freely flowing rivers demonstrate significant day-to-day variations in the concentration of suspended solids. The suspended load may vary with different reaches of a stream, particularly when the bed is resuspended or deposited due to changes in the hydraulic regime.

The question of how much suspended sediment is deleterious to surface waters cannot be precisely delineated, and standards are only available for turbidity. However, detrimental effects of sediment loads on the ecosystems can be characterized, and the importance of each effect outlined.

Suspended sediment alters aquatic environments, primarily by inhibiting light, changing heat radiation, blanketing the stream bottom, and retaining organic materials and other substances that create unfavorable conditions for benthic organisms.

Sedimentation and blanketing of the stream bottom concern both organisms living in the bottom of the receiving water as well as higher organisms that depend upon bottom organisms for food. It has been long recognized that sediment deposition is primarily harmful to fish eggs buried in the bottom. Heavy or irritating concentrations of solids might interfere sufficiently with the gill movement of fish to affect circulation. However, some studies indicate rather significant resistance of adult fish to turbidities. Regardless of the uncertainties as to the effect of sediment on fish populations, the truth remains that indirect damage to fish through destruction of the food supply, eggs, or alevins, or changes in the habitat probably occur long before any adult fish can be directly harmed.

Increasing turbidity reduces the light penetration, which in turn limits the primary productivity by limiting the column of water in which light intensity is sufficient (about 1% of incident light) for the rate of photosynthesis to exceed the respiration rate. Since photosynthetic organisms form the base of the food web, any significant change in their population would have a widespread effect on the organisms depending upon them for food.

The surface of particulate matter may act as a substrate for bacteria and other microorganisms, thus changing their habitat. Sediment also traps nutrients, phosphorus in particular, and other toxic materials, which may be buried with the sediment in places of sediment deposition. Nutrient release and possible release of toxic materials may follow when resuspension of bottom materials occurs during periods of high flow or local scour.

Sediment deposition in reservoirs limits their useful lifetime as is shown in Fig. 2-4. Similarly, many navigation waterways must be continuously dredged to keep the navigation channels passable, a process that is greatly polluting water by resuspending sediments and associated pollutants and nutrients.

Soil loss, which is the primary source of sediment, contains one or more percent of organic matter, nutrients in amounts of about a fraction of a percent, and often significant amounts of pesticides, toxic metals, and PCBs.

Deposition of organics, which have lower specific density than sand, silt, and

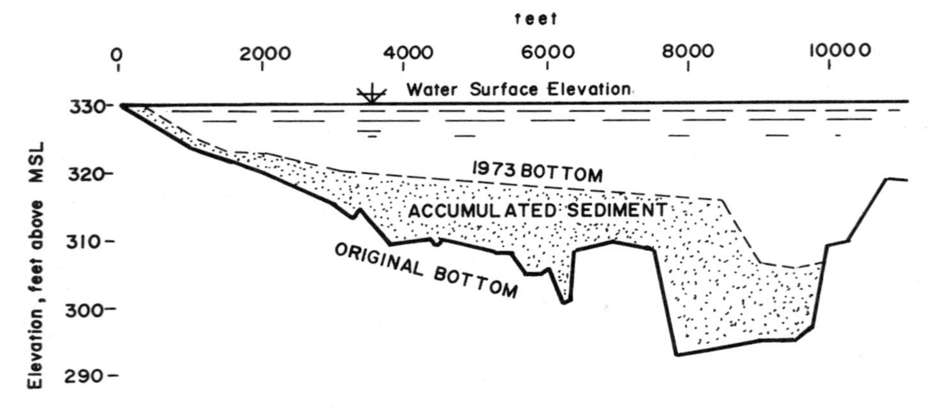

Fig. 2-4. Sediment accumulation in Clark Hill reservoir on Savannah River, Georgia. (Source U.S. Corps of Engineers.)

clay particles, occur at lower velocities. Most of the organics stay as part of the washload.

As will be seen in the next section, one of the most serious water quality problems is deposits containing large amounts of organics. Their oxygen demand deprives water of oxygen. Furthermore, processes taking place in the sediment layers may also, under certain conditions, release large amounts of nutrients back into the water.

Dissolved Oxygen Problem. Dissolved oxygen concentration of surface water is the primary parameter on which suitability of water for fish and wildlife is determined. The D.O. concentration standard is also the major parameter used in the waste assimilative capacity determination. The dissolved oxygen problem caused by nonpoint discharges such as runoff from manure applications, animal feedlots, and urban storm or combined sewers, can be devastating to the receiving water. The problem is not limited to runoff discharges containing high amounts of organics–for example, feedlot runoff. Large amounts of relatively "clean" storm water may resuspend bottom deposits and increase their oxygen demand by several orders of magnitude.

In streams, the dissolved oxygen concentration is a response to various oxygen sinks and sources. The sinks of oxygen, that is, the biochemical and biological processes that use oxygen, include:

1. Deoxygenation of biodegradable organics whereby bacteria and fungi utilize oxygen as an electron acceptor in the biooxidation process.
2. Benthal oxygen demand, where oxygen is utilized by the upper layers of the bottom sediment deposits.

3. Nitrification, in which oxygen is utilized during the oxidation of ammonia and organic nitrogen to nitrates.
4. Respiration by algae and aquatic vascular plants which use oxygen during night hours to sustain their living process.

Major oxygen sources are:

1. Atmospheric reaeration, where oxygen is transported from the air into the water by turbulence at the water surface.
2. Photosynthesis, where chlorophyll-containing organisms (algae and aquatic plants) convert CO_2 (or alkalinity of water) to organic matter with a consequent production of oxygen.

The basic concept of the oxygen balance in streams was proposed by Streeter and Phelps[6] and was later summarized by Phelps.[7] This concept, which is primarily used for evaluation of the point source impact on water quality, is still in use with some modifications. However, the original model, which assumed a point sewage discharge from a single source, cannot be used without modifications for nonpoint source evaluation. Smith and Eilers,[8] Meadows et al.,[9] and Wanielista[10] among others attempted to modify the Streeter-Phelps model for evaluation of nonpoint source impact on receiving streams.

The basic assumptions of the model are:

a. The biochemical oxygen demand (BOD) load to the stream can take place either as a point source (combined or storm sewer or a major tributary) or as a distributed source.
b. The flow conditions do not change considerably during the time of analysis—an assumption that is often violated, but as shown in reference 9, no significant error will result as long as $X \leqslant 0.05\ Q_0/q$, where X is the length of the reach, Q_0 is the flow at the head of the reach, and q is the lateral inflow per unit length of the reach.
c. Effects of nitrification and photosynthesis are negligible.
d. Effect of dispersion, such as in tidal streams, is negligible.
e. The reach is uniform with no appreciable changes in cross-sectional dimensions, flow velocity, and depth.

Under these assumptions, the equations, for the BOD and dissolved oxygen concentrations, become:[8, 9, 11]

BOD variations

$$L = L_0 e^{-K_r (X/u)} + \frac{L_r}{K_r}(1 - e^{-K_r (X/u)}) \qquad (2\text{-}1)$$

dissolved oxygen variations

$$D = D_0 e^{-K_a(X/u)} + \frac{K_r}{K_a - K_r}\left(e^{-K_d(X/u)} - e^{-K_a(X/u)}\right)L_0$$
$$+ \left[\frac{K_d}{K_a K_r}\left(1 - e^{-K_a(X/u)}\right) - \frac{K_d}{(K_a - K_r)K_r}\left(e^{-K_r(X/u)} - e^{-K_a(X/u)}\right)\right]L_r$$
$$+ \left(1 - e^{-K_a(X/u)}\right)\frac{S_B}{K_a H} \qquad (2\text{-}2)$$

where:

$D = c_s - c$ is the oxygen deficit at the end of the reach (mg/liter)
D_0 = initial oxygen deficit (mg/liter)
c_s = oxygen saturation (mg/liter)
c = oxygen concentration at the end of the reach (mg/liter)
L_0 = initial ultimate BOD concentration (mg/liter)
L_r = ultimate BOD concentration in the lateral nonpoint pollution input (mg/liter)
K_r = overall BOD removal coefficient (day^{-1})
K_a = reaeration coefficient (day^{-1})
K_d = BOD deoxygenation coefficient (does not include BOD removal by sedimentation) (day^{-1})
X = length of the reach (m)
u = average stream flow velocity (m/day)
S_B = benthic oxygen demand (g/m^2 - day)
H = average depth of flow (m)

The most important task in the dissolved oxygen analysis of nonpoint pollution effects on receiving water is determining the design flow, appropriate coefficients, and stream characteristics, and temperature conditions for Equations 2-1 and 2-2.

Flow. Traditionally, for point source evaluations and waste assimilative capacity determination, the flow on which the analysis is performed is statistically derived from historical flow records that are in some cases available from the U.S. Geological Surveys. The low-flow characteristic is usually a dry weather flow that would statistically occur once in ten years with a magnitude which for seven days during the critical period, would not be exceeded (so called 7 days–10 years expectancy flow). In Europe, the low-flow magnitude is evaluated from flow cumulative frequency curves.

This low-flow characteristic can hardly be used for nonpoint source pollution evaluation since its occurrence presumes a long period of drought and excludes any surface runoff—the primary contributor of nonpoint pollution.

The most critical nonpoint pollution impact may occur when a local storm enters a receiving water body after an extensive drought period, a condition that must be statistically analyzed from flow and rain data and that will result in a flow characteristic higher than the 7 days–10 years expectancy discharge.

Coefficient of Deoxygenation. The coefficient of deoxygenation measured in streams is a composite of three processes that may not occur simultaneously. These processes are: deoxygenation of BOD in free-flowing water, effects of benthic slimes on BOD absorption and removal, and sedimentation of particulate organics. Hence[2]

$$K_r = K_1 + K_3 + B \tag{2-3}$$

and

$$K_d = K_1 + B$$

where:

K_1 = deoxygenation rate constant for free-flowing water approximately equaling the laboratory BOD rate

K_3 = BOD removal rate by sedimentation which does not result in the oxygen demand

B = effect of absorption of BOD by benthic slimes which results in oxygen demand

The magnitudes of the laboratory BOD removal coefficient, K_1, have been reported in ranges of 0.1 to 0.6 day^{-1} (K_1 base e), while the overall ranges of the deoxygenation rate, K_r, can be between 0.1 and 5 day^{-1} depending on the type of the stream, flow conditions, benthic slime density and activity, and sewage or wastewater conditions.[2,12,13]

Atmospheric Reaeration. The gas transfer theory demonstrates that stream reaeration by atmospheric oxygen is proportional to the turbulence intensity at the water surface and the ratio of the surface area to the water volume. The magnitude of the reaeration coefficient has been subjected to intensive investigations from purely theoretical, such as that of O'Connor and Dobbins,[14] to experimental field investigations such as those by Churchill et al.[15] These studies have usually resulted in a relationship of the following type:

$$K_a = \frac{Cu^n}{H^m} \tag{2-4}$$

Reported values of the coefficients for some of the most common reaeration formulas are in Table 2-4.

TABLE 2-4. Summary of Coefficients for the Reaeration Formula.[a]

Investigator	C	n	m
O'Connor, Dobbins[14]	3.92	0.5	1.5
Churchill et al.[15]	5.05	0.969	1.673
Owens et al.[16]	3.0	0.67	1.65
Langbein and Durum[17]	5.15	1.0	1.33

[a] $K_a = C(u^n/H^m)$, where u = velocity (m/sec), H = depth (m), K_a = base e.

Benthal Oxygen Demand. Stream bottoms may be covered by highly active organic sediments, biological slimes, and residual sludges in streams polluted by sediments from nonpoint sources and by residual solids from point sources. The growth and accumulation of these materials result from deposition of suspended sediments and from the transfer of soluble organics from flowing water into the bottom biological matter. Deposition of suspended organics occurs during low flows when velocities are low and/or in reaches where, due to high depth, the flow velocity slows down (see Chapter 5 for more detailed discussion on sedimentation and scour).

Only the upper layer of deposited sludges and sediments remains aerobic and exerts the benthic oxygen demand. The remainder of the sludge banks and bottom layers are anaerobic. The thickness of the upper aerobic layer is in fractions of a centimeter. However, during a sudden increase of flow velocity, which is typical for storm events, the settled sediment can be resuspended. The organic matter, which under anaerobic conditions would not consume oxygen, may, when resuspended, increase oxygen demands of bottom sediment by several orders of magnitude. This disturbed sediment oxygen demand is typical especially following a discharge from storm or combined sewers into an otherwise slow-moving stream, harbor, or estuary with substantial deposits of organic sediments on its bottom.

Table 2-5 shows dissolved oxygen demand of benthic sediments typical of those from nonpoint pollution. Note that disturbed sediment demand for oxygen, a situation common in areas of concentrated overflows from storm and combined sewers, can increase the oxygen demand by several orders of magnitude as compared to packed undisturbed sediment deposits. Disturbing and scouring of sediment is one of the most severe secondary effects of nonpoint pollution caused by surface runoff heavily laden by organic rich soils or urban particulate matter. The consequences of such phenomena may be detrimental to the oxygen regime of the stream in spite of relatively higher flows that result from the surface runoff inputs.

TABLE 2-5. In Situ Measured Oxygen Demand of Bottom Sediments.

Benthic Deposits	Benthic Oxygen Demand S_B (g/m^2-day) Undisturbed	Disturbed	Temp. (°C)	Investigator
Rivers				
River sludge	0.9			Oldaker et al.[a]
River muds 2 cm	3.4			McDonnel and Hall[a]
25 cm	6.4			McDonnel and Hall[a]
Sandy bottom	0.2–1.0			Thomann[11]
Mineral soils	0.05–0.1			Thomann[11]
R. Venaviken	1.44		2	Edberg and Hofsten[19]
	0.68		10	Edberg and Hofsten[19]
Sjomosjon	0.31		0	Edberg and Hofsten[19]
Jaders Bruk	0.84		9	Edberg and Hofsten[19]
	1.4		2	Edberg and Hofsten[19]
Lower Milwaukee R.	2.8–6.7	360–1200	20	Kreutsberger et al.[20]
Potomac estuary	2.5		15	O'Connel and Wales[a]
Lakes				
Lake Erken	2.6		18	Edberg and Hofsten[19]
	0.43		4	Edberg and Hofsten[19]
Lake Ramsen	2.3		17	Edberg and Hofsten[19]
Lake Michigan– Green Bay	1.6–1.9		12	Paterson et al.[21]

[a]Quoted in reference 18.

Figure 2-5 shows the effect of the disturbed oxygen demand on the D.O. conditions in the Milwaukee River estuary following a storm-water discharge from the Milwaukee metropolitan area. Large deposits of sediments, mostly from urban nonpoint sources and combined sewer overflows, have accumulated in the harbor, and the rapid decline of the D.O. concentration is caused almost exclusively by local scouring of sediment by concentrated flows near the storm and combined sewer outfalls.

The benthic oxygen demand depends on the composition of the sediment, on oxygen conditions in the overlying water, and on mixing and turbulence in the bottom boundary layer.[18,19]

The anaerobic processes taking place in the lower portion of bottom sediments cause production of methane (CH_4) and hydrogen sulfide (H_2S) gases, which penetrate up into water and into the atmosphere.

Temperature Effects. Almost all of the reaction rates in the oxygen balance process are temperature dependent. The relationship is expressed in the following form:[2,12,13]

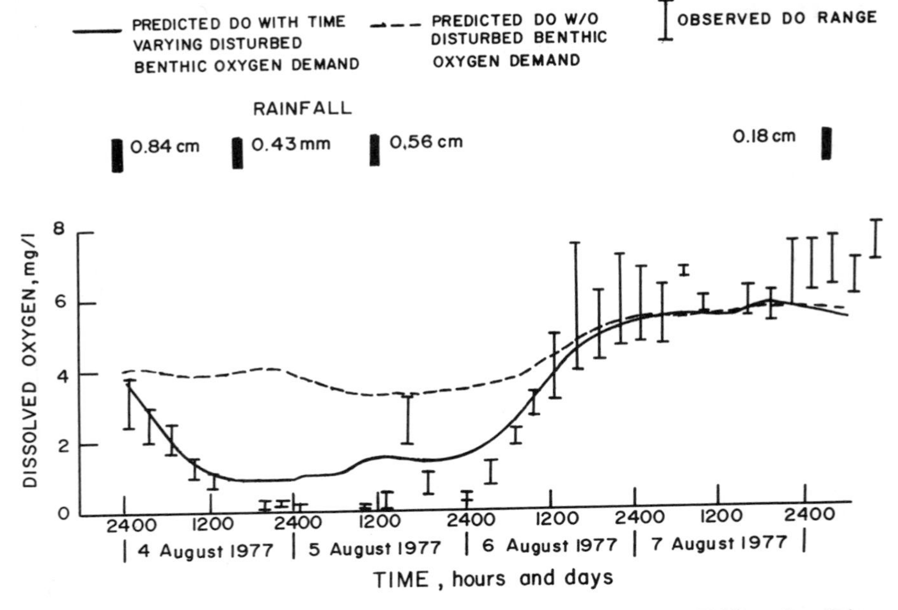

Fig. 2-5. Dissolved oxygen conditions in the harbor portion of Milwaukee River (Wisconsin) following discharges of storm water from storm and combined sewer overflows. (Source Kreutsberger et al.[20])

$$K_T = K_{20}\,\theta^{(T-20)} \tag{2-5}$$

where:

K_T = reaction rate at temperature T (T in °C)
K_{20} = reaction rate at temperature 20°C
θ = thermal factor, which has the following accepted values:

Deoxygenation rates K_r, K_d, K_1	$\theta = 1.047$
Reaeration rate K_a	$\theta = 1.025$
Benthic oxygen demand S_B	$\theta = 1.065$

Temperature also affects the oxygen saturation value, which can be approximated as follows:[2]

$$c_s = 14.652 - 0.41022T + 0.007991T^2 - 0.000077774T^3 \tag{2-6}$$

The temperature conditions selected for the waste assimilative capacity evaluation should correspond to maximum average monthly summer temperatures, since statistically rare temperatures in combination with low flow would lead to excessively rare and unrealistic conditions.

It must be realized that these coefficients are statistical quantities and should be preferably measured by in situ surveys rather than estimated. The water quality models that use such coefficients must always be verified by field data and measurements.

Example 2-1: Dissolved Oxygen Computation

A medium-size city with a population of 50,000 and a city area of 20 km^2 *is discharging its treated sewage and storm runoff by separate sewer systems into an impounded river. The length of the impounded reach is* $X = 10$ *km.*

Compare the dissolved oxygen conditions in the river during a critical drought period and following a storm event which resulted in runoff volume from the city area $R = 1.5$ *cm, lasting 2 h. The lateral rural inflow in the stream is* $q = 0.1$ m^3/sec *per 1 km of the receiving stream. The measured (undisturbed) benthal oxygen demand was 2.5 g/m*^{2}*-day at 25°C.*

Stream characteristics:

Low flow (7 days duration–10 years expectancy)	$Q_R = 15$ m^3/sec
Average depth during the low-flow conditions	$H = 3$ m
Average low flow velocity	$u = 0.1$ m/sec
Average water temperature	$T = 25°$C

Deoxygenation rate $K_R = K_d = 0.25$ day^{-1} at 25°C
Upstream BOD_5 concentration = 1.2 mg/liter
Upstream D.O. concentration $c_0 = 7.6$ mg/liter

From Equation 2-6 the oxygen saturation concentration at 25°C becomes $c_s = 8.4$ *mg/liter.*

Sewage characteristics:

$$BOD_5 \text{ discharge} = 54 \text{ g/cap-day}$$

$$= 54 \text{ (g/cap-day)} \times 50{,}000 \text{ (pop)} \times \frac{1 \text{ (day)}}{86{,}400 \text{ (sec)}}$$

$$= 31.25 \text{ g/sec.}$$

Assuming 90% removal in the sewage treatment plant, the BOD_5 *loading from sewage is*

$$0.1 \times 31.25 = 3.12 \text{ g/sec}$$

Storm characteristics:

Flow from storm sewers

$$Q_s = 1 \text{ (cm)} \times 20 \text{ (km}^2) \times 10^6 \frac{(\text{m}^2)}{(\text{km}^2)} \times \frac{1}{2 \text{ (hr)}} \times \frac{1 \text{ (hr)}}{3600 \text{ (sec)}}$$

$$= 27.78 \text{ m}^3/\text{sec}$$

From Table 1-3 the average BOD_5 concentration of urban storm water is 30 mg/liter.
The BOD_5 concentration of the lateral rural runoff can be assumed as 3 mg/liter.

Dry weather D.O. conditions
Initial ultimate BOD concentration

$$L_0 = 1.4 \left(\frac{\text{sewage BOD}_5\ \text{loading}}{Q_R} + \text{upstream BOD}_5 \right)$$

$$= 1.4 \left(\frac{3.12}{15} + 1.2 \right) = 1.97 \text{ mg/liter}$$

Initial D.O. deficit

$$D_0 = c_s - c_0 = 8.4 - 7.6 = 0.8 \text{ mg/liter}$$

Time of flow in the reach

$$t = \frac{X}{u} = \frac{10{,}000\ (\text{m})}{0.1\ (\text{m/sec}) \times 86{,}400\ (\text{sec/day})} = 1.16 \text{ days}$$

Coefficient of reaeration (from O'Connor-Dobbins formula)

$$K_a = \frac{3.92\, u^{1/2}}{H^{3/2}} = \frac{3.92 \times 0.1^{1/2}}{3^{3/2}} = 0.24 \text{ day}^{-1}$$

and converted to 25°C (Equation 2-5)

$$K_{a25^\circ C} = 0.24 \times 1.025^{(25-20)} = 0.27 \text{ day}^{-1}$$

Use Eq. 2-2 to compute D.O. deficit assuming that $L_r = 0$ and $S_B = 2.5$ g/m²-day

$$D = 0.8 e^{-0.27 \times 1.16} + \frac{0.25 \times 1.97}{(0.27 - 0.25)} \left(e^{-0.25 \times 1.16} - e^{-0.27 \times 1.16} \right)$$

$$+ \left(1 - e^{-0.27 \times 1.16}\right) \frac{2.5}{0.27 \times 3} = 1.84 \text{ mg/liter}$$

and then the D.O. concentration becomes

$$c = c_s - D = 8.4 - 1.84 = 6.56 \text{ mg/liter}$$

Other D.O. concentrations for distances between 0 and 10 km are plotted in Fig. 2-6.

Wet weather D.O. conditions
Average flow

$$\overline{Q} = \overbrace{Q_R + Q_S}^{Q_0} + \tfrac{1}{2} q \times X = 15 + 27.78 + \tfrac{1}{2}\, 0.1 \times 10 = 43.28 \text{ m}^3/\text{sec}$$

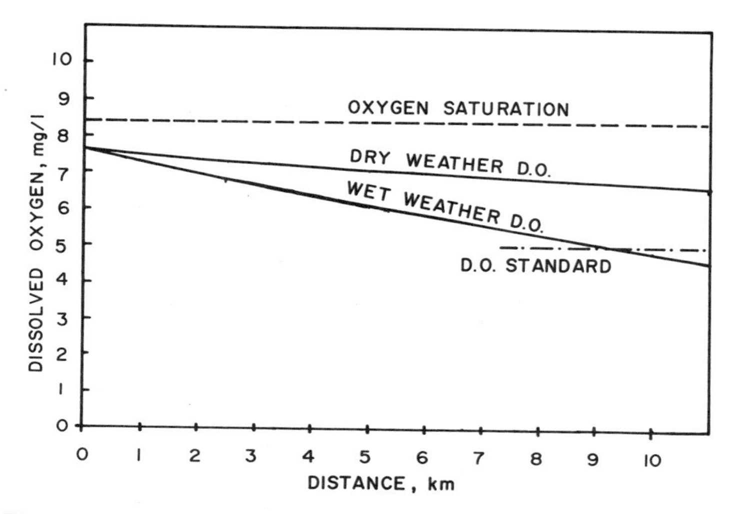

Fig. 2-6. Plot of dissolved oxygen concentrations for Example 2-1.

Since

$$X = 10\text{ km} < 0.05 \times \frac{Q_0}{q} = 0.05\,\frac{42.78}{0.1} = 21.3\text{ km}$$

no appreciable error occurs if dispersion is omitted.

The depth of flow, velocity, and detention time must be adjusted for $\overline{Q}$. In impounded waters the depth remains about the same; however, velocity increases. Let us assume

$$u_{\text{wet}} = u_{\text{dry}}\,\frac{Q}{Q_{\text{dry}}} = 0.1\,\frac{43.28}{15} = 0.29\text{ m/sec}$$

and the time flow

$$t = \frac{X}{u} = \frac{10{,}000}{0.29 \times 86{,}400} = 0.40\text{ day}$$

The initial ultimate BOD concentration is then

$$L_0 = 1.4\left(\frac{\text{sewage BOD}_5\text{ load} + \text{storm-water load}}{Q_0} + \text{upstream BOD}_5\right)$$

$$\cdot\, 1.4\left(\frac{3.125 + 27.7830}{42.78} + 1.2\right) = 29.06\text{ mg/liter}$$

The coefficient of reaeration becomes

$$K_a = \frac{3.92\ 0.29^{1/2}}{3^{3/2}} = 0.41\text{ day}^{-1}$$

corrected for 25° C

$$K_{a25°C} = 0.41 \times 1.025^{(25-20)} = 0.46 \text{ day}^{-1}$$

Compute the oxygen deficit by Equation 2-2 assuming that $L_r = 1.4 \times 3 = 4.2$ mg/liter and $S_B = 2.5$ mg/liter. Then

$$D = 0.8\,(e^{-0.46\times0.4}) + \left[\frac{0.25}{0.46 - 0.25}\,(e^{-0.25\times0.4} - e^{-0.46\times0.4})\right] 29.06 + \left[\frac{0.25}{0.46 \times 0.25}\,(1 - e^{-0.46\times0.4}) - \frac{0.25}{(0.46 - 0.25)\,0.25}\,(e^{-0.25\times0.4} - e^{-0.46\times0.4})\right] 4.2 + (1 - e^{-0.46\times0.4})\,\frac{2.5}{0.46 \times 3} = 3.57 \text{ mg/liter}$$

and the concentration is then

$$c = c_s - D = 8.4 - 3.57 = 4.83 \text{ mg/liter}$$

Note that the undisturbed benthic oxygen demand was assumed in this analysis. If, as indicated in Table 2-6, the benthic oxygen demand increased as a result of concentrated flow and scour near the outfall and/or increased flow velocity in the channel, the oxygen deficit would increase by 0.12 mg/liter for every 1 g/m^2-day increase in the magnitude of S_B.

As a result of the foregoing storm-water discharge the dissolved oxygen standard would be violated.

Nutrient Problems and Eutrophication. Although many elements and chemical compounds are essential for plant and algal growth, only nitrogen and phosphorus are considered the limiting nutrients controlling their growths. On the average, about 50% of phosphorus and an even greater proportion of nitrogen originates from nonpoint sources.

The studies undertaken by the National Eutrophication Survey in the early 1970s clearly indicated that nationwide there is a distinct correlation between general land use and nutrient concentrations in streams. Streams draining agricultural watersheds had on the average considerably higher nutrient concentrations than those draining forested watersheds. The nutrient concentrations were proportional to the percent of land in agricultural and urban land use.[22] It should be pointed out that these conclusions were based on surveys of watersheds that had no known point sources. Figures 2-7 and 2-8 summarize the relationship of the average nutrient concentrations to the general land use in three geographical regions of the United States. No significant correlation

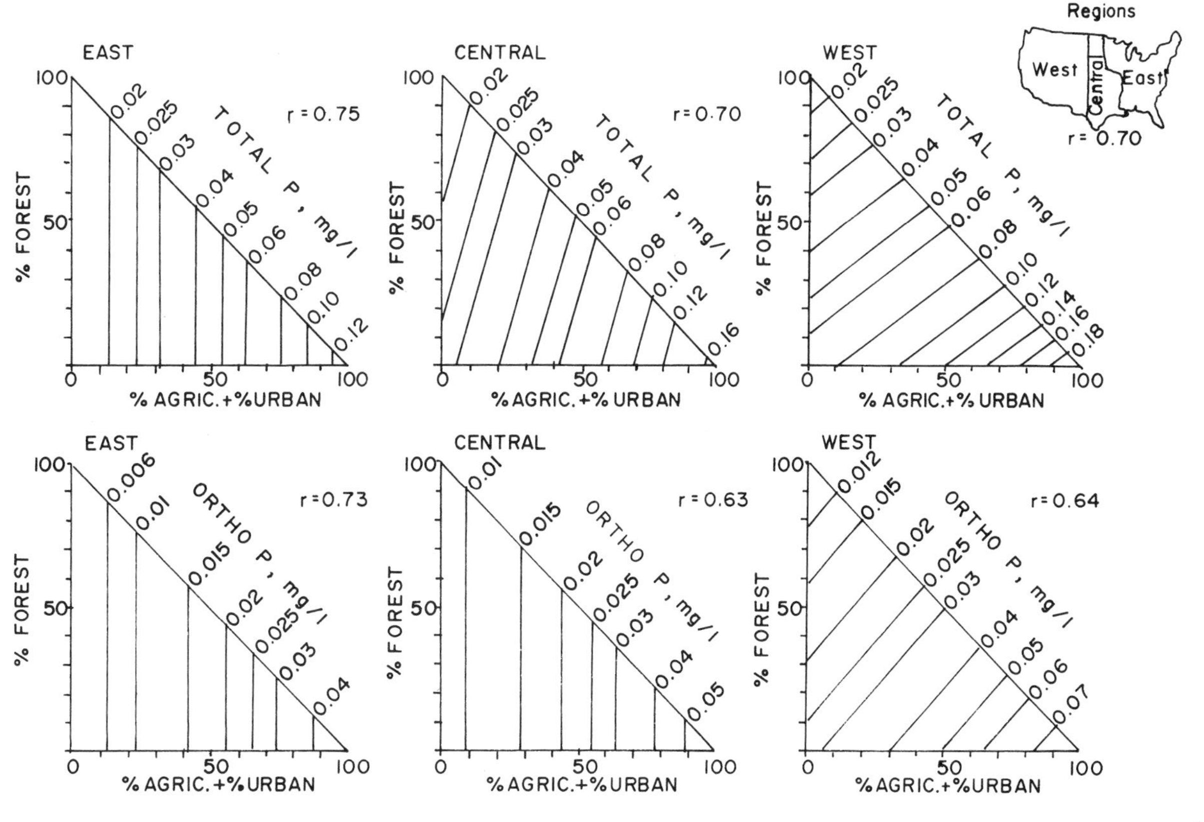

Fig. 2-7. Regional relationships between land use and phosphorus concentrations in streams. (Redrawn from Omernik.[22])

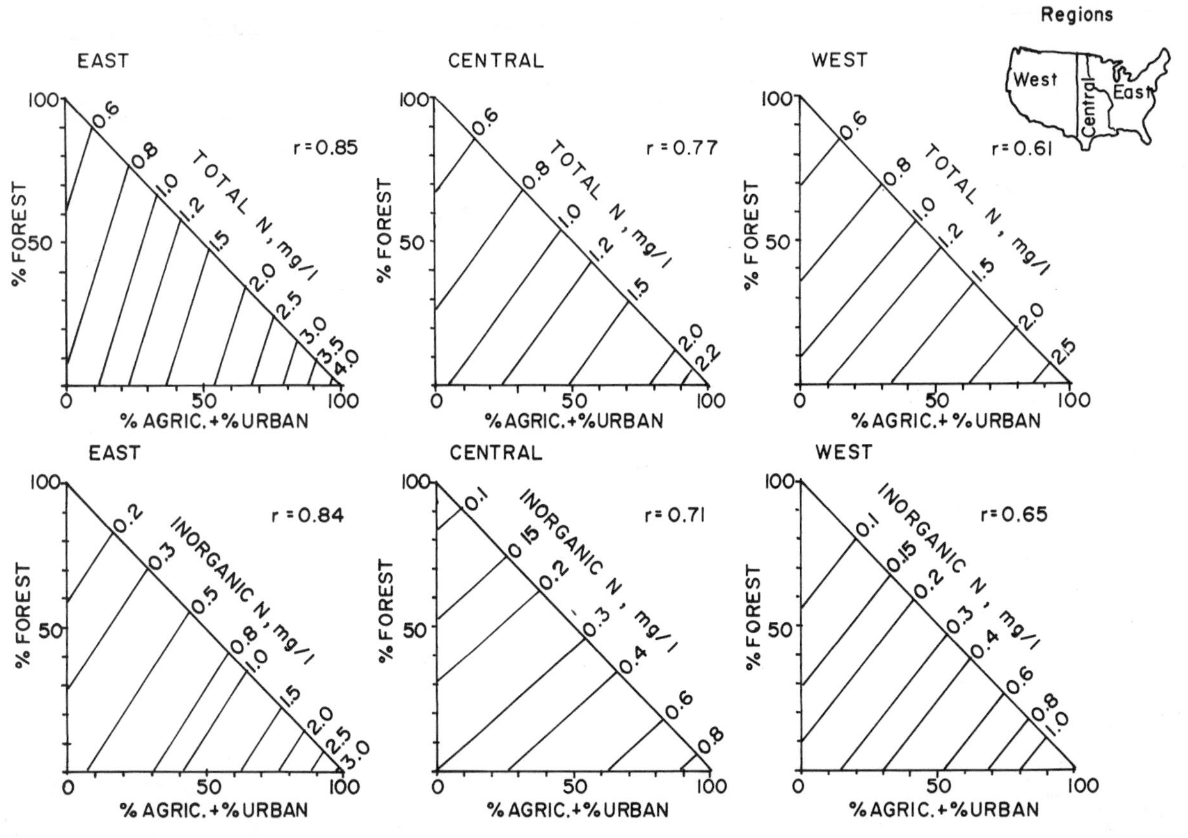

Fig. 2-8. Regional relationships between land use and nitrogen concentrations in streams. (Redrawn from Omernik.[22])

was found between the nutrient concentrations and the geology or drainage density of the watersheds.

The nutrient problem is especially important for lakes, reservoirs, and estuaries. In these water bodies, the classical dissolved oxygen concept may not work in evaluating the waste assimilative capacity, and the production of organic matter by phytoplankton and large plants–macrophytes–may greatly exceed the BOD contribution from point sources, runoff, and tributaries. Oxygen levels are also affected by photosynthesis, and respiration and BOD concentrations are affected by the planktonic organisms and their residues.

The production of organic matter in lakes and reservoirs depends on the trophic state of the impoundment. Many lakes and reservoirs have experienced accelerated eutrophication in the last few decades. The term *eutrophication* is usually associated with excessive growths of autotrophic organisms, primarily planktonic algae (*phytoplankton*) and aquatic weeds (*macrophytes*). Eutrophication is not synonymous with pollution, however. A body of water may experience excessive algal growths if it receives pollutants containing certain kinds of nutrients, primarily phosphorus and nitrogen. The eutrophication process takes place in a surface water body in which organic mass production nourished by nutrients exceeds its loss by respiration, decay, and outflow. In today's context, "Eutrophication refers to the natural and artificial addition of nutrients to bodies of water and to the effect of these added nutrients on water quality."[23]

Eutrophication is the process by which a water body progresses from its origin to its extinction according to the level of nutrient and organic matter accumulation. The rate of eutrophication depends on many environmental factors, many of them uncontrollable. It is a dynamic process with highly variable rates that differ from year to year, season to season, and even hour to hour.

The process of eutrophication and trophic states of surface waters has recently attained much public attention. Lakes Erie and Ontario and portions of Lakes Michigan and Huron have been deteriorating rapidly due to accelerated production of organic matter with consequent limitations on their beneficial uses. Many other lakes and reservoirs have become highly eutrophicated in the last few decades as a result of increased nutrient loadings from intensive farming operations and urban developments. Eutrophication problems in some reservoirs occurred after a few years of operation.

The term *oligotrophic* is used to describe the youngest stage of a surface water body–lake or reservoir–and is characterized by water with very low mineral and organic content. As the nutrient content is increased by runoff or by wastewater disposal, the photosynthetic (autotrophic) organisms increase in number. These organisms are called producers, and they initiate the entire cycle of production of organic matter in the lake. At a later state of the development process, the lake becomes *mesotrophic*, and when the nutrients and organic matter productivity are high, the water body becomes *eutrophic*. The final

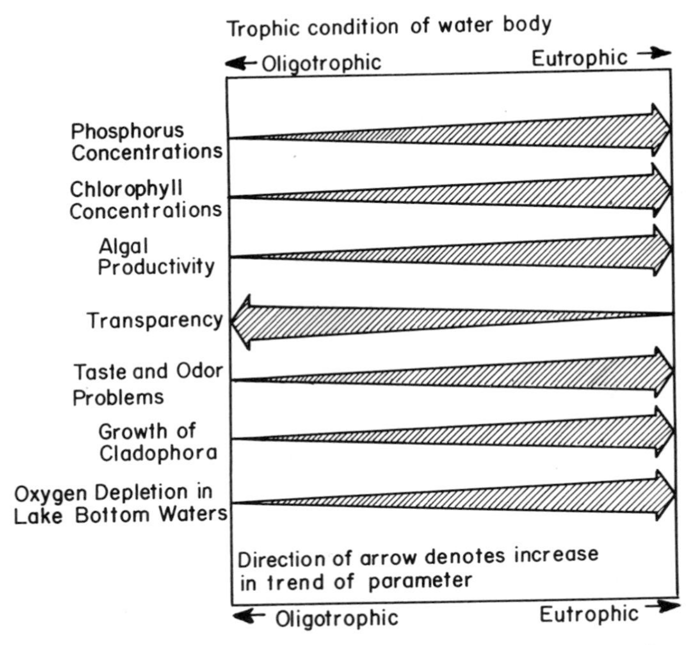

Fig. 2-9. Symptoms of eutrophication in lakes. (Adapted from reference 1.)

stages of the surface body's existence are pond, swamp, marsh, or wetland. Since eutrophication is related to the aging of water bodies, some disagreement exists as to the applicability of the concept to other nonstagnant water bodies. Streams and estuaries do not age in the same sense as lakes, although added nutrients may cause similar water quality problems. Figure 2-9 shows symptoms of eutrophication while Fig. 2-10 clearly demonstrates aesthetic impairment of water quality caused by excessive growths of planktonic algae and aquatic weeds—two of the symptoms of eutrophication.

The photosynthetic process of algae and macrophytes in eutrophic lakes and reservoirs (as well as streams) can be recognized by the cyclic fluctuations of the dissolved oxygen concentrations in the euphotic zone (a zone where light can penetrate). Oxygen is produced during the day and consumed by respiration during the night. On bright sunny days during the production season, this often results in supersaturation of upper (epilimnion) water zones by oxygen during afternoon hours and a significant drop of D.O. concentrations during the late night and early morning hours.

The related physical and chemical changes caused by advanced eutrophication (pH variations, oxygen fluctuations or lack of them in lower—hypolimnium—zones, organic substances) may interfere with recreational and aesthetic uses of

Fig. 2-10. Excessive planktonic algae growths in a reservoir located in southeastern Wisconsin receiving nutrient inputs from agricultural nonpoint sources. (Photo authors.)

water and may also cause a shift in fish population from game to rough fish. In addition, possible taste and odor problems caused by algae can make water less suitable or desirable for water supply and human consumption.

When the concentration of phytoplankton algae during the late summer period exceeds a certain threshold nuisance value, the situation is termed *algal bloom.*

Hutchinson[24] pointed out that the terms eutrophic, mesotrophic, and oligotrophic should not only be used to describe water quality but should also be related to the drainage area and the sediment transport. In this context, the term *available nutrients* should be used to describe the eutrophication potential.

The necessity of including watersheds in eutrophication systems brought about categorizing the lakes and reservoirs into *autotrophic*, that is, lakes that receive a major portion of the nutrients from internal sources (sediment, atmosphere), and *allotrophic*, which receive a major portion of the nutrients from external sources. *Allochthonous* nutrients are those originating from the watershed, that is, from the point and nonpoint sources.

Trophic Index and Trophic Status. Determination of the trophic status of lakes and other water bodies based on limnological observations, taxonomy or distribution of organisms and their productivity, and chemical water quality, is very difficult and requires great experience and often subjective judgments. A lack of a precise definition of "trophic status" makes it difficult to develop an accurate engineering tool that would enable estimation of the stage of the eutrophication process of a given water body. Prediction of the effects of various remedial measures to reduce nutrient loadings is still more or less guess work.

Some methods of estimating the trophic status (i.e., to determine whether the water body is oligotrophic, mesotrophic, or eutrophic) have evolved and have been published and/or used for classifying lakes. Many of the techniques are relative systems in which lakes are classified and ranked only in respect to each other or to an average quality of a given group of lakes and not according to some objective independent scale. These systems often give different weights to various parameters similar to those shown in Fig. 2-9, and the sum of these is then called "trophic status index." The most frequently used ones are dissolved oxygen (in the lower hypolimnetic zone), total phosphorus, transparency by Secchi disk (the depth at which a white disk can no longer be seen from the surface), inorganic nitrogen, and chlorophyll *a* concentration.

Trophic Index by Carlson.[25] This index was developed for lakes that are phosphorus limited. Carlson based his index on the fact that there are intercorrelations between the transparency expressed by the Secchi disk depth, algal concentration expressed by chlorophyll *a*, and vernal (spring) or average annual phosphorus concentrations. The trophic status index was then defined as

$$TSI(SD) = 10\left(6 - \frac{\ln SD}{\ln 2}\right) \tag{2-7}$$

where:

SD = Secchi disk depth (m)

From the correlations between the chlorophyll *a* concentrations, total phosphorus, and Secchi disk depth, the other two expressions for the *TSI* became

$$TSI(Chl) = 10\left(6 - \frac{2.04 - 0.68 \ln Chl}{\ln 2}\right) \tag{2-8}$$

and

$$TSI(TP) = 10\left(6 - \frac{\ln 48/TP}{\ln 2}\right) \tag{2-9}$$

where

Chl = concentration of chlorophyll *a* (μg/liter)
TP = concentration of total phosphorus (μg/liter).

This method offers three indices instead of one single value. The best indicator of the trophic status varies from lake to lake and from season to season. Secchi disk values may be erroneous in lakes where turbidity is caused by factors other than algae.

Based on the observation of several lakes,[26,27] most of the oligotrophic water bodies had TSI below 40, mesotrophic lakes had TSI between 35 and 45, while most eutrophic lakes had TSI greater than 45. Hypereutrophic lakes may have values above 60.

Permissible Nutrient Loadings. During investigations of eutrophication problems in Madison, Wisconsin lakes, Sawyer[28] noted that algal blooms occurred when concentrations of inorganic nitrogen (NH_4^+, NO_2^-, NO_3^-) and inorganic phosphorus exceeded respective values of 0.3 mg N/liter and 0.01 mg P/liter. It should be noted that due to the uptake of nutrients by algae, very low concentrations would be measured during the productive summer period; therefore, the critical nutrient concentrations should be evaluated during the spring overturn.

Based on the work by Vollenweider[29] and Sawyer,[28] Schindler (quoted in reference 30) developed loading functions for nitrogen and phosphorus related to the mean depth of the lake as follows:

Admissible loadings

$$\log_{10} P_A = 0.6 \log_{10} H - 1.60 \tag{2-10}$$

$$\log_{10} N_A = 0.6 \log_{10} H - 0.42 \tag{2-11}$$

Dangerous loadings

$$\log_{10} P_D = 0.6 \log_{10} H - 1.30 \tag{2-12}$$

$$\log_{10} N_D = 0.6 \log_{10} H - 0.12 \tag{2-13}$$

The phosphorus loadings, P_A and P_D, and nitrogen loadings, N_A and N_D, are in grams per square meter-year and the mean depth, H, is in meters. Oligotrophic lakes are supposed to occur at loadings below the admissible levels, while eutrophic lakes occur above the dangerous levels and mesotrophic lakes lie between the admissible and dangerous levels.

Vollenweider[31,32] then developed an input-output phosphorus model that included sedimentation as follows:

$$\text{Change in total P} = \text{inflow} - \text{sedimentation} - \text{outflow}$$

This leads to the determination of "admissible" and "dangerous" loadings

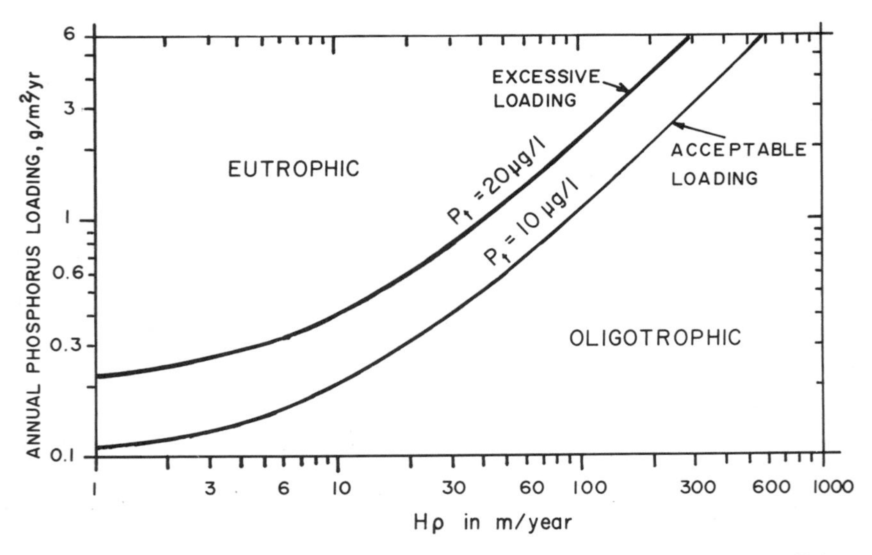

Fig. 2-11. Relationships between nutrient loading and lake trophic conditions. (Adapted from Vollenweider.[31] Courtesy *Schweizerische Zeitschrift für Hydrologie*.)

(Fig. 2-11) that are related to the depth of the lake, H, and annual flushing rate ρ = annual inflow/lake volume.

Assuming that the phosphorus is the limiting nutrient, Lee et al.[33] have expanded the Vollenweider relationships to include chlorophyll *a*, Secchi disk, and hypolimnetic oxygen depletion, thus closing the circle between the nutrient loading, water quality parameters, and trophic status. Figues 2-12 through 2-14 show the result of Lee et al. On the abscissa:

$L(P)$ = areal phosphorus loading (mg P/m^2-yr)
q_s = hydraulic loading (m/yr) = $H/\tau_w = H\rho$
H = mean depth (m) = water body volume/surface area
$\tau_w = 1/\rho$ = hydraulic residence time (yr) = water body volume/annual inflow volume

The studies indicate that nonpoint source control is essential in many lakes if phosphorus is the limiting nutrient.

Limiting Nutrients. Many components and elements must contribute to the growth of producers, but according to Liebig's law of minimum, the component that is in shortest supply will control the growth rate. This also applies to the process of eutrophication where either available nitrogen or phosphorus concentrations will control. The nutrient that is in shortest supply is then the limiting

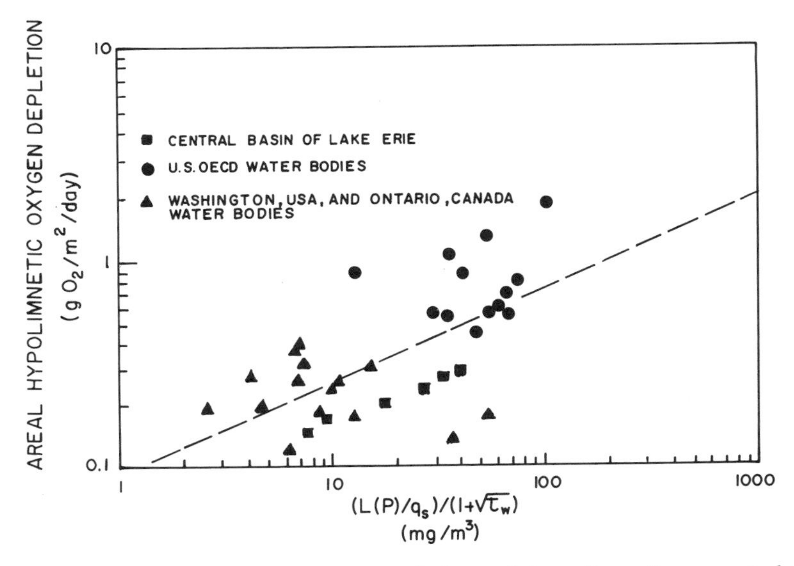

Fig. 2-12. Phosphorus loading characteristics and hypolimnetic oxygen depletion in lakes. (Redrawn from Lee et al.[33] with permission. Copyright by the American Chemical Society.)

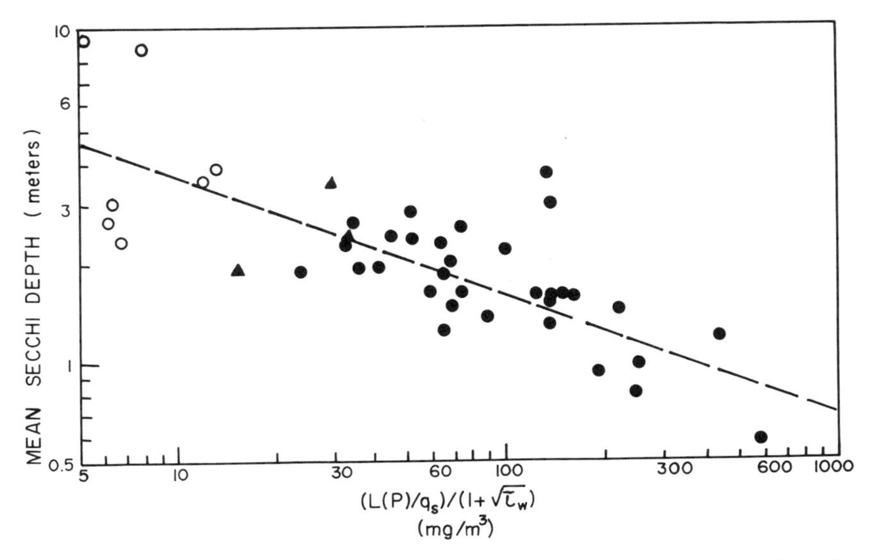

Fig. 2-13. Phosphorus loading characteristics and mean Secchi depth for lakes. (Redrawn from Lee et al.[33] with permission. Copyright by the American Chemical Society.)

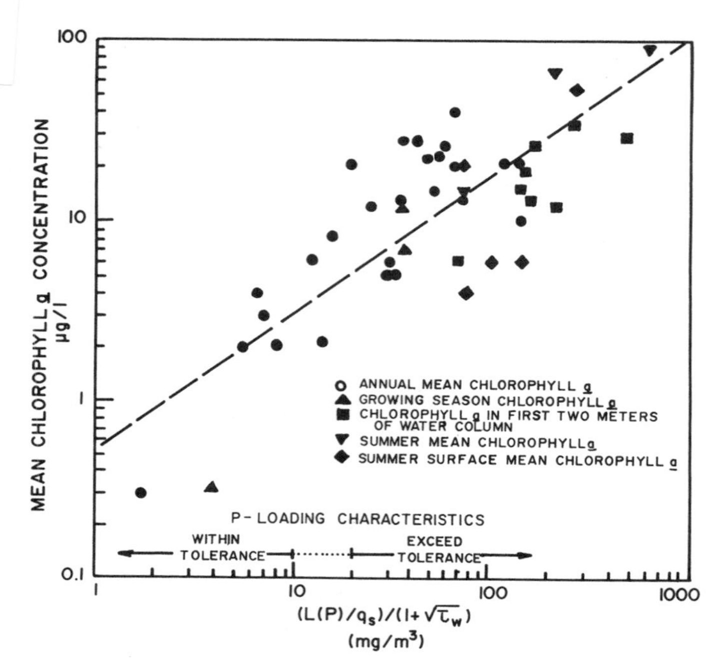

Fig. 2-14. Phosphorus loading characteristics and mean chlorophyll *a* relationship for lakes. (Redrawn from Lee et al.[33] with permission. Copyright by the American Chemical Society.)

nutrient, and controls should be focused on reducing this particular nutrient level. Most of the inland lakes are phosphorus limited, while many oceanic coastal waters including estuaries are limited by nitrogen.

To determine whether nitrogen or phosphorus is limiting the eutrophication of a particular water body, one can plot the nitrogen concentration versus consequent phosphorus concentrations on an arithmetic plot. The straight line approximation will intercept either the nitrogen or phosphorus abscissa. The nutrient that at low concentrations is exhausted first (zero or negative intercept on the abscissa) is then the limiting one.

A test that in laboratory growth conditions can determine the limiting nutrient has been recommended by the U.S. Environmental Protection Agency. In this test, the water sample containing added standard algae is spiked by nutrients and the increase of growth rates is then observed for several days. The nutrient that results in the highest growth of the culture is then considered limiting.

Example 2-2: Eutrophication Potential

A lake with 28 km^2 surface area and an average depth of 13 m has the following annual loadings:

Source	*N (kg/yr)*	*P*	*Flow (mil m³/yr)*
Urban sewage	*23,000*	*8,000*	*1.0*
Urban runoff	*15,000*	*4,000*	*7.0*
Rural runoff	*26,000*	*10,000*	*10.0*
Precipitation	*47,000*	*1,000*	*21.0*
Groundwater	*125,000*	*300*	*47.0*
Evaporation loss	–	–	*-8.0*
Total	*236,000*	*23,300*	*78.0*

Estimate the trophic level of the lake based on the Vollenweider and Lee et al. models.

$$\text{Annual flushing rate } \rho = \frac{\text{annual inflow}}{\text{lake volume}} = \frac{78{,}000{,}000\ (\text{m}^3/\text{yr})}{28\ (\text{km}^2) \times 13\ (\text{m}) \times 10^6\ (\text{m}^2/\text{km}^2)}$$

$$= 0.21\ \text{yr}^{-1}$$

Annual nutrient loading

Nitrogen

$$L_N = \frac{236{,}000\ (\text{kg/yr}) \times 10^3\ (\text{g/kg})}{28\ (\text{km}^2) \times 10^6\ (\text{m}^2/\text{km}^2)} = 8.43\ \text{g/m}^2\text{-yr}$$

Phosphorus

$$L_P = \frac{23{,}300\ (\text{kg/yr}) \times 10^3\ (\text{g/kg})}{28\ (\text{km}^2) \times 10^6\ (\text{m}^2/\text{km}^2)} = 0.83\ \text{g/m}^2\text{-yr}$$

From Equations 2-10 through 2-13:

Admissible loading

$$\log_{10} P_A = 0.6 \log_{10} (13) - 1.6 \qquad P_A = 0.12\ \text{g/m}^2\text{-yr}$$

$$\log_{10} N_A = 0.6 \log_{10} (13) - 0.42 \qquad N_A = 1.77\ \text{g/m}^2\text{-yr}$$

Dangerous loading

$$\log_{10} P_D = 0.6 \log_{10} (13) - 1.3 \qquad P_D = 0.23\ \text{g/m}^2\text{-yr}$$

$$\log_{10} N_D = 0.6 \log_{10} (13) - 0.12 \qquad N_D = 3.53\ \text{g/m}^2\text{-yr}$$

Loadings are above the dangerous levels and the lake is expected to be eutrophic.

From Fig. 2-11, the admissible and dangerous loadings for $H\rho = 13 \times 0.21 = 2.73$ *would be about the same.*

Lee et al. model: Use for example Fig. 2-13 (Secchi disk) and apply the following values:

$$q_s = H\rho = 2.73 \text{ m/yr}$$

$$\tau_w = \frac{1}{\rho} = 4.78 \text{ yr}$$

$$(L_p/q_s)/(1 + \sqrt{\tau_w}) = \frac{830 \text{ (mg/m}^2\text{-yr)}}{2.73 \text{ (m/yr)}} \Big/ (1 + \sqrt{4.78}) = 100$$

The expected Secchi disk depth would be around 2 m. Then from Equation 2-7

$$TSI(SD) = 10\left(6 - \frac{\ln 2}{\ln 2}\right) = 50$$

indicating again eutrophic conditions.

The above method renders itself also to evaluating the effect of management practices resulting in the reduction of nutrient loadings on the trophic status of receiving water bodies.

Toxic Metal Problems. Almost any element in the earth's crust can potentially be found, at least in trace quantities, in surface water bodies. Most of these elements originate from natural sources. However, increased industrialization and urbanization have raised the levels of the trace metals in surface waters, especially the levels of toxic, sometimes improperly called "heavy," metals.

The periodic table includes over 90 elements from hydrogen to trans-uranians, and all but 20 can be classified as metals. As many as 59 of these elements can be considered "heavy metals."[34]

However, only 17 of the heavy metals are considered both very toxic and relatively accessible. Of these 17 toxic metals (Table 2-6), nine are being mobilized into the environment by man at rates greatly exceeding those of natural geological processes.[34,36]

Most likely each of these nine metals occurs naturally in water to some extent, and all organisms are naturally exposed to some level. In addition, some metals, including Cd, Cu, and Zn, have known biochemical roles and others may be required as trace elements by the cell.

Since any metal may have a positive biological role, it follows that the total absence of a metal may be detrimental, that some concentrations may be optimal, and that some concentrations will be toxic. The toxicity levels will depend on the type of the metal, its biological role, and on the type of the organism and its ability to regulate its body concentrations of the metal. Table 2-7 shows a general ranking of toxicity of the nine toxic metals to aquatic biota.

PLUARG, in studying the impact of pollution of the Great Lakes, determined that the following elements should be considered potentially hazardous and requiring further attention:

TABLE 2-6. Toxic Metals of Particular Environmental Concern.[a]

Very Toxic and Readily Accessible	Man-Induced Mobilization Higher than Natural Rate
Co, Bi	Ag[b]
Ni, Cu	Cd
Zn, Sn[b]	Cu
Se,[b] Te[b]	Hg[b]
Pd,[b] Ag[b]	Ni
Cd, Pt[b]	Pb[b]
Ag,[b] Hg[b]	Sb
Tl,[b] Pb[b]	Sn[b]
Sb	Zn

[a]From Wood, 1975, and Ketchum, 1972, quoted in reference 34.
[b]Metal alkyls stable in aqueous systems and reported to be biomethylated.

1. Mercury, lead
2. Arsenic, cadmium, selenium
3. Copper, zinc, chromium, vanadium

Radioactive strontium, which in the era of atmospheric nuclear tests was of primary concern, is now below dangerous levels and for the time being is of no major concern in spite of occasional nuclear atmospheric tests by a few irresponsible nations.

The ranking is based on environmental potential or an existing hazard to the Great Lakes. The elements of concern and their environmental hazard may change for other geographical locations.

Toxic metals are most commonly added to streams as salts—sulfides, carbonates, and phosphates—all of which are quite insoluble in waters with appreciable hardness and travel mostly with the sediment.

The primary sources of toxic metal pollution are mining and smelting operations, and transportation (lead).

TABLE 2-7. General Ranking of Toxicity of Metals in the Aquatic Environment.[34]

Metal	Ag, Cd, Hg	>	Cu	>	Ni, Pb, Zn	>	Sb	>	Sn
Toxic level	10^{-8} *M*		10^{-7} *M*		10^{-6} *M*		10^{-5} *M*		10^{-4} *M*

TABLE 2-8. Background Levels of Toxic Metals (μg/liter) in Pristine Soft-Water Lakes in Areas not Receiving Highly Polluted Precipitation.[a]

Region	No. of lakes	Zn	Pb	Cu	Cd
Central and northern Norway	110	0.5–12	0.5–2.0	0.5–2	0.1–0.5
Northern Sweden	–	10–30	–	–	0.05–0.2
Sierra Nevada, California	170	0.7–5	0.3–2	0.5–3	3
Northwestern Ontario	102	1	1	1–3	0.1

[a]Courtesy *Water Research*, Pergamon Press.[37]

A large portion of toxic metals come from atmospheric fallout. Measurements of toxic metals in snow and ice from Greenland indicated that the concentrations of Pb in snow is today about 0.2 μg/liter versus only 0.001 μg/liter 3000 years ago.[35] Similarly, levels of Cu have risen from 0.15 μg/liter in 1900 to 0.85 μg/liter and Zn from 0.4 to 1 μg/liter.[36] The present-day levels in snow from Greenland are similar to those measured in pristine soft-water lakes in the Northern Hemisphere and could represent "background" levels for most of North American and European surface waters (Table 2-8).

The modification and transformation of toxic metals in aquatic environments is quite complex and depends on many factors such as pH, hardness, and presence of sediment. Many of the elements including mercury, lead, selenium, and arsenic can undergo methylation by bacteria whereby elements are converted from their metallic state into a potent toxic organometallic component.[38,39] For example, mercury can be converted into a nerve poisoning compound, monomethyl (CH_3-Hg) and bimethyl mercury (CH_3-Hg-CH_3), which has caused human poisoning in Japan (Minamata disaster). The itai-itai disease in Japan resulted from people ingesting fish contaminated with Cd; the cadmium replaced calcium in bones.[2] Both disasters, however, resulted from point pollution of receiving waters by industrial complexes, and no incidents of such magnitude are known to have resulted from nonpoint pollution.

The process of methylation takes place in lake and stream sediments.[1,38,39] Algae and aquatic vegetation have a capacity to trap and concentrate toxic metals in their tissues. For example, quantities of lead accumulated in aquatic vegetation in lead mining areas have reached several percent on a dry weight basis.[40] This can initiate the food web effect when metal contamination from producers (algae, vegetation) is passed to consumers (benthic invertebrates, fish). Concentrations of metals in fish in the Great Lakes are reported in Table 2-9.

In addition to their toxic effect in higher concentrations, toxic metals in lower concentrations can affect the metabolism of bacteria, increasing their uptake and thus altering the microbiological composition of aquatic bodies.

TABLE 2-9. Concentrations of Metals in Great Lakes Fish.[1]

Lake	Hg[a] (mg/kg of wet weight)	Pb[b] (mg/kg of wet weight)
Superior	0.07–0.78	0.012–0.066
Michigan	0.22–0.54	0–0.54
Huron	0.06–0.18	0.04–0.1
St. Clair	0.06–3.8	0.47–0.63
Erie	0.03–1.53	0.04–0.12
Ontario	0.06–0.49	1.0

[a]Accepted guideline concentration for Hg is 0.5 mg/kg.
[b]Accepted guideline concentration for Pb is 10 mg/kg.

Bacteria have been known to pass their metal contamination to invertebrates feeding on these cells, and the invertebrates can then pass these metals on to predatory fish.[41]

Table 2-10 lists some toxic metals, their nonpoint sources, and recommended stream criteria.

TABLE 2-10. Some Toxic Metals that can be Found in Surface Waters.[2]

Metal	Nonpoint Source	Alleged Effects	Recommended Drinking Water Standards	Limits by EPA Criteria Aquatic Life
Arsenic	Pesticides, soil	Chronic, cumulative, carcinogenic, cardiovascular effects	50 μg/liter	0.04 mg/liter
Cadmium	Soil, industrial pollution	Arteriosclerosis, cancer, itai-itai disease	10 μg/liter	0.01–1.2 μg/liter[a]
Chromium	Industrial pollution	Cancer, ulceration	50 μg/liter	0.29 μg/liter
Lead	Transportation, industrial pollution	Cumulative, plumbism	50 μg/liter	0.1–3 μg/liter[a]
Mercury	Soil, industrial pollution, fungicides	Methylated cause Minamata disease	2 μg/liter	0.0005 μg/liter
Zinc	Soil	Taste problem	5 mg/liter	47 μg/liter

[a]Depends on hardness of water according to U.S. EPA criteria for toxic substances.

Organic Chemical Problems. The past few decades, especially the period of the 1970s, have been an era of controversy and concern over benefits and environmental cost of the use of organic chemicals, namely pesticides and polychlorinated biphenyls (PCBs).

The monitoring of DDT and dieldrin in fish species in the Great Lakes added greater evidence to the pesticide problem.[42] High salmon mortalities were observed in Lake Michigan, where DDT in fish tissues averaged two to five times higher than those from Lake Superior and 60 times higher than those from the Pacific west coast.

Of most concern to those interested in water quality is the potential concentration increase of toxic materials that can occur via the food web. The process called *biomagnification* is of particular significance to the pesticide and PCB problem since large quantities of these toxicants can be found in higher trophic level organisms (end of the food web), while lower trophic organisms and water itself exhibit lower, often undetectable concentrations. The process of biomagnification is demonstrated in Table 2-11. It is interesting to note that DDT, for example, was magnified 1350 times through four trophic levels.

There are tremendous amounts of organic chemicals that can be potentially dangerous to the aquatic environment. By their very purpose, all pesticides are toxic and can cause damage. The amount and nature of pesticides reaching surface waters from agricultural lands are primarily functions of the persistence of the compound used, intensity and length of time pesticides have been applied, and transport mechanisms from the area of application to receiving waters.[44]

Many organic chemicals, including chlorinated hydrocarbons, are potentially carcinogenic. Some of the potentially dangerous chemicals such as persistent

TABLE 2-11. Concentrations of *p,p'*-DDT in a Marine Environment.[43]

Species	Trophic Level	*p,p'*-DDT (mg/kg)
Oar weed	1	0.003
Microzooplankton	2	0.03
Macrozooplankton	3	0.16
Lobster	3	0.024
Shag liver	4	2.87
Herring gull liver	4	0.456
Cormorant liver	5	4.14
Common dolphin blubber	5	1.28

TABLE 2-12. Organic Chemicals of Present Concern to Great Lakes.

Chemicals	Recommended Limits by EPA
Phenolics	3 μg/liter aquatic life
PCB	0.014 μg/liter aquatic life
Phthalate	3 μg/liter aquatic life
DDT	0.001 μg/liter aquatic life
DDD	
DDE	
Heptachlor	0.0038 μg/liter aquatic life
Aldrin	3.0 μg/liter aquatic life
Lindane	4 μg/liter drinking water
	0.08 μg/liter aquatic life
Heptachlor epoxide	
Dieldrin	
Methoxychlor	100 μg/liter drinking water
	0.03 μg/liter aquatic life
MIREX	

*DDT derivatives.

pesticides and PCBs have become a part of the environment and can travel by air, by water, and with the sediment. Many of them, for example DDT and most PCBs, are highly insoluble, hydrophobic (water repellent) substances and travel mostly with air aerosols and aquatic particulate matter.

Table 2-12 shows some organic chemicals of present concern in the Great Lakes. Continued research is needed to understand the significance of pesticides and other organic chemicals to fish and aquatic biota. The contamination of surface waters by pesticides and other organic chemicals can have a great impact on humans, commercial fishing, fish propagation, and survival of aquatic species in general. For example, commercial fishing in the Great Lakes has been drastically reduced, and interstate sale of fish has been virtually banned, resulting in great economic losses to commercial and sport fishing.

Polychlorinated biphenyl concentrations in the Great Lakes fish seem to give the final blow to commercial fishing. PCBs have been in wide industrial use since the 1930s. The chemical properties that make PCBs desirable industrial materials are their excellent thermal stability, their strong resistance to both acidic and basic hydrolysis, and their general inertness. Despite their overall use and known toxic effects, their presence in the environment and in surface waters in particular was not discovered until the late 1960s in Sweden and the United States.[45] Subject to these findings, the production of PCBs in the United States was voluntarily limited by the producers. In spite of the production curtailment in 1972, the PCBs are now in the environment. Although most of the produced PCBs are in closed systems as media in transistors and large capacitors or in heat transfer and hydraulic systems that can be controlled and/or recovered, leakages and losses do occur. In addition, PCBs used in lubricating and cutting oils, in some insecticides, as sealers in waterproofing compounds, in asphaltic materials, etc., are in direct contact with the environment, and their entry into environmental systems is very difficult to control.

In aquatic environments, PCBs are highly hydrophobic and travel solely with sediment. They can be biomagnified, and their present concentrations in the Great Lakes significantly exceed the recommended limits (Table 2-13). In addition to the industrial sources, PCBs can be found in the air particulate matter, soil, urban dust, and bottom sediments of many surface water bodies. In most cases (with a few notable exceptions such as industrial losses of PCBs in Waukeegan, Illinois, or Sheboygan, Wisconsin), the entry of PCBs into the environmental systems, including surface waters, can be considered a nonpoint source. The total annual worldwide loss of PCBs into the environment has been estimated to be of the order of 1.5 to 2×10^3 tonnes/yr into the atmosphere, 4 to 5×10^3 tonnes/yr into fresh and coastal waters, and 1.8×10^3 tonnes/yr into dumps and landfills.[46]

In spite of the limitation of the PCB manufacturing, these components are still finding their way into certain streams and lakes (especially the Great Lakes, which are characterized as a PCB sink).

Figure 2-15 shows the distribution of PCB concentrations in the sediment of the lower Great Lakes. It can be clearly seen that the highest PCB concentrations occur near larger urban centers. This indicates that urban areas are the primary PCB sources.

Data gathered by the National Water Monitoring Program of the Environmental Protection Agency indicated that the occurrence of PCBs in surface waters and bottom sediments is widespread throughout the major drainage basins of the United States. Median residue levels of the positive detections for the years 1971 to 1974 ranges between 0.1 to 3 μg/liter for unfiltered water samples and from 1.2 to 160 μg/kg for bottom sediments. The highest levels were found in basins east of the Mississippi River.[47]

pH and Acidity. The pH, which expresses the molar concentration of the hydrogen ion as its negative logarithm ($pH = -\log[H^+]$), is one of the primary indicators used for evaluation of surface water quality and suitability for various beneficial uses. Most aquatic biota are sensitive to pH variations. Fish kills and reduction and change of other aquatic species result when the pH is altered outside their tolerance limits. Most of the aquatic species prefer a pH near neutral but can withstand a pH in the range of about 6 to 8.5.

TABLE 2-13. Summary of PCB Levels in Great Lakes Fish.[1]

Lake	Years of Sampling	Mean PCB Concentration All Samples (mg/kg)	Range (mg/kg)	Percentage of Major Species Exceeding Guidelines
Superior	1968–1975	0.61 (2.0)[a]	<0.1–3.7	75%–whitefish (1974)–Marathon 20%–chub (1975)
Michigan	1972–1974	10.2 (5.0)	2.1–18.9	13%–chubs 50%–lake trout
Huron	1968–1976	0.82 (2.0)	<0.1–7.0	50%–coho salmon (1971) 75%–rainbow trout (1974)–Douglas Point 33%–rainbow trout (1974)–Goderich 10%–rainbow trout (1974)–Nottawasaga River
Erie	1968–1976	0.88 (2.0)	<0.1–9.3	27% of all samples analyzed had, in the range of concentrations for a sample, an upper PCB concentration >2 mg/kg. Of these, 75% were from the western and central basins. Species with the largest number of individuals exceeding the guideline were coho salmon, freshwater drum, white bass and channel catfish.
Ontario	1972–1977	2.37 (2.0)	<0.1–21.1	30%–carp (1972–74) 100%–coho salmon (1972–73) 27%–rock bass (1972–74) 88%–catfish (1972–74) 0%–sunfish (1972–74) 36%–white perch (1972–74) 42%–northern pike (1972–73) 0%–cisco 77%–smelt (1972–74) Smelt, coho salmon and lake trout had PCB levels in whole fish ranging from 0.4 to 16.2 mg/kg. Mean concentrations in fish from the eastern basin were highest at 5.31 mg/kg (1977).

[a]Accepted guideline for fish.

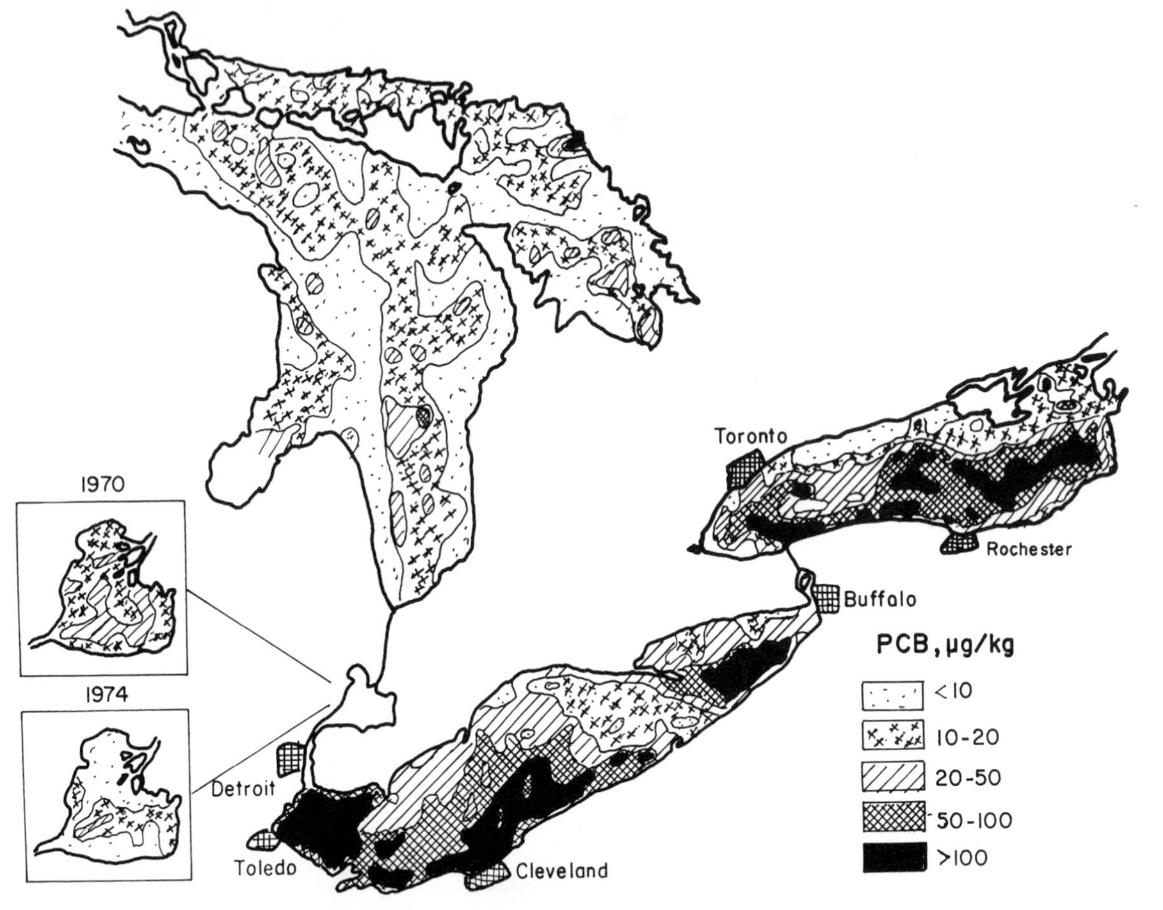

Fig. 2-15. PCB concentrations in surface sediments of Lakes Huron, Erie, and Ontario. (From reference 1.)

The toxicity of other toxic components can also be altered if the pH is changed. The solubility of many metals as well as other compounds (ammonia) is affected by the pH, resulting in increased toxicity in the lower pH range.

Change of pH and acidity of surface waters resulting from nonpoint inputs can occur mainly from two sources of acidity: (1) acid mine drainage water, and (2) acid precipitation.

Both sources have similar origins. Acid mine drainage is a result of mine water being in contact with sulfur-bearing minerals, while acidity of precipitation is caused primarily by atmospheric sulfur. Oxidation of these sulfuric compounds in surface or atmospheric water produces sulfuric acid, which then dissociates to H^+ and SO_4^{2-} ions.

Acid rain, which is defined as rain with a pH less than 5.6, is a result of sulfuric and nitrate emissions from urban, industrial, and electric utility burning operations that use sulfur- and nitrogen-containing fuel (primarily coal or, to a lesser degree, low-quality oil).

Oxidation of SO_2 to SO_3 (sulfates) is greatly enhanced by the presence of metallic catalysts. Chapter 4 contains a more detailed analysis of the acid rain phenomenon and its occurrence.

Lethal and sublethal effects of acid rain or acid mine drainage have been noticed both in the United States and in Europe (Scandinavia).[48-51] Undesirable "oligotrophication" (a severe loss of productivity by the low pH conditions) and fish kills are the most visible and dangerous consequences of acidification. Loss of natural fish populations due to acidic rain and snow input in New York's Adirondack Mountains and in many pristine lakes of North America and Scandinavia has been documented and widely publicized. The damage to the fish populations of these lakes was brought about by both a long-term decrease of pH and short-term pH shocks by runoff from large storms and snow-melt.

Many watersheds and surface water systems have a natural ability to neutralize the excess acidity. During the overland flow, rain and snow-melt water dissolves calcium- and magnesium-containing rocks (limestone and dolomite) or soils, and it is enriched by mineral and organic salts such as phosphates and humates. These constituents often provide enough buffering capacity to maintain the pH of surface waters within acceptable ranges. At this time, rain acidity does not seem to have a great adverse effect on larger water bodies such as the Great Lakes that have elevated hardness and salinity contents.

The ability of surface waters to neutralize acidic inputs depends primarily on the carbonate (CO_3^{2-}) and bicarbonate (HCO^-) content that is expressed as alkalinity. However, many North American and Scandinavian lakes are particularly sensitive to acid inputs. These lakes have watersheds underlain by siliceous hard rocks such as granite, some gneisses, quartzite and quartz sandstone. These materials are highly resistant to weathering and produce waters that contain very low concentrations of neutralizing components (alkalinity less than 30 mg

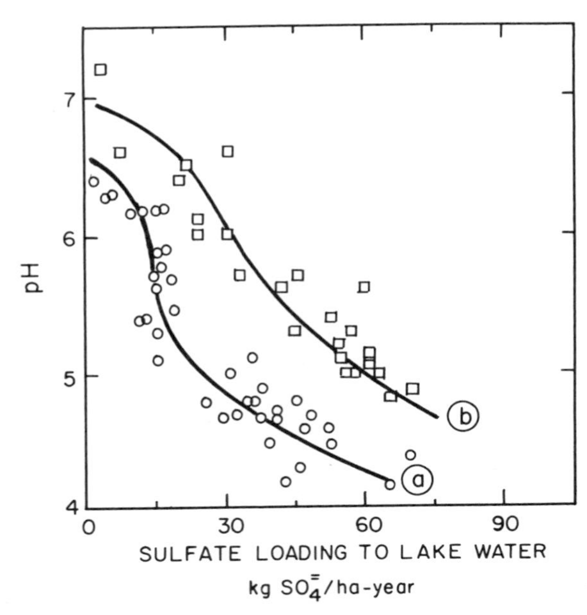

Fig. 2-16. Sulfate loading effect on pH of Swedish lakes. (a) Very sensitive lake systems. (b) Less sensitive lake systems. (Replotted from Glass et al.[52] Data from National Swedish Environmental Protection Board, Jolna, Sweden.)

$CaCO_3$/liter). When acid rain falls on such watersheds the acids are not neutralized during the overland flow, and streams and lakes become acidified (Fig. 2-16).[51]

Areas of highly siliceous bedrock are widespread on the Precambrian Fennoscandian Shield in Scandinavia, the Canadian Shields, the Rockies, New England, the Adirondack Mountains of New York, the Appalachians, and smaller areas elsewhere (Fig. 2-17).

As reported by Likens et al.,[51] acidification of thousands of freshwater lakes and streams in southern Norway has affected fish populations in an area of 33,000 km^2. In addition to damages to fish population, other adverse effects can be attributed to acidification. Bacterial decomposition is reduced at a lower pH, nitrification does not exist in low-pH waters, and changes occur at all levels of the food web.

Although not yet documented in the literature, urban streams and lakes may also be highly amenable to acidification especially when their drainage area is highly impervious and has a developed storm sewer drainage system. In these cases, urban dust and dissolution of calcium-containing building materials (concrete, limestone, rocks, etc.) are the only buffering agents available to neutralize the elevated acidity of urban precipitation.

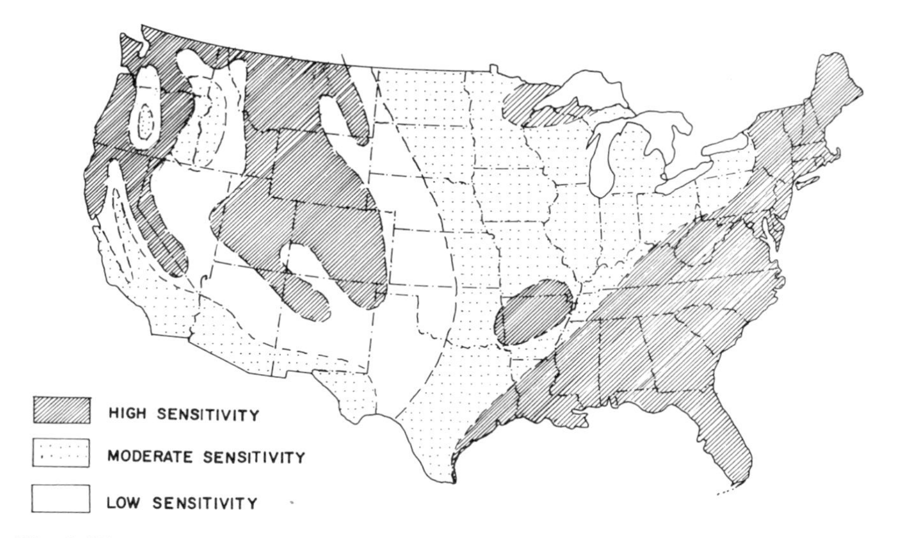

Fig. 2-17. Areas of the continental United States sensitive to acid deposition based on soils, climatic patterns, geology, and types of vegetation. (Source U.S. EPA).

Biological Availability of Pollutants

A significant portion of pollutants from nonpoint sources is carried by—or is a part of—suspended solid loads. Most of the pollutants in the particulate form, however, are not readily available to the initiators of the food chain—the aquatic algae and green photosynthetic organisms. On the other hand, most of the dissolved pollutants can be readily picked up by the algae, contaminate their tissues, and initiate the food web process. By biomagnification, concentrations of the pollutants in higher trophic levels can increase by several orders of magnitude. Thus, assessing the effect of nonpoint pollution on aquatic biota requires an understanding of the biological availability of different pollutants entering the water bodies.

The biologically available forms of pollutants—primarily nutrients, toxic metals, and organic chemicals—are those that are dissolved or can be extracted from the sediment by desorption, dissolution, or elutriation. Therefore, the adsorption-desorption and precipitation-dissolution reactions are the primary driving mechanisms determining the availability of a pollutant.[53] These reactions can be affected by various environmental factors including pH, temperature, and presence or absence of some organic compounds or inorganic catalysts.

Availability of Phosphorus. Almost 75% of the total phosphate input to the Great Lakes is in particulate form. However, only dissolved phosphate is available

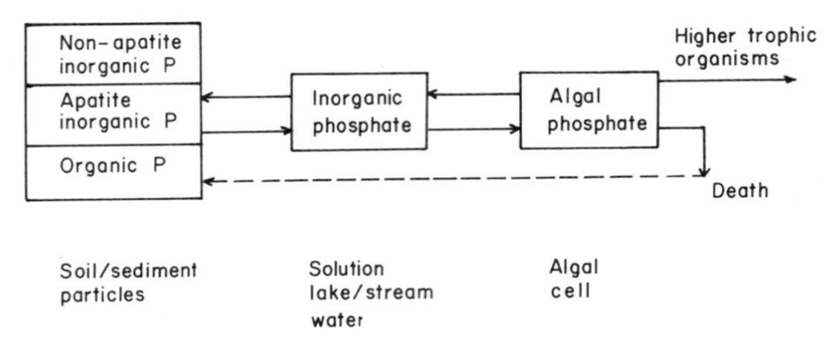

Fig. 2-18. Relation between particulate, dissolved, and algal phosphate. (Adapted from Armstrong et al.[53])

to algae and determines the rate of eutrophication. Consequently, the biological availability of phosphorus in suspended sediments (particulate P) is determined by the rate and extent of conversion to dissolved inorganic phosphate.[53]

The relationship between particulate phosphorus, dissolved inorganic phosphate, and algal phosphorus is illustrated in Fig. 2-18. Particulate P, in the sediment, can be divided into nonapatite (largely Fe- and Al-associated), apatite, and organic forms. The inorganic forms tend to control the phosphate concentration in solution through adsorption-desorption and precipitation-dissolution reactions. Particulate organic P can release dissolved organic P to solution, but dissolved organic P must be converted to inorganic P before it becomes available.

The factors controlling the availability of particulate P in suspended sediments include:[53]

1. The forms and amounts of phosphorus in the particulate fraction.
2. The residence time of the particle in the lake water.
3. The phosphorus status of the algal population.
4. The solution phosphate concentration maintained by the algal or other phosphorus sinks.
5. Other factors such as pH.

The available phosphorus can be measured by extraction of P from the sediment by NaOH (nonapatite inorganic fraction) and HCl (apatite P). Nonapatite P is much more readily available (within 48 h) than the other two fractions of the particulate P.[53] Available P (NaOH-P), expressed as a percent of total P, averaged from 14 to 35% in five investigated tributaries to the Great Lakes.[53]

Nitrogen Availability. The quantities of nitrogen in suspended sediments range from 0.02 to 10% and represent mostly nitrogen from eroded soils and sewage sludge. Nitrogen in aquatic environments can exist as organic N (particulate and dissolved), ammonia (dissolved or adsorbed on sediments), nitrite and nitrate N, and dissolved nitrogen gas, N_2. Due to the fast conversion of nitrite form to nitrate by *nitrobacter*, nitrites can be found only in trace quantities. Dissolved atmospheric N_2 is of little relevance.

Available nitrogen is defined as that fraction of the total N that can be readily or moderately assimilated by either phytoplankton (small photosynthetic aquatic organisms) or macrophytes (larger rooted or floating aquatic plants). The most important of these are the inorganic nitrogen and simple hydrolyzable organic compounds containing free amino or amide groups. The inorganic nitrogen forms in solution (ammonium and nitrate) are considered directly available, while organic forms are made available through conversion to inorganic nitrogen by mineralization. Ammonia can exist either as fixed by soil and organic particles or as exchangeable ammonium. Fixed ammonia is considered unavailable.[53]

Simple organic compounds can undergo rapid ammonification to ammonia and further by nitrification to nitrate. The process of mineralization of aquatic organic nitrogen is somewhat similar to that occurring in soils. (Chapter 6).

Measured available nitrogen in sediments of the Great Lakes tributaries ranged from 52 to 73% (mean values) of the total nitrogen in the suspended sediments. The available nitrogen was measured as exchangeable ammonium, nitrate, and a portion of hydrolyzable organic nitrogen.[53]

Toxic Metals and Organics Availability. The availability of these compounds is determined mainly by precipitation-dissolution reactions (metals) and adsorption-desorption processes (organics). Precipitation is strongly affected by the pH of the water; therefore more metallic ions can be dissolved, and hence they become more available, at lower pH values. Investigations also indicate that humic acids, for example, from decaying leaf sediments in forest areas, increase dissolution of toxic metals by decreasing the pH and by formation of toxic metals–humic acid complexes.[54] Availability of heavy metals to first trophic level organisms can be measured as the fraction extracted by a hydroxylamine hydrochloride reagent or by a chelating cation exchange resin.[53]

Schneider, Jeffries, and Dillon[55] noticed increased mercury content in fish in Ontario lakes as a result of the decrease in pH due to acid precipitation inputs.

Mean values of the available metal fractions for the Great Lakes tributaries[53] ranged from 25 to 45% of the sediment total toxic metal content (Cu, Pb, and Zn).

Many pesticides and organic chemicals, including the most dangerous ones, can be readily adsorbed–fixed–by suspended sediment. In the particulate form they are not readily available and are not taken up by algae and macrophytes.[44] The adsorption-desorption process and general desorptivity of the chemical then determines the extent of availability of the organic pollutants to aquatic orga-

TABLE 2-14. Approximate Regional Natural Water Quality–Average Annual Concentrations.[56]

Parameter	U.S. Region				
	Eastern	Midwest	Great Plains	Mountains	Pacific
Suspended sediment (mg/liter)	5–10	10–50	20–100	5–20	2–5
BOD_5 (mg/liter)	1.0	1–3	2–3	1–2	1
Nitrate N (mg/liter)	0.05–0.2	0.2–0.5	0.2–0.5	0.1	0.05–0.1
Total P (mg/liter)	0.01–0.02	0.02–0.1	0.1–0.2	0.05	0.05–0.1
Total coliforms (MPN/100 ml)	100–1000	1000–2000	500–2000	100	100–500

nisms. However, not much is known about the process and further research is necessary.

Background Water Quality

As defined in Chapter 1, background water quality represents the chemical and biological composition of surface waters that would result from natural causes and factors. Natural or background water quality should be measured rather than estimated. Almost every river basin has headwater reaches that are representative of undisturbed drainage areas. Background water quality represents the limit of the waste assimilation process to which polluted waters should eventually approach during the recovery.

The natural or background water quality is also the goal of water pollution abatement measures, including nonpoint pollution control practices. It is impossible to eliminate all sediment from rivers as well as it is impossible to bring BOD down to zero or oxygen concentrations to saturation values.

A study by the Midwest Research Institute compiled the data of the average concentrations of various pollutants from undisturbed streams.[56] The values from 57 stations comprising the National Hydrologic Benchmark Network showed distinct regional distribution as summarized in Table 2-14. However, it should be noted that the ranges for BOD_5, nitrate, suspended solids, phosphate, and bacteria reported in Table 2-14 are based on approximately only one station per state.

Table 2-8 reported background values of some toxic metals, and background mineral contribution from groundwater will be discussed in Chapter 7. It must be remembered that there are no background natural levels for organic chemicals (pesticides, PCBs, organic chlorinated hydrocarbons, etc,) since these substances are man-made and are alien to nature. The apparent background levels of these substances mostly originate from atmospheric fallout.

The natural levels of organic pollution as well as sediment loads depend on the type of stream. Even within the same geographical region, background natural quality will vary between mountain streams and lowland rivers, between small creeks and large rivers, and between streams draining wetlands and those draining prairies. Mountain streams, especially those draining glaciers and snowfields, exhibit very low mineral and organic contents, whereas lowland sluggish-moving streams are rich in organic life, including bacteria, and usually have high mineral content. The water quality of stagnant water bodies also depends on the trophic state of the natural eutrophication process.

The establishment of water quality goals in accordance with the natural water quality levels is imperative for all water pollution control studies.

REFERENCES

1. PLUARG. 1978. Environmental management strategy for the Great Lakes systems. International Joint Commission, Windsor, Ontario.
2. Krenkel, P. A., and Novotny, V. 1980. "Water Quality Management." Academic Press, New York.
3. Patrick, R. 1975. Some thoughts concerning correct management of water quality. In: W. Whipple, Jr. (ed.), "Urbanization and Water Quality Control." AWRA, Minneapolis, Minnesota.
4. LeFoll, Y., Pinot, R., and Lesouef, A. 1977. A multi-dimensional analysis of the results of the French 1971 surface water quality network control in the river basin, Seine-Normandie. *Prog. Water Tech.*, **9**:89–102.
5. Anon. 1968. Water quality criteria. Rep. of the Nat. Tech. Advisory Committee, FWPCA, Washington, D.C. Reprinted EPA, Washington, D.C., 1972.
6. Streeter, H. W., and Phelps, E. B. 1925. A study of the pollution and natural purification of the Ohio River. Public Health Bull. 146, U.S. Public Health Service, Washington, D.C.
7. Phelps, E. B. 1944. The oxygen balance. *Stream Sanitation*. John Wiley & Sons, Inc., New York.
8. Smith, R., and Eilers, R. G. 1978. Effect of stormwater on stream dissolved oxygen. *J. Env. Eng. Div.*, ASCE, **104(EE4)**:549–560.
9. Meadows, M. E., Weeter, D. W., and Green, J. M. 1978. Assessing nonpoint water quality for small streams. *J. Env. Eng. Div.*, ASCE, **104(EE6)**:1119–1133.
10. Wanielista, M. 1978. Stormwater management; Quantity and Quality. Ann Arbor Science, Ann Arbor, Michigan.
11. Thomann, R. V. 1972. System analysis and water quality management. Environmental Research Application, New York.

12. Novotny, V., and Krenkel, P. A. 1975. A waste assimilative capacity model for a shallow, turbulent stream. *Water Res.*, **9**:233–241.
13. Krenkel, P. A., and Novotny, V. 1979. River water quality model construction. In: H. Shen (ed.), "Modeling of Rivers." Wiley-Interscience, New York.
14. O'Connor, D. J., and Dobbins, W. E. 1958. Mechanism of reaeration in natural streams. *Trans. ASCE*, **123**:641.
15. Churchill, M. A., Elmore, H. L., and Buckingham, R. A. 1962. The prediction of stream reaeration rates. "Advances in Water Pollution Research, Vol. 1, Proceedings of the First International Conference, IAWPR." Pergamon Press, London.
16. Owens, M., Edwards, R. W., and Gibbs, J. W. 1964. Some reaeration studies in streams. *Int. J. Air Water Poll.*, **8**:469.
17. Langbein, W. B., and Durum, W. H. 1967. The aeration capacity of streams. U.S.G.S. Circular No. 542, U.S. Dept of Interior, Washington, D.C.
18. Filos, J., and Molof, A. 1972. Effect of benthal deposits on oxygen and nutrients economy of flowing waters. *J. WPCF*, **44(4)**:644–662.
19. Edberg, N., and Hofsten, B. V. 1973. Oxygen uptake of bottom sediments studied in situ and in the laboratory. *Water Res.*, **7(9)**:1285–1294.
20. Kreutsberger, W. A., Meinholz, T. L., Harper, M., and Ibach, J. 1980. Predicting the impact of sediments on dissolved oxygen concentrations following combined sewer overflow events in the Lower Milwaukee River. *J. WPCF*, **52**:192–201.
21. Paterson, D. J., Epstein, E., and McEvoy, J. 1975. Water pollution investigation: Lower Green Bay and lower Fox River. U.S. EPA, 905/9-74/D17, Washington, D.C.
22. Omernik, J. M. 1977. Non-point source-stream nutrient level relationship: A nation wide study. EPA-600/3-77/105, U.S. EPA, Corvallis, Oregon.
23. Rohlich, G. A. (ed.) 1969. Eutrophication: Causes, consequences, correctives. National Academy of Science, pp. 3–7, Washington, D.C.
24. Hutchinson, G. E. 1969. Eutrophication, past and present. In. G. A. Rohlich (ed.) "Eutrophication: Causes, Consequences, Correctives." NAS, pp. 17–26, Washington, D.C.
25. Carlson, R. E. 1977. A trophic state index for lakes. *Limnol. Oceanog.*, **22**:361–369.
26. Sloey, W. E., and Spangler, F. L. 1978. Trophic status of the Winnebago pool lakes. Proc. 2nd Ann. Conference, Wis. Sec. AWRA, WRC-Univ. of Wisc., Madison.
27. Uttormark, P. D., and Hutchins, M. L. 1978. Input/output models as decision criteria for lake restoration. Tech. Rep. Wisc. WRC-78-03, Water Res. Center, Univ. of Wisc.
28. Sawyer, C. H. 1947. Fertilization of lakes by agricultural and urban drainage. *New England Water Works Assoc.*, **61**:109–127.
29. Vollenweider, R. A. 1968. Scientific fundamentals of the eutrophication of lakes and flowing water with particular reference to nitrogen and phosphorus as factors in eutrophication. Organ. Econ. Coop. Dev., Paris, Technical Rep. No. DAS/CSI/68-27, 159 pp.

30. Porcella, D. B., Bishop, A. B., Andersen, J. C., Asplund, O. W., Crawford, A. B., Greeney, W. J., Jenkins, D. I., Jurinek, J. J., Lewis, W. D., Middlebrooks, E. J. and Walkingshaw, R. W. 1973. Comprehensive management of phosphorus water pollution. EPA-68-01-0728, Washington, D.C.
31. Vollenweider, R. A. 1975. Input-output models with special reference to the phosphorus loading concept in limnology. *Schweiz. Z. Hydrol.*, **37**:53–83.
32. Vollenweider, R. A. 1976. Advances in defining critical loading levels for phosphorus in lake eutrophication. *Mem. Ist. Ital. Idrobiol.*, **33**:53–83.
33. Lee, G. F., Rast, W., and Jones, R. A. 1978. Eutrophication of water bodies: Insights for an age-old problem. *Env. Sci. Tech.*, **12**:900–908.
34. Chapman, G. 1978. Toxicological considerations of heavy metals in the aquatic environment. In: "Toxic Materials in the Aquatic Environment," Oregon State Univ., WRI, Corvallis, Oregon.
35. Murozumi, M., Chow, T. J., and Patterson, C. 1969. Chemical concentrations of pollutant lead aerosols, terrestrial dust, and sea salts in Greenland and Antarctic snow strata. *Geochim. Cosmochim. Acta.*, **33**:1277–1294.
36. Weiss, H., Bertine, K., Koide, M., and Goldberg, E. D. 1975. The chemical composition of Greenland glacier. *Geochim. Cosmochim. Acta.*, **29**:1–10.
37. Henriksen, A., and Wright, R. F. 1978. Concentration of heavy metals in small Norwegian lakes. *Water Res.*, **12**:101–112.
38. Krenkel, P. A. 1973. Mercury: Environmental considerations, Part I. CRC critical reviews in environmental control, pp. 303–373.
39. Jernelov, A. 1976. Microbial alkylation of metals. Report B332, Institute for Vatten-Och Luft-Vardforshkming, Stockholm, 269 pp.
40. Bogges, W. R., and Wixson, B. G. 1977. Lead in the environment. RANN Report-NSF, U.S. Govt. Printing Office.
41. Guthrie, R. K., Singleton, F. L., and Cherry, D. S. 1977. Aquatic bacterial populations and heavy metals–II. Influence of chemical content of aquatic environments on bacterial uptake of chemical elements. *Water Res.*, **11**: 643–646.
42. Johnson, H. E., and Ball, R. C. 1972. In: S. D. Faust (ed.), "Organic Pesticide Pollution in an Aquatic Environment." Am. Chem. Soc., Washington, D.C.
43. Robinson, J., Richardson, A., Crabtree, A. N., Coulson, J. C., and Potts, G. R. 1967. Organochlorine residues in marine organisms. *Nature*, **214**: 1307–1311.
44. Chesters, G. and Simsiman, G. V. 1974. Impact of agricultural use of pesticides on the water quality of the Great Lakes. Task-A-5, Water Resources Center, University of Wisconsin, Madison.
45. Anon. 1976. Environmental health criteria 2-polychlorinated biphenyls and terphenyls. World Health Organization, Geneva, Switzerland.
46. Nisbet, C. T., and Sarofin, A. F. 1972. Rates and routes of transport of PCBs in the environment. *Env. Health Perspectives*, **1**:21–38.
47. Dennis, D. S. 1976. Polychlorinated biphenyls in the surface waters and bottom sediments of the major drainage basins of the United States. Na-

tional Conference on Polychlorinated Biphenyls. U.S. EPA QV63JN277C, Washington, D.C.

48. Almer, B., Ekstrom, C., Hornstorm, E., and Miller, V. 1974. Effects of acidification on Swedish lakes. *Ambio*, **3**:30–36.
49. Beamish, R. J., and Harvey, H. H. 1972. Acidification of the La Cloche Mountain Lakes resulting fish mortalities. *J. Fish Res. Bd. Can.*, **29(8)**: 1131–1143.
50. Likens, G. E., and Borman, F. H. 1974. Acid rain: A serious regional environmental problem. *Science*, **184**:1176–1179.
51. Likens, G. E., Wright, R. F., Galloway, J. N. and Butler, T. J. 1979. Acid rain. *Sci. Amer.*, **241**:43–51.
52. Glass, N. R., Glass, G. E., and Rennie, P. J. 1979. Effects of acid precipitation. *Env. Sci. Tech.*, **13**:1350–1355.
53. Armstrong, D. E., Perry, J. R., and Platness, D. E. 1979. Availability of pollutants associated with suspended or settled river sediments which gain access to the Great Lakes, Vol. 11, Menomonee River Pilot Watershed Study, International Joint Commission, Windsor, Ontario.
54. Bolter, E., and Butz, T. R. 1977. Mobility of heavy metal humates in soils and surface waters. Project No. B-114-MO, Missouri Water Res. Research Center, University of Missouri, in Rolla.
55. Schneider, W. A., Jeffries, D. S., and Dillon, P. J. 1979. Effect of acid precipitation on precambrian freshwaters in Southern Ontario. *J. Great Lakes Res.*, **5**:45–51.
56. McElroy, A. D., Chiu, S. Y., Nebgen, J. W., Aleti, A., and Bennett, F. W. 1976. Loading functions for assessment of water pollution from non-point sources. EPA-600/2-76/151, U.S. EPA, Washington, D.C.

3

Hydrologic considerations

INTRODUCTION

Pollution from nonpoint sources is a hydrologic problem. It is a known fact that there is a close relationship of pollutant loadings from areal sources to the rain volume and intensity, infiltration and storage characteristics of the watershed, and other hydrologic parameters. Unlike pollution from point sources, which bears little relation to the watershed hydrology, pollution from nonpoint sources has its beginning in the atmospheric transport of pollutants, and its occurrence and magnitude are closely related to the hydrologic cycle. Subsequently, the pollutant load from nonpoint sources has a random character. In addition, uncontrolled hydrologic modifications of watersheds caused by man can increase the nonpoint pollutant loads. On the other hand, hydrologic modifications aimed to reduce the hydrologic activity of lands are often effective for controlling excessive pollution.

Most of the models used for simulating nonpoint pollutant loadings are basically models of the watershed hydrology or

Fig. 3-1. High flows and floods carry most of the pollution from nonpoint sources.

are closely related to it. Rainfall energy and the splashing effect of rain droplets liberate soil particles which are then amenable to the transport by surface runoff. If agricultural chemicals or organic fertilizers are placed on the land and surface runoff is generated from these areas, a significant portion of these pollutants can be lost into surface waters. Mobile pollutants can be leached into the groundwater zone and cause groundwater contamination and pollution.

The highest pollutant loadings and highest concentrations of pollutants from non-point sources occur usually during high-flow and flood conditions (Fig. 3-1). Point source impact is most severe during critical low-drought periods—when nonpoint source contribution is minimal or nonexistent.

To control and understand nonpoint pollution generation and transport it is necessary to study the hydrologic processes causing and contributing to nonpoint pollution and to consider the various paths the pollutants travel before they reach surface waters.

PRECIPITATION-RUNOFF RELATIONSHIP

A classical representation of the rainfall-runoff transformation and the components of the hydrologic cycle are shown in Fig. 3-2.

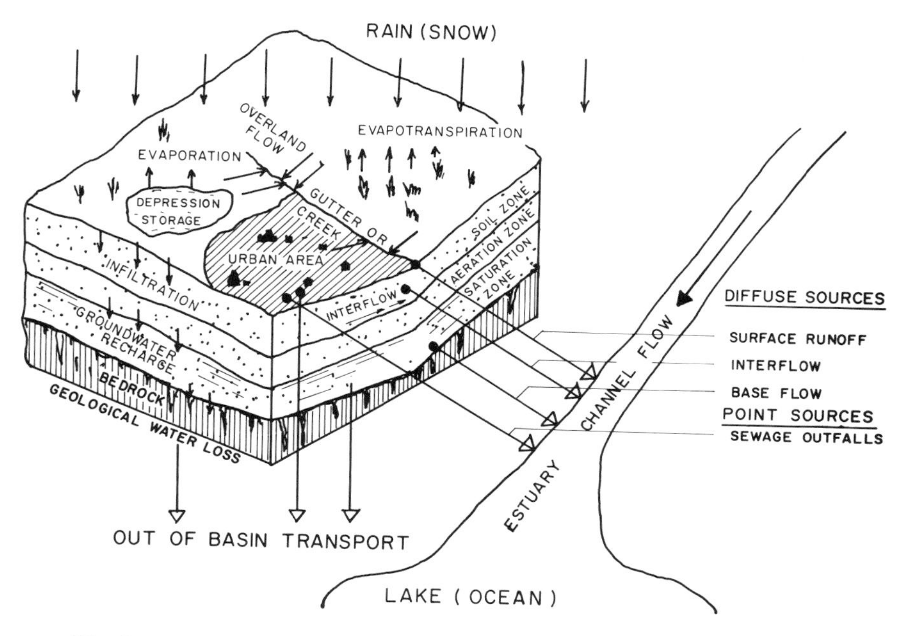

Fig. 3-2. Schematic representation of a watershed hydrologic system.

The first stage of the runoff formation is condensation of atmospheric moisture into rain droplets or snowflakes. During this process, water is in contact with atmospheric pollutants, and, in fact, the pollution content of rainwater can often reach high levels. In addition, rainwater dissolves atmospheric carbon dioxide, sulfur, and nitrogen oxides, and as a weak acid it then reacts with soil and limestone or dolomite geological formations.

The runoff formation begins after the rain (snow) particles reach the surface. The runoff formation, during the winter months, may be delayed by snow storage and melting. During the initial phase of the runoff formation, the rain energy impact liberates the soil particles and picks up the pollutants deposited on the surface.

The runoff generated by precipitation has three components:

1. Surface runoff is a residual of the precipitation after all losses have been satisfied. Numerical subtraction of the losses from precipitation will yield so-called *excess* or *net rain*. The losses include interception by surface vegetative cover, depression storage and ponding, infiltration, and evaporation from the surface. Since the surface runoff is a residual from the precipitation, a linear relationship between the precipitation volume and runoff does not exist. Most pollution from nonpoint sources is generated and carried by surface runoff.

2. Interflow is that portion of water infiltrating into the soil zone, which moves in a horizontal direction due to lower permeability of subsoils. The amount of interflow is again a residual from infiltration after subtraction of the groundwater recharge, soil moisture storage, and evapotranspiration from the upper soil zones.

3. Groundwater runoff (base flow) is defined as that part of runoff contribution which originates from springs and wells. Most stream flows during prolonged drought periods can be characterized as groundwater runoff. In some arid or semiarid regions, the natural base flow may be zero during some times of the year, and the measured flow may originate from sewage outfalls.

The magnitude of runoff and its three components has a stochastic nature. In many areas, it is subject to modifications and control by man; in fact, few important streams have truly natural flows. Runoff quantity can be controlled by the change of use of land by man, urbanization, flood storage reservoirs, etc.

The quality of runoff can be related to the erosion intensity by precipitation and to the amount of pollutants accumulated on the surface. Interflow and groundwater pollution can be correlated to the amount of mobile pollutants present in the soil reflecting also the basic chemical composition of soils, subsoils, and bedrocks. Very often, interflow and groundwater flow pollution may result from excessive contamination of soils: for example from overloaded septic tank seepage or overfertilized farms and urban lawns.

Mathematically, the runoff relation to precipitation can be expressed as:

Surface runoff

$$R_s = P - S_i - S_d - f\Delta t \tag{3-1}$$

Interflow (lateral soil water movement)

$$R_i = (f - ET)\Delta t - S_s - q_g \tag{3-2}$$

Groundwater (base) flow

$$R_g = q_g - S_g - q_d \tag{3-3}$$

where:

R_s = volume of the surface runoff in cm during a time interval Δt
P = precipitation volume (cm)
S_i = change in the available interception storage (cm)
S_d = change in the available depression surface storage (cm)
f = infiltration rate (cm/hr)
ET = evapotranspiration rate from the soil zone (cm/hr)
S_s = soil moisture storage change (cm)
q_g = groundwater recharge (cm)
R_i = interflow volume (cm)
R_g = groundwater flow contribution (cm)

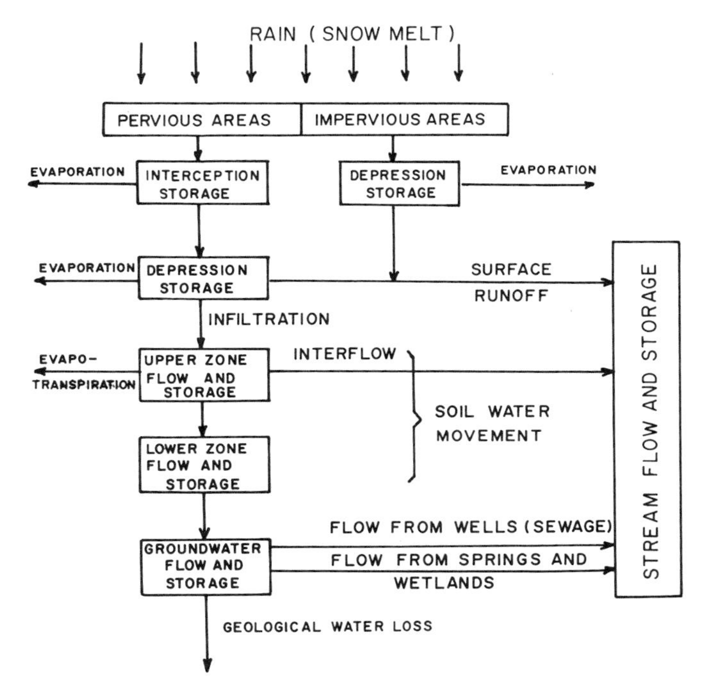

Fig. 3-3. Block diagram of watershed hydrologic model. (Adapted from reference 3.)

S_g = groundwater storage change (cm)
q_d = geological water loss (cm)
Δt = time interval (hr)

A block diagram of the rainfall runoff transformation process is shown in Fig. 3-3.

Components of the Rainfall-Runoff Transformation Process

Interception. A part of the precipitation volume is intercepted by vegetation and adheres to the vegetation surface until a sufficiently heavy film is formed, when gravity begins to prevail over adhesion. Interception storage is that part of precipitation that adheres or wets the aboveground objects and is returned to the atmosphere by evaporation. The amount that is intercepted depends on the type and intensity of vegetation, intensity and volume of rainfall, roughness of the surface, and season of the year or growth stage of the vegetation.

The few models reported in the literature for interception storage are crude and inaccurate. Generally, about 0.5 to 2.5 mm of rain can be held on foliage before appreciable drip can take place.[1] Similarly, about 1.2 to 1.8 mm of precipitation can be intercepted by grass and dense herbs. Interception does not have an important influence on runoff from agricultural areas and is usually lumped together with depression storage into a single surface storage parameter.[2,3]

Depression Storage. Water reaching the surface must first fill the surface depressions, forming small puddles, ponding, or general wettness of the area. Water stored in the depression storage either evaporates or percolates into the soil zone. Only when the precipitation rate exceeds infiltration and all surface storage is exhausted will surface runoff result.

The character of the depression storage as well as its magnitude depends largely on the surface characteristics that can be generally related to land use. The primary factors determining the depression storage are the surface character, roughness, and slope. Accurate estimations of depression storage are not possible and little information is available that could serve as a guide for choosing values of the depression storage that would be based on physical measurements in the field. Tholin and Keifer[4] estimated the surface storage (evidently including also the interception) for Chicago's urban areas as being 6.25 mm ($\frac{1}{4}$ in.) on pervious areas and 1.56 mm ($\frac{1}{16}$ in.) on impervious areas. Figure 3-4 relates the surface storage to the slope for various agricultural land uses.

Since an accurate estimation of the surface storage parameter is not available, the appropriate values for various lands are usually determined from calibration of watershed hydrologic models to measured field data. The outputs of such models are often very sensitive to the magnitude of surface storage, and its proper selection is crucial for further modeling.

Depression storage can be increased by engineering and agronomic practices such as plowing, raking of the surface, and/or by seeding and planting of vegetation. The depression storage on agricultural fields is at a maximum during planting but decreases afterwards.

Infiltration. Although acknowledged as very important, the rate of infiltration in runoff models has been subject to inadequate mathematical formulations. By way of example, it should be noted that the well known and widely used "rational formula," which relates the runoff to rainfall, incorporates the infiltration effect into the runoff coefficient, *C*. Several other methods avoid the evaluation of infiltration by relating surface runoff to the net rain only.

Infiltration and permeability are not synonymous. Permeability is defined from Darcy's law and denotes the rate of water movement through the soil column under saturated conditions and a unit slope. Infiltration, on the other hand, is the rate at which water percolates from the surface storage into the soil zone.

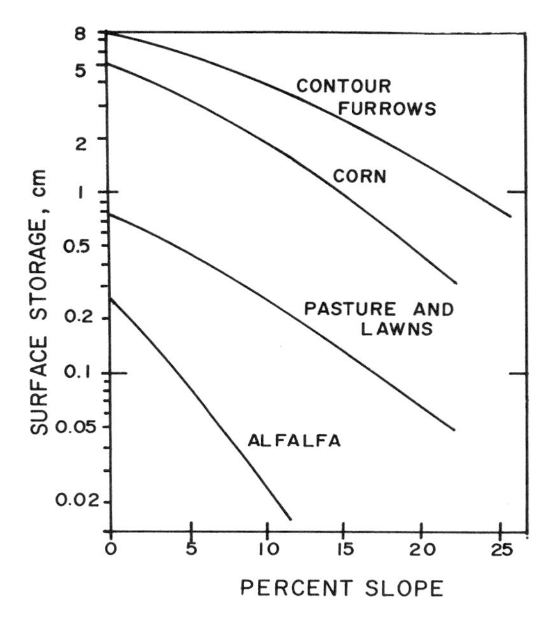

Fig. 3-4. Relationship of surface storage parameter to the slope of the land. (From reference 5.)

Permeability of soils depends on such soil characteristics as texture, compactness, and organic and chemical composition. Infiltration is a function of permeability of soils and subsoils, soil moisture, vegetation cover, temperature, and possibly other parameters.

Water enters from the surface storage due to the combined effects of gravity and capillary forces. When the process continues, the capillary pore space becomes filled and the capillary tension decreases. As water percolates to greater depths, the gravitational forces encounter increased resistence due to the increase in the length of the channels, decreases pore size from swelling of clay particles, or the pressure on an impermeable barrier such as rock or clay. Subsequently, the infiltration rate decreases with time from the commencement of the storm as shown on Fig. 3-5.

The soils of the United States have been classified by the Soil Conservation Service as to their permeability and runoff potential into four hydrologic groups.[7]

1. Group A are soils with low total surface runoff potential due to high infiltration rates even when thoroughly wetted. These soils consist chiefly of deep, well to excessively drained sands and gravel.

2. Group B are soils of low to moderate surface runoff potential that have

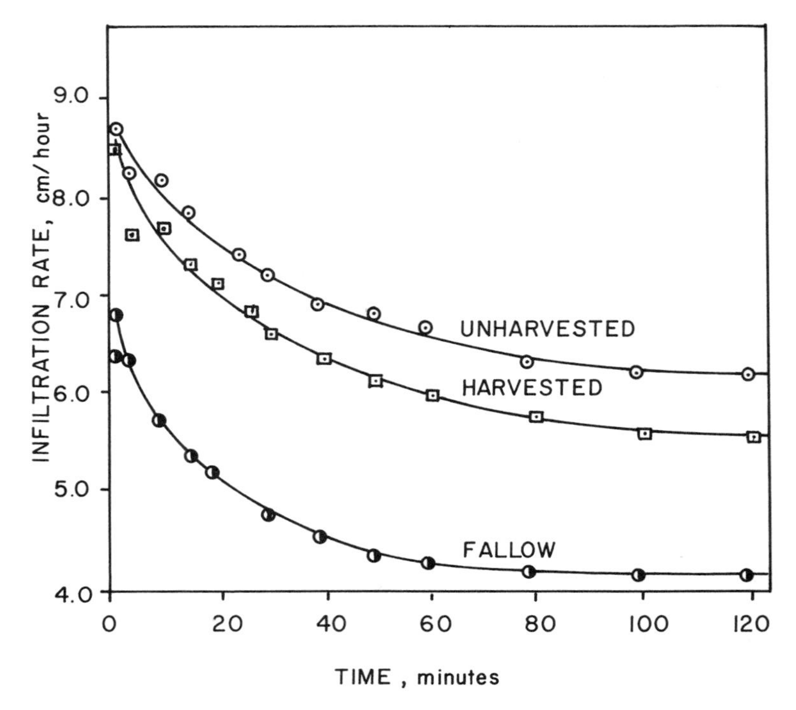

Fig. 3-5. Characteristic relationship of infiltration rate vs time. (From V. Larson, J. H. Axley, and G. L. Miller, Univ. Md, WRC.[6])

moderate infiltration rates and include moderately fine to moderately coarse texture.

3. Group C soils have high to moderate surface runoff potential and slow infiltration rates, and consist chiefly of soils with a layer that impedes downward movement of water, or soils with moderately fine to fine texture.

4. Group D soils have high surface runoff potential and very slow infiltration rates, and consist chiefly of clay soils with a high swelling potential, soils with a permanently high water table, soils with a clay pan or clay layer at or near the surface, and shallow soils over nearly impervious material.

The soil classification and approximate permeabilities can be obtained from the SCS soil maps available for most of the conterminous United States.

Particle size distribution (texture), arrangement of soil particles, organic matter content, clay mineral content and composition, exchangeable sodium percentage, and total concentration of salts are the most important factors affecting permeability.[8] In addition, permeability rates can be affected by soil compaction, cultivation, vegetation, and land cover.

Most guides developed as aids for estimating soil permeability are based on the relationship of permeability to the soil texture (Fig. 3-6). Table 3-1 shows the permeability ranges of the hydrologic soil groups recognized by the U.S. Soil Conservation Service.

Soil Water Storage. Storage of soil moisture can be divided into two moisture classes: that held between saturation and 0.3 bar tension, and that held between 0.3 bar and 15 bar tensions, respectively. The former moisture content can be drained from the soil by gravity, whereas the latter moisture content is retained by the soils and is generally considered available to plants. Moistures below 15 bar tensions are not available to most crops and plants, and the water content below these levels can be reduced only by evaporation. Gravitational water (G) is then determined by subtracting 0.3 bar moisture volume percentage from the total porosity. Available soil water capacity (AWC) is the difference between the moisture contents at 0.3 bar tensions and 15 bar tensions. Figure 3-7 shows the moisture characteristics related to the soil texture.

Infiltration Models. In modeling infiltration, one must realize that the process depends on the antecedent and/or current soil moisture characteristics. As was shown in Fig. 3-5, under dry soil conditions the infiltration rate is significantly higher than the soil permeability due to the soil water tension effect that prevails over gravity. As the moisture approaches saturated conditions, the tension decreases and gravity begins to control.

The models presented in the literature describe the infiltration rate either by using the empirical concept or by equations based on the physical concept of water entry into soils or porous media in general. Equations proposed by Horton[9] and Holtan[10,11] are examples of the first approach, whereas Philip[12] presented a semiempirical soil physics model of water penetrating in porous media.

Horton's equation simply describes the exponential decrease of the infiltration rate as

$$f = f_c + (f_0 - f_c)e^{-Kt} \tag{3-4}$$

where

f = rate of infiltration
f_c = the infiltration rate assumed similar to the saturation permeability
f_0 = the initial rate of infiltration
K = a constant

Equation 3-4 does not relate the infiltration rate to the soil moisture conditions and as such it would be of lesser importance to the modeling of watershed

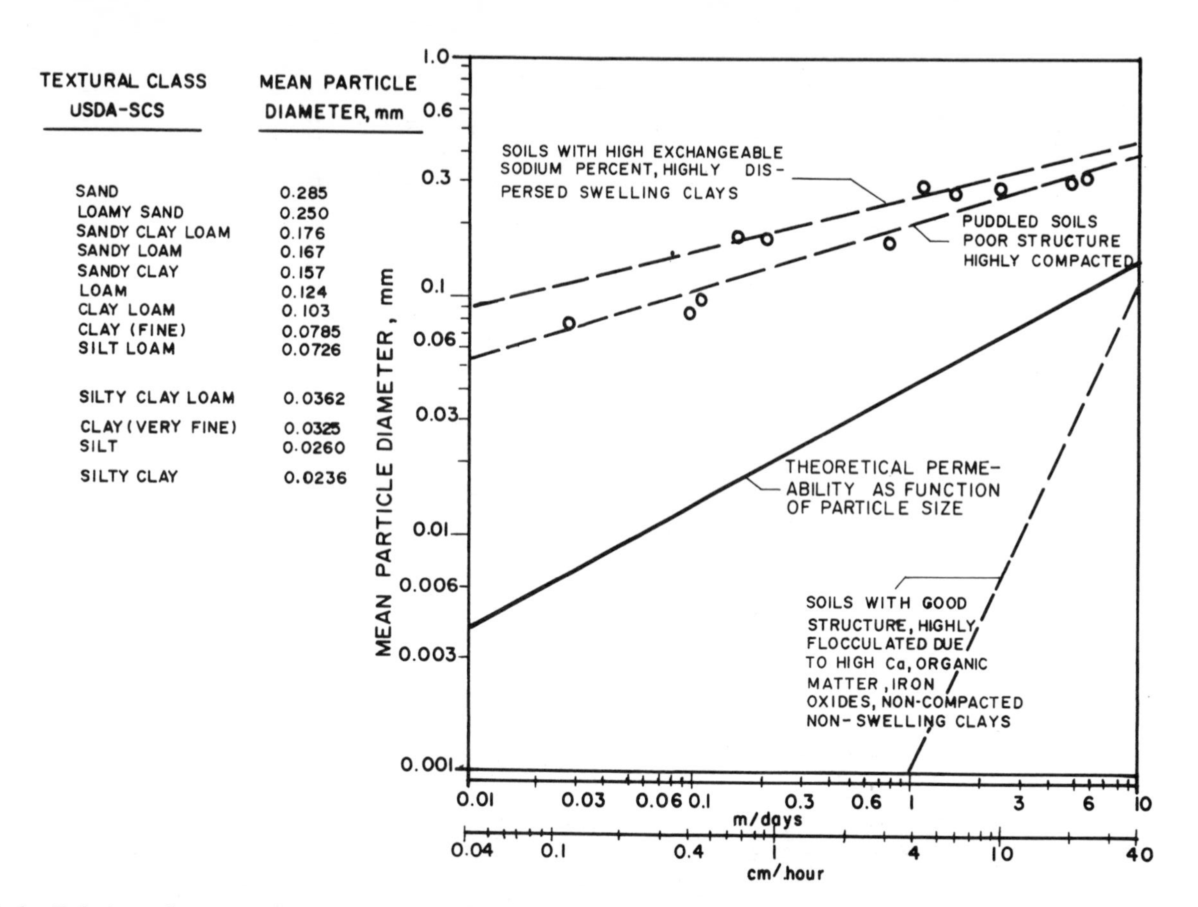

Fig. 3-6. Relation of permeability to soil texture (reference 8). The circled points represent permeability of septic seepage fields after a few years' operation.

TABLE 3-1. Permeability Classes According to the Soil Conservation Service, U.S. Department of Agriculture.

Permeability Class	In./hr	Cm/hr	M/day
A Very rapid	>10	>25	>6.0
+B Rapid	5.00–10.00	12.5–25.0	3.0–6.0
B Moderately rapid	2.50–5.00	6.3–12.5	1.5–3.0
+C Moderate	0.80–2.50	2.0–6.3	0.5–1.5
C Moderately slow	0.20–0.80	0.5–2.0	0.12–0.5
+D Slow	0.05–0.20	0.12–0.5	0.03–0.12
D Very slow	<0.05	<0.12	<0.03

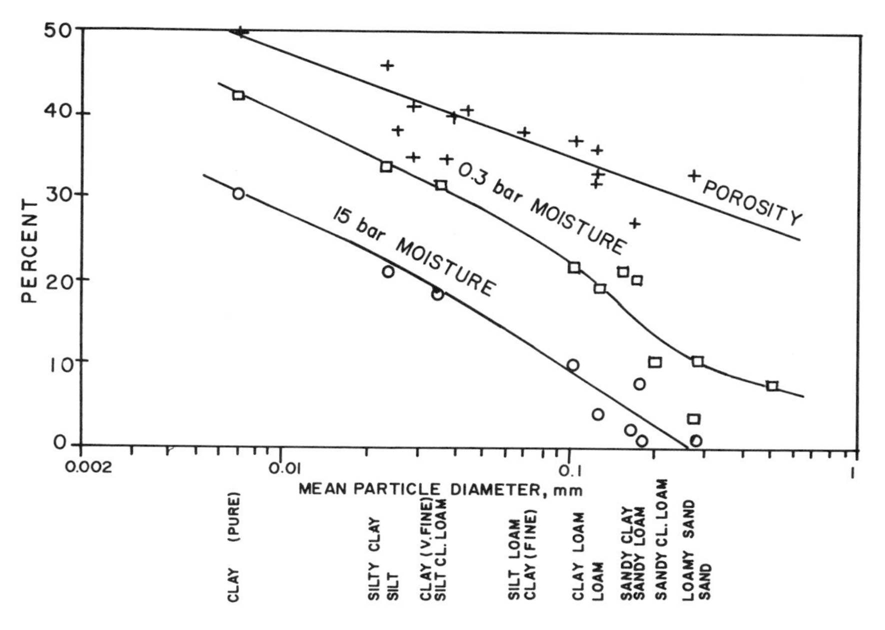

Fig. 3-7. Moisture characteristics of soils.

hydrology. However, in spite of difficulties with determining the parameters f_0 and K, it has been used in earlier models of rainfall-runoff transformation.

Recognizing the serious difficulties of relating infiltration rates to soil characteristics and soil moisture distribution, Holtan[10] proposed a formula that would relate the infiltration rate to the exhaustion of the available soil moisture storage. The formula was presented as follows:

$$f = a(S - F)^n + f_c = aF_p^n + f_c \quad (3\text{-}5)$$

where:

S = the volume of soil water storage above the control horizon
F = cumulative infiltration
F_p = a measure of the soil moisture remaining in the soil column at any time
a, n = coefficients

The coefficients a and n have average suggested values of $a = 0.42$ and $n = 1.387$, if the soil moisture is expressed in centimeters.

The value of n was similar for all experiments, but the coefficient a varied between 0.05 to 0.70 and its suggested values are given in Table 3-2.

Equation 3-5 was modified later and included in the USDAHL watershed model in the form[11]

$$f = GIaF_p^{1.4} + f_c \quad (3\text{-}5a)$$

where:

a = a vegetation parameter
GI = the growth index in percent of maturity

TABLE 3-2. Estimates of the Vegetation Factor *a* in Holtan's Infiltration Equation.[a]
for f in centimeters per hour and F_p in centimeters.

Land Use and Cover	Basal Area Rating[b] Poor Conditions	Basal Area Rating[b] Good Conditions
Fallow[c]	0.07	0.2
Row crops	0.07	0.14
Small grains	0.14	0.20
Hay (legumes)	0.14	0.28
Hay (sod)	0.28	0.40
Pasture (bunchgrass)	0.14	0.28
Temporary pasture (sod)	0.28	0.40
Permanent pasture (sod)	0.55	0.68
Woods and forest	0.55	0.68

[a] Source: U.S. Department of Agriculture, Agricultural Research Service, 1975.
[b] Adjustments needed for "weeds" and "grazing."
[c] For fallow land only, "poor conditions" means "after row crop," and "good conditions" means "after sod."

Information on the estimated magnitudes of the growth index for various crops and growing seasons is available from U.S. Department of Agriculture publications.[11]

The depth of the control horizon is supposed to coincide with the top soil zone between the soil surface and subsoils, that is, the soil horizon A (see Chapter 6 for definitions of soil horizons).

Example 3-1: Infiltration Rate

Estimate the infiltration rate curve for ponded soil with a saturation permeability of 2 cm/hr and soil moisture characteristics as follows:

porosity – 45%
0.3 bar moisture – 30%
1.5 bar moisture – 21%

The depth of the control horizon is assumed to be 50 cm and the antecedent moisture is equal to the gravity water content (30%). Then the initial available storage capacity becomes

$$F_p(0) = (\text{porosity} - 0.3 \text{ bar moisture})50 \text{ cm}$$

$$= (0.45 - 0.3)50 = 7.5 \text{ cm}$$

If the land surface is fallow, then the approximate magnitude of the vegetation parameter will be between 0.07 and 0.14. Select $a = 0.1$ and $GI = 1.0$. Then the initial infiltration rate at $t = 0$ becomes

$$f(0) = 0.1(7.5)^{1.4} + 2 = 3.68 \text{ cm/hr}$$

When $t \rightarrow \infty$ the infiltration rate would approach 2 cm/hr. However, due to the exponent n equaling 1.4, an exact solution of Equation 3-5 is not possible. Overton and Meadows[13] report that satisfactory results can be obtained with $n = 2$. A simple numerical solution shown below will yield adequate results as well.

Since the infiltration capacity is a function of the available water storage, a simultaneous solution of the storage equation must accompany the infiltration equation. Hence

$$\frac{dF_p}{dt} = f_r - f \tag{3-6}$$

where:

f_r = storage recovery rate

If the storage recovery rate $f_r = f_c$ (which occurs when the permeability of subsoils is more than or equal to the permeability of the top layer, and moisture *content is above the gravitational water) then*

$$\frac{dF_p}{dt} = -aF_p^n \tag{3-7}$$

If the soil moisture rate is below the 0.3 bar tension, the recovery rate f_r equals the evapotranspiration rate.

The above differential equation can be solved by simple numerical techniques such as Runge-Kutta, Euler, or Heund's methods. Heund's method would yield an equation

$$F_{p_{t+\Delta t}} = F_{p_t} - \frac{\Delta t}{2}\left(a(F_{p_t})^n + a(F_{p_t} - a(F_{p_t})^n \Delta t)^n\right) \tag{3-8}$$

and the solution for $\Delta t = 1.0$ hr is given in Table 3-3 and plotted in Fig. 3-8.

Philip's infiltration model[12] is based on soil physics of water movement in porous media. The infiltration rate is expressed as

$$f = \frac{1}{2} St^{-0.5} + A \tag{3-9}$$

where:

S = sorptivity of the soil
A = conductivity at the wetting front

Sorptivity is computed from the soil moixture distribution and is, generally, very difficult to define.

The three infiltration models are most commonly used by the current watershed models. For example, Horton's model was incorporated in the Unversity of

TABLE 3-3. Solution of Holtan's Infiltration Equation.

t (hr)	F_p (cm)	f (cm/hr)
0	7.5	3.68
1.0	5.94	3.21
2.0	4.89	2.92
3.0	3.84	2.66
4.0	3.26	2.52
5.0	2.79	2.42
6.0	2.41	2.34
8.0	1.89	2.24
10.0	1.49	2.17
15.0	0.93	2.09
20.0	0.62	2.05

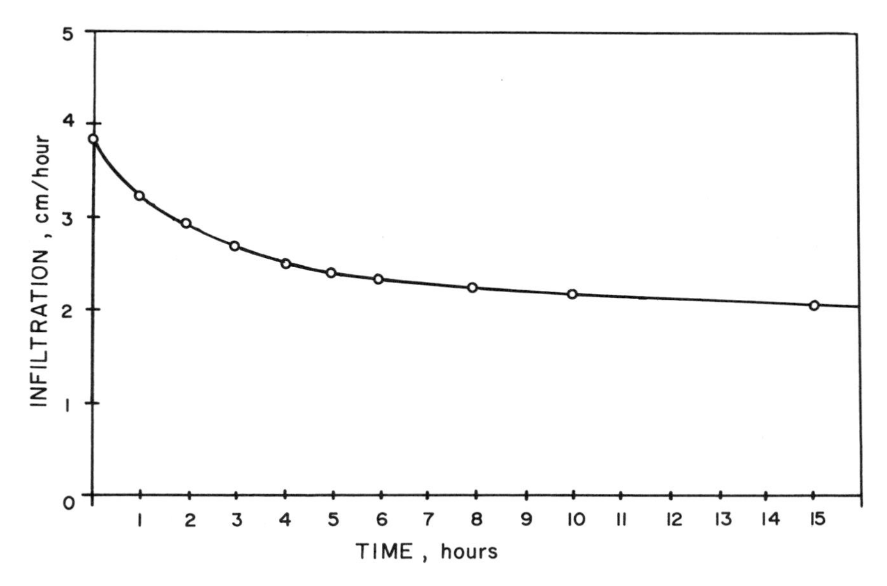

Fig. 3-8. Infiltration rate curve for Example 3-1.

Cincinnati storm water model and the EPA's Storm Water Management Model; Holtan's model was used by the U.S.D.A-SCS model ACTMO and USDAHL, and the Stanford Watershed Model; a modified Philip's model is incorporated in the newer versions of the Hydrocomp watershed model, which evolved from the Stanford model. Both Holtan's and Philip's models are included in the LANDRUN watershed model by Marquette University. All models are relatively sensitive to the magnitude of the permeability parameter, which must be calibrated and verified.

Infiltration rates can be partially controlled by engineering and agricultural practices, such as tillage, raking of the surface, enrichment of soils by organic residues and root systems, and chemical treatment of soils. Compaction by heavy machinery will reduce permeability and, hence, the infiltration.

Frozen soil usually has lower infiltration rates. If the soil is frozen dry, the decrease of permeability is not as significant as if the soil is wet (Fig. 3-9).

Evaporation and Evapotranspiration. Knowledge of evaporation and evapotranspiration is important for estimating the rates of recoveries of interception, depression, and soil moisture storage capacities. The most significant changes in the storage capacities take place between rains, when the air moisture content (ambient vapor pressure) is low and below saturation.

By definition, evapotranspiration represents water loss into the atmosphere by evaporation from both open water surfaces and soils, and transpiration refer-

Fig. 3-9. Permeability of frozen soils can be significantly reduced resulting in high surface runoff potential. (Photo U.S.D.A.–Soil Conservation Service.)

ring to water drawn from the soil zone by root systems of plants and vegetation and released to the atmosphere as a part of the life cycle of plants.

The potential evaporation can be either measured or computed. Pan evaporation data are available from many U.S. Weather Service Bureau stations; however, it must be realized that pan evaporation may differ from evaporation from large open water surfaces and open lands.

Most of the evaporation formulas are based on the simple aerodynamic equation:

$$E_0 = (a + bU)(e_s - e_a) \tag{3-10}$$

where:

E_0 = the evaporation rate
U = wind speed
e_s = the saturation vapor pressure at the mean air temperature
e_a = vapor pressure of air
a, b = constants

Table 3-4 lists the most common empirical formulas for the estimation of potential evaporation.

Several methods have been developed for estimating the evaporation requirements of crops and forested areas. The methods are sufficiently described in the

TABLE 3-4. Some Evaporation Formulas of the Type $E_0 = (a + bU)(e_s - e_a)$.[a]

Author	Formula
Zaikov (from Braslavskii and Vikulina[14])	$\frac{E_0}{e_s - e_a} = 0.11(1 + 0.72U)$
Lake Hefner (from Edinger and Geyer[15])	$\frac{E_0}{e_s - e_a} = 0.10U$
Lake Colorado (from Edinger and Geyer[15])	$\frac{E_0}{e_s - e_a} = 0.15U$
Harbeck[16] or WHO measurements	$\frac{E_0}{e_s - e_a} = 0.18A^{-0.05}U$

[a]Units: evaporation rate, E_0 (mm/day)
wind velocity, U (m/sec)
vapor pressure, e (m bars)
surface area, A (hectares)
U measured 8 m above the water surface.

engineering handbooks.[7,17,18] The models can be separated into two groups. The first group estimates evapotranspiration from the heat balance at the ground surface. The method most widely used throughout the world was developed by Penman.[19,20] It involves the use of an equation developed in 1948 for estimating evaporation from wind velocity and vapor pressures at the mean and dew point air temperatures. It estimates the crop evapotranspiration by multiplying the evaporation values by empirical constants depending on latitude and the length of daylight.

The second group utilizes a crop stage coefficient based on the crop type and period of the growing season, by which the potential evaporation is multiplied, for example, the evaporation index method described by McDaniel.[21]

$$ET = E_0 \times KU \tag{3-11}$$

where:

ET = crop evapotranspiration requirement (consumptive use) on a monthly or shorter period basis

E_0 = climatic index, which is identical to the evaporation from a shallow hypothetical lake situated at the locality under consideration

KU = crop-use coefficient, which reflects the growth stage of crops. Average values for crop-use coefficient are presented in Table 3-5.

A similar approach has been suggested by the Agricultural Research Service, which proposed the following equation for the potential evapotranspiration:[10]

TABLE 3-5. Crop-Use Coefficients for Use in Evaporation-Index Method.[a]

Perennial Crops (Northern Hemisphere)

Crop	Average *KU* values by months											
	Jan.	Feb.	Mar.	Apr.	May	June	July	Aug.	Sept.	Oct.	Nov.	Dec.
Alfalfa	0.83	0.90	0.96	1.02	1.08	1.14	1.20	1.25	1.22	1.18	1.12	0.86
Grass pasture	1.16	1.23	1.19	1.09	0.95	0.83	0.79	0.80	0.91	1.91	0.83	0.69
Grapes	–	–	0.15	0.50	0.80	0.70	0.45	–	–	–	–	–
Citrus orchards	0.58	0.53	0.65	0.74	0.73	0.70	0.81	0.96	1.08	1.03	0.82	0.65
Deciduous orchards	–	–	–	0.60	0.80	0.90	0.90	0.80	1.50	0.20	0.20	–
Sugarcane	0.65	0.50	0.80	1.17	1.21	1.22	1.23	1.24	1.26	1.27	1.28	0.80

Annual Crops

Crop	*KU* values at listed percent of growing season										
	0	10	20	30	40	50	60	70	80	90	100
Field corn	0.45	0.51	0.58	0.66	0.75	0.85	0.96	1.08	1.20	1.08	0.70
Grain sorghum	0.30	0.40	0.65	0.90	1.10	1.20	1.10	0.95	0.80	0.65	0.50
Winter wheat[b]	1.08	1.19	1.29	1.35	1.40	1.38	1.36	1.23	1.10	0.75	0.40
Cotton	0.40	0.45	0.56	0.76	1.00	1.14	1.19	1.11	1.83	0.58	0.40
Sugar beets	0.30	0.35	0.41	0.56	0.73	0.90	1.08	1.26	1.44	1.30	1.10
Cantaloupes	0.30	0.30	0.32	0.35	0.46	0.70	1.05	1.22	1.13	0.82	0.44
Potatoes (Irish)	0.30	0.40	0.62	0.87	1.06	1.24	1.40	1.50	1.50	1.40	1.26
Papago peas	0.30	0.40	0.66	0.89	1.04	1.16	1.26	1.25	0.63	0.28	0.16
Beans	0.30	0.35	0.58	1.05	1.07	0.94	0.80	0.66	0.53	0.43	0.36
Rice[c]	1.00	1.06	1.13	1.24	1.38	1.55	1.58	1.57	1.47	1.27	1.00

[a] From "Handbook of Applied Hydraulics," by Davis and Sorensen.[17] Copyright © 1969 by McGraw-Hill. Used with the permission of the McGraw-Hill Book Company.
[b] Data given only for springtime season of 70 days prior to harvest (after last frost).
[c] Evapotranspiration only.

$$ET = GI \times k \times E_0 \left(\frac{S - SA}{AWC}\right)^x \tag{3-12}$$

where:

ET = evapotranspiration potential (cm/hr or in./hr)
GI = growth index of crops defined previously
k = ratio of ET to pan evaporation, usually 1.0 to 1.2 for short grasses, 1.2 to 1.6 for crops up to shoulder height, and 1.6 to 2.0 for forests
E_0 = pan evaporation (cm/hr or in./hr)
S = total porosity
SA = available porosity
AWC = available water storage capacity defined as 0.33 bar moisture minus 15 bar moisture
$x = AWC/G$
G = gravitational water storage.

Methods for estimating evapotranspiration also include the Blaney-Criddle method and the Lowry-Johnson method. The reader is referred to references 7, 17, and 18 for a more detailed discussion.

The evapotranspiration rate, which includes both evaporation and transpiration, is a primary factor in determining the water loss of the rainfall-runoff balance. The water lost by infiltration is only temporary, since most of the groundwater runoff will eventually appear on the surface, but the loss by evapotranspiration is considered permanent by all watershed models.

Snow-pack Formation and Snow-melt. Precipitation in snow form has much less energy than rain droplets. The snow accumulates on the surface until it is transformed into snow-melt. Due to the lowered erosion, snow-melt pollution is not as high as it would be from a rain of the same magnitude. In most cases only pollutants that accumulate on the snow surface or have been scavenged by the snow from the atmosphere will contribute to the pollution; however, the amount of accumulated pollutants can be quite high in areas with a high atmospheric fallout and/or street salting.

The major factors determining the amount of snow-melt are meteorological factors such as temperature, solar radiation, and wind velocity. The simplest model of the snow-melt process is called temperature index or degree-day method, in which the air temperature is the only variable. The formula was presented by Gray[18] as follows:

$$P_s = C(T_a - T_b) \qquad \text{for } T_a > T_b$$

and

$$P_s = P \qquad \text{for } T_a \leqslant T_b \tag{3-13}$$

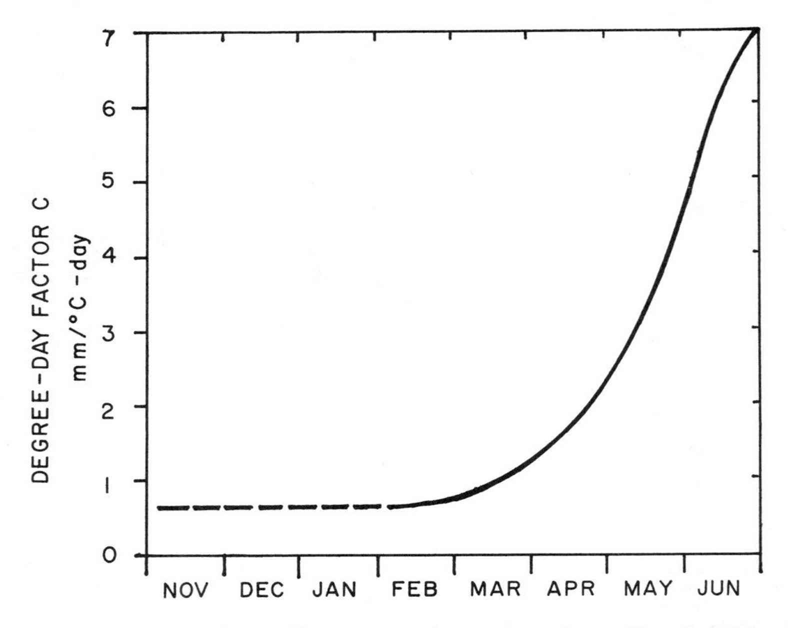

Fig. 3-10. Degree-day factor for snow-melt computation. (Reprinted by permission of the National Research Council of Canada. "Handbook on the Principles of Hydrology," by D. M. Gray.[18])

where:

P = water equivalent of precipitation (mm/day)
P_s = change of the snow-pack water content (mm/day)
T_a = mean or maximum daily air temperature (°C)
T_b = base temperature close to 0°C
C = a coefficient determined by trial and error, assuming lower values in early melt season and higher at a later time (Fig. 3-10)

A more complex model, based on the heat budget of the snow-pack, was developed by the U.S. Corps of Engineers in cooperation with other agencies. A massive program of research resulted in the treatise "Snow Hydrology,"[22] which gives the most detailed and comprehensive coverage of the subject currently available.

Hydrologically Active Area Concept

Rainfall Excess. From the foregoing discussion, it is now clear that the surface runoff, which is often the most polluted component of the total runoff, will be generated from a surface only when precipitation is greater than the losses, that is:

$$R_s = P - S_i - S_d - f\Delta t \qquad \text{if } P > S_i + S_d + f\Delta t$$

and

$$R_s = 0 \qquad \text{for } P \leqslant S_i + S_d + f\Delta t \tag{3-14}$$

The term rainfall excess or net rain is used to denote simple numerical subtraction of losses from the precipitation volume. This differentiates it from surface runoff, which refers to the part of flow in the receiving water body that was generated by the rainfall excess. The unit for rainfall excess is depth of water on the surface from the excess rain generated during a time interval, while the unit of surface runoff is volume/time. A time lag between maximum rain excess and the peak of the runoff is typical for larger drainage areas due to the flow routing overland and in channels.

The estimation of the rainfall excess can be shown in the following example:

Example 3-2: Rainfall Excess Determination

Determine the rainfall excess from a storm where the hyetograph is given in the top portion of Fig. 3-11. A dry period preceding the storm lasted 5 days. The evaporation rate, during the dry period, averaged 0.3 cm/day. The area is covered by grass and has an average slope of 2%. For this slope and surface, the combined interception and depression storage can be read from Fig. 3-4 as 0.62 cm. The soil and infiltration characteristics are similar to Example 3-1.

Gravitational moisture content is G = 50(porosity - 0.3 bar moisture) = 7.5 cm.

The initial soil moisture and available storage capacity can be computed assuming that the soil was saturated during the preceding rain and that the crop-use factor, KU, is close to 1. Since the saturation permeability is 2 cm/hr, the soil moisture has reached 0.3 bar tension shortly after the last rain and the initial storage capacity becomes

$$F_p(0) = G + ET \times 5 \text{ days} = (\text{porosity} - 0.3 \text{ bar moisture})50$$

$$+ 5 \text{ (days)}0.3 \text{ (cm/day)}1.0 = 7.5 + 1.5 = 9\text{cm}$$

For $F_p(0) = 9.00$ cm, the initial infiltration rate is

$$f(0) = 0.1(9.0)^{1.4} + 2 = 4.17 \text{ cm/hr}.$$

The excess rain is computed in 15-min time intervals by the simultaneous solution of the following equations (based on equation 3-1):

$$R_s = P - S_{id} - f\Delta t \qquad \text{if } P > S_{id} + f\Delta t$$

$$S_{id} = 0, \qquad \Delta F_p = (f_r - f)\Delta t \tag{3-15}$$

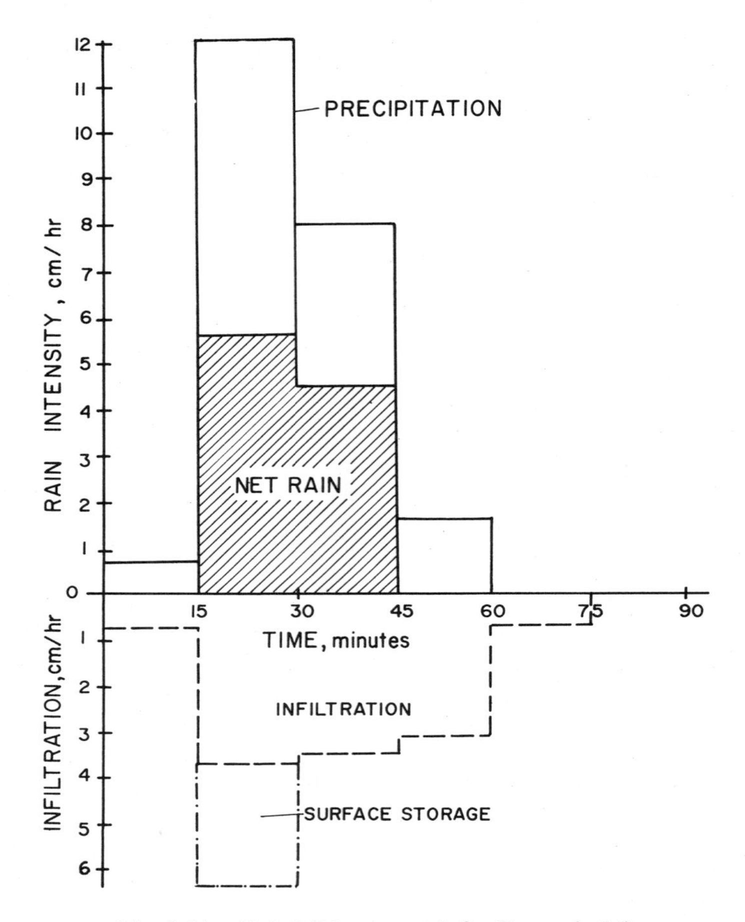

Fig. 3-11. Rainfall hyetograph for Example 3-2.

and

$$R_s = 0 \qquad \text{if } P < S_{id} + f\Delta t$$

with

$$-\Delta S_{id} = P - f(\Delta t)$$

$$f(\Delta t) = P + \Delta S_{id} \qquad \text{if } (aF_p^{1.4} + f_c) > (P + \Delta S_{id})/\Delta t$$

$$\Delta F_p = (f_r - f)\Delta t$$

f_r = 0 if soil moisture is below 0.3 bar tension (or $F_p > G$).

During the rain period the recovery of the surface and soil water storages by

TABLE 3-6. Rainfall Excess Computation.

Time Interval	Precipitation Intensity During the Time Interval	Precipitation Volume, P	Available Depression and Interception Storage S_{id}	ΔS_{id}[a]	Infiltration Volume[b]	Soil Moisture Storage Capacity, F_p	Excess Rain Volume, R_s	Net Rain Intensity, R_s/t
min	cm/hr	cm	cm	cm	cm	cm	cm	cm/hr
1	2	3	4	5	6	7	8	9
0–15	3.0	0.75	0.62	0.0	0.75	9.0	0.0	0.0
			0.62			8.25		
15–30	12.0	3.00	0.0	−0.62	0.96	7.40	1.42	5.68
30–45	8.0	2.00	0.0	0.0	0.90	7.00	1.10	4.40
45–60	1.6	0.40	0.47	+0.47	0.87	6.63	0.00	0.00
60–75	0.0	0.0	0.62	+0.15	0.15	6.98	0.00	0.00

[a] $\Delta S_{id} = f\Delta t - P$, min $S_{id} = 0$, max $S_{id} = 0.62$ cm.
[b] Infiltration volume $= f\Delta t$ or $= P + S_{id}$, whichever is less.
[c] $\Delta F_p = (f_r - f)\Delta t$, $f_r = f_c$ if $F_p < G$, $f_r = 0$ if $F_p > G$; rate of exhaustion of the soil moisture storage capacity reduced at $F_p = 7.5$ cm.

evaporation is minimal. The computation and results are given in Table 3-6 and Fig. 3-11.

Rainfall Excess Estimation by the SCS Method. The Soil Conservation Service developed a method for estimating rainfall excess that does not require computation of infiltration and surface storage. The method evolved from analysis of numerous storms under a variety of soil and cover conditions. The excess rain-precipitation relationship suggested by the SCS is shown in Fig. 3-12.

According to the SCS procedure, the excess rain is estimated from precipitation with the aid of runoff curves. Selection of the appropriate curve is dependent upon the soil type, antecedent soil moisture conditions, and cover type. Soil classification into hydrologic soil groups A to D has been discussed in the previous portion of this chapter.

The antecedent soil moisture conditions for determining the runoff curve have been classified as follows:[23]

AMC I: A condition of watershed soils where the soils are dry but not to the wilting point, and when satisfactory plowing or cultivation takes place.

AMC II: The average case for annual floods, that is, an average of the conditions that have preceded the occurrence of the annual flood on numerous watersheds.

AMC III: If heavy rainfall or light rainfall and low temperatures have occurred during 5 days prior to the given storm and the soil is nearly saturated.

For average soil moisture conditions (AMC II), the runoff curve number (CN) can be found in Table 3-7. The corresponding curve numbers for conditions

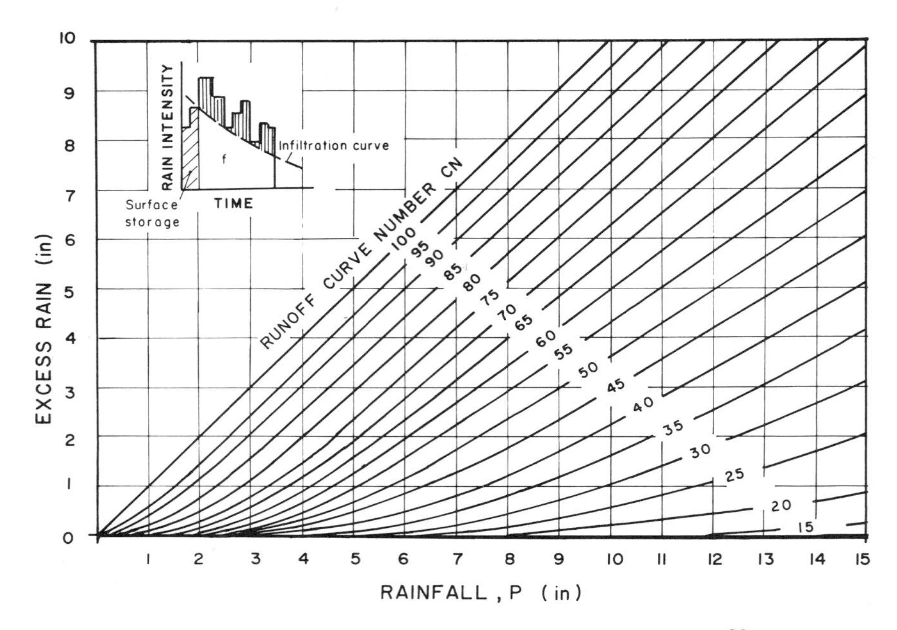

Fig. 3-12. Precipitation-excess rain relationship for SCS method.[23] To convert from inches to centimeters, multiply by 2.54.

AMC I and AMC III can be obtained from Table 3-8, if the CN for AMC II is known.

Example 3-3: Excess Rain Estimation by the SCS Method

Estimate rainfall excess for the storm using the SCS method. The following information is given:

Total precipitation P = 6.15 cm = 2.42 in.
Hydrologic soil group: C
Antecedent soil moisture conditions: AMC II
Surface cover: grass.

From Table 3-7 the runoff curve number for grass (meadows) and soil hydrologic group C is CN = 71. This curve number represents the antecedent soil moisture conditions of AMC II.

Enter Fig. 3-12 with P = 2.42 in. and read excess rain R_s = 0.6 in. (=1.52 cm) for CN = 71.

Rainfall Excess from Impervious Areas. Due to the surface impermeability and small depression storage, the impervious areas seem to be 100% active (i.e., they

TABLE 3-7. Runoff Curve Numbers for Hydrologic Soil-Cover Complexes (Antecedent Moisture Condition II).[23]

Cover			Hydrologic Soil Group			
Land Use or Cover	Treatment or Practice	Hydrologic Condition	A	B	C	D
Fallow	Straight row		77	86	91	94
Row crops	Straight row	Poor	72	81	88	91
	Straight row	Good	67	78	85	89
	Contoured	Poor	70	79	84	88
	Contoured	Good	65	75	82	86
	Contoured and terraced	Poor	66	74	80	82
	Contoured and terraced	Good	62	71	78	81
Small grain	Straight row	Poor	65	76	84	88
		Good	63	75	83	87
	Contoured	Poor	63	74	82	85
		Good	61	73	81	84
	Contoured and terraced	Poor	61	72	79	82
		Good	59	70	78	81
Close-seeded legumes[a] or rotation meadow	Straight row	Poor	66	77	85	89
	Straight row	Good	58	72	81	85
	Contoured	Poor	64	75	83	85
	Contoured	Good	55	69	78	83
	Contoured and terraced	Poor	63	73	80	83
	Contoured and terraced	Good	51	67	76	80
Pasture or range		Poor	68	79	86	89
		Fair	49	69	79	84
		Good	39	61	74	80
	Contoured	Poor	47	67	81	88
	Contoured	Fair	25	59	75	83
	Contoured	Good	6	35	70	79
Meadow		Good	30	58	71	78
Woods		Poor	45	66	77	83
		Fair	36	60	73	79
		Good	25	55	70	77
Farmsteads		–	59	74	82	86
Roads (dirt)[b]		–	72	82	87	89
(hard surface)[b]		–	74	84	90	92

[a] Close-drilled or broadcast.
[b] Including right of way.

TABLE 3-8. Curve Numbers (*CN*) for Wet (AMC III) and Dry (AMC I) Antecedent Moisture Conditions Corresponding to an Average Antecedent Moisture Condition.[23]

CN for AMC II[a]	Corresponding *CN*s	
	AMC I[b]	AMC III[c]
100	100	100
95	87	98
90	78	96
85	70	94
80	63	91
75	57	88
70	51	85
65	45	82
60	40	78
55	35	74
50	31	70
45	26	65
40	22	60
35	18	55
30	15	50
25	12	43
20	9	37
15	6	30
10	4	22
5	2	13

[a] AMC II. The average condition.
[b] AMC I. Lowest runoff potential. Soils in the watershed are dry enough for satisfactory plowing or cultivation.
[c] AMC III. Highest runoff potential. Soils in the watershed are practically saturated from antecedent rains.

generate surface runoff even during small rains). However, not all of the rainfall excess will appear as surface runoff. The rainfall excess generated on impervious surfaces that are not directly connected to a drainage channel or storm sewer will overflow on nearby pervious areas and/or into soils. Such cases include roof drains and driveways overflowing on lawns, roadways and other impervious surfaces with poor or no apparent drainage systems, and parking lots separated by pervious areas. The amount overflowing must be added to the net rain balance of the accepting pervious surface, that is, the depth of precipitation should be increased by the corresponding amount of the overflow from adjacent impervious surfaces.

It is apparent that the fraction of the impervious areas directly connected to

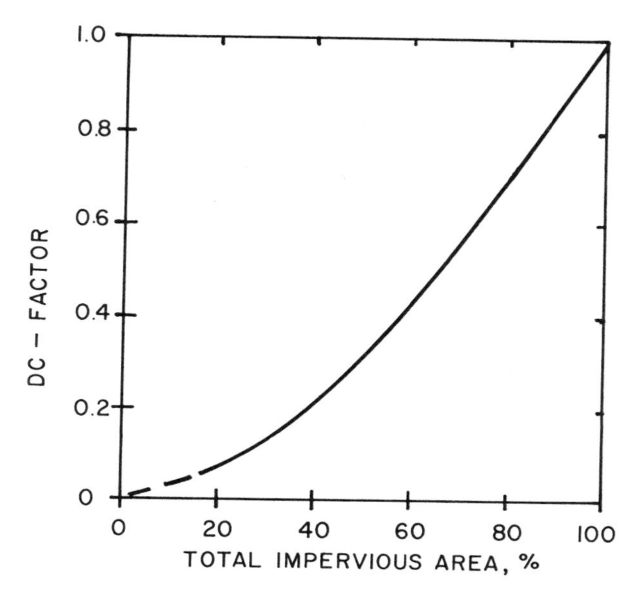

Fig. 3-13. Relation of fraction of impervious surface directly connected to channel drainage systems (DC factor) to total imperviousness of the area (Adapted from Ref. 24).

a drainage system increases with the degree of urbanization reflected in the total imperviousness of the area. The rain excess from impervious surfaces in rural areas will predominantly overflow on adjacent pervious surfaces or pervious road ditches. In densely built-up urban zones, there may not be pervious surfaces available and all runoff will be connected to a drainage system. An approximate relation of the fraction of directly connected impervious surfaces, DC, to the total imperviousness of the area is shown in Fig. 3-13.

In rural zones, during medium- and low-intensity storms, when most of the surface runoff will originate from impervious surfaces, the volume of the surface runoff might be sensitive to the magnitude of the parameter DC. It is, therefore, recommended that the value of the DC factor be estimated and verified by comparison of the computed surface runoff volumes to measured field observations. Areas with storm or combined sewers will have a much higher DC factor than unsewered areas.

Hydrologically Active Areas. Based on the net rain estimation, it is evident that not all areas within the watershed will generate surface runoff and subsequently nonpoint pollution loadings associated with them. The areas that will produce surface runoff are called hydrologically active, while the rest of the watershed contributes only to interflow and base flow. The areas showing the highest hydrologic activity are connected impervious surfaces (even here the depression storage must be substituted), clayey soils with low permeability, frozen soils with high moisture content, soils with a high groundwater table, and highly compacted surfaces.

Areas with high surface storage such as wood-and-flat cropland and, generally, soils with high permeability rates have the lowest hydrologic activity and often generate surface runoff only during extreme storms and/or frozen ground conditions.

It must be remembered that the hydrologic activity of an area is a stochastic phenomenon depending on the magnitude and intensity of the storm, soil conditions, and surface characteristics of the area. The contributing hydrologically active area will be smaller for storms with small recurrence intervals and will increase with the magnitude of the storms and their recurrence intervals. Figure 3-14 shows an example of contributing areas for 2-yr and 10-yr recurrence storms.

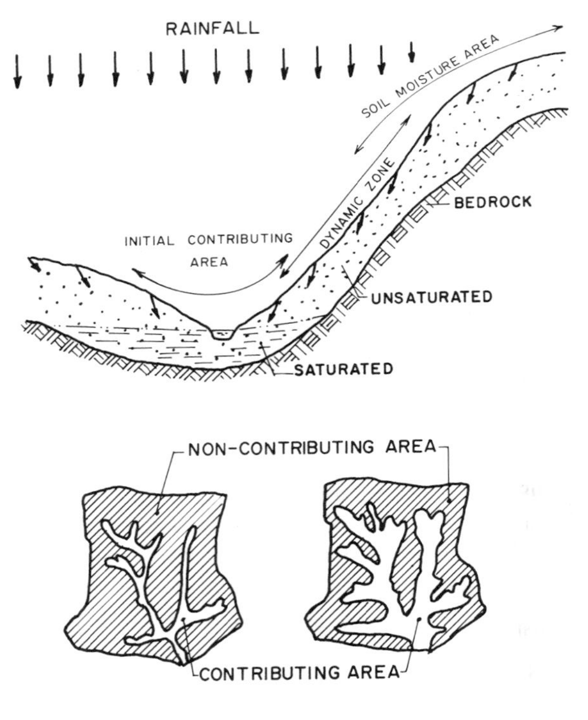

Fig. 3-14. Hydrologically active area concept. (Reference 25.)

Identification of the areas that have a tendency to be hydrologically active is a necessary step in the abatement of nonpoint pollution. These areas contribute to sediment and all surface runoff pollution. Hydrologically active areas can be determined by field surveys, plane and satellite photogrammetry and imagery, and/or hydrological modeling techniques.

Reducing the hydrological activity, for example by increasing depression and interception storage or permeability, or by draining high groundwater levels, is one of the most effective measures to abate nonpoint pollution.

OVERLAND ROUTING OF THE PRECIPITATION EXCESS

The excess or net rain can be imagined as a depth of water on the surface contributing directly to the surface runoff. Overland flow routing is a process by which the excess rain is transformed into surface runoff. The term channel flow routing is used mostly for estimation of the flood or high flow wave modification in stream channels. It is defined as a procedure whereby the time and magnitude of a flood wave, at a point on a stream, is determined from the known or assumed data at one or more points upstream.[7]

The watershed size and length of the overland flow, along with the roughness and slope characteristics, volume and intensity of precipitation, and percent imperviousness, seem to be the most important factors affecting the shape and the magnitude of the runoff hydrograph. For larger drainage areas, the hydrograph curve is affected by channel routing. The channel portion of the routing process may not be as significant for smaller watersheds up to 5 km^2. Channel routing is a process that depends on the hydraulic characteristics of the channel.

Although the hydrograph shape seems to be of lesser importance in nonpoint pollution control than the determination of rainfall excess, the hydrograph evaluation is necessary if the impact of nonpoint sources on receiving waters is studied. Furthermore, as will be demonstrated in Chapter 5, the delivery of pollutants from the source area to a receiving water body is affected by the slope of the receding portion of the hydrograph.

The literature on storm-water and flow routing is quite extensive, and the most recent references include those by Overton and Meadows,[13] Eagelson,[26] Wooding,[27] and Wanielista.[28]

Early overland flow routing models are known as the "rational formula" and the "unit hydrograph." Both concepts are oversimplified and some assumptions on which the models are based can not be satisfied under actual conditions. In spite of these shortcomings, and mainly because of their simplicity, the models are still used and are quite popular.

Rational Formula. The origin of the rational formula can be dated back into the second part of the 19th century. The formula has been widely used for the design of storm and combined sewer systems. It relates the peak flow of the runoff hydrograph to the rain intensity as

$$Q_p = CIA \tag{3-16}$$

where:

Q_p = peak discharge (liters/min or cfs)
C = a runoff coefficient depending on the characteristics of the drainage basin
I = the average rainfall intensity during a specified interval called the time of concentration (mm/min or in./hr)
A = area (m^2 or acres)

In the SI unit system, the units are consistent: however, in the U.S. unit system, the quantities of the units are only approximate.

The time of concentration is the time required for the surface runoff, from the remotest part of the drainage basin, to reach the point of consideration (the watershed outlet). For small watersheds with simple drainage patterns, the time of concentration may be very close to the lag time of the peak flow on the hydrograph resulting from a short intensive rain. For larger watersheds, the time of concentration is generally greater than the lag time of the peak flow.[7] The SCS procedure for hydrograph determination in medium-sized watersheds[23] estimates the runoff peak lag time as $t_p = 0.7t_c$, where t_c is the time of concentration.

The runoff coefficient, C, is a highly empirical quantity. The magnitudes of the coefficient are presented in Table 3-9. Note that the coefficient, C, is a quantity of proportionality that relates the runoff directly to the precipitation. Due to the fact that the runoff is a residual of precipitation and not proportional to the precipitation, the rational method itself and its reliability have been questioned. It should be used only for very crude preliminary estimations.

The Unit Hydrograph. The unit hydrograph concept was developed and proposed by Sherman[29] in early 1930. The unit hydrograph of a drainage basin is defined as a hydrograph of direct runoff resulting from a unit (1 in. or 1 cm) of the excess rain generated uniformly over the basin area at a uniform rate during a specified period of time of duration. The unit hydrograph theory is based on the following assumptions:[7]

1. The effective rainfall is uniformly distributed during the time of duration.
2. The areal distribution of the precipitation excess is uniform over the entire watershed.
3. The base or time duration of the surface runoff hydrograph due to the precipitation excess of unit duration is constant.
4. The ordinates of the surface runoff hydrograph of a common base time are

TABLE 3-9. Runoff Coefficients.[a]

Urban Areas	
Description of Area	Runoff Coefficient
Flat, residential, with about 30% of area impervious	0.40
Moderately steep, residential, with about 50% of area impervious	0.65
Moderately steep, built-up, with about 70% of area impervious	0.80

Table from Horner and Flynt-Transactions ASCE 1936.

Deductions from unity to obtain runoff coefficients for agricultural areas

Type of Area	Value of c'[b]
Topography	
Flat land, with average slopes of 0.02–0.06%	0.30
Rolling land, with average slopes of 0.3–0.4%	0.20
Hilly land, with average slopes of 3–5%	0.10
Soil	
Tight, impervious clay	0.10
Medium combination of clay and loam	0.20
Open, sandy loam	0.40
Cover	
Cultivated lands	0.10
Woodland	0.20

Table from Bernard-Transactions ASCE 1935.

[a] Reprinted from "Handbook on the Principles of Hydrology" by D. M. Gray[18] by permission of the National Research Council of Canada.

[b] The magnitude of the runoff coefficient, C, is obtained by adding values of c s for each of the three factors: topography, soil, and cover, and by subtracting the sum from unity.

directly proportional to the total volume of the surface runoff represented by each hydrograph.

5. For a given drainage basin, the hydrograph of runoff due to a given period of rainfall reflects all the combined physical characteristics of the basin.

Figure 3-15 shows the graphical convolution of the excess rain into the flow using the unit hydrograph concept.

The Instantaneous Unit Hydrograph. The instantaneous unit hydrograph is a hydrograph that would result from a unit rainfall excess pulse, that is, when the

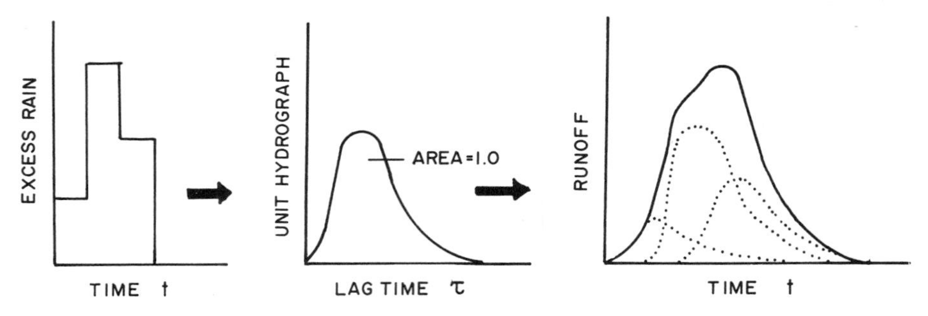

Fig. 3-15. Graphical convolution of rainfall into runoff using unit hydrograph concept.

time of duration of the rain excess becomes infinitesimally small. In this case, the rainfall-runoff relationship can be expressed using a so-called convolution integral

$$Y(t) = \int_0^t h(\tau) X(t - \tau)\, d\tau \tag{3-17}$$

where:

$h(\tau)$ = the ordinate of the instantaneous unit hydrograph
τ = lag time
X = the precipitation excess input expressed as the rain intensity
Y = flow output expressed as volume/time
t = time

The above input-output relationship is typical for linear systems, that is, for systems where the shape of the unit hydrograph does not depend on the magnitude of the input or output of the system. The unit hydrograph concept, as defined herein, corresponds to the linear system representation.

The principle of linearity has long been questioned. Horton[30] and Izzard[31] showed that the ordinates of the unit hydrograph, as well as the time of concentration in the rational formula, depend on the intensity of the rainfall excess. This deficiency of the instantaneous unit hydrograph concept could be overcome assuming a nonlinear input-output relationship so that the convolution integral becomes[32]

$$Y(t) = \int_0^t h(X(t - \tau); \tau) X(t - \tau)\, d\tau \tag{3-18}$$

The function describing the instantaneous unit hydrograph can be based on two simplified watershed representations (Fig. 3-16). The first representation is

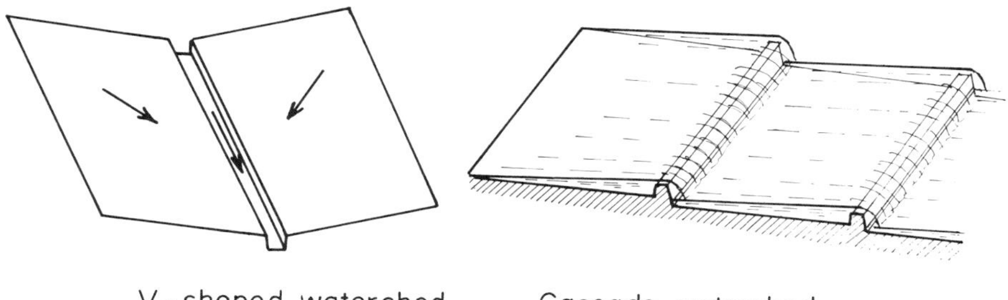

Fig. 3-16. Representations of watersheds in simple rainfall-runoff models.

a two-plane V-shaped watershed, as shown on the left side of the Fig. 3-16. The hydrograph solution for this watershed can be obtained by kinematic wave approximation,[13,26,27,33] which neglects the dynamic terms in the momentum equation for overland flow. Hence, the equations of continuity and momentum are simplified to the following form

$$\frac{\partial H}{\partial t} + \frac{\partial q}{\partial x} = i - f$$

and

$$q = \alpha H^n \tag{3-19}$$

where:

H = water depth
q = the discharge rate/unit width
$i - f$ = rainfall excess
x = distance measured downstream from the top of the catchment
α, n = empirical coefficients

The kinematic wave model can be applied to small impervious drainage areas of uniform slope and surface characteristics such as parking lots and driveways. The solutions and various types of hydrographs are discussed in reference 13.

The second watershed representation breaks the watershed into a series of overflowing pool-cascades and uses the continuity equation for routing the flow. The cascade models have been successfully applied to larger nonuniform watersheds, and as stated by Overton and Meadows,[13] the concept "has been a workhorse in watershed runoff computations." The cascade watershed instantaneous hydrograph function was developed and proposed by Nash[34] as

$$h(t) = \frac{1}{K_N} \frac{e^{-t/K_N}}{\Gamma(n)} \left(\frac{t}{K_N}\right)^{n-1} \tag{3-20}$$

where:

K_N = the storage constant
n = a watershed characteristic representing approximately the number of reservoirs
$\Gamma(n)$ = the gamma function of n

If n approaches 1.0, the above hydrograph function can be replaced by a single reservoir model given by

$$h(t) = \frac{1}{K} e^{-t/K} \tag{3-21}$$

where:

K = the single reservoir constant

In a watershed modeling process, the principal question is how many equal linear "pools" are needed to the model. Based on the authors' experience, n = 1.0 gives satisfactory results for small watersheds up to 10 km^2. For larger watersheds, the ranges of n may be higher; however, as stated by Overton and Meadows[13] the model rapidly approaches translation ($n \to \infty$) for $n > 5$. An analysis of a very large number of storm hydrographs by Holtan and Overton (quoted in reference 13) indicates that n = 2 produces optimum results in fitting computed runoff to measured hydrographs.

Both constants can be related to the time of travel of water from the most remote point on the watershed to the watershed outlet. This time is called the time of equilibrium or time of concentration. According to Rao et al.[35]

$$t_e = K = nK_N \tag{3.22}$$

Henderson and Wooding[33] used the kinematic wave approximation for a V-shaped watershed and developed an equation for t_e which when converted to metric units becomes

$$t_e = 6.9 \frac{L^{0.6} n_M^{0.6}}{i^{0.4} S^{0.3}} \tag{3-23}$$

where:

t_e = time to equilibrium (min)
L = length of the overland flow (m)
i = rain intensity (mm/hr)
S = slope (m/m)
n_M = Manning's surface roughness factor (from Table 3-10)

An almost identical formula was developed and independently published by Morgali and Linsley.[37]

Rao et al.[35] statistically analyzed the hydrograph curves for several urbaniz-

TABLE 3.10. Manning's Roughness Factor, n_M, for Overland Flow.

Ground Cover	Manning's n_M
Urban Zones[a]	
Smooth asphalt	0.012
Street pavement	0.013
Asphalt or concrete paving	0.014
Packed clay	0.03
Light turf	0.20
Dense turf	0.35
Dense shrubbery or forest litter	0.40
Agricultural Zones[b]	
Fallow field	
Smooth–rain packed	0.01–0.03
Medium–freshly disked	0.1–0.3
Rough–freshly turn-plowed	0.4–0.7
Cropped field	
Grass and pasture	0.05–0.15
Clover	0.08–0.25
Small grain	0.1–0.4
Row crops	0.07–0.2

[a] From reference 3.
[b] From reference 36.

ing watersheds. The authors investigated the effects of many variables on the shape of the runoff hydrographs and only those which were found statistically significant were included in the final formulas. Based on the above work, t_e and n can be estimated from

$$t_e = 304 \frac{(AW)^{0.458}(TR)^{0.104}}{(1+U)^{1.662}(i)^{0.269}} \tag{3-24}$$

and

$$n = \frac{2.23(AW)^{0.069}}{(1+U)i^{0.155}} \tag{3-25}$$

where:

t_e = the lag time or time to equilibrium (hr)
AW = watershed area (km^2)
i = rain intensity (mm/hr)
U = fraction of the impervious area of the total watershed area
TR = rain duration (hr)

Gray[38] developed a relationship for the watershed constant, K_N, applicable to small rural watersheds (0.6 to 80 km^2) in Wisconsin, Illinois, central Iowa, and Missouri as

$$K_N = \frac{t_e}{n} = 7.33\left(\frac{L}{S_c}\right)^{0.562} \quad (3\text{-}26)$$

where:

L = length of watershed (km) that includes overland flow length and the length of the longest watershed channel
S_c = average watershed slope (%)
t_e = time of concentration (min)

Using Equation 3-23 or 3-24, the watershed characteristic, n, can then be estimated by Equation 3-22 knowing that min $n = 1$.

The unit hydrograph solution can be conveniently programed for a digital computer model. Compared to conventional numerical finite difference solutions, the savings on computer time are substantial, and the accuracy is adequate with no stability and convergence problems. According to Equation 3-18, each time component of the rainfall excess will be convoluted with different unit hydrographs due to the dependence of the parameters, n and K_N, on the rain input. However, the loss of accuracy is minimal and a close agreement of measured and computed flows can be obtained as shown in Figure 3-17.

Example 3-4: Runoff Estimation

For the rainfall excess computed in Example 3-2 and reported in Table 3-6, estimate the magnitude and shape of the surface runoff hydrograph if the drainage area is 1 km^2. The slope is 2% and the surface is covered by grass; 3% of the watershed is impervious.

Since the watershed size is less than 10 km^2, the reservoir characteristic, n, is assumed to be one. This can be checked from Equation 3-25 using the smaller rain intensity reported in Table 3-6, column 9. Hence, i = 4.4 cm/hr = 44 mm/hr. Then

$$n = \frac{2.23(1.0)^{0.069}}{(1 + 0.03)44^{0.155}} = 1.20$$

Therefore, Equation 3-21 can be used for representing the instantaneous unit hydrograph. The Manning factor for grass is close to 0.35. For i = 57 mm/hr

$$K = t_e = 6.9\,\frac{(0.5 \times 1000)^{0.6}\,0.35^{0.6}}{57^{0.4}\,0.02^{0.3}} = 98.2 \text{ min} = 1.64 \text{ hr}$$

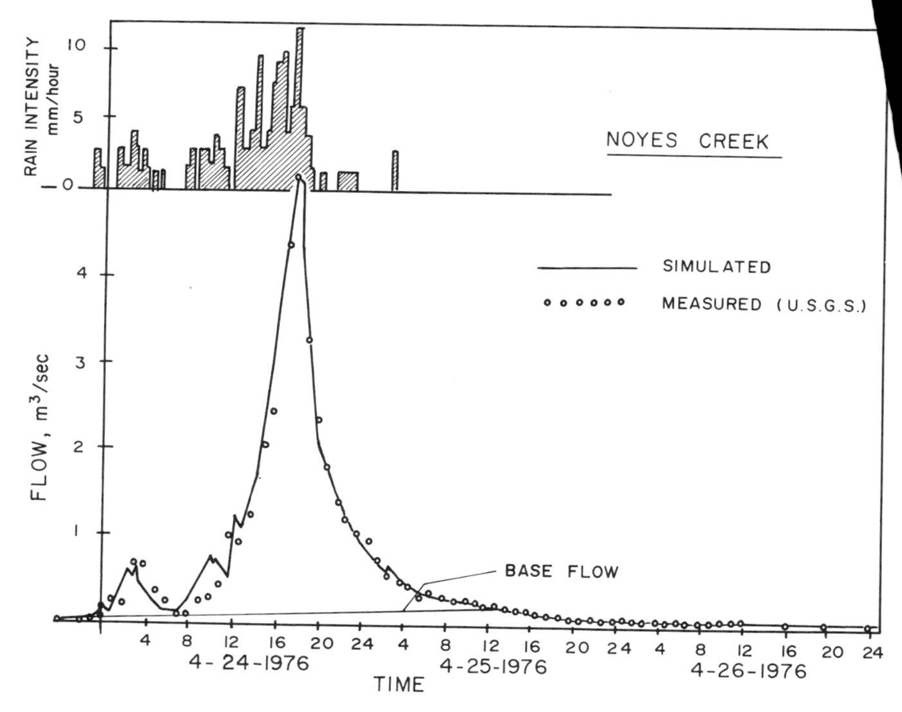

Fig. 3-17. Comparison of measured and computed flows by a rainfall-runoff model for a small urban watershed in southeastern Wisconsin.

For i = 44 mm/hr

$$K = t_e = 6.9 \frac{(0.5 \times 1000)^{0.6} 0.35^{0.6}}{44^{0.4} 0.02^{0.3}} = 99.3 \text{ min} = 1.65 \text{ hr}$$

Since both storage constants are similar, linear response can be assumed. The hydrograph computation is shown in Table 3-11 and plotted in Fig. 3-18.

INTERFLOW

Although the quantity of interflow may represent only a small portion of total runoff, for some mobile pollutants its pollution effect may be of the same order of magnitude as the surface runoff. The interflow pollution originates mainly from salts deposited in soils. On the other hand, interflow may be free of particulate pollutants.

The occurrence of interflow may be observed in areas where the permeability of the subsoils is less than that of the upper soil zone, causing a horizontal move-

ˌLE 3-11. Hydrograph Computation for Example 3-4.

ɹe ˌin) or τ	Unit Hydrograph at Time τ	Ordinates of the Hydrograph (m^3/sec)		
		P57 mm/hr	P44 mm/hr	Total
		At Time t		
0	0	0	0	0
15	0.56	0	0	0
30	0.48	2.22	0	2.22
45	0.41	1.90	1.71	3.61
60	0.35	1.62	1.47	3.09
75	0.31	1.38	1.26	2.64
90	0.26	1.22	1.07	2.29
105	0.23	1.03	0.95	1.98
120	0.19	0.91	0.79	1.70
150	0.15	0.67	0.64	1.31
180	0.11	0.49	0.46	0.95
210	0.08	0.38	0.34	0.72

$$\text{Flow (m}^3\text{/sec)} = \text{Area (m}^2) \times \frac{\text{m}}{1000 \text{ mm}} \times \frac{\text{hr}}{3600 \text{ sec}}$$

$$\times \sum_{\tau=0}^{t} [X_{t-\tau}(\text{mm/hr}) h(\tau) \Delta t]$$

X = rain intensity

h = hydrograph ordinate

ment of water in the upper zone. Lateral water movement in soils is especially significant during the snow-melt period, when subsoils are still frozen.

Theoretically, the amount of interflow could be computed from Darcy's law if the depth of the saturated upper zone storage, the saturation permeability, and the slope of the soil water piezometric surface were known.

The Stanford Watershed Model[3] approximates the outflow from the interflow storage as

$$INTF = \alpha \cdot SRGX \tag{3-27}$$

where:

$SRGX$ = current volume of water in interflow storage

For a 15-min computational interval used in the model

$$\alpha = 1 - (IRC)^{1/96} \tag{3-28}$$

where:

IRC = recession constant

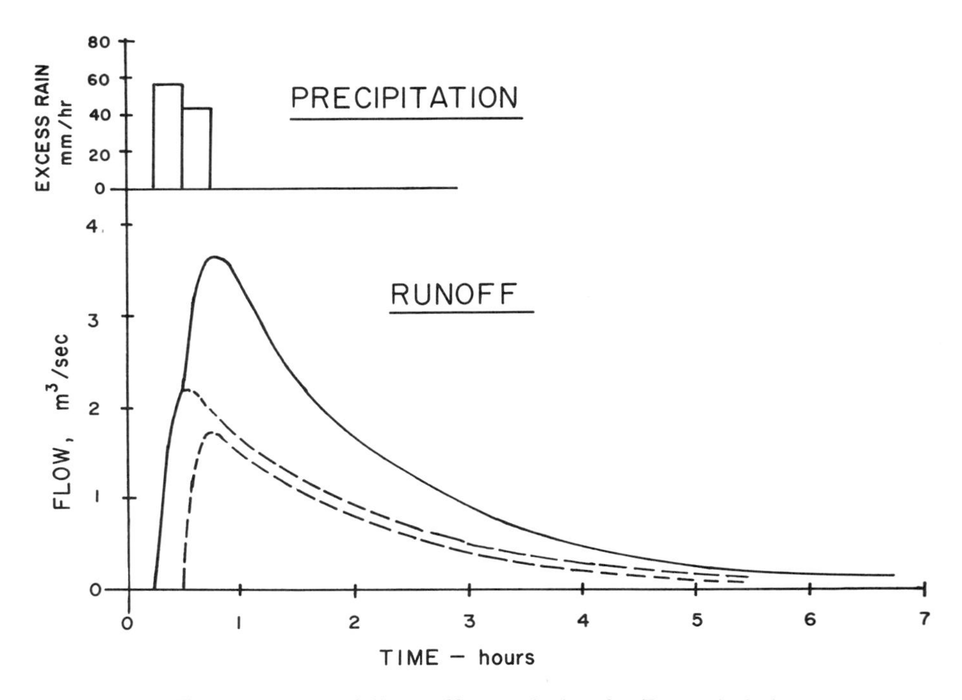

Fig. 3-18. Rainfall-runoff convolution for Example 3-4.

GROUNDWATER FLOW

Groundwater movement and occurrence are an integral part of the hydrologic cycle. Almost all groundwater originates from infiltrated precipitation, after subtraction of surface losses–surface runoff, evapotranspiration, and interflow. During prolonged dry periods, most of the natural flow in surface waters originates from groundwater systems and is referred to as base flow or groundwater runoff.

Aquifers and Aquitards

Water below the ground surface occurs in four zones (Fig. 3-19): soil zone, intermediate zone, capillary zone, and saturation zone. The soil, intermediate, and capillary zones are called vadose zones (in Latin *vadosus* means shallow) or zones of aeration. The zone of aeration contains voids and cracks occupied partially by water and air. In the saturated zone, all interstices are filled by water.

The top of the saturated zone–the groundwater table–can be revealed in wells and borings that penetrate into the saturation zone. It represents the water surface under atmospheric pressure. Due to capillary forces acting at the ground-

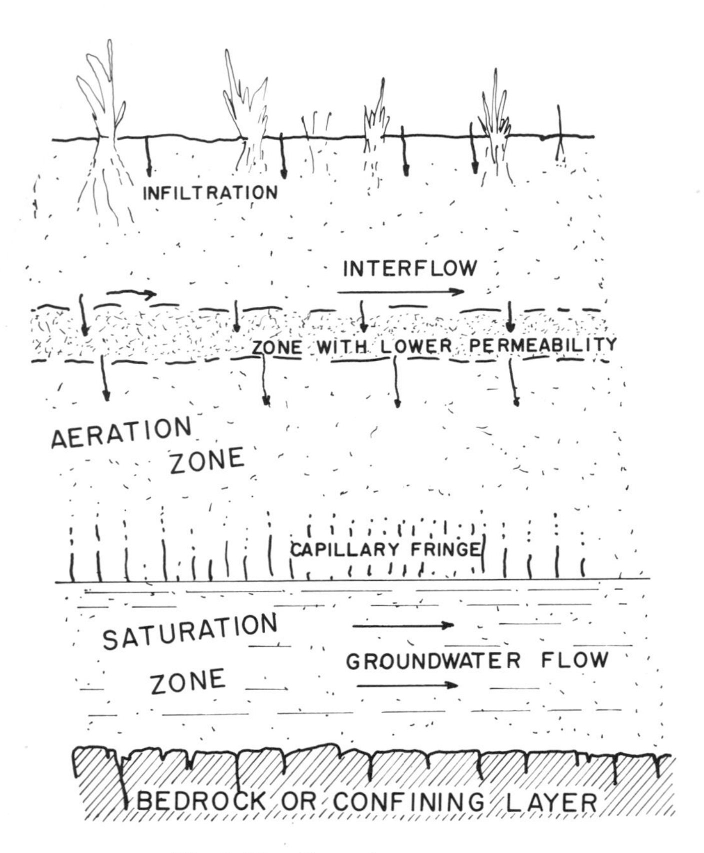

Fig. 3-19. Groundwater zones.

water surface, saturated conditions may exist above the groundwater table; however, water is held there at less than atmospheric pressure.[39]

Water occurring in the saturated zone is referred to as groundwater. Its upper boundary is the water table or an overlying impermeable stratum. In the latter case, the maximum water elevation in a well penetrating the groundwater zone is called the piezometric water table. The lower boundary extends to the underlying impermeable stratum such as bedrock (e.g., shale or crystalline rocks) or clayey layers. The groundwater zone, with both upper and lower boundaries consisting of impermeable layers, is confined, while the unconfined zone has an upper boundary (water table) overlain by the zone of aeration.

The intermediate zone extends from the bottom of the soil water zone to the top of the capillary fringe. Water in the intermediate zone moves vertically downward under the influence of gravity. This zone may be absent under high

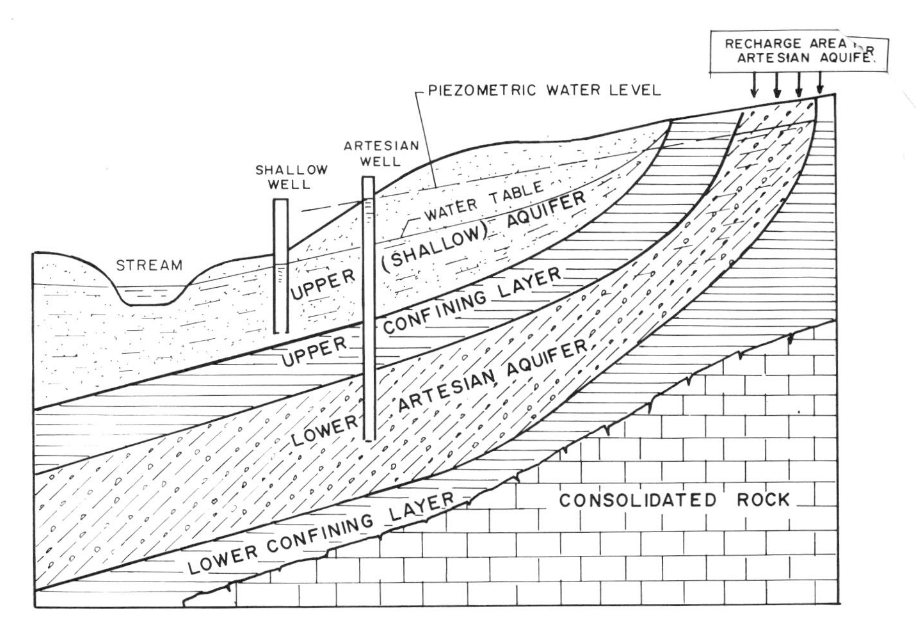

Fig. 3-20. Representation of aquifers.

groundwater conditions or may be several hundred meters thick in arid and semiarid regions. Water in the intermediate zone is held by hygroscopic and capillary forces. The excess moves downward by gravity.

A geological formation saturated by water that will yield appreciable quantities of water that can be economically used and developed has been defined as an aquifer.[39] An aquifer can be either confined or unconfined (Fig. 3-20).

An unconfined aquifer is one in which the upper boundary of the saturation zone is the same as the water table. Confined aquifers, also known as artesian or pressure aquifers, are overlain by an impermeable geological stratum that keeps water under pressure.

Water enters an aquifer through a recharge area, which is an area where the water-bearing stratum is exposed to the atmosphere or is overlain by a permeable zone of aeration.

Aquitards are geological formations that are not permeable enough for economical development as a groundwater source. An aquiclude is a formation that stores water but is incapable of transmitting (e.g., clays). A rock that neither transmits nor stores water is an aquifuge.

Aquifers and aquitards can exist in layers with an unconfined saturated zone occurring on top, underlain by one or more confined zones. The top unconfined aquifer, often called a shallow aquifer, is most susceptible to nonpoint pollution.

groundwater that can be recovered by springs and wells represents the groundwater or base runoff.

Groundwater Systems

According to Toth[40] (quoted in reference 41), groundwater flow systems can be divided into three types: local, intermediate, and regional. A local flow system has its recharge area in a topographic high and discharges into an adjacent topographic low. The intermediate flow system is viewed as having recharge and discharge areas that are not in adjacent topographic highs and lows. The discharge area, for the intermediate system, may be located several subbasins downstream. The regional system has its recharge in the regional topographic high, while its discharge occupies the topographic low for the basin (Fig. 3-21).

The flow in groundwater zones is generally slow, and the response to surface and subsurface pollution loadings are often gradual. The residence time of water in local groundwater systems may range from days to months, intermediate zone flows may have residence times ranging from months to years, and the residence of water in regional groundwater systems is years. In most cases, surface contamination by pollutants affects only local groundwater systems and shallow aquifers.

The three systems have well-defined boundaries that theoretically identify changes in flow patterns, water quality, and relative rate of groundwater move-

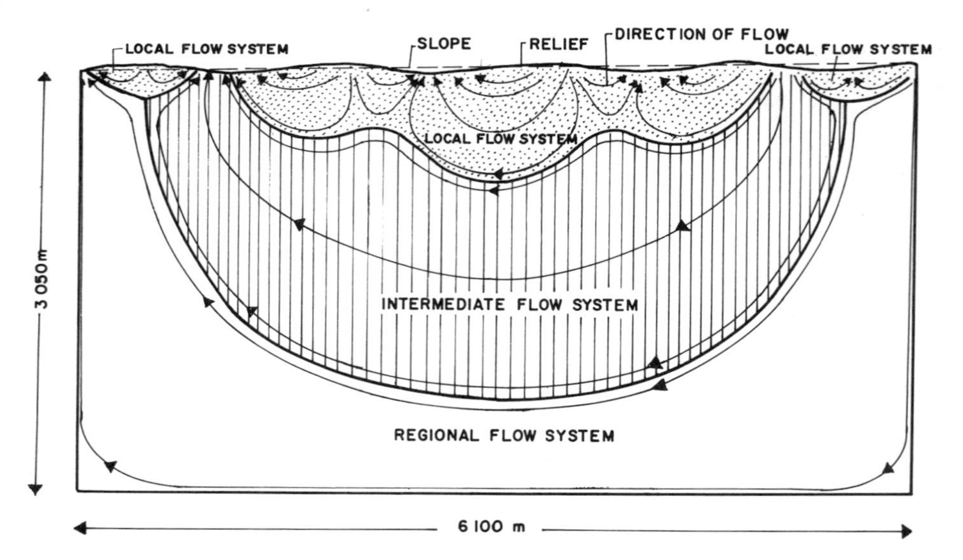

Fig. 3-21. Theoretical patterns for local, intermediate, and regional flow systems. (Adapted from reference 40.)

ment.[40,41] The water movement in a local system is commonly faster, which will produce water quality different from the regional system. Due to the longer residence time and slower flow rates, the regional systems will produce groundwater with higher concentrations of dissolved ions than groundwater moving through a local system.

The depths at which the boundaries between flow systems develop are dependent upon the following geomorphological factors: local topographic relief and its steepness, distribution and depths of bedrock and impermeable geological strata, aquifer thickness, and general topographic and geomorphological characteristics of the basin.

The computer simulations of Toth[40] indicated that the development, intensity, and depth of local flow systems could be related directly to increased local topographic relief and inversely to increased regional slope. Strong regional flow systems developed where local relief was negligible and where the ratio of total aquifer thickness to local relief was high. A weak local flow system will have only a portion of the total aquifer contributing discharge to that subbasin. Moreover, a strong local flow system has most or all of the aquifer discharging into that subbasin. Figure 3-22 shows typical flow patterns for a glacial aquifer overlying dolomite and the flow patterns in the two aquifers.

Relation Between Surface and Groundwater Systems. Ground and surface waters are interrelated through two processes: *recharge* and *discharge*. There are two major sources of natural recharge to an aquifer. The first is the residual of precipitation which infiltrates through the unsaturated zone, the second is freshwater inflow from surface water bodies such as streams, rivers, and lakes. In addition, aquifers may be recharged by septic tanks, irrigation, artificial recharge, and sewer leakage. Natural discharge from aquifers may occur through springs, streams, lakes, and oceans, and through evapotranspiration from larger plants known as phreatophytes, whose roots extend to the water table. In addition, man extracts water from aquifers by wells and other more sophisticated collection systems. Under natural conditions, natural recharge and discharge are commonly in balance.

Groundwater Hydrologic Models. The state-of-the-art of groundwater hydrology is well advanced, and numerous models have been developed to represent the flow and quality conditions of aquifers. An EPA publication surveyed and summarized over 200 available models.[42] Kisel and Duckstein[43] presented the theoretical background of groundwater models, their advantages, and their deficiencies.

The models can be either distributed parameter flow models or lumped parameter aquifer models. The lumped system considers aquifers as homogeneous, and the lumped values can be obtained by averaging the aquifer characteristics at a

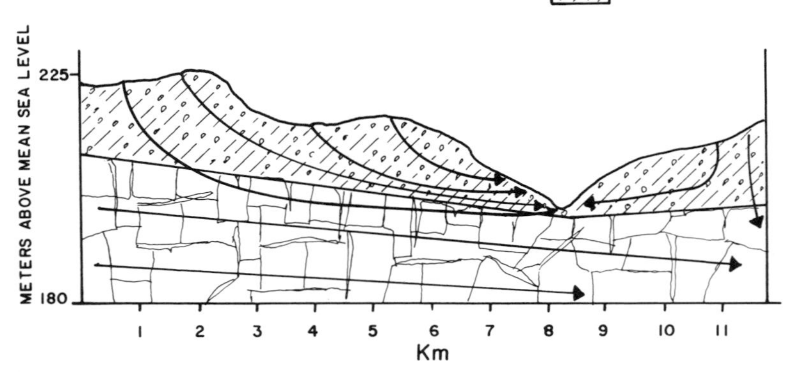

Fig. 3-22. Cross section showing generalized flow patterns for local flow system in a shallow glacial aquifer underlain by a dolomite aquifer. (From reference 41.)

few discrete points or by estimating the parameters from the response of the aquifer to a known input, for example, a dye injection.

Many models are three-dimensional box-grid representations using finite difference approximations of the flow equations. Due to the linear nature of the basic flow equations, many groundwater flow problems can be solved by electric-analog simulations.

For nonpoint pollution studies, groundwater flow models can provide basic information on residence times of water in groundwater zones, dispersion patterns, and basic direction of flow.

REFERENCES

1. Lull, H. W. 1964. Ecological and silvicultural aspects, Chapter 6 in: V. T. Chow (ed.), "Handbook of Applied Hydrology." McGraw-Hill, New York.
2. Steward, B. A., Woolhiser, D. A., Wischmeier, W. H., Caro, J. H., and Frere, M. H. 1976. Control of water pollution from cropland, Vol. II–An overview. U.S. EPA and ARS U.S. Dept. of Agriculture, EPA 600/2-75/026b, Washington, D.C.
3. Crawford, N. H., and Linsley, R. K. 1966. Digital simulation in hydrology, Stanford Watershed Model IV. Technical Rep. No. 39, Dept. of Civil Engineering, Stanford University, Palo Alto, California.
4. Tholin, A. L., and Keifer, C. J. 1960. Hydrology of urban runoff. *Trans. ASCE*, Pap. No. 3061.

5. Hiemstra, L. 1968. Frequencies of runoff for small basins. Ph.D. Thesis, Colorado State University, Fort Collins, Colorado.
6. Larsen, V., Axley, J. H., and Miller, G. L. 1972. Agricultural wastewater accommodation and utilization of various forages. WRRC A-006-Md, 14-01-0001-790, University of Maryland, College Park, Maryland.
7. Chow, V. T. (ed.). 1964. "Handbook of Applied Hydrology." McGraw-Hill, New York.
8. Horn, M. E. 1971. Estimating soil permeability rates. *J. Irrigation Drainage Div., Proc. ASCE*, **97**:263–274.
9. Horton, R. E. 1939. Approach toward a physical interpretation of infiltration capacity. *Proc. Soil Sci. Soc. Amer.*, **5**:399–417.
10. Holtan, H. N. 1961. A concept for infiltration estimates in watershed engineering. USDA Agricultural Res. Service, ARS pp. 41–51, Washington, D.C.
11. Holtan, H. N., and Lopez, N. C. 1973. USDAHL-73 Revised model of watershed hydrology. U.S. Dept. of Agriculture, Plant Physiology Institute, Rep. No. 1.
12. Philip, J. R. 1969. Theory of infiltration. In: V. T. Chow (ed.), "Advances in Hydroscience." Academic Press, New York.
13. Overton, D. E., and Meadows, M. E. 1976. "Stormwater Modeling." Academic Press, New York.
14. Braslavskii, A. P., and Vikulina, Z. A. 1963. "Evaporation norms from water reservoirs." Gidrometeorologicheskoe izdatelstvo, Leningrad, 1954, English translation: Israel Program for Scientific Translation, Jerusalem, 1963.
15. Edinger, J. E., and Geyer, J. C. 1965. Heat exchange in the environment. Cooling Water Studies for Edison Electric Institute, John Hopkins University, Baltimore, Maryland.
16. Harbeck, G. E., Jr. 1962. A practical field technique for measuring reservoir evaporation utilizing mass transfer theory. U.S. Geological Survey, Prof. Pap. 272-E, U.S. GPO, Washington, D.C.
17. Davis, C. V., and Sorensen, K. E. 1969. "Handbook of Applied Hydraulics." McGraw-Hill, New York.
18. Gray, D. M. 1973. "Handbook on the Principles of Hydrology." Water Information Center, Port Washington, New York.
19. Penman, H. L. 1948. Natural evaporation from open water, bare soil and grass. *Proc. Royal Soc. London*, Ser. A, **193**:120–145.
20. Penman, H. L. 1958. Evaporation: An introductory survey. *Neth. J. Agr. Sci.*, **4**:9–29.
21. McDaniel, L. L. 1960. Consumptive use of water by major crops in Texas. Texas Water Dev. Board Bull. 6019, Austin, Texas.
22. Anon. 1956. "Snow Hydrology." U.S. Army Corps of Engineers, North Pacific Div., Portland, Oregon.
23. Anon. 1968. Hydrology. Suppl. A to Sec. 4. "Engineering Handbook." USDA, Soil Cons. Service.
24. Anon. 1972. Hydrocomp simulation programming operation manual. Hydrocomp International, Palo Alto, California.

25. Engman, E. T. 1974. Partial area hydrology and its application to water resources. *Water Resources Bull.*, **10**(3):512–521.
26. Eagelson, P. S. 1970. "Dynamic Hydrology." McGraw-Hill, New York.
27. Wooding, R. A. 1965. A hydraulic model for the catchment-stream problem, Parts I and II. *J. Hydrol.*, **3**:254–282.
28. Wanielista, M. P. 1978. "Stormwater Management—Quantity and Quality. Ann Arbor Science, Ann Arbor, Michigan.
29. Sherman, L. K. 1932. Streamflow from rainfall by unit-graph method. *Eng. News Rec.*, p. 501.
30. Horton, R. E. 1938. The interpretation and application of runoff plot experiments with reference to soil erosion problems. *Proc. Soil Sci. Soc. Amer.*, **3**:340–349.
31. Izzard, C. F. 1946. Hydraulics of runoff from developed surfaces. *Proc. Highway Res. Board*, **26**:129–150.
32. Amorocho, J., and Brandstetter, A. 1971. Determination of non-linear functional response function in rainfall-runoff process. *Water Resources Res.*, **7**:1087–1101.
33. Henderson, F. M, and Wooding, R. A. 1964. Overland flow and groundwater flow from a steady rainfall of finite duration. *J. Geophys. Res.*, **69**:1531–1540.
34. Nash, J. E. 1957. The form of the Instantaneous Unit Hydrograph. *Bull. Int. Assoc. Sci. Hydrol.*, **111**:114–121.
35. Rao, R. A., Delleur, J. W., and Sarma, B. S. P. 1972. Conceptual hydrologic model for urbanizing basins. *J. Hydraulics Div., ASCE*, **98**:1205–1220.
36. Huggins, L. F., Podmore, T. H., and Hood, C. F. 1976. Hydrologic simulation using distributed parameters. Purdue Univ. Water Resources Research Center Tech. Rep. No. 82, West Lafayette, Indiana.
37. Morgali, J. R., and Linsley, R. K. 1965. Computer analysis of overland flow. *J. Hydraulics Div., Proc. ASCE*, **91**:81–100.
38. Gray, D. M. 1961. Synthetic unit hydrographs for small watersheds. *J. Hydraulic Div., Proc. ASCE*, **87**:33–54.
39. Todd, D. K. 1959. "Groundwater Hydrology." Wiley, New York.
40. Toth, J. 1963. A theoretical analysis of groundwater flow in small drainage basins. *J. Geophys. Res.*, **68**:4795–4812.
41. Eisen, C. E., and Anderson, M. P. 1978. Field data quantifying groundwater surface water interaction. Final Rep., Menomonee River Pilot Watershed Study, Internatl. Joint Commission, Winsdor, Ontario, and U.S. EPA, Washington, D.C.
42. Anon. 1978. Utilization of numerical groundwater models for water resources management. EPA 600/8-78/012, U.S. EPA, Washington, D.C.
43. Kisel, C. C., and Duckstein, L. 1976. Groundwater models. In: A. K. Biswas (ed.), "Systems Approach to Water Management." McGraw-Hill, New York.

4

Pollution from the atmosphere

ATMOSPHERIC INPUTS

The atmosphere is the portion of the environment where some of the most severe nonpoint pollution problems originate, and, in fact, the intensity and magnitude of nonpoint pollution often can be correlated with contamination of the atmosphere and atmospheric and meteorological conditions. Emissions of sulfur dioxide and nitric oxides by coal burning processes and traffic cause the phenomenon of "acid rain," which is having a devastating effect on many water bodies throughout the world. Particulate aerosols in the atmosphere contain appreciable quantities of sulfur, toxic and trace metals, pesticides, organic compounds, fungi, pollen, and soil; other particulate matter including fly ash, nutrients, and tar; and a variety of other chemical compounds such as oxides, nitrites, nitrates, chlorides, fluorides, ozone, and silicates. Several extensive treatises have been devoted to air pollution problems and the reader is referred to reference 1 for the most extensive coverage of the topic.

Sources of Pollutants

Sources of atmospheric pollution and contamination include a variety of natural and cultural emissions that can affect the composition of the air on a global or a local scale.

Major global sources of pollution are:[1]

A. Urban and industrial emissions including those from power generation, industrial processes, and domestic heating.
B. Agricultural and forest emissions resulting from human activities such as:
   a. Soil erosion by wind during dry weather,
   b. Slash burning, which in many parts of the world is recommended to prevent the spread of disease and to reduce fire hazard.
   c. Fertilizer components reaching the atmosphere through wind erosion and/or volatilization.
   d. Pesticides entering the atmosphere from drift during application and by wind erosion and by volatilization.
   e. Decomposing farm wastes and animal operations releasing ammonia, hydrogen sulfide, methane, and mercaptans to the atmosphere.
C. Naturally-occurring emissions on the global scale including:
   a. Dust blown from arid and desert areas.
   b. Forest, bush, and grass fires.
   c. Volcanic eruptions, which are a source of sulfuric compounds and ash.
   d. Volatile hydrocarbons emitted from forests and other silvicultural activities.
   e. Sea spray, a significant source of particulate matter such as salt particles.
   f. Evaporation from large water bodies, which can contribute significant quantities of volatile compounds and trace gases.

Table 4-1 summarizes the input of particulate matter into the troposphere, that is, the first 15-km layer of the atmosphere above the ground. It should be noted that so-called "background" atmospheric pollution often referred to in the literature on nonpoint sources is actually a composite of natural and man-made sources. Background levels of pollution measured in the snow of Greenland or the Arctic contain large quantities of pollutants that emanate from distant urban and nonurban man-made sources.

Local sources of pollution include most sources mentioned earlier plus traffic. The magnitude of pollution–and, hence, atmospheric deposition–is magnified by several orders of magnitude in the vicinity of some sources and falls off rapidly to background levels as distance from the source is increased. For example, most of the lead deposited from automobile exhausts is found within 100 m of the roadway.

Atmospheric pollution consists of gases, and aerosols or atmospheric particulates. The particulate matter ranges in size from 6×10^{-4} to 10^3 μm. The term

TABLE 4-1. Estimates of Tropospheric Aerosol Production Rates.*

Source	Production Rate (Metric Tons/Day)	Weight (% of Total)
A. Natural sources		
1. Primary		
Wind-blown dust	2×10^4–10^6	9.3
Sea spray	3×10^6	28
Volcanoes	10^4	0.09
Forest fires	4×10^5	3.8
2. Secondary		
Vegetation	5×10^5–3×10^6	28
Sulfur cycle	10^5–10^6	9.3
Nitrogen cycle	2×10^6	14.8
Volcanoes (gases)	10^3	0.009
Subtotals	10×10^6	94
B. Man-made sources		
1. Primary		
Combustion and industrial	1×10^5–3×10^5	2.8
Dust from cultivation	10^2–10^3	0.009
2. Secondary		
Hydrocarbon vapors	7×10^3	0.065
Sulfates	3×10^5	2.8
Nitrates	6×10^4	0.56
Ammonia	3×10^3	0.028
Subtotals	6.7×10^5	6
Total	10.7×10^6	100

*Data by G. M. Hidy and J. R. Brock (1970) Proc. 2nd Int. Clean Air Congress, pp. 1088–1097–Table reprinted from Ref. 1.

aerosol should be differentiated from dust. Dust contains particles that are mostly insoluble, while aerosols contain water-soluble materials (about 50% according to reference 2) and they may react with or adsorb other substances.

Besides direct emissions from terrestrial sources, aerosols can be formed in the atmosphere by precipitation, absorption, and chemical reactions.

Atmospheric transport and contamination is part of the overall cycle of elements in the air, water, and land (Fig. 4-1). Oceans and large lakes are often the final sinks of pollutants transported by air and water.

Entry of Atmospheric Pollutants in Surface Waters

Removal of particles from the atmosphere is due to:

A. Removal by rainfall and snowfall.
B. Dry deposition by sedimentation.
C. Dry deposition by impact on vegetation and rough surfaces.

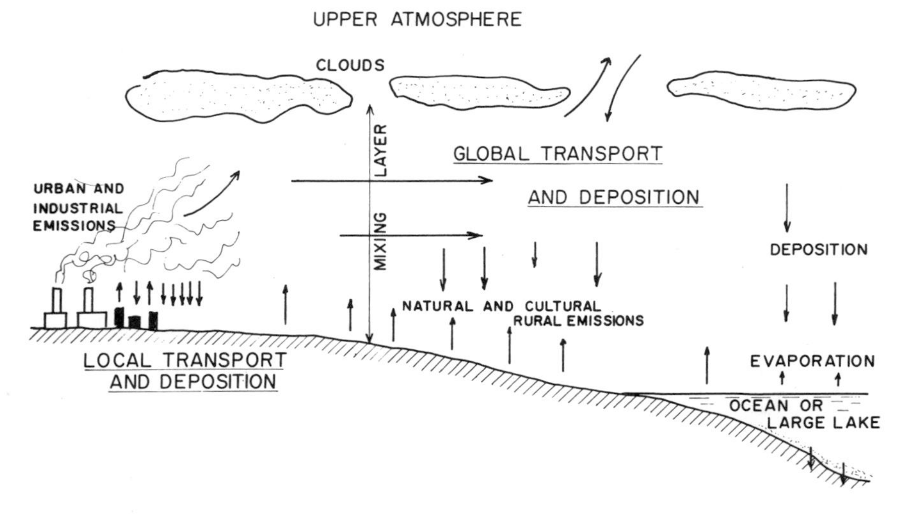

Fig. 4-1. Transport and deposition of pollutants through the atmosphere.

Removal of gases occurs primarily by:

A. Removal during periods of precipitation.
B. Absorption at earth's surface.
C. Adsorption on aerosolic particles and subsequent deposition.

Wet removal or scavenging of pollutants by moisture contained in the atmosphere is the primary mechanism for transferring atmospheric pollutants to surface waters. Sedimentation affects contaminants in particles of sizes greater than 10 μm.[1,3]

Local–and to some extent global–levels of atmospheric contamination can be reduced by air pollution control measures. These measures are aimed mostly at reducing emission rates of aerosols (stack effluent control), at reducing pollutant emissions from automobile exhausts, or at banning certain compounds from the market. In the United States, these efforts were mandated by the Clean Air Act enacted and passed by the Congress and signed by President Carter in 1977.

In a parallel fashion to the Clean Water Act (see Chapter 2), the Clean Air Act relies on the two sets of standards:[4]

A. The national ambient air standards designed to protect public health and welfare.
B. Emission standards based on pollution control technology that are imposed on the sources (stationary or moving) to bring pollutant concentrations below ambient standards and keep them there.

The agency responsible for implementation of the Clean Air Act is the U.S. Environmental Protection Agency (EPA).

With the exception of acidity and some toxic metals, the atmospheric contribution to pollution of surface waters is less than from terrestrial sources. However, for lakes with a small ratio of drainage area to lake surface area, the atmospheric contribution of many pollutants is significant. The major example of this relationship is the Great Lakes.

Wet Deposition and Composition of Precipitation. The notion that "pure rainwater" exists should be abandoned at the beginning of the discussion. Due to the fact that precipitation scavenging is one of the most effective processes for cleansing the atmosphere, rain and snow contain many pollutants in quantities that may be harmful to terrestrial and aquatic ecosystems. Pollutants contained in precipitation include acidity, toxic metals, organic chemicals, phosphates, and nitrogen compounds. In some areas pollution from rain has been devastating to surface water biota and oftens leads to acidification of lakes, fish kills, and severe reduction in lake productivity.

Contaminants can be incorporated into precipitation within or below clouds. In-cloud scavenging is called rainout or snowout, the below-cloud process of enrichment is called washout.[1,3]

When a water particle falls (precipitation) or moves (cloud water) through the atmosphere it sweeps out a volume of air. Due to the forces of inertia, particles and gas molecules collide with the water drops and are absorbed by or otherwise react with the water. Scavenging efficiency is greater for snow than rain due to the larger specific surface exposed by snowflakes.[4] Precipitation is an excellent carrier and scavenger of such atmospheric contaminants as gases (NO_x, CO_2, SO_2), aerosols (e.g., $(NH_4)_2SO_4$) and large particles (e.g., $CaCO_3$, silicates).[5] The typical composition of precipitation measured in urban and rural areas is given in Table 4-2.

From Table 4-2 it can be seen that rainwater contains surprisingly high amounts of acidity (expressed as pH), nutrients, and toxic metals. The nutrient content of rain at times can be greater than the recommended limits for preventing lake eutrophication (see Chapter 2). Most of the acidity and toxic metals originate from coal-fired power plants, but the problem is by no means purely local, since precipitation with high acidity and metal content falls over a large portion of Europe and North America, and other parts of the world (Table 4-2). In the absence of strong terrestrial point or nonpoint sources of pollution, rainwater may be the major input of pollutants to some surface waters. For example, atmospheric inputs of nutrients are the prevailing source of pollution in undisturbed watersheds (Table 4-3).

Some measures intended to control air pollution levels near the stack actually spread the problem over larger areas. The practice of building tall stacks—some over 150 m high—changes a local air pollution problem into a regional nonpoint

TABLE 4-2. Average Precipitation Characteristics in Urban and Rural Areas (mg/liter).

	Urban					Rural							
											Harp Lake, Ontario[10]		
Constituent	Knoxville, Tenn.[6]	St. Louis, Mo.[6]	Houston, Tex.[7]	London, England[8]	Oslo, Norway (snow)[11]	Gatlinburg, Tenn.[6]	Oak Ridge, Tenn.[6]	Hubbard Br., N.H.[5]	Houston, Tex.[7]	Chadron, Neb.[9]	Snow	Rain	Rural South Norway[11]
pH[a]	5.1	4.9		5.1–5.6	3.98	4.2		4.1				3.9	4.6
COD (TOC)	65		14.1					(1.31)	15.4				
Org. N	2.5												
NH_3–N	0.41		0.30		0.7	0.18	0.13		0.22		0.26	0.44	0.30
NO_x–N	0.47		0.52		0.97	0.28	0.25	1.39	0.31		0.7	0.76	0.37
PO_4–P	1.1		0.012		0.002		0.12	0.025	0.04		0.01	0.03	0.003
Lead	0.05	0.03		0.024–1.04	0.056					0.004			
Mercury	0.0009												
Cadmium				0.001–0.1	0.004					0.0003			
Susp. solids	16												

[a]In pH units.

TABLE 4-3. Sources of Nitrogen and Phosphorus for Various Lakes as Percentages of the Total Input.

Sources	Disturbed Watershed					Undisturbed Watershed		
	Lake Mendota[5]	Lake Rotona[5]	Lake Maldren[5]	Cayuge Lake[5]	Lake Michigan[12]	Clear Lake[5]	Rouson Lake[5]	Mirror Lake[5]
Phosphorus								
Precipitation	6	2	4	2	26	61	50	17
Urban drainage	35	14	39		44			
Rural drainage and groundwater	59	83	57	98[a]				
Nitrogen								
Precipitation	17	6	4	7		75	50	78
Urban drainage	11	11	25					
Rural drainage and groundwater	66	83	71	93[a]		25	50	22
Fixation	7	?	?	?		?	?	?

[a]All terrestrial sources.

pollution problem such as is now evident with the magnitude and areal extent of the acid precipitation.

Global effects on the distribution of nitrogen loadings by wet fallout are shown in Fig. 4-2. Since loading is a product of the volume of the rain and the constituent concentrations, the lowest loadings are observed in the arid regions of the western United States while highest loadings are typical of the more humid Midwest, which has relatively high pollutant concentrations in the atmosphere.

The major sources of nitrogen are fuel combustion and traffic in the form of nitrogen oxides (NO_x) and decomposition of organic materials contained in soils (ammonia and NO_x). Calcareous soils (soils with $pH > 7$) of the midwestern United States are primary sources of volatilized ammonia.

Acid Rain. Acid rain is a serious global pollution problem caused by wet deposition. Acidity* of water is commonly measured as a molar concentration of H^+ ion and expressed as $pH = \log(1/[H^+])$. In relatively unpolluted atmosphere, acidity of rainwater is caused mainly by dissolved CO_2 (carbonate acidity). The pH of rainwater in equilibrium with CO_2 is in the range of 5.6 to 5.7. However, investigators in the early 1950s noticed a surprisingly low pH of precipitation in southern Scandinavia, and a similar phenomenon was reported in the late 1960s

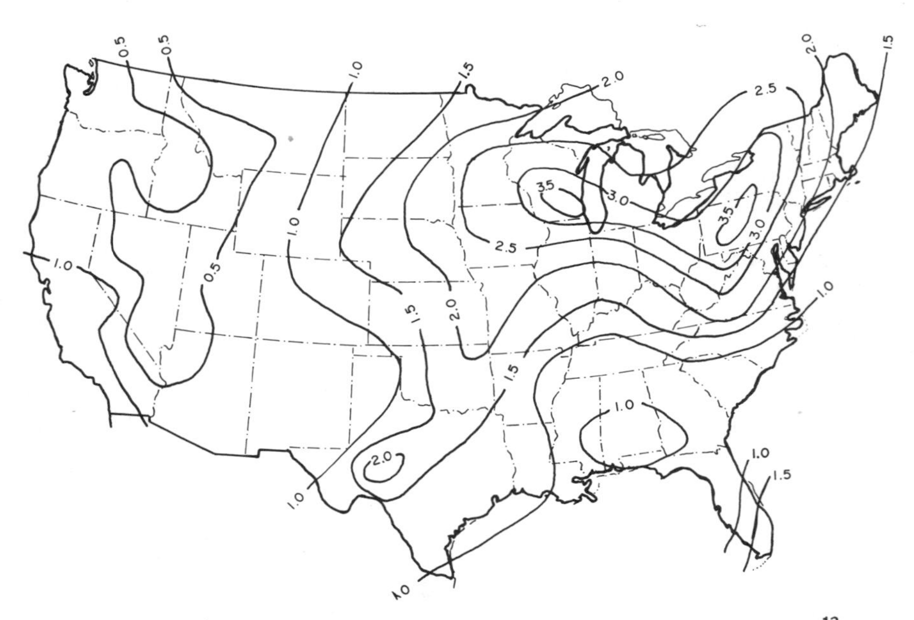

Fig. 4-2. Nitrogen deposition (NO_3^--N and NH_4^+-N) by precipitation.[13]

*Besides pH, acidity is measured by standard laboratory procedures as an amount of a titrant base solution necessary to neutralize the sample.

in the northeastern United States and southern portions of Ontario and Quebec. The pH values were reported near 4 or less.[14,15] From this discussion it follows that the term "acid" rain or "acid" precipitation should be used for precipitation having a pH < 5.6.

The extent and severity of the acid rain problem and the potential damage to the environment are only just being realized, although some serious surface water problems have been reported, including fish kills and oligotrophication (see Chapter 2).

The origin of acid rain is attributed primarily to the presence of sulfates and sulfites (SO_4^{2-} and SO_3^{2-}) and nitrates (NO_3^-) in the atmosphere. Sulfur is one of the elements always found in the atmosphere and occurs as SO_4^{2-} and SO_3^{2-} in aerosols and SO_2 and H_2S gases. Hydrogen sulfide in air is normally oxidized to SO_2, which is then oxidized to SO_3. The oxidation reaction proceeds quickly if such metallic catalysts as iron and manganese oxides are present.[1] These metallic compounds are commonly emitted by burning processes in fly ash. Forma-

Fig. 4-3. Formation of sulfuric acid from the SO_2 emission is enhanced greatly when stack effluent combines with the vapor drift from cooling towers. (Photo authors.)

tion of sulfuric acid is greatly enhanced by the moisture emitted from cooling towers (Fig. 4-3).

The SO_4^{2-} and NO_3^- anions in the air are balanced by cations, primarily NH_4^+, Ca^{2+}, Mg^{2+}, and Na^+. The major sources of these compounds are sea spray, soil dust, and ammonia volatilization from soils. Since the Na^+ from sea spray is already balanced by Cl^- in the absence of other buffering agents in the air, there may not be other cations available to balance additional SO_4^{2-} and NO_3^- ions and they can only react with H^+ to produce acid rain (Fig. 4-4).

Although the elevated acidity of precipitation is commonly attributed to energy production, several other sources contribute to the sulfur and nitrogen oxide contents of the atmosphere. Global estimates indicate that about 70% of atmospheric sulfur originates from marshland or sea and about 30% is from cultural anthropogenic sources.[16] However, Swedish investigators have calculated that more than 70% of the sulfur in the atmosphere over southern Sweden is from human activities, and 77% of it is believed to originate from sources outside Sweden, mainly from burning of low-quality "brown" coal in Central Europe and from coal burning operations in Great Britain. There are no appreciable coal deposits in Scandinavia, and most of the energy is generated from hydropower or from imported oil and high-quality coal. Similarly, in the United States the origin of acid rain is attributed to coal burning operations (primarily from power plants) in the northeastern and midwestern United States (Fig. 4-5).

The origin of nitrates is not so well known, but nitrates are attributed largely to agricultural and traffic sources. Data from New York state and parts of New England revealed that approximately 60 to 70% of the acidity is ascribed to sul-

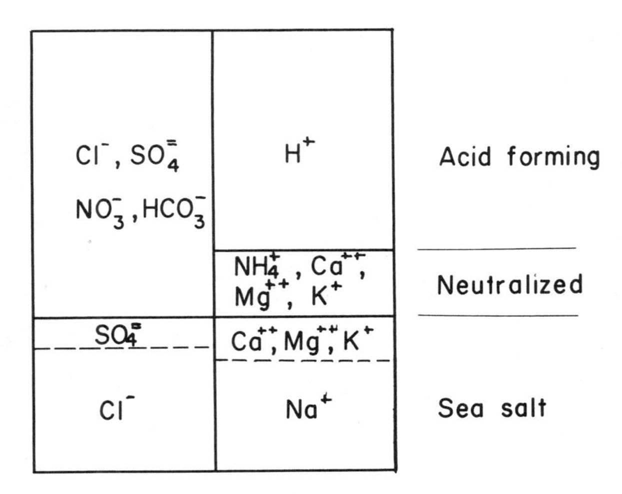

Fig. 4-4. Atmospheric cation-anion balance resulting in an increase of H^+ ions in rainwater. (From reference 14.)

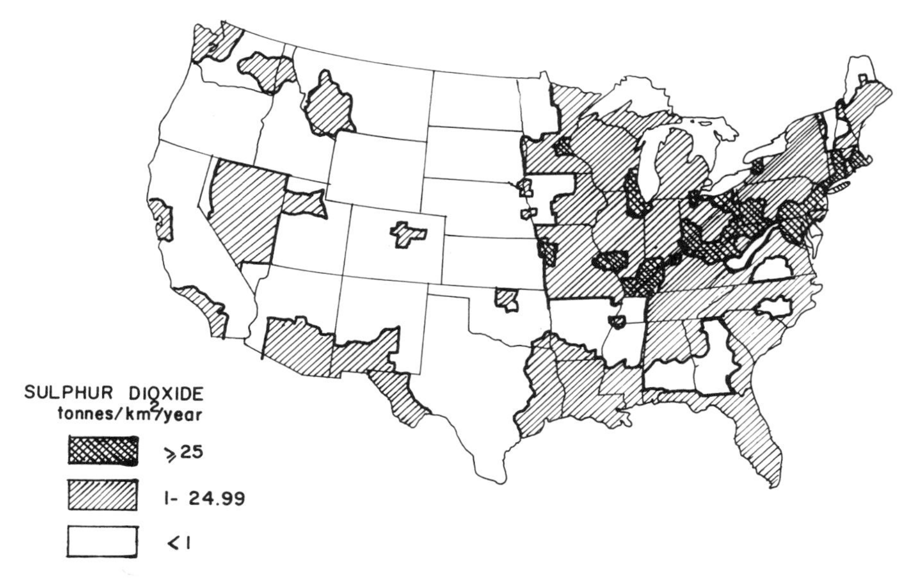

Fig. 4-5. Sulfur emissions in the conterminous U.S. (From reference 17.)

furic acid, and 30 to 40% is due to nitric acid.[18] The proportion may be reversed in areas of heavy traffic (e.g. southern California) where high amounts of nitric oxides are emitted from automobiles.

Acidity is a local as well as (and primarily) a regional problem. The lowest pH values of precipitation are measured usually in the vicinity of large coal-fired power plants, near smelting operations, and in heavy traffic corridors. The pH values are often measured locally near or below 3. It is interesting to note that in England, in the early part of this century, the pH of rain in the vicinity of Leeds (a heavy coal-use region) was measured on occasions to be below 3.0 (reference 18). The lowest pH value reported for an individual storm in Europe up to now is 2.4 on April 10, 1974, at Pitlochry, Scotland. In the same month values of 2.7 and 3.5 were reported on the west coast of Norway and at a remote station in Iceland, respectively.[19]

In the United States there is a noticeable trend of increased acidity of precipitation in the period from 1950 to the early 1970s (Fig. 4-6). As stated by Likens et al.,[19] an irony in this matter is that the trend to relieve local air pollution problems by building taller and taller stacks has turned the local problems into regional ones. Thus, the coal-burning operations in the Appalachian region may be causing acidification of lakes in New York state, New England, and Ontario that are over a thousand kilometers away. Besides the eastern United States and northeastern Europe, acid rain has been measured in other parts of the world.

Some alarming facts are associated with the acid rain problem:[5,14]

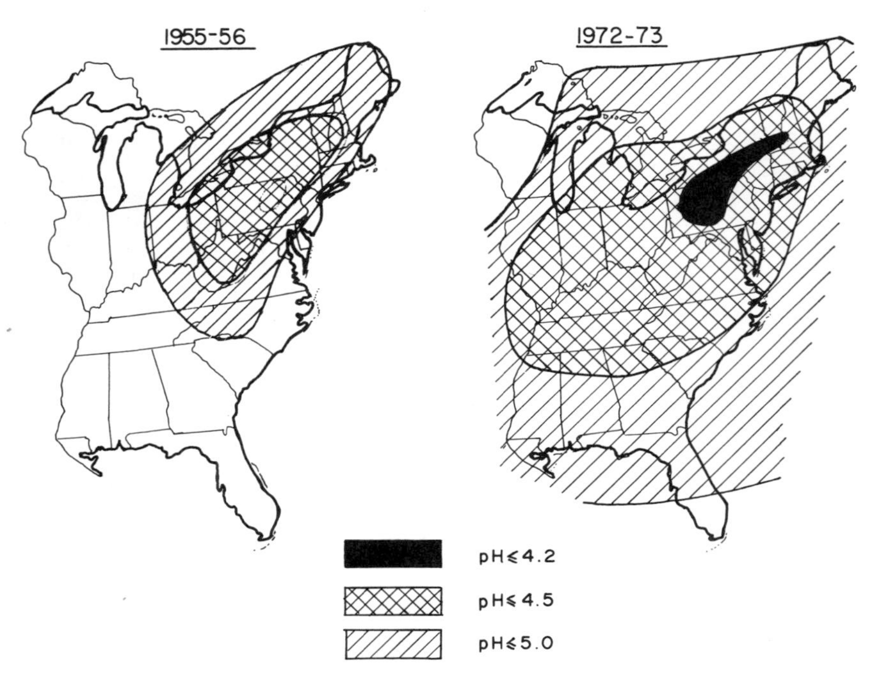

Fig. 4-6. The weighted annual average pH of precipitation in the eastern United States in 1955–56 and 1972–73. (References 5, 20.)

A. The pH of precipitation has been decreasing over most of the northeastern United States and northwestern Europe for the past few decades as energy demand has increased.
B. Acid precipitation affects almost all parts of the ecosystem, including forests and surface waters. Some aquatic systems of low alkalinity have already manifested such decreases in pH that fish populations are no longer self-sustaining.
C. Low pH values (<5.6) are experienced over large portions of North America and Europe.
D. The atmospheric content of sulfur—the primary source of acidity—is likely to increase unless drastic and expensive measures are undertaken globally to reduce sulfur emissions from power plants and urban and industrial operations. With diminishing resources of low-sulfur oil, the prospects for reducing SO_2 emission are not good.

Washout Function. In many cases the amount of pollutants deposited by the wet fallout can be estimated only from known atmospheric concentrations because the rainwater concentrations either are not known or are unreliable. The

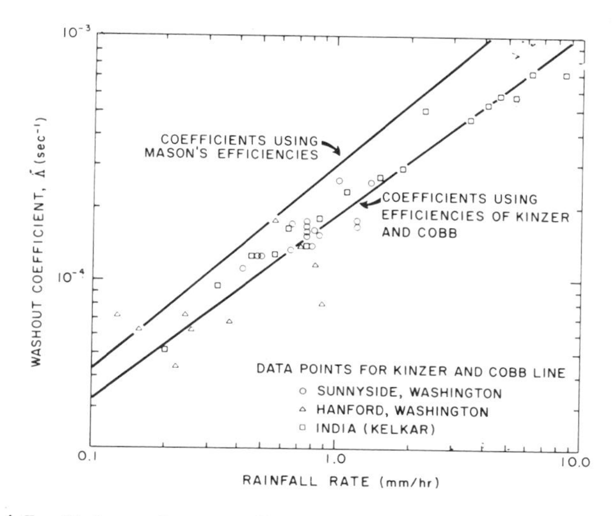

Fig. 4-7. Rain washout coefficient λ related to rainfall intensity.[3]

process of scavenging of pollutants by raindrops during washout or rainout (snowout) is basically an exponential function:[3]

$$C_w = C_{wo} \exp(-\lambda t) \tag{4-1}$$

where:

C_w = the atmospheric concentration of the contaminant after the rain
C_{wo} = the atmospheric concentration of the contaminant before the rain
t = duration of the rain
λ = the washout coefficient

The wet fallout per unit area, D_w, is then:

$$D_w = (C_{wo} - C_w)H = C_{wo}(1 - \exp(-\lambda t))H \tag{4-2}$$

where:

H = the depth of atmosphere through which the pollutant plume is mixed

The magnitude of the washout coefficient, λ, is of the order 10^{-4} sec^{-1} and is a function of rain intensity (Fig. 4-7). The magnitudes of λ are similar both for gases and particulates.

Example 4-1: Concentration of Pollutants in Precipitation

Atmospheric concentration of phosphate before a rain was estimated or measured as $C_{wo} = 10\ \mu g/m^3$. The depth of the mixed atmospheric layer extended about

1000 m above the ground surface. Estimate the amount of wet fallout during a storm with a volume of 10 mm lasting 2 hr.

From Fig. 4-5 the washout coefficient for the storm intensity $i = 10\ mm/2\ hr = 5\ mm/hr$ *is* $\lambda = 7 \times 10^{-4}\ sec^{-1} = 2.52\ hr^{-1}$ *Then the mass of the deposited phosphate becomes:*

$$D_w = C_{wo}(1 - e^{-\lambda t})H = 10(1 - e^{-2.52 \times 2})1000 = 9935\ \mu g/m^2$$

The phosphate concentration in rainwater (rain volume $V_r = 10\ mm = 0.01\ m^3/m^2$*):*

$$C_r = \frac{D_w}{V_r} = \frac{9.93\ mg/m^2}{0.01\ m^3/m^2 \times 1000\ liters/m^3} = 0.99\ mg/liter$$

Example 4-2: Acidity of Precipitation

The ambient concentration of sulfur trioxide and sulfates (SO_3 *and* SO_4^{2-}*) is about 30% of the ambient* SO_2 *concentrations. Estimate the approximate pH of rainwater resulting from 5 mm of rain lasting 5 hr if the ambient* SO_2 *concentration is 20* $\mu g/m^3$*. The mixed atmospheric depth* H = *1000 m.*

From Fig. 4-5 the washout coefficient $\lambda = 2 \times 10^{-4}\ sec^{-1} = 0.72\ hr^{-1}$

Rainfall volume $V_r = 5\ mm\ 5\ liters/m^2$.

Mass of sulfates washed out:

$$D_w = 0.3 \times C_{wo}(1 - e^{-\lambda t})H = 0.3 \times 20(1 - e^{-0.72 \times 5})1000 = 5836\ \mu g/m^2.$$

Sulfate concentration in rainwater:

$$C_r = \frac{D_w}{V_r} = \frac{5.836\ mg/m^2}{5\ liters/m^2} = 1.17\ mg/liter$$

Change C_r *to equivalent* SO_4^{2-} *(equivalent weight of* $SO_3 = [32 + (3 \times 16)]/2 = 40\ g/eq = 40{,}000\ mg/eq$*).*

$$\{SO_4^{2-}\} = \frac{C_r}{EW(SO_3)} = \frac{1.17\ mg/liter}{40{,}000\ mg/eq} = 2.92 \times 10^{-5}\ eq/liter$$

Each equivalent of SO_4^{2-} *must be balanced by 1 equivalent of* H^+ *or any other available atmospheric cation (such as* NH_4^+*). The* $[H^+]$ *concentration of rainwater not contaminated by* SO_x *(pH = 5.6) is* $[H^+] = 10^{-pH} = 10^{-5.6} = 2.5 \times 10^{-6}$ *moies/liter, which is more than one order of magnitude less than the equivalent weight of* H^+ *required to balance* SO_4^{2-} *ion. Since almost all of the sulfuric acid in water is dissociated, the resulting pH is roughly:*

$$\text{pH} = \log \frac{1}{[\text{H}^+]} = \log \frac{1}{\{\text{SO}_4^{2-}\}} = \log \frac{1}{2.92 \times 10^{-5}} = 4.5$$

Dry Fallout of Aerosol Particles. Since the rainout, snowout, and washout of aerosols is the most effective atmospheric cleansing mechanism, dry fallout of particulates is only of significance between rains and the deposition rate usually is lower than for wet fallout. However, overall loadings may be of the same order of magnitude. For example, in St. Louis, Missouri,[22] the dry deposition rate usually is <10% of the wet deposition rate, but the total loadings from wet and dry fallouts are about the same.

Gravity force is the primary mechanism determining the rate of dry fallout from the atmosphere, but other effects such as surface impaction, electrostatic attraction, adsorption, and chemical interaction may explain that the deposition rate of small particles (order of magnitude of 1 μm or less) onto the ground is often greater than can be expected by appropriate gravitational fall velocity.[3]

The rate of deposition of aerosol particles can be related to their average above-ground concentration:[3]

$$D_d = v_d\, C(x, y, z) \tag{4-3}$$

where:

D_d = amount of aerosols removed per unit area per unit time (e.g., g/m²-day or tons/km²-month)
$C(x, y, z)$ = average concentration of the aersols at x, y, and z location from the source or coordinate origin (g/m³)
v_d = deposition velocity of particles (m/day or m/month)

The atmospheric aerosols do not contain only dust particles, that is, particles that are mostly insoluble, but they also include mist, which contains liquid particulates; smoke, a mixture of solid and liquid particles from incomplete combustion of materials; and fog, a high concentration of water droplets in the air.[1]

The particle size of atmospheric aerosols ranges from a fraction of 1 μm to a fraction of 1 mm. The gravitational fall velocity can be described by the Stokes law:

$$v_g = \frac{2r^2 g\rho}{9\mu} \tag{4-4}$$

where:

v_g = gravitational settling velocity in the air
r = particle radius

g = gravity acceleration
ρ = specific density of the particle
μ = air viscosity

It must be realized that the deposition rate v_d may not be the same as the gravitational settling, v_g. Air turbulence and other factors mentioned previously also control the deposition velocity and v_d must be determined largely from field data, namely ground concentrations and deposition rates.

Example 4-3: Fall Velocity of Aerosol Particles

Compute fall velocity of lead aerosols with mean diameter of 0.4 µm ($r = 2 \times 10^{-5}$ cm. Specific density of lead is about 11 g/cm³ and air viscosity at 18°C is $\mu = 182.7 \times 10^{-6}$ g cm^{-1} sec^{-1} ($g = 981$ cm/sec²).

From Stokes' law:

$$v_g = \frac{2r^2 g\rho}{9\mu} = \frac{2 \times (2 \times 10^{-5})\, 981 \times 11}{9 \times 182.7 \times 10^{-6}} = 0.005 \text{ cm/sec}$$

The above gravitational settling velocity of lead particles is about two orders of magnitude less than the measured deposition velocity of lead for the southern portion of Lake Michigan[23] ($v_d \simeq 0.3$ cm/sec).

The discrepancy between v_g and v_d is due to the fact that at fall velocities <1 cm/sec the effect of sedimentation is negligible, and the vertical movement of the particle is largely controlled by larger turbulent and mean air motions.[3] The gravitational settling becomes important when the particle diameter is >10 µm. For particles <10 µm the deposition velocity can be read from Fig. 4-8. Note that the concept given by Equation 4-3 also can be extended to gases in spite of the fact that in this case it would not be physically correct. The deposition velocities of gases have been reported to range from 0.5 to 2.5 cm/sec, depending on the type of ground surface.[24]

Estimating Mean Aboveground Concentrations of Pollutants

Concentration of atmospheric pollutants at a location often must be estimated. There are several important factors that affect the concentration of a pollutant and, hence, its deposition:

A. Frequency and characteristics of winds between the source and the area of deposition.
B. Source strength and type of emission (continuous, temporal, instantaneous).
C. Loss of the pollutant during transport by dry and wet fallout.
D. Atmospheric conditions that affect dispersion.

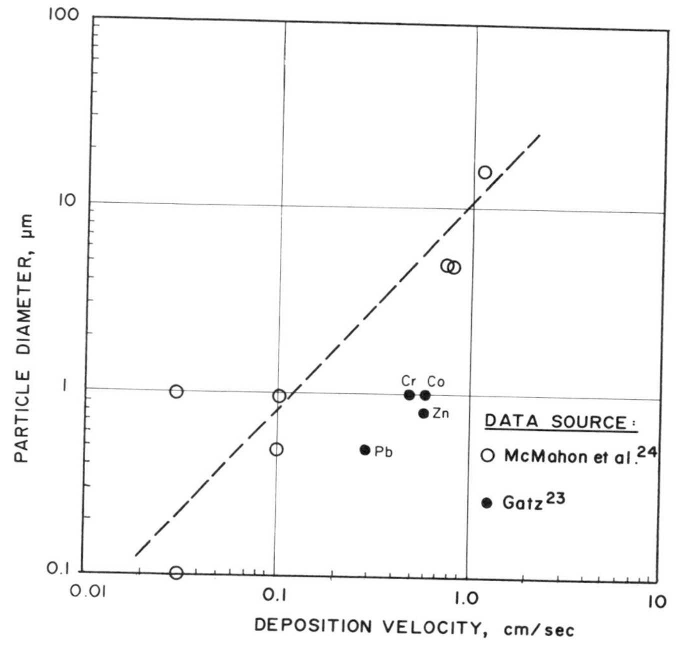

Fig. 4-8. Deposition velocity for particles <10 μm (for particles >10 μm use Equation 4-4).

Sources can be characterized according to their geometrical configuration or temporal patterns of emission. In a parallel fashion to sources of water pollution, sources of atmospheric pollution can be of the following types:

A. Point sources (industrial and power plant stacks, volcanoes, explosions).
B. Diffuse sources (urban areas, landfills, agricultural fields, forests).
C. Line sources (highways).

Sources also can be elevated (stacks) or surface (ground).
As to the temporal patterns of emissions sources can be:

A. Continuous–emission rate does not vary significantly during a certain period of time.
B. Instantaneous as exemplified by explosions and eruptions.

Most of the models that approximate the concentration of atmospheric pollutants at a given point are based on the so-called Gaussian plume dispersion concept (Figs. 4-9 and 4-10), which for a continuous point source is given as:[3]

Fig. 4-9. Plume dispersion. (Photo Department of Agricultural Journalism, University of Wisconsin.)

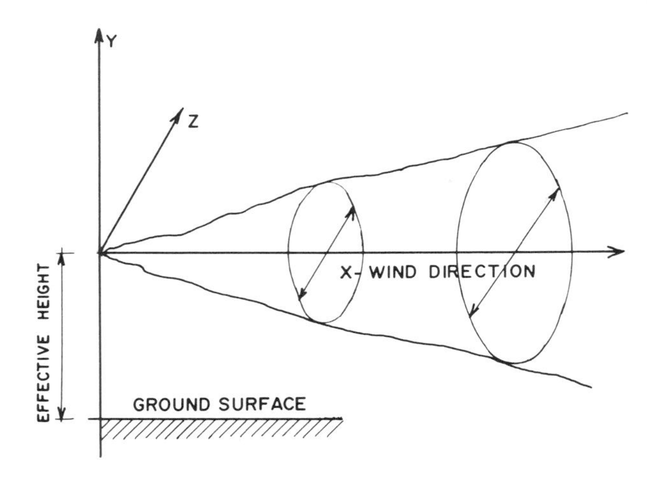

Fig. 4-10. Gaussian plume dispersion concept.

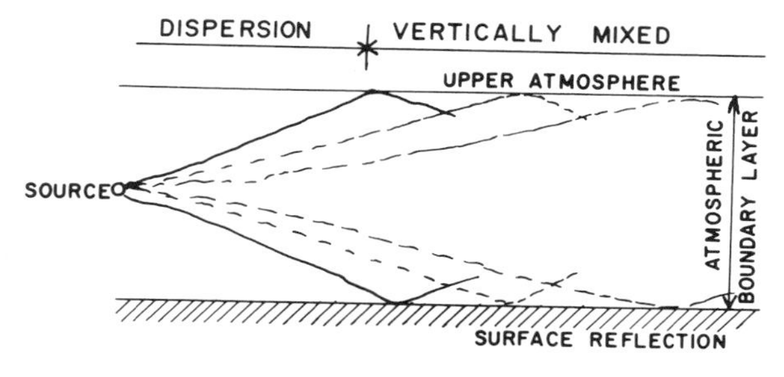

Fig. 4-11. Zones of application of the Gaussian and completely mixed models.

$$\frac{C(x,y,z)}{Q} = (2\pi\sigma_y\sigma_z\bar{u})^{-1} \exp\left[-\left(\frac{y^2}{2\sigma_y} + \frac{z^2}{2\sigma_z}\right)\right] \tag{4-5}$$

where:

$C(x, y, z)$ = air concentration of pollutant at a location x, y, z from the source
σ_y, σ_z = the respective standard deviations characterizing the plume spread in y and z directions
Q = source strength in mass of pollutant per unit time
$\bar{u}$ = average wind velocity in x direction

The Gaussian plume model describes the spread of pollutants near the source. With increased distance of the receptor area from the source, the plume may be reflected by the lower (ground level) and upper (inversion) boundaries and become trapped inside the so-called atmospheric mixing layer (Fig. 4-11).

For nonpoint pollution studies, the original form of the Gaussian dispersion model is of lesser importance; however, its integrated simplified versions can be used for estimating atmospheric loadings from longer distances and point sources. Two modifications of the Gaussian model will be introduced herein. Both models can be applied for longer but not global distances from a point source to the receptor area.

The crosswind integrated concentration model describes downwind concentrations from a continuous line source of a large length such as a major highway or a large urban area. The downwind concentration, C_d, at a distance x from the source becomes:[3]

$$\bar{C}_d = \frac{2Q'}{\pi\sigma_y\bar{u}} \exp\left(-\frac{h^2}{2\sigma_y{}^2}\right) \tag{4-6}$$

where:

h = effective height of the source above ground (includes stack height plus buoyancy effects)

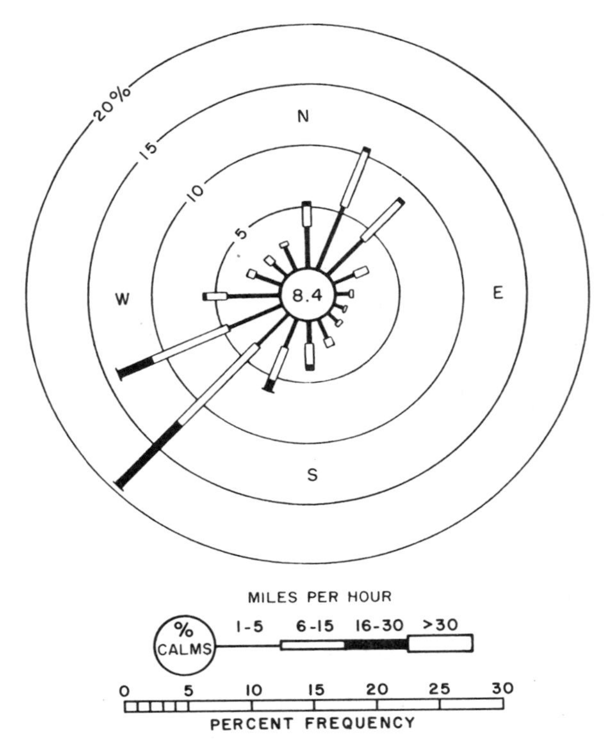

Fig. 4-12. A typical wind rose with wind-speed information.[3]

Q' = the source strength of the line source in mass per unit length of the source per unit time.

Averaged Long-period Concentrations. Over a period of time the direction of the mean wind shifts. In a statistical sense, there is a certain probability that wind of certain magnitudes directed from the source towards the receptor area occurs. These probabilities or frequency distributions are best expressed by so-called wind rose (Fig. 4-12).

To obtain an estimate of the average concentration of pollutants, C_{LT}, over a period that is very long compared with that over which the mean wind is computed, multiply Equation 4-5 by the frequency with which the wind blows from the source towards the receptor area or a given sector and divide by the width of the sector at a distance x from the source (Fig. 4-13):

$$C_{LT} = \left(\frac{2}{\pi}\right) \frac{0.01fQ}{\sigma_y \bar{u}(2\pi x/n)} \exp\left(-\frac{h^2}{2\sigma_y^2}\right) \quad (4\text{-}7)$$

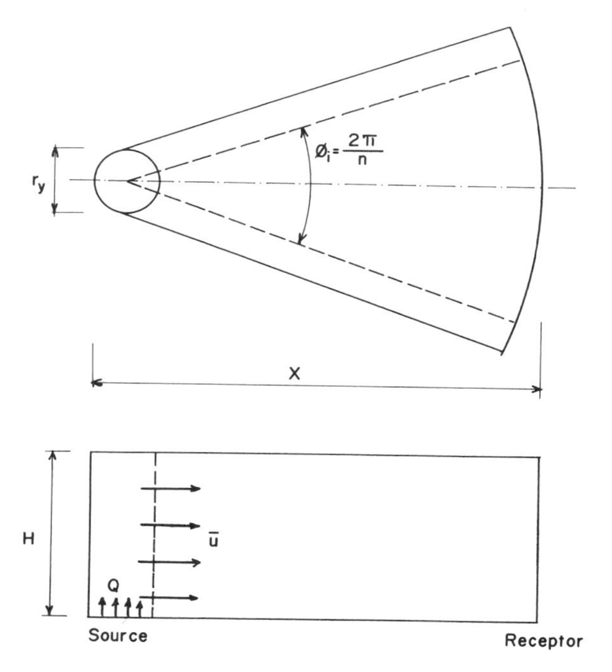

Fig. 4-13. Circular box approach for estimating atmospheric transport of pollutants.[24]

where:

f = frequency (%)
Q, σ, n, and $\bar{u}$ = averages over the long time
$2\pi x/n$ = the sector width

The above functions do not account for wet and dry deposition between the source and the receptor area.

A further simplified formula based on the so-called "circular box" approach has been used for estimating atmospheric loadings to the Great Lakes. The formula is:[24]

$$C_{LT} = \frac{QL}{\bar{u}(\phi_i x + r_y)H_i} \tag{4-8}$$

where:

L = loss function equaling the proportion of pollutant remaining after wet and dry en-route losses (dimensionless)
$\phi_i = 2\pi/n$ - box angle of dispersion in radiants
r_y = diameter of the circle of area equaling the emission area

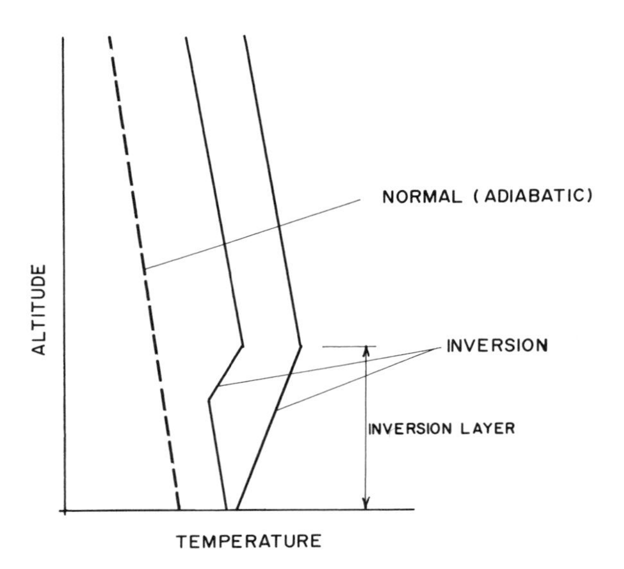

Fig. 4-14. Vertical temperature profile under adiabatic and inversion conditions.

The circular box approach and representation are also shown in Fig. 4-13.

Note that Equations 4-7 and 4-8 neglect horizontal dispersion of the plumes. This assumption introduces insignificant errors only when the frequency of wind direction is about the same in all directions.

Atmospheric Conditions That Affect Dispersion. Under almost all meteorological conditions there is a distinct demarcation between the lower atmospheric zone, where most of the pollution is contained and vertically mixed, and the upper "cleaner" air. The lower mixed layer is usually 300 to 3000 m deep and its thickness depends on the vertical temperature profile.

Under normal temperature conditions, air temperature decreases roughly 0.65°C/100 m altitude distance from the surface. These meteorological conditions are termed adiabatic or neutral conditions. Very often, due to radiation inputs or surface cooling, the vertical temperature change does not follow the adiabatic conditions. The most severe atmospheric pollution (and deposition) occurs when the temperature profile in the lower zone is reversed, that is, the temperature increases vertically with altitude. This phenomenon is called inversion, and the zone of reversed temperature profile is the "inversion layer" (Fig. 4-14). Under inversion conditions the vertical mixing of pollutants is limited to the inversion layer.

The thickness of the mixing layer under noninversion conditions can be roughly

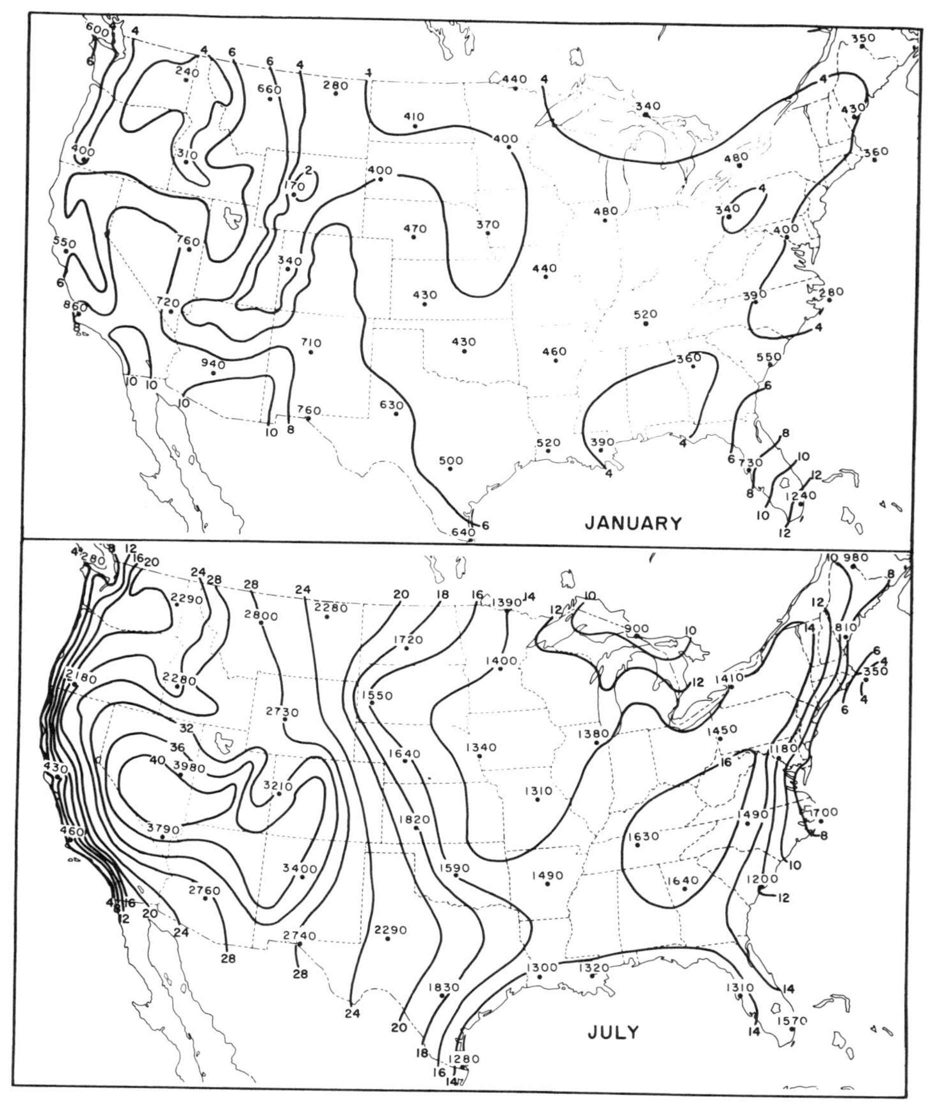

Fig. 4-15. The mean maximum mixing depth (m) for January and July according to Holsworth. (Reprinted from Ref. 3.)

estimated as the height of cumulus clouds above the ground. Even under cloudless conditions there is a distinct upper boundary of the mixing layer. Figures 4-15 and 4-16 show the average winter and summer thickness of the mixing layer and the time percentages when the inversion layer is <150 m (severe air pollution

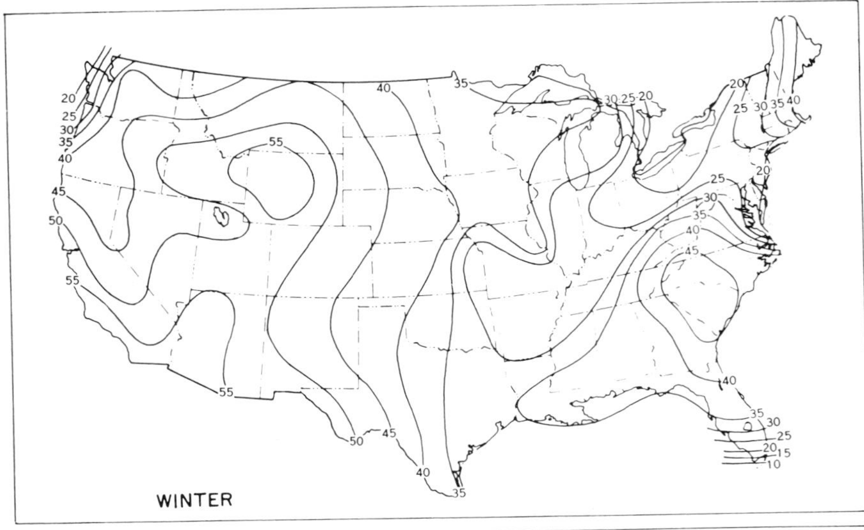

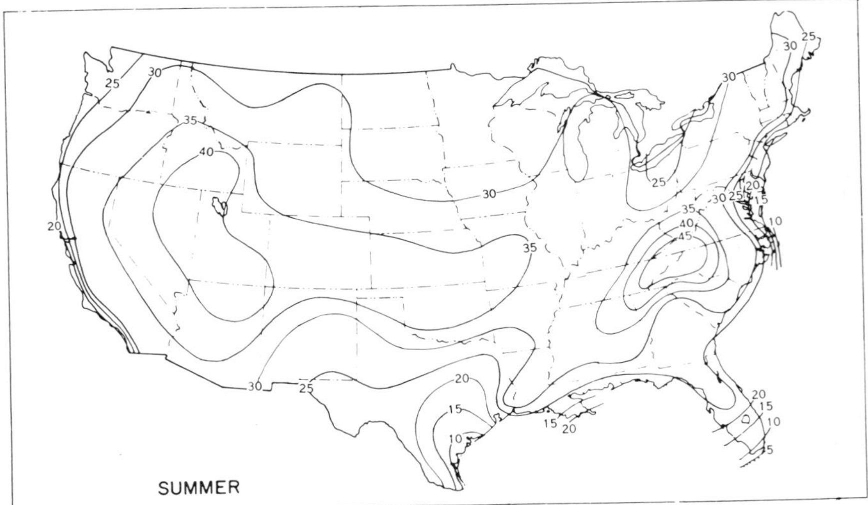

Fig. 4-16. Percentage frequency (percent of total hours) of the occurrence of inversion based below 150 m during the winter and summer according to Hosler. (Reprinted from Ref. 3.)

and deposition). Pollution from ground sources (with the exception of such sources as volcanoes) is supposed to be limited to the mixing layer.

In order to use Equations 4-5 to 4-7 one also must know the standard deviations σ_y and σ_z. Both variables are proportional to the distance x from the

TABLE 4-4. Relation of Atmospheric Dispersion to Weather Conditions (after reference 3).

Surface Wind Speed (m/sec)	Daytime Isolation			Nighttime Conditions	
	Strong	Moderate	Slight	Thin Overcast or ⩾4/8 Cloudiness[a]	<3/8 Cloudiness
2	A	A–B	B		
2	A–B	B	C	E	F
4	B	B–C	C	D	E
6	C	C–D	D	D	D
6	C	D	D	D	D

A–Extremely unstable
B–Moderately unstable
C–Slightly unstable
D–Neutral conditions[b]
E–Slightly stable
F–Moderately stable

[a]The degree of cloudiness is defined as that fraction of the sky above the local apparent horizon which is covered by clouds.
[b]Applicable to heavy overcast, day or night.

source. Pasquill (quoted in reference 3) suggested a practical scheme for estimating atmospheric dispersion expressed in the coefficients σ_y and σ_z. The method was verified by field experiments and proven to be accurate. The magnitudes of the coefficients depend on the atmospheric weather conditions defined in Table 4-4. Figures 4-17 and 4-18 then provide the magnitudes of the dispersion coefficients based on these atmospheric conditions.

Example 4-4: Dispersion of Pollutants

Estimate the atmospheric phosphate concentrations near a lake located 50 km downwind from an urban area. Assuming that the urban area is a sole source, use circular box concept during a prevailing wind.

Given:	
Mean wind speed	$\bar{u} = 7.5\ km/hr = 2.08\ m/sec$
Source diameter of the urban area	$r_y = 10{,}000\ m$
Atmospheric mixing height	$H = 1{,}000\ m$
Average phosphate concentration in the urban atmosphere	$C_0 = 100\ \mu g/m^3$
Assume a loss function	$L = 1.0$

A. Determine the source strength.

The source strength is obtained from the product of urban air concentration and wind speed, that is,

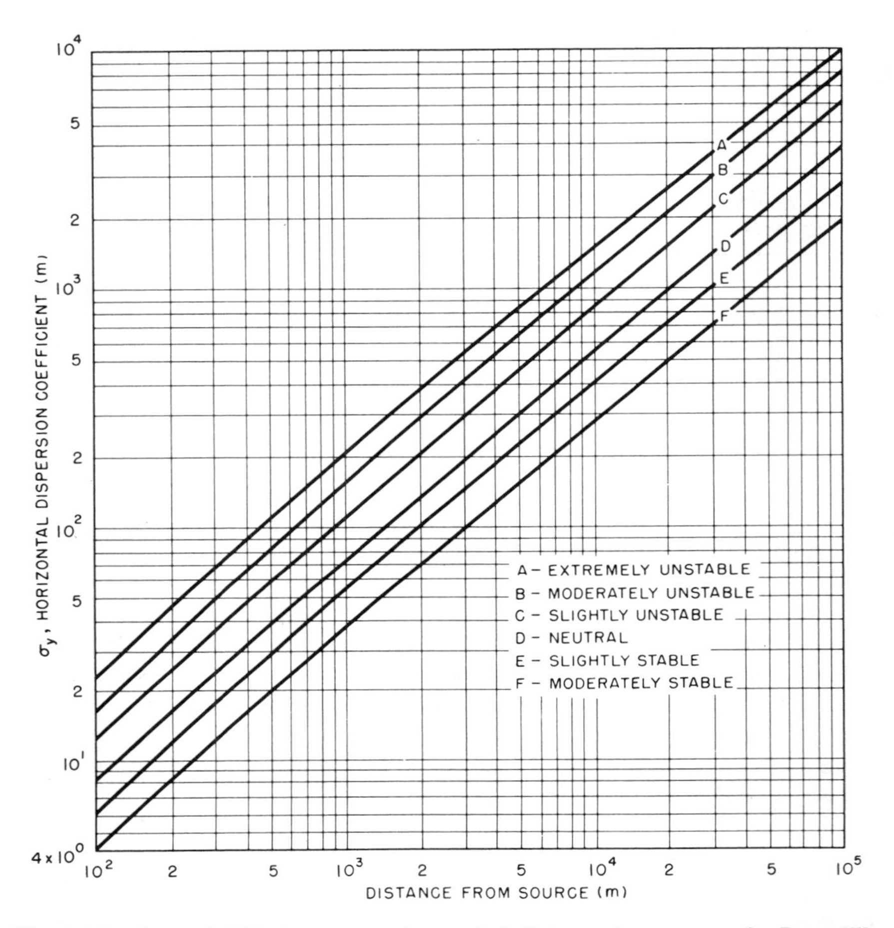

Fig. 4-17. Lateral diffusion, σ_y, vs downwind distance from source for Pasquill's turbulence types. (Reprinted from reference 3.)

$$Q = C_0 \bar{u} H r_y$$
$$= 100(\mu g/m^3) \times 2.08(m/sec) \times 1000(m) \times 10{,}000(m)$$
$$= 2.08 \times 10^9 \ \mu g/sec$$

B. Determine ambient concentrations.

In order to use the model, divide the circle around the urban area into arbitrarily chosen 5° sectors. Then:

$$\phi = \frac{2\pi 5}{360} = \frac{2\pi}{72}$$

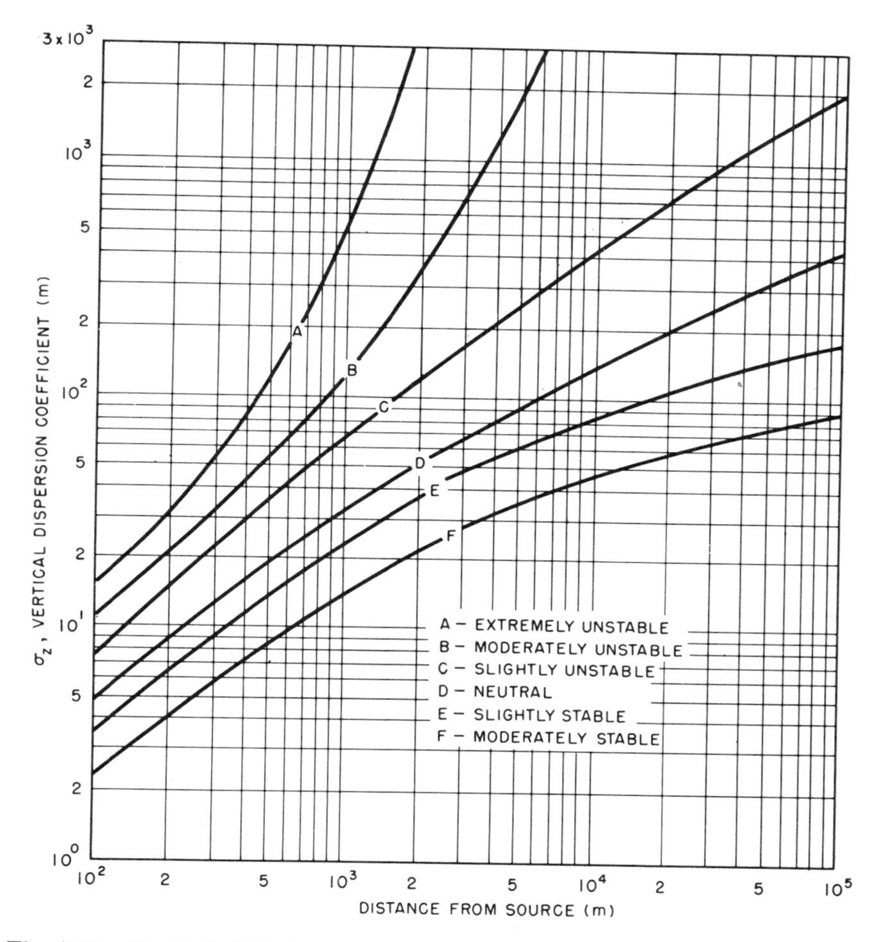

Fig. 4-18. Vertical diffusion, σ_z, vs downward distance from source for Pasquill's turbulence types. (Reprinted from reference 3.)

The circular box model from Equation 4-8:

$$C_{LT} = \frac{QL}{\bar{u}(\phi x + r_y)} = \frac{2.08 \times 10^9 \times 1.0}{2.08 \left(\frac{2\pi}{72} \times 50{,}000 + 10{,}000\right) \times 1000}$$

$$= 69\ \mu g/m^3$$

In order to obtain deposition, multiply concentrations by appropriate deposition velocities.

Example 4-5: Phosphate Deposition

Estimate atmospheric (dry deposition) for the previous example assuming that most of the phosphate-containing particles are in the 1-μm range.

From Fig. 4-8 and particle diameter of 1 μm, the deposition velocity is v_d = *0.15 cm/sec = 0.0015 m/sec.*

For an atmospheric concentration of phosphates of 69 μg/m³, the deposition rate is:

$$D_d = v_d \times C_{LT} = 0.0015 \times 69 = 0.10\ \mu g/m^2\text{-sec} = 0.86\ mg/m^2\text{-day}$$

Global Transport of Pollutants

The Gaussian dispersion models are applicable to local atmospheric pollution-deposition problems. However, measurements at the most remote points on the earth such as Antarctica indicate that many pollutants, including unnatural components—pesticides and PCBs—enter the global cycle and are deposited in appreciable quantities anywhere on the earth. Most of the discussion on global transport is taken from reference 25.

The concentration of pollutants in the atmosphere is determined by the mass balance between global sources and sinks of the pollutant. Mathematically, one can write that:

$$\frac{dM}{dt} = Q - S(M) \qquad (4\text{-}9)$$

where:

M = the global mass of the pollutant in the atmosphere
Q = the global source strength of the pollutant
$S(M)$ = the global sink of the pollutant

Under a steady-state assumption, which can be applied only to time intervals of 1 yr or more, and fairly steady inputs, the left side of Equation 4-9 becomes zero and:

$$Q_{ss} = S(M) \qquad (4\text{-}10)$$

The sinks of atmospheric pollutants include:

A. Deposition (wet and dry) on land and sea surfaces.
B. Adsorption on land and sea surfaces.
C. Decomposition by atmospheric chemical and photochemical processes.
D. Emission into the stratosphere.

The global removal (sink) rate always is a function of mass of pollutant present in the atmosphere (or its concentration), and if deposition prevails as in the cases of some "inert" components (DDT, PCBs), it can be approximated by a relationship similar to Equation 4-3.

The average residence time of a pollutant in the atmosphere under steady-state conditions is given by:

$$T = \frac{M}{Q} = \frac{M}{S(M)} \text{ (yr)} \tag{4-11}$$

The relationship between the deposition velocity, v_d, and the residence time, T, is:

$$T = \frac{M}{S(M)} = \frac{C_M V_A}{v_d C_M A_G} = \frac{H}{v_d} \tag{4-12}$$

where:

C_m = average global (background) concentration of the pollutant
A_G = global surface area
V_A = volume of the atmosphere within the mixing layer
H = average depth of the mixing layer

The most effective removal process is attachment of pollutants to atmospheric aerosols and their subsequent removal by dry and wet fallout on land and sea surfaces.[25] For many pollutants, the sea is the final sink since pollutants deposited on land can be resuspended or can reenter the atmosphere by volatilization, wind erosion, and other processes described previously.

If the input of a pollutant into the global transport system is instantaneous as occurs during an explosion, volcanic eruption, or one-time pesticide application, and if the sink function is linearly proportional to the mass of the constituent in the atmosphere ($S(M) = S_0 \times M$), Equation 4-9 can be solved to yield:

$$M(t) = \frac{Q_0}{V_A} e^{-S_0 t} = \frac{Q_0}{V_A} e^{-t/T} \tag{4-13}$$

where:

Q_0 = the mass of the instantaneous input

Example 4-6: Global Pollution Transport

Background (steady-state) concentrations of an "inert" pesticide in the atmosphere are measured at about $C_M = 0.1\ ng/m^3 = 10^{-10}\ g/m^3$. *The worldwide production of the pesticide is about 1000 tons/yr* $= 10^9$ *g/yr from which about 40% is lost to the atmosphere during and after application. Estimate the average residence time and deposition velocity of the pesticide. Assume average mixing height of* H = *3000 m.*

Atmospheric input of the pesticide:

$$Q = 0.4 \times 10^9 \text{ (g/yr)} = 4 \times 10^8 \text{ (g/yr)}$$

To determine the amount of the pesticide in the atmosphere it is necessary to know:

Volume of the atmospheric mixing layer (earth radius $r = 6.3 \times 10^6$ m)

$$V_A = 4\pi r^2 H = 4\pi (6.3 \times 10^6)^2 \times 3000 = 1.5 \times 10^{18} \text{ m}^3$$

Mass of the pesticide in the atmosphere:

$$M = C_M V = 10^{-10} (\text{g/m}^3) \times 1.5 \times 10^{18} (\text{m}^3) = 1.5 \times 10^8 \text{ g}$$

Average residence time:

$$T = \frac{M}{Q} = \frac{1.5 \times 10^8}{4 \times 10^8} = 0.375 \text{ yr} = 137 \text{ days}$$

and the deposition velocity:

$$v_d = \frac{H}{T} = \frac{3000}{0.375} = 8000 \text{ m/yr} = 0.025 \text{ cm/sec}$$

Dust Pollution

Dust, that is, particles consisting largely of matter originating from soil and urban litter, is always present in the air. In arid zones of the western United States and other parts of the world, the flux of dust particles may increase by several orders of magnitude during periods of dry soil conditions and high winds. The term "dust storm" is used when blowing dust reduces visibility below 1 km.[26] A theoretical visibility-concentration relation can be written as:[27]

$$C = \frac{57.2}{\text{Visibility (km)}} \tag{4-14}$$

where:

C = concentration of particles (mg/m^3)
57.2 = an empirical constant

This relationship indicates that the dust content of air during a dust storm is $\geqslant$60 mg/m^3. These values are fairly high, although measured concentrations of dust have exceeded this value. Measured dust concentrations during dust storms have ranged from 0.1 to 176 mg/m^3, depending on the stage of the storm.[26]

Dust storms during the 1930s devastated large portions of the United States (Fig. 4-19). Due to the implementation of erosion control measures by farmers to reduce soil loss, the occurrences and magnitudes of dust storms have been greatly reduced although not altogether eliminated. Hagen and Woodruff[27] esti-

Fig. 4-19. Dust storm in late 1930s in Colorado. Wind speed reached approximately 50 km/hr (30 mph). (Photo USDA–Soil Conservation Service.)

mated that the average amount of dust suspended by wind erosion in Great Plains locations of the United States during the 1950s was about 244 million tons. Since the dust particles are rather large and conservative, their deposition is equal to their erosion. Estimated average wind erosion-atmospheric deposition of dust particles in the Great Plains from Nebraska to Texas averages about 500 kg/ha-yr.

Factors that may cause particle translocation by wind include soil properties, surface protection from wind, and meteorological factors.[26] Soils that have a potential to form dust are common throughout the world, and other factors such as climate and surface protection are the primary determinants of wind erosion.[26]

At this time, the occurrence of dust storms is relatively infrequent. Their percentage frequency for various regions of the United States is given in Table 4-5. Note that the data show dust conditions with concentrations >3 to 5 mg/m^3, which corresponds to visibility of 11 km. Figure 4-20 shows the regions of the contiguous United States presented in Table 4-5.

Table 4-5 indicates that the south central region is the largest source of dust

TABLE 4-5. Average Regional Dust Storm Frequency for Seven Regions of the United States[a] (after Reference 26).

Region of the United States[b]	Annual RADF (%)[c]
South Central	0.61
Rocky Mountain	0.17
North Central	0.17
Pacific Coast	0.07
Southeastern	0.06
Great Lakes	0.04
Northeastern	0.01

[a]Dust concentrations of more than 3 to 5 mg/m^3 or visibility <11 km.
[b]Region location given in Fig. 4-20.
[c]RADF—regional average dust storm frequency defined as number of hours with dusty conditions divided by total number of hours.

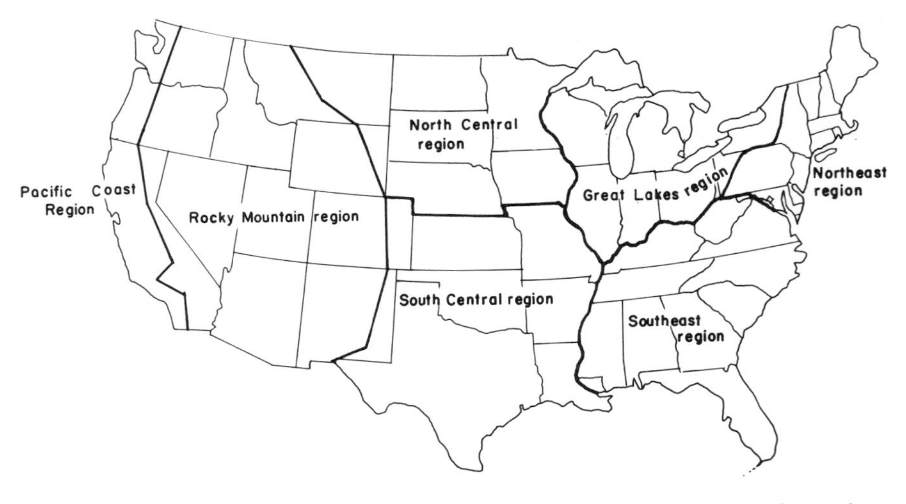

Fig. 4-20. Regions of the United States for dust storm frequency information in Table 4-5.

particles. As stated before, most of the dust particles are rather large and deposit near the source. However, global effects are possible. During the 1883 eruption of the volcano Krakatoa and 1980 eruption of Mt. St. Helens fly ash and dust particles were deposited thousands of kilometers away from the eruption.

Fly ash from industrial coal burning operations and disintegration of urban litter is another significant source of atmospheric deposition, especially in or

TABLE 4-6. Average Annual Dust Deposition Rates in Various Cities.

City	Period of Observation	Dustfall ($tons/km^2$-month)
Seattle[28]	1954–1956	8.8
Seattle[28]	1963–1964	13.5
New York[28]	1951–1955	27.8
Chicago[28]	1954–1956	19.3
Cincinnati[28]	1950–1956	7.7
Pittsburgh[28]	1951–1953	17.7
Detroit[28]	1951–1953	20.5
Milwaukee[29]	1951–1969	11.1
Munich[30]	1970–1979	8.7
Paris[30]	1970–1978	8.6
Cairo[30]	1960–1962	18–45
Hamburg[30]	1969–1978	33

near urban and industrial centers. Table 4-6 shows average dustfall deposition rates for several United States cities. The portion of the dustfall originating from stationary fuel burning operations has been significantly reduced in the last 20 yr by implementing air pollution control measures and by switching from coal to natural gas and oil as domestic and industrial fuels. This trend may be reversed in the future if energy requirements become more reliant on coal. Table 4-7 shows the past trends for Milwaukee, Wisconsin, where dust fallout rate was reduced in central sections of the city by 25 to 50% in the period from 1951 to 1969. Note that in outlying areas, where soil erosion is the source of atmospheric particulate fallout, no such trend is exhibited.

TABLE 4-7. Average Annual Particulate Deposition Rate in Milwaukee County.[29]

	Year							
	1951	1957	1963	1965	1966	1967	1968	1969
Agricultural and rural suburbs	4.87	5.36	7.10	8.51	9.54	10.8	8.21	6.70
Residential	7.80	6.89	7.38	8.28	8.14	7.93	7.83	6.75
Local business	12.7	9.43	10.4	8.54	9.08	10.2	10.3	8.04
Commercial	15.9	16.7	12.8	14.5	12.8	15.9	14.1	12.2
Industrial	28.6	19.6	14.4	15.8	14.5	15.0	14.8	14.2

TABLE 4-8. Chemical Composition of Dustfall in German Urban Areas.[31]

Parameter	Amount (% of dry weight)
Volatile susp. solids	30–40
PbO	0.05–0.2
ZnO	0.01–0.5
NH_3	0.5–1
NO_3	0.5–1

Chemical analyses of dust particles from rural areas can be correlated with soil composition. However, enrichment due to higher contents of pollutants adsorbed on fine soil particles must be taken into account. The composition of dustfall in urban areas is affected by local air pollution. Typical composition of urban dustfall in German cities is given in Table 4-8. Composition of urban street refuse is discussed in Chapter 8.

A significant portion of dust particles that are deposited in urban areas may be water soluble.[30] This means that the control measures employing sedimentation may be less effective.

Air Pollution by Traffic

Highway traffic has the following effects:

A. Abrasion of road surfaces by motor vehicles generates dust.
B. Traffic resuspends deposited dust particles and transfers them to near the curb or median barrier and/or onto adjacent areas.
C. Traffic contributes many inorganic and some organic pollutants, especially lead, asbestos, oil, and grease.

The emission rates of lead from automobiles using leaded gasoline vary from 30 to 90% of the lead content, depending on vehicle speed, operation, and exhaust system. Average emission rates for lead are 60% for urban driving and 80% for highway driving.[32]

The lead content of gasoline is presently being reduced on a worldwide scale. Table 4-9 shows trends in lead emissions based on increased use of cars requiring unleaded gasoline and improved mileage.

The average median diameter of lead particulates emitted from automobile exhausts in the Los Angeles area is about 5 μm, and a significant portion of the lead emissions is contained within 40 to 60 m of the highways.[33]

Cantwell et al.[34] sampled lead emissions by particle size distribution and

TABLE 4-9. Trends in Lead Emissions in the United States.

Year	Maximum Allowable Average Lead Content of Gasoline[a] (g/liter)	Average Fuel Economy Standards for Passenger Cars in Each Model Year (mpg)	(km/liter)	Lead Emission[b] (mg/km-car) Urban	Highway
1975	0.40	14.8	6.3	38	57
1980	0.13	20.0	8.5	9	12
1985	0.13	27.5	11.7	7	9

[a]Reference 32.
[b]Based on 60% emission for urban driving and 80% emission for highways.

found emissions of 19 mg/km-car in particles >9 μm, 11 mg/km-car in particles between 1 and 9 μm, and 19 mg/km-car in particles <1 μm.

Lead aerosols are dispersed rapidly by air diffusion and depleted by deposition. Close to the highway, deposition is affected by sedimentation, which is significant only for particles >5 to 10 μm. Since lead deposition is a function of particle size and its atmospheric concentration, both declining rapidly away from the highway,[35] lead deposition rate shows a rapid decrease over a short distance. This phenomenon can be revealed in soil and vegetation concentrations of lead. Wheeler and Rolfe[36] found that lead from highway sources in roadside soil and vegetation follows a double exponential function of the type:

$$Pb_s = 13.6 + ADT\,(0.187e^{-1.43x} + 0.0147e^{0.08x}) \quad (4\text{-}15)$$

and

$$Pb_v = 8.9 + ADT\,(0.0356e^{-2.75x} + 0.00618e^{-0.64x}) \quad (4\text{-}16)$$

where

Pb_s = lead concentration in soil (μg/g)
Pb_v = lead concentration in plants (μg/g)
ADT = average daily traffic volume (axles/day)
x = distance from the highway (m)

A large portion of lead appearing in the submicrometer range remains airborne for larger distances, that is, at least 10 km.[33]

Typical deposition rates for lead have been reported[33] as 2.3 $\mu g/m^2$-day on a busy highway. Background deposition rates are 0.046 to 0.186 $\mu g/m^2$-day, and natural deposition rates have been observed to be 0.004 $\mu g/m^2$-day.

In urban areas lead becomes incorporated into street dust and accumulates within 1 m of the curb (see Chapter 9 for a more detailed discussion).

Lead aerosols remote from highways have been shown to have a mass mean diameter of 0.4 μm and average deposition velocity v_d = 0.3 cm/sec.[23,37]

Higher concentrations of asbestos aerosols near highways originate from the wearing of brake linings. The average emission rates were determined as 17.8 μg/km-car for passenger vehicles and 55 μg/km-car for trucks.[38] Most of the asbestos is in the <2 μm range. Concentrations in areas of considerable vehicular traffic range from 10 to 60 μg/m³ of air.[39]

Atmospheric Transport of Organic Chemicals

Pesticides and other organic chemicals become airborne, travel in the atmosphere for long distances, and reside there for a long time. The process of atmospheric transport of organic chemicals is not well understood. Major articles on the subject appeared in the literature as early as the mid-1960s.

Most pollutants in the category of organic chemicals are unnatural components of the environment. The atmosphere becomes contaminated with pesticides by drift, wind erosion, and post-application volatilization.[40,41]

Drift is that portion of the spray that is moved away from the target area by wind or other meteorological factors. Aerial spraying (Fig. 4-21) contributes a large portion of the atmospheric drift input. Drift losses ranging from 25 to 75%

Fig. 4-21. View of pesticide spray operations by helicopter. (Photo USDA–Soil Conservation Service.)

of the quantity applied were estimated for pesticides applied aerially.[41] The extent of drift and subsequent dispersion and transport of pesticides in the air are governed primarily by prevailing atmospheric conditions, chemical constitution of the pesticide, and method of application. High temperatures and windy conditions (unstable atmosphere) accentuate drifting. The quantity and extent of drift losses can be reduced by considering a combination of interrelated factors such as spray formulation, type of equipment used, meteorological factors, and spray method used.

Volatilization has been recognized as a pathway of loss of pesticides from soil, plant, and water surfaces. Once in soil, pesticide losses may differ depending on whether the pesticide is surface applied or incorporated into the soil. Field measurements indicate that significant volatilization loss may occur if pesticides are not incorporated.[41]

Wind erosion of pesticides is typical for those materials adsorbed on particles contained in topsoil.

Measured atmospheric concentrations of pesticides vary considerably, indicating both local and global sources and transport. Figure 4-22 shows the basic processes occurring during local and global atmospheric transport. Although any organic chemical can cause a local nonpoint pollution problem, only persistent compounds such as organochlorine pesticides, DDT and its derivatives, aldrin,

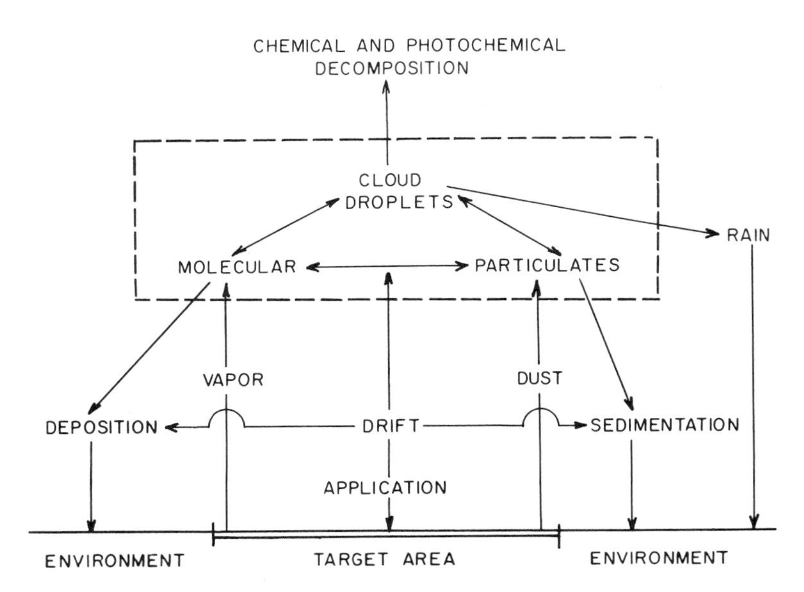

Fig. 4-22. Pesticide input and removal from the atmosphere. [Reprinted with permission from the *Journal of the Air Pollution Control Association*, 28:918 (1978).[40]]

dieldrin, and PCBs have global significance. Discovery of DDT and PCBs in Antarctic seals and snow[42-45] suggest that both components have become a part of global atmospheric pollution and travel through the air to the most remote points of the earth's surface. In contrast to toxic and trace metals, almost all organic chemicals are unnatural. No natural background loadings or concentrations exist. Due to the environmental problems and damages associated with these components, the use of both DDT and PCBs has been banned in some parts of the world, and their production has been severely limited (DDT has been banned in the United States and Canada since the early 1970s and limited use of PCBs is allowed only in closed systems).

Ample evidence exists that the atmosphere is contaminated by pesticides and PCBs from drifting spray and airborne particulate matter, and through volatilization from soil. Precipitated dust collected in Cincinnati after a dust storm in the 1960s indicated the presence of DDT, chlordane, Ronnel, and DDE as the major pesticide components of the dust;[46] heptachlor epoxide, 2,4,5-T, and dieldrin were present in lesser amounts. The concentrations ranged from 3 ng/g dieldrin to 600 ng/g DDT based on the air-dried weight of the dust. It was believed that the dust originated in the southeastern United States where agricultural fields were treated heavily with these pesticides. A country-wide attempt to determine atmospheric contamination by pesticides was performed by collection of samples in nine locations including urban and rural sampling sites.[47] Nineteen pesticides and metabolites were sought in the samples. Detected pesticide levels ranged from 0.1 ng/m^3 of air to 1560 ng/m^3 *p,p'* DDT, 2520 ng/m^3 toxaphene, and 465 ng/m^3 parathion. Only DDT was found at all locations. Levels of DDT were higher in agricultural areas, and the atmospheric content was more closely correlated with spraying activities than with meteorological parameters. These pesticide levels reflected the situation before DDT was banned. Present ambient concentrations of DDT are much lower (<0.1 ng/m^3).

Table 4-10 presents mean concentrations of PCBs and some pesticides found in the precipitation to the Great Lakes.

Residence Time of Organic Chemicals in the Atmosphere and Their Deposition. In the absence of a strong local source, the transport of persistent chemicals is of global nature. Despite its simplistic approach, the global model (Equations 4-9 to 4-13) can be applied with some accuracy to estimate global (background) loadings, deposition rates, and residence time. The Gaussian models should be used when a strong point source is known. Atmospheric layers up to the mid-latitudes become mixed relatively quickly; air parcels move around the world in 3 to 4 weeks; and a quasi-steady state is reached in months.

The residence time of a chemical component in the atmosphere depends on its form and its sinks. There is a great deal of controversy as to the average global residence time and, hence, deposition velocity of the most important, persistent

TABLE 4-10. Mean Concentrations of PCBs and Some Organochlorine Pesticides in Great Lakes Precipitation.*

	Mean Concentration (ng/liter)			
	L. Ontario	L. Erie	L. Huron	L. Superior
Total PCB	32	9	11	26
Lindane	4.7	6.1	6.0	4.9
α-BHC	19.1	10.3	13.3	4.6
ΣDDT Residues	5.6	3.8	2.7	0.8
α-Endosulfan	12.0	2.0	2.1	1.0
Dieldrin	1.3	2.6	1.0	0.5
Methoxychlor	8.5	13.1	9.5	1.6
HCB	nd	nd	nd	2.8

*From Strachan and Huneault[48] –Table 6.

organic chemicals. Due to the fact that the vapor pressure of these chemicals indicates that they most likely exist in the vapor phase, the residence times were compared to those for gases, which are in years. For example, it was estimated that the residence time of DDT is 4 yr.[49] However, ambient background concentrations of DDT are quite low,[40,50] indicating that at least part of the vaporized chemical becomes sorbed by atmospheric aerosols and is deposited at a much faster rate and with much shorter residence times.

Ambient concentrations of pesticides and PCBs found in the mid-Atlantic Ocean are in the order:[51] toxaphene > PCBs > chlordane > DDT > dieldrin. Ambient concentrations of PCBs in air range from 1 ng/m^3 to 50 ng/m^3 according to measurements in the United States. Similar values have been found in Sweden and Japan.[51] Lower PCB concentrations up to about 0.05 ng/m^3 have been measured over the Atlantic Ocean.[52]

The overall average residence time for atmospheric aerosols is about 7 days,[25] and if ϕ denotes the fraction of an organic chemical present in the air and adsorbed on aerosols, the residence time is approximated as $7/\phi$. For DDT the value of ϕ was estimated in the range 0.28 to 0.48, and ϕ for PCBs seems to be <0.01. This would indicate that while the average residence time for DDT and similar organic components could be in months, the residence time of PCBs may be >1 yr. Reference 51 reports that the deposition velocity of DDT, lindane, and dieldrin from the atmosphere is roughly $v_d = 0.04$ cm/sec. From Equation 4-12, and assuming that the average thickness of the atmospheric mixing layer is about 2000 m, the residence time is $T = 200{,}000$ (cm)/0.04 (cm/sec) = 5×10^6 sec = 2 months, which agrees closely with the hypothesis introduced in reference 25.

References 50 and 53 provide data for another check of the global residence time of DDT. In the summer of 1974, the U.S. Forest Service conducted a DDT spraying program for selected Pacific Northwest forests. This program represented the only DDT input in North America since the restriction of its use in the United States (1972) and Canada (1971). Altogether about 200,000 ha were sprayed at application rates of 0.75 kg/ha. The program resulted in airborne concentrations of DDT up to 18 ng/m^3. The DDT concentrations up to 1.8 mg/liter were detected in rainwater for about 5 months after the spray in New York state, some 5000 km from the area of application.[53] This indicates that the residence time of DDT is longer than weeks but shorter than 6 months. Measured ambient (background) concentrations of DDT[50] were of the order of 0.01 ng/m^3.

Although the presence of organic chemicals in the atmosphere is well documented, little is known about the ultimate fate of these compounds after they enter the atmosphere. More research is needed to evaluate photochemical alterations, adsorption on aerosols, and fallout and washout of airborne organic chemicals.

REFERENCES

1. Stern, A. C. (ed.). 1976. "Air Pollution," Vol. 1–5. Academic Press, New York.
2. Peterson, J. T., and Junge, C. E. 1971. Sources of particulate matter in the atmosphere. Pp. 310–320. In: W. H. Matthews, W. W. Kelogg, and G. D. Robinson (eds.), "Man's Impact on the Climate." MIT Press, Cambridge, Massachusetts.
3. Slade, D. H. (ed.). 1968. Meteorology and atomic energy—1968. U.S. Atomic Energy Commission, Div. of Technical Information, Washington, D.C.
4. Easton, E. B., and O'Donnel, F. J. 1977. The clean air act amendments of 1977. *J. Air Pollut. Assoc.*, **27**:947–949.
5. Galloway, J. N., and Cowling, E. B. 1978. The effect of precipitation on aquatic and terrestrial ecosystems: A proposed precipitation chemistry network. *J. Air Pollut. Assoc.*, **28**:229–235.
6. Betson, R. P. 1978. Bulk precipitation and streamflow quality relationship in an urban area. *Water Resources Res.*, **14**:1165–1169.
7. Characklis, W. G., Ward, C. H., King, J. M., and Roe, F. L. 1979. Rainfall quality, land use and runoff quality. *J. Env. Eng. Div., Proc. ASCE*, **105**:416–421.
8. Harrison, R. M., Perry, R., and Wilings, R. A. 1975. Lead and cadmium in precipitation: Their contribution to pollution. *J. Air Pollut. Assoc.*, **25**:627–630.
9. Struempler, A. W. 1976. Trace metals in rain and snow during 1973 at Chadron, Nebraska. *Atmos. Env.*, **10**:33–37.

10. Nicholls, K. H., and Cox, C. M. 1978. Atmospheric nitrogen and phosphorus loading to Harp Lake, Ontario, Canada. *Water Resources Res.*, **14**:589–592.
11. Johannessen, M., and Henriksen, A. 1978. Chemistry of snow meltwater: Changes in concentration during melting. *Water Resources Res.*, **14**:615–619.
12. PLUARG. 1978. Environmental management strategy for the Great Lakes system. International Joint Commission, Windsor, Ontario.
13. Uttormark, P. D., Chapin, J. D., and Green, K. M. 1978. Estimating nutrient loadings of lakes from nonpoint sources. EPA-660/3-74-020, U.S. EPA, Washington, D.C.
14. Cogbill, C. V., and Likens, G. E. 1974. Acid precipitation in the northeastern United States. *Water Resources Res.*, **10**:1133–1137.
15. Likens, G. E., and Borman, F. H. 1974. Acid rain: A serious regional environmental problem. *Science*, **184**:1176–1179.
16. Junge, C. E. 1960. Sulphur in the atmosphere. *J. Geophys. Res.*, **65**:227–237.
17. Gatz, D. F., and Changnon, S. A. 1976. Environmental status of the Lake Michigan region, Vol. 8. Atmospheric environment of the Lake Michigan region. ANL/ES-40 Vol. 8; Argonne National Laboratory, Argonne, Illinois.
18. Glass, N. R., Glass, G. E., and Rennie, P. J. 1979. Effects of acid precipitation. 4th Annual Energy Research and Development Conf., Amer. Chem. Soc., Washington, D.C.
19. Likens, G. E., Wright, R. F., Galloway, J. N., and Butler, T. J. 1979. Acid rain. *Sci. Amer.*, **241**:43–51.
20. Likens, G. E. 1976. Acid precipitation. *Chem. Eng. News*, **54**:29–35.
21. Scott, W. D. 1978. The pH of cloud water and the production of sulfate. *Atmos. Env.*, **12**:917–927.
22. Huff, F. A. 1976. Relation between atmospheric pollution, precipitation, and streamwater quality near a large urban-industrial complex. *Water Res.*, **10**:945–953.
23. Gatz, D. F. 1975. Pollutant aerosol deposition into southern Lake Michigan. *Water Air Soil Pollut.*, **5**:239–251.
24. MacMahon, T. A., Dension, P. J., and Fleming, R. 1976. A long-distance air pollution transportation model incorporating washout and dry deposition component. *Atmos. Env.*, **10**:751–761.
25. Junge, C. E. 1977. Basic considerations about trace constituents in the atmosphere as related to the fate of global pollutants. In: I. H. Suffett (ed.), "Fate of Pollutants in the Air and Water Environments–Part I." Wiley-Interscience, New York.
26. Orgill, M. M., and Sehnel, G. A. 1976. Frequency and diurnal variation of dust storms in the contiguous USA. *Atmos. Env.*, **10**:813–825.
27. Hagen, L. J., and Woodruff, N. P. 1973. Air pollution from dust storms in the Great Plains. *Atmos. Env.*, **7**:323–332.
28. Johnson, R. E., Rossano, Jr., A. T., and Sylvester, R. O. 1976. Dustfall as a source of water quality impairment. *J. Sanitary Eng. Div., Proc. ASCE*, **92**:245–268.

29. Anon. 1970. Ambient air quality (particulates and sulfur oxides) in Milwaukee County. Milwaukee County Dept. of Air Pollut. Control, Milwaukee, Wisconsin.
30. Goettle, A. 1978. Atmospheric contamination, fallout and their effects on storm water quality. *Prog. Water Technol.*, **10**:455–467.
31. Kettner, H. 1974. Beeinträchtigumg oberirdischer Gewässer durch anorganische Stöffe (Enrichment of surface runoff by inorganic matter). Föd. Europäischer Gewässerschutz, Informationsblatt 21, Zürich, Switzerland.
32. Provenzano, G. 1978. Motor vehicle lead emission in the USA: An analysis of important determinants, geographic patterns, and future trends. *J. Air Pollut. Assoc.*, **28**:1193–1199.
33. Hoggan, M. C., Davidson, A., Brunelle, M. F., Nevitt, J. S., and Gins, J. D. 1978. Motor vehicle emissions and atmospheric lead concentrations in the Los Angeles area. *J. Air Pollut. Assoc.*, **28**:1200–1206.
34. Cantwell, E. N., Jacobs, E. S., Gunz, W. G., and Libery, V. E. 1972. Control of particulate lead emissions from automobiles. Pp. 95–107. In: M. G. Curry and G. M. Gigliotti (eds.). "Cycling and Control of Metals." U.S. EPA, Cincinnati, Ohio.
35. Laxen, D. P. H., and Harrison, R. M. 1977. Review paper—The highway as a source of water pollution: An appraisal with the heavy metal lead. *Water Res.*, **11**:1–11.
36. Wheeler, G. L., and Rolfe, G. L. 1979. The relationship between daily traffic volume and the distribution of lead in roadside soil and vegetation. *Env. Pollut.*, **18**:265–274.
37. Stolzenburg, T. R., and Andren, A. W. 1978. Atmospheric deposition of lead and phosphorus on the Menomonee River watershed. International Joint Commission, Windsor, Ontario.
38. Jacko, G. J., DuCharne, R. T., and Sommers, J. H. 1973. How much asbestos do vehicles emit? *Auto. Eng.*, **81**:38–40.
39. Alste, J., Watson, D., and Bagg, J. 1976. Airborne asbestos in the vicinity of a freeway. *Atmos. Env.*, **10**:583–589.
40. Glotfelty, D. E. 1978. The atmosphere as a sink for applied pesticides. *J. Air Pollut. Assoc.*, **28**:917–921.
41. Chesters, G., and Simsiman, G. V. 1974. Impact of agricultural use of pesticides on the water quality of the Great Lakes. Task A-5, Water Resources Center, University of Wisconsin, Madison, Wisconsin.
42. Peterle, T. J. 1969. DDT in Antarctic snow. *Nature*, **224**:629–630.
43. Sladen, W. J. L., Menzie, C. M., and Reichel, W. L. 1966. DDT residues in Adelic penguins and a crabeater seal from Antarctica. *Nature*, **210**:670–671.
44. Cramer, J. 1973. Model of the circulation of DDT on earth. *Atmos. Env.*, **7**:241–256.
45. Resebrough, R. W., Walker, II, W., Schmidt, T. T., deLappe, B. W., and Connors, C. W. 1976. Transfer of chlorinated biphenyls to Antarctica. *Nature*, **264**:738–739.
46. Cohen, J. M., and Pinkerton, C. 1966. Widespread translocation of pesticides by air transport and rainout. Pp. 163–176. In: R. F. Gould (ed.).

"Organic Pesticides in the Environment." Adv. Chem., Series 60, Am. Chem. Soc., Washington, D.C.

47. Stanley, E. W., Barney, II, J. E., Helton, M. R., and Yobs, A. R. 1971. Measurement of atmospheric levels of pesticides. *Env. Sci. Technol.*, **5**:430–435.
48. Strachan, W. M. J., and Huneault, H. 1979. Polychlorinated biphenyls and organochlorine pesticides in Great Lakes precipitation. *J. Great Lakes Res.*, **5**:61–68.
49. Woodwell, G. M., Craig, P. P., and Johnson, H. A. 1971. DDT in the biosphere. Where does it go? *Science*, **174**:1101–1107.
50. Orgil, M. M., Sehnel, G. A., and Petersen, M. R. 1976. Some initial measurements of airborne DDT over Pacific Northwest forest. *Atmos. Env.*, **10**:827–834.
51. Anon. 1976. Environmental health criteria–Polychlorinated biphenyls and terphenyls. World Health Org., Geneva, Switzerland.
52. Harvey, G. R., and Steinhauer, W. G. 1973. Atmospheric transport of polychlorobiphenyls to the North Atlantic. *Atmos. Env.*, **8**:777–782.
53. Peakal, D. B. 1976. DDT in rainwater in New York following application in the Pacific Northwest. *Atmos. Env.*, **10**:899–900.

5

Erosion and sediment yields

SEDIMENT PROBLEM

Sediment is the most visible pollutant originating from nonpoint sources. Effects of excessive sediment loading on receiving waters include deterioration in aesthetic values, loss of storage capacity in reservoirs, changes in aquatic populations and their food supplies, and accumulation of bottom deposits, which impose additional oxygen demand and inhibit some advantageous benthic processes.

Sediment per se is a major pollutant of waterways. Furthermore, sediment—especially its fine fraction—is a primary carrier of such other pollutants as organic components, metals, ammonium ions, phosphates, and other toxic materials. For example, persistent organochlorine compounds such as dieldrin and PCBs have low solubility in water but are adsorbed readily by soil particles. Fertilizer phosphorus also is adsorbed strongly by soils and thus moves with sediments.

Large amounts of particulate sediment materials originate from pervious and impervious urban areas. Urban erosion is known to contribute significant amounts of sediment containing toxic metals, organochlorine pesticides and other chemicals emanating from traffic, elevated air pollution levels, and general urban litter accumulation.

Erosion, transportation, and deposition of sediment are natural processes that have been occurring throughout geological times. In many areas, the top surface layers are relatively young, that is, years to thousands of years. These areas—usually mildly sloping lowlands—are located where sediment is being deposited by surface runoff or wind erosion. Areas on exposed high slopes are the principal sources of sediment. It is, therefore, unrealistic to expect or require complete control or elimination of sediments from nonpoint sources. Such "wall-to-wall" control measures would be technically and economically impossible. However, it is feasible to control or manage excessive sediment loadings from the most critical (hazardous) sediment sources and to limit undesirable soil losses caused by man's activities.

Brant et al.[1] have listed the most important sources of sediment:

1. Natural erosion occurs from the weathering of soils, rocks, and uncultivated land; geological abrasion; and other natural phenomena. The average rate of natural erosion is estimated to be about 40 Tonnes/km^2/yr, but very high amounts of soil loss can be measured locally. Landslides or erosion in badlands result in sediment loading far exceeding any other sediment source.

2. Agricultural erosion is a major sediment source due to the large area involved and the land-disturbing effects of cultivation. Most of the sediment loading from agricultural sources takes place during spring, especially when spring rains fall on still-frozen soil. Reported erosion rates range from 100 to 4000 Tonnes/km^2/yr.

3. Urban erosion originates mainly from exposed bare soil in areas under construction and from street dust and litter accumulation on impervious surfaces. Sediment yields from urban developing areas can be extremely high, sometimes reaching values of 50,000 Tonnes/km^2/yr.

4. Highway erosion is associated with the stripping of large areas of their vegetative protection during road construction. Uncompacted or unsettled fills may be subject to intensive rill erosion and, eventually, to landslides. The magnitudes of the sediment yields from these areas are similar to those from developing urban areas.

5. Stream bank, channel, and shoreline erosion result from concentrated water flows and wave action in channels and flood plains. Channel erosion usually is a source of large-grain-size sediments and in most cases represents a smaller portion of the total sediment loading. However, sediment loading, originating from bank and channel erosion of some ephemeral streams in arid and semiarid regions, may be quite high.[2]

An understanding of erosion processes is fundamental if sediment movement is to be controlled. It is a fact that sediment control is easier at its source and it is, of course, less detrimental to surface water quality if the sediment never reaches the watercourse. In-stream control measures—such as dredging—are only temporary cosmetic techniques that often are damaging to downstream water quality and the aquatic ecosystem.

Although sediment per se and its associated pollutants originate mostly from nonpoint sources, it should be noted that large portions of the sediment loading may arise from a small portion of the drainage area, that is, "hydrologically active areas." Sources or conditions that cause excessive sediment movement are listed by Stewart et al.[3] as:

Farming on long slopes without terraces or runoff diversions.
Row cropping up and down moderate or steep slopes.
Bare soil following seeding of crops.
Bare soil between harvest and establishment of a new crop canopy.
Intensive cultivation close to a stream.
Intensive runoff from upslope pasture or rangeland that traverses areas of row crops.
Poor crop stands.
Gully formation.
Residential or commercial construction.
Poorly managed idle or wooded land.
Unstable stream banks.
Unstable road banks.
Surface mining.
Feedlots close to a stream.
Long exposure of bare soil resulting from any land use.

It also should be realized that sediment produced at the source is not quantitatively or qualitatively the same as the sediment measured in the receiving body of water. Many factors such as the distance of the source from the receiving body of water, vegetative buffers, slope and roughness characteristics of the land, and ponding and presence of depositional areas during overland flow can affect the delivery of the sediment from the source to the receiving water body.

DEFINITIONS

Erosion is defined as the abrading of the land.[4] Agents of erosion are water, wind, ice, gravity, and human activities. Water accounts for the bulk of sediment erosion and transport. Erosion processes can be divided into sheet and rill erosion (Fig. 5-1), gully erosion (Fig. 5-2), stream and floodplain scour (Fig. 5-3),

Fig. 5-1. Rill and interrill erosion on exposed slopes. (Photo authors.)

Fig. 5-2. Gully erosion. (Photo authors.)

Fig. 5-3. Stream bank erosion and deposition. (Photo Soil Conservation Service.)

shoreline erosion (Fig. 5-4), roadside erosion (Fig. 5-5), and erosion from mines, industrial areas, and construction sites (Fig. 5-6). Hydrologically, these processes can be categorized into sheet (upland) erosion and stream or channel erosion and transport. Thus, the total amount of on-site sheet and channel erosion is the gross (potential) erosion.

The term sediment yield denotes the amount of sediment measured in the waterway at the watershed outlet, and the delivery ratio (DR) is the ratio of sediment yield (Y) to gross erosion in the watershed (A):

$$DR = \frac{Y}{A} \tag{5-1}$$

Theoretically, DR can range from zero to one. Very often sediment yield is related to upland sheet erosion, but since channel erosion is included in the DR, the factor can have apparent values >1.

Eroded soil materials are classified according to their grain sizes into clay, silt, sand, and gravel fractions. Clay particle sizes range from 10^{-4} to 10^{-3} mm, silt

Fig. 5-4. Shoreline erosion—Lake Michigan near Racine Wisconsin. (Photo Soil Conservation Service.)

ranges from 10^{-3} to 10^{-2} mm, sand from 10^{-2} to 10^{0} mm, and gravel particles are $>10^{0}$ mm (Table 5-1).

Soil materials are classified as cohesive and noncohesive sediments. Cohesive sediments include clays and other fine particles including organic materials. These fractions can form bonds between particles which resist scour, cause flocculation during overland and channel flows, and sorb pollutants. Noncohesive sediments, mostly particles of sand and gravel size, are transported as individual particles, the movement of which depends solely on such particle properties as mass, shape, size, and relative position of the particles with respect to surrounding particles. Their sorption potential is small compared to cohesive sediments.

Once in the channel phase of transport, sediments are categorized as washload or bedload. Washload is carried by streams in suspension, while bedload movement consists of shifting bottom sediments due to turbulence intensity and mean stream velocity. It has been found by Gottschalk[4] that most of the sediment carried by streams in humid agricultural areas originates from sheet erosion. Since sheet erosion is responsible for production of fine-grained sediments, it follows that the bulk of the sediment carried by these streams is in the form of washload. As to pollution potential, washload seems to be more damaging than bedload. Since most of the clays, fine silt, and organic materials are contained in the washload portion of the sediment (with the exception of slow-moving,

Fig. 5-5. Roadside erosion. No temporary seeding was applied and tremendous amounts of sediment and pollutants were washed into an adjacent stream. (Photo Soil Conservation Service.)

sediment-laden streams), the analysis of washload movement is important in determining delivery of sediment and sediment-associated pollutants.

Enrichment ratio refers to the difference in particle size distribution and sorbed pollutant content of washload particles and the soils from which the sediment originated. Erosion is a selective process by which clays and fine particles are transported by surface runoff.

ESTIMATING SEDIMENT YIELD

Sediment yield from a watershed can be estimated by one of the following methods:[4,5,6]

1. Stream flow sampling methods, which yield sediment discharge/flow relationships rating curve (Fig. 5.7). Once the long-term sediment/flow relation-

Fig. 5-6. Soil erosion from construction. Two large gullies were formed after heavy rains, and considerable erosion has taken place. (Photo Soil Conservation Service.)

TABLE 5-1. Scale of Sediment Particle Sizes.[a]

	Particle Size (mm)	
Class	AGU Scale	USDA Scale
Cobbles and boulders	>64	>10
Gravel	2–64	2–10
Very coarse sand	1–2	1–2
Coarse sand	0.5–1.0	0.5–1.0
Medium sand	0.25–0.5	0.25–0.5
Fine sand	0.125–0.25	0.1–0.25
Very fine sand	0.062–0.125	0.05–0.1
Coarse silt	0.031–0.062	0.002–0.05
Medium silt	0.016–0.031	
Fine silt	0.008–0.016	
Very fine silt	0.004–0.008	
Coarse clay	0.002–0.004	<0.002
Medium clay	0.001–0.002	
Fine clay	0.0005–0.001	
Very fine clay	0.00024–0.0005	

[a]According to the American Geophysical Union (AGU) and U.S. Department of Agriculture (USDA).

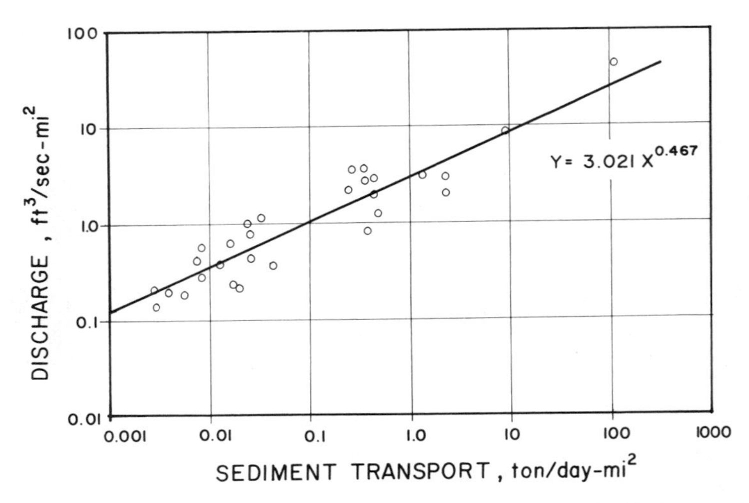

Fig. 5-7. Relationship between sediment transport and discharge for the Menomonee River in Wauwatosa, Wisconsin. (Compiled from U.S. Geological Survey, Wisconsin Department of Natural Resources, and Southeastern Wisconsin Regional Planning Commission data.)

ship is established it is combined with a long-term flow-frequency curve to obtain average annual yields.

2. The reservoir sedimentation survey method, which allows sediment input to be determined if accumulated sediment and reservoir trap efficiencies are measured.

3. The sediment delivery method, in which sediment yield to some downstream cross section or deposition point is based on an estimate of total upstream erosion factored down to account for loss (or gain) of sediment during overland and channel transport. This method requires expressing the ratio between sediment yield and gross upstream erosion, usually determined or based on a semiempirical formula. Both *DR* determination and upland gross erosion estimation are still unreliable if calibrated and verified field data are not available.

4. Bedload function methods, which use of mathematical equations developed for calculating rate and quantity of sediment materials in the bed portions of alluvial streams. Application of these equations requires information on sediment particle size, channel gradients and cross sections, and flow duration curves. The equations can be used when sediment transport is not limited by upstream supply but depends solely on the transport capacity of the flow. These models are used mostly for noncohesive, larger grain size sediments.

5. Methods using empirical equations relating sediment yields (directly measured by methods 1 or 2) to watershed hydrologic or morphologic characteristics. Most of these empirical equations have severely limited applications even in the region of origin.[5]

6. Simulated watershed sediment models, which usually are attached to a watershed hydrologic model. These watershed models are capable of simulating individual storm events or seasonal water and sediment yields. The hydrologic portion is necessary for determining the so-called hydrologically active areas, that is, areas from which most intensive surface runoff and erosion occurs. Chapter 9 includes an expanded dicussion of mathematical modeling.

ESTIMATING UPLAND EROSION

Variables influencing upland (sheet) erosion are: climate, soil properties, vegetation, topography, and human activities. Rainfall, snowfall, and temperature are the primary climatic factors. Soil particles are detached and transported by the impact of rainfall energy resulting in eroded materials being carried by surface runoff. Freezing temperatures and snow cover affect permeability and reduce the energy of precipitation. Conversely, spring rains occurring when subsoils are still frozen may cause high sediment yields due to the reduced soil permeability.

The major soil properties related to the erosion are soil texture and composition. Soil texture determines the permeability and erodibility of soil. Permeability and infiltration are the primary factors affecting the hydrologic activity of an area. Erosion occurs only when surface runoff is generated or when wind picks up loose soil particles. Vegetation influences sediment yield of dissipating some rainfall energy, binding the soil by its root system, and reducing soil moisture, thereby increasing infiltration.

Topographic factors of greatest importance are slope and the path length traversed by sediment-generating flow. Human activities relate mostly to agricultural and construction practices.

Universal Soil Loss Equation

The universal soil loss equation (USLE) is the most common estimator of soil loss caused by upland erosion. The equation and its development utilized more than 40 yr of experimental field observations gathered by the Agricultural Research Service of the USDA. The USLE was formulated by Wischmeier and Smith[7] as:

$$A = (R)\,(K)\,(LS)\,(C)\,(P) \qquad (5\text{-}2)$$

where:

A = computed soil loss in Tonnes/ha for a given storm
R = rainfall energy factor
K = soil erodibility factor
LS = slope-length factor
C = cropping management (vegetative cover) factor
P = erosion control practice factor

The equation expresses soil loss/unit area due to erosion by rain. It does not include wind erosion and it does not give a direct sediment yield estimation.

The ***rainfall energy factor*** (R_r) is equal to the sum of the rainfall erosion indices for all storms during the period of prediction. For a single storm it was defined[7] as follows:

$$R_r = \Sigma\,[(2.29 + 1.15 \log X_i)D_i]\,I \tag{5-3}$$

where:

i = rainfall hyetograph time interval
D_i = rainfall during time interval i (cm)
I = maximum 30-min rainfall intensity of the storm
X_i = rainfall intensity (cm/hr)

Average yearly rainfall energy factors were determined for eastern portions of the United States and later were developed for the remainder of the country. The distribution of average R_r factors for the 48 continental states are shown in Fig. 5-8. These curves can aid as first estimates of gross erosion potential, but as shown in Fig. 5-9, significant yearly and seasonal differences in magnitudes of the rainfall energy factor R_r may be typical for seasons and years with high-intensity storms or low rainfall intensity with extensive dry periods. In the mid-western part of the United States, summer rains have the highest erosion potential, while the effect of winter precipitation on sediment yields is minimal.

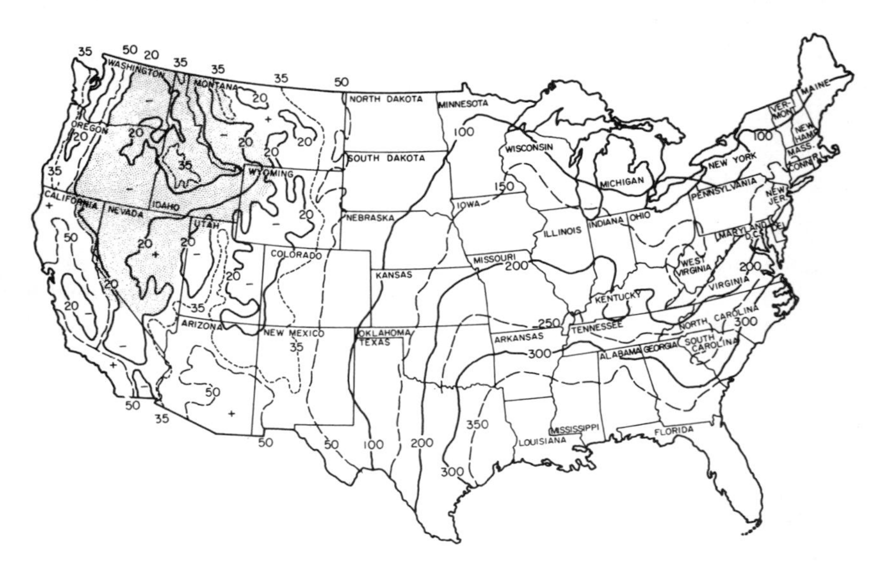

Fig. 5-8. Values of rainfall factor (R) in tons/acre. To convert to Tonnes/ha multiply by 2.24. (Taken from reference 3.)

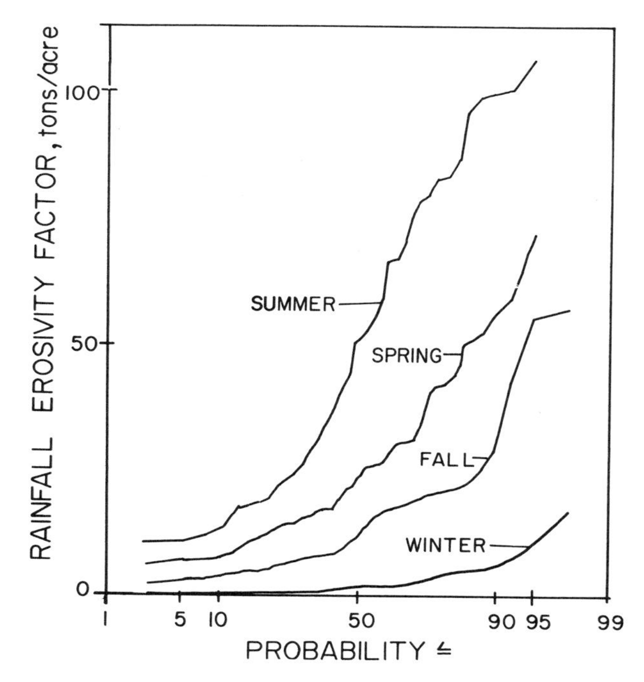

Fig. 5-9. Seasonal cumulative frequency of rainfall factor (R) for southeastern Wisconsin in tons/acre. To convert to Tonnes/ha multiply by 2.24.

The distribution of erosive rains differs significantly for different sections of the country, as shown in Fig. 5-10. In the Western Plains and Great Lakes regions, from 40 to 50% of the erosive rainfall normally occurs during a 2-month period following spring planting when soils have least protection. In most Corn Belt areas and eastern parts of Kansas, Oklahoma, and Texas, the value is about 35%; for the lower Mississippi Valley and southeastern United States, the value is about 20 to 25%. In the dry-land grain-growing region of the Pacific Northwest, about 80 to 90% of the annual erosion occurs in the winter months when the soil has little crop cover because grain is seeded late in the fall.[3]

Both rainfall energy and detachment of soil particles by overland runoff contribute to soil loss. Thus, the rainfall factor (R) also should include the effect of runoff. A modification to Equation (5-3) was proposed:[8]

$$R = aR_r + bcQq^{1/3} \tag{5-4}$$

where:

a and b = weighting parameters ($a + b = 1$)
c = an equality coefficient

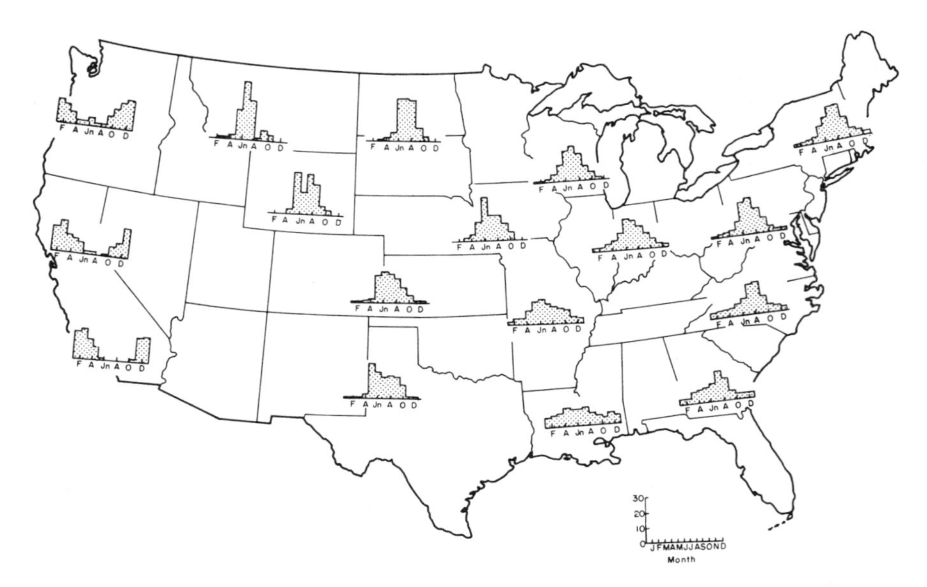

Fig. 5-10. Monthly distribution of erosive rainfall as a percentage of total rainfall. (Reprinted from reference 3.)

Q = runoff volume (cm)
q = maximum runoff rate (cm/hr)

The weighting factors compare the relative amounts of erosion caused by rainfall and runoff under unit conditions. It was suggested that the detachment of particles by runoff and rain energy is about evenly divided ($a = b = 0.5$). The equality coefficient in SI units is about 15.0. Substituting values for a, b, and c into the USLE, the overall rainfall factor (R) becomes:

$$R = 0.5R_r + 7.5\,Qq^{1/3} \tag{5-5}$$

The *soil erodibility factor* (K) is a measure of potential erodibility of soil and has units of tons/unit of erosion index. The soil erosion nomograph shown in Fig. 5-11 is used to find the appropriate values of the soil erodibility factor using five estimated soil parameters. These parameters are: percent silt + fine sand, that is, 0.05 to 0.1 mm fractions; percent sand >0.1 mm; percent organic matter; textural class and permeability. General magnitudes of soil erodibility factors are shown in Table 5-2.

The *slope-length factor* (LS) is a function of overland runoff length and slope. The general magnitudes of the LS factor are given in Fig. 5-12. For slopes >4%, the LS factor can be estimated as follows:

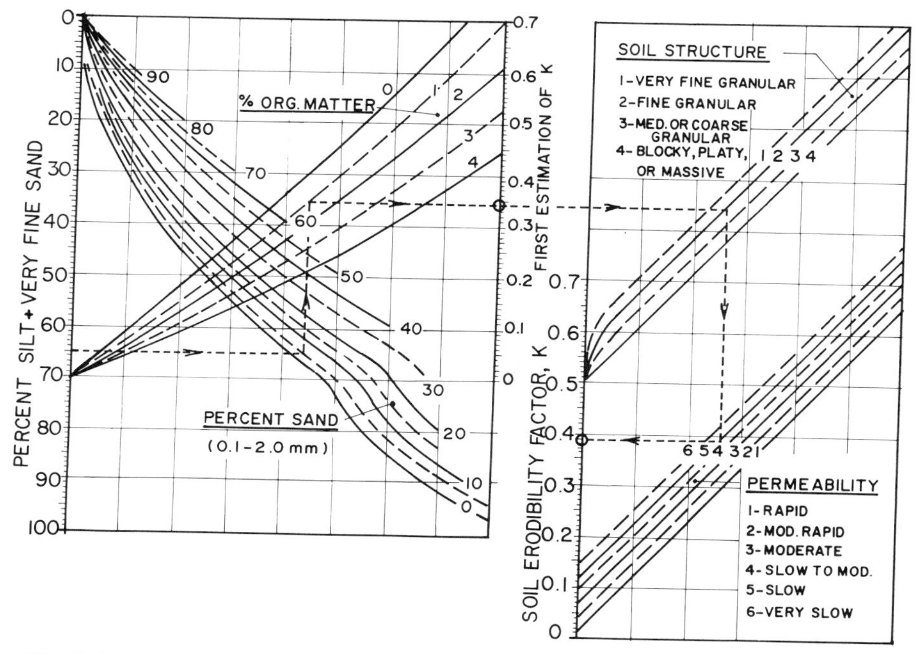

Fig. 5-11. Soil erodibility nomograph for determining (K) factor for U.S. mainland soils. (Reprinted with permission from Wischmeier et al., *J. Soil Water Cons.*, **26**:189–193.[9] Copyright © 1971 by the *Journal of Soil and Water Conservation*.)

$$LS = L^{1/2}(0.0138 + 0.00974S + 0.00138S^2) \tag{5-6}$$

where:

L = length in meters from the point of origin of the overland flow to the point where the slope decreases to the extent that deposition begins or to the point at which runoff enters a defined channel

S = the average slope (%) over the runoff length

Values of the LS factor estimated for lengths >100 m or slopes >18% are extrapolated beyond the experimental data from which the magnitude of the factor was determined.

If average slope is used in calculating the LS factor, predicted erosion differs from actual erosion when the slope is not uniform. The equation for LS factors shows that when the actual slope is convex, the average slope will underestimate predicted erosion. Conversely, for a concave slope, the equation will overestimate actual erosion. To minimize these errors, large areas should be broken up

Table 5-2. Magnitude of Soil Erodibility Factor *K*.[a]

Textural Class	*K* for Organic Matter Content (%)		
	<0.5	2	4
Sand	0.05	0.03	0.02
Fine sand	0.16	0.14	0.10
Very fine sand	0.42	0.36	0.28
Loamy sand	0.12	0.10	0.08
Loamy fine sand	0.24	0.20	0.16
Loamy very fine sand	0.44	0.38	0.30
Sandy laom	0.27	0.24	0.19
Fine sandy loam	0.35	0.30	0.24
Very fine sandy loam	0.47	0.41	0.33
Loam	0.38	0.34	0.29
Silt loam	0.48	0.42	0.33
Silt	0.60	0.52	0.42
Sandy clay loam	0.27	0.25	0.21
Clay loam	0.28	0.25	0.21
Silty clay loam	0.37	0.32	0.26
Sandy clay	0.14	0.13	0.12
Silty clay	0.25	0.23	0.19
Clay		0.13–0.2	

[a]Adapted from Stewart et al.[3] The values shown are estimated averages of broad ranges of specific soil values. When a texture is near the border line of two texture classes, use the average of the two *K* values.

into areas of fairly uniform slope. If sediment moves from an area with steep slope to an area of less steep slope, the smaller *LS* factor will control the amount eroded and the excess sediment is likely to be deposited.

The *cropping management factor* (*C*)—also called the vegetation cover factor—estimates the effect of ground cover conditions, soil conditions, and general management practices on erosion rate. The *C* factor is set equal to one for continuous fallow ground, which has been defined as land that has been tilled up and down the slope and maintained free of vegetation and surface crusting. The effect of vegetation on erosion rates results from canopy protection, reduction of rainfall energy, and protection of soil by plant residues, roots, and mulches.

Table 5-3 shows the general magnitudes of *C* for agricultural land, permanent

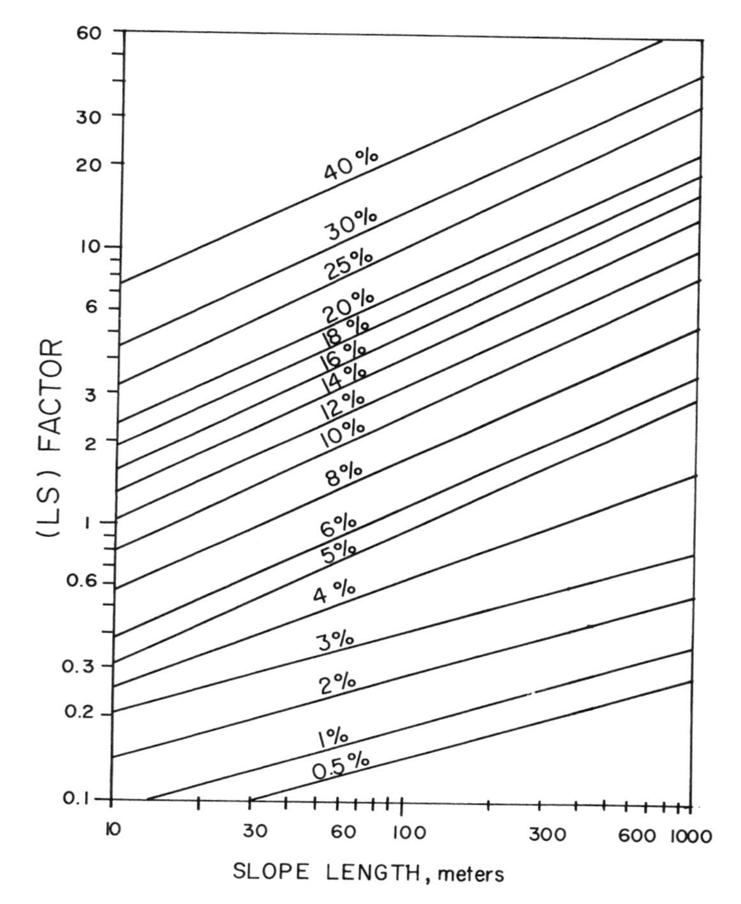

Fig. 5-12. Length-slope factor (*LS*) for different slopes. (Taken from reference 3.)

pasture, woodland, and idle rural land. Grassed urban areas have *C* factors similar to those for permanent pasture.

The *C* factor for construction sites can be reduced if the surface is protected by seeding or application of hay, asphalt, wood chips, or other protective covers. The effect of these protective practices on *C* is given in Table 5-4.

The ***erosion control practice factor*** (*P*) accounts for the erosion control effectiveness of such land treatments as contouring, compacting, establishing sediment basins, and other control structures. Terracing does not affect *P* because soil loss reduction by terracing is reflected in the value of *LS*. Generally, *C* reflects protection of the soil surface against the impact of rain droplets and

TABLE 5-3. Values of *C* for Cropland, Pasture, and Woodland.[a]

Land Cover or Land Use	*C*
Continuous fallow tilled up and down slope	1.0
Shortly after seeding or harvesting[b]	0.3–0.8
For crops during main part of growing season	
Corn[c]	0.1–0.3
Wheat[c]	0.05–0.15
Cotton	0.4
Soybeans[c]	0.2–0.3
Meadow[c]	0.01–0.02
For permanent pasture, idle land, unmanaged woodland	
Ground cover 95–100%	
As grass	0.003
As weeds	0.01
Ground cover 80%	
As grass	0.01
As weeds	0.04
Ground cover 60%	
As grass	0.04
As weeds	0.09
For managed woodland	
Tree canopy of 75–100%	0.001
40–75%	0.002–0.004
20–40%	0.003–0.01

[a]Adapted from references 3, 7, and 10.
[b]Depending on root and residue density.
[c]Depending on yield.

subsequent loss of soil particles. On the other hand, *P* involves treatments that retain liberated particles near the source and prevent further transport.

Values of *P* for various farm and urban practices are given in Tables 5-5 and 5-6, respectively.

Example 5-1: Estimation of Sediment Yields

A 50-ha land area is to be developed into a single-family residential area. The map indicates that the soil is loam with the following composition:

Clay 20%
Silt 35%
(Silt + fine sand) 55%

TABLE 5-4. *C* **Values and Slope-Length Limits (*SL*) for Construction Sites.**[a]

Mulch				
Type	Tonnes/ha	Slope (%)	*C*	*SL*
No mulch or seeding		All	1.0	
Straw or hay tied down	2.25	<5	0.2	60
by anchoring and	2.25	6–10	0.2	30
tracking equipment	3.4	<5	0.12	90
used on slope	3.4	6–10	0.12	45
	4.5	<5	0.06	120
	4.5	6–10	0.06	60
	4.5	11–15	0.07	45
	4.5	16–20	0.11	30
	4.5	21–25	0.14	23
Crushed stone	300	<15	0.05	60
	300	16–20	0.05	45
	300	21–33	0.05	30
	540	<20	0.02	90
	540	21–33	0.02	60
Wood chips	15	<15	0.08	23
	15	16–20	0.08	15
	27	<15	0.05	45
	27	16–20	0.05	23
	56	<15	0.02	60
	56	16–20	0.02	45
	56	21–33	0.02	30
Asphalt emulsion 12 m^3/ha			0.03	
Temporary seeding with grain or fast-growing grass with				
No mulch			0.70[b] 0.10[c]	
Straw	2.25		0.20[b] 0.07[c]	
Straw	3.4		0.12[b] 0.05[c]	
Stone	300		0.05[b] 0.05[c]	
Stone	540		0.02[b] 0.02[c]	
Wood chips	15		0.08[b] 0.05[c]	
Wood chips	27		0.05[b] 0.02[c]	
Wood chips	56		0.02[b] 0.02[c]	
Sod			0.01[b] 0.01[c]	

[a]Adapted from reference 11.
[b]During first 6 weeks of growth.
[c]After the 6th week of growth.

TABLE 5-5. Values of *P* for Agricultural Land.[a]

Slope (%)	Contouring	Strip Cropping and Terracing	
		Alternate Meadows	Closegrown Crops
1.1–2.0	0.6	0.30	0.45
2.1–7.0	0.5	0.25	0.40
7.1–12.0	0.6	0.30	0.45
12.1–18.0	0.8	0.40	0.60
18.1–24.0	0.9	0.45	0.70
>24.0 1.0			

[a]Adapted from reference 7.

TABLE 5-6. Values of *P* for Construction Sites.[a]

Erosion Control Practice	*P*
Surface Condition with No Cover	
Compact, smooth, scraped with bulldozer or scraper up and down hill	1.30
Same as above, except raked with bulldozer and root-raked up and down hill	1.20
Compact, smooth scraped with bulldozer or scraper across the slope	1.20
Same as above, except raked with bulldozer and root-raked across slope	0.90
Loose as a disked plow layer	1.00
Rough irregular surface, equipment tracks in all directions	0.90
Loose with rough surface >0.3 m depth	0.80
Loose with smooth surface >0.3 m depth	0.90
Structures	
Small sediment basins:	
0.09 basins/ha	0.50
0.13 basins/ha	0.30
Downstream sediment basins:	
With chemical flocculants	0.10
Without chemical flocculants	0.20
Erosion control structures:	
Normal rate usage	0.50
High rate usage	0.40
Strip building	0.75

[a]Adapted from reference 12.

Fine sand 20%
Coarse sand and gravel 25%

The organic content of the soil is 1.5%.

The lot has a square shape with a drainage ditch in the center. A future storm sewer is proposed to replace the ditch. The average slope of the lot towards the ditch is 2.4%.

Determine soil loss (potential erosion) for a storm whose hyetograph is given in Fig. 5-13. Soil loss should be determined from the pervious areas for the two periods, namely, during construction when all vegetation is stripped from the soil surface (100% pervious) and subsequent to construction when 25% of the area is impermeable (streets, etc.).

The rainfall energy factor R_r is determined from the hyetograph shown in Fig. 5-13. From this information it can be determined that the maximum 30-min rainfall intensity I is 2.5 cm/hr.

Utilizing Equation 5-3

$$R_r = \Sigma\,[(2.29 + 1.15 \log X_i)D_i]\,I$$
$$= 15.4$$

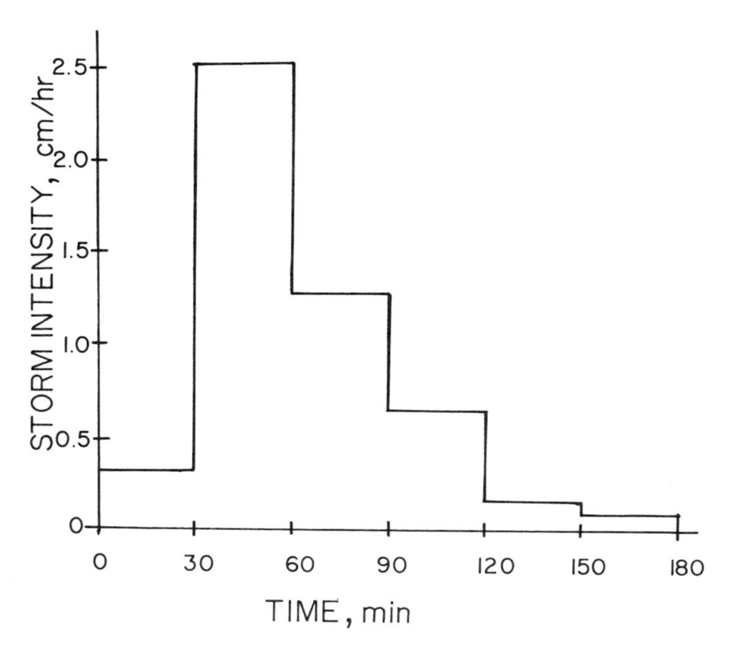

Fig. 5-13. Storm hyetograph for Example 5-1.

The soil erodibility factor is determined from Fig. 5-11 assuming soil texture to be fine grained and permeability to be moderate, giving a K value of 0.33.

To determine the LS factor for a 50-ha area with a ditch or storm sewer in the middle, the length of the overland flow $L = 0.5\sqrt{50 \times 100 \times 100} = 353.5$ *m. With the use of Fig. 5-12 or Equation 5-6, the LS factor for L = 353.5 m and S = 2.4% can be determined as*

$$LS = (353.5)^{1/2}\,[0.0138 + (0.00974 \times 2.4) + (0.00138 \times 2.4^2)]$$

$$= 18.8\,[0.0138 + 0.023376 + 0.00795] = 0.47$$

Factors R, K, and LS are the same for both alternatives. The remaining two factors, C and P, must be evaluated for each alternative. For the period during construction (alternative 1), C is estimated assuming no vegetative protection. In this case, C is approximately the same as for bare fallow ground, that is, C = 1. In the absence of erosion control practices, P = 1. Thus soil loss/ha for this particular storm is:

$$A = 15.4 \times 0.33 \times 0.47 \times 1 \times 1 = 2.4 \text{ tons/ha}$$

then, for 50 ha, total soil loss for the storm is:

$$50 \times 2.4 = 120 \text{ tons}$$

For the period after construction (alternative 2), and assuming that the pervious areas are covered by lawns, C is reduced to C = 0.01 and soil loss/ha is

$$A = 15.4 \times 0.33 \times 0.47 \times 0.01 \times 1 = 0.024 \text{ tons/ha}$$

Given that 75% of the area is subject to soil loss, the total sediment generation from pervious areas is:

$$0.75 \times 50 \times 0.024 = 0.9 \text{ tons}$$

In order to complete the sediment analysis, sediment generated from connected impervious areas would have to be added to the above amount.

Sediment Generation Models

The Negev sediment generation model[13] simulates generation and transport of soil by raindrop impact and overland flow. The production of fine soil particles by raindrop splash is determined for each time interval and unit area:

$$A(t) = (1 - COV) \times K_N \times P(t)^{RER} \tag{5-7}$$

where:

A = fine soil particles produced during time interval t

COV = fraction of vegetative cover as a function of the relative value during the growing season

K_N = the coefficient of soil properties
P = precipitation during the time interval Δt
RER = an exponent

The model should represent the production of fine soil particles, that is, silt and clay fractions, of the total washload.

The fine particles produced by raindrop impact are available immediately for transport by overland flow if overland flow occurs during the time interval. If overland flow does not occur, for example, during initial or final stages of the storm, the fine particles accumulate at the soil surface and represent a reservoir of liberated fine materials available for transport by subsequent overland flow. The mechanism is modeled by:

$$SER(t) = KSER \times SRER(t-1) \times ROSB(t)^{SR} \tag{5-8}$$

where:

SER = fine particles transported during time interval Δt
$KSER$ = coefficient of transport
$SRER(t-1)$ = reservoir of deposited fine particles existing at the beginning of the time interval Δt
$ROSB$ = the overland flow occurring during time interval Δt
SR = an exponent

The David and Beer model[14] divides the erosion loss into rill and interrill contributions. In the interrill zone the soil particles are liberated mainly by the effect of raindrop splash. The following equation was proposed to determine the amount of soil splash:

$$E_d = (SC_F)(LS_F)I^a \exp(-ky) \tag{5-9}$$

where:

SC_F = soil cover factor
LS_F = land-slope factor
I = rainfall intensity
y = overland flow depth
a and k = coefficients

In addition to soil splash, soil detachment by overland flow—which takes place along well-defined paths or rills—may be significant:

$$E_r = C'y^{\beta} \tag{5-10}$$

where:

E_r = amount of overland flow scour
C' = a constant representing the soil characteristics and the overland flow surface slope
β = a constant

The detached material that does not directly enter a stream may be redeposited on the ground or trapped by vegetation or its residues, or be transported if overland flow occurs. The redeposited soil particles are left loosely on the ground for some time as detachment storage until the next overland flow occurs. Then

$$D = D_0 \exp(-Rt) \tag{5-11}$$

where:

D_0 = total detachment storage at the beginning of the time interval
D = total detachment storage at the end of the time interval
$R = a_s/a_c$ where a_s is a soil factor and a_c a climate factor
t = the time interval

The Negev[13] and David and Beer[14] models require a knowledge of several empirical factors and coefficients. Unlike the USLE, these models are not substantiated by extensive experimental field data and measurements, and the coefficients must be estimated by calibrating the model against the extensive set of field data. Possible conversions between the USLE coefficients and some factors of the Negev model have been suggested.[15]

SEDIMENT DELIVERY

Sediment yield measured at a watershed outlet is not equal to upland erosion. The state-of-the-art for estimating delivery ratio (DR) has not progressed beyond the initial stages, and available models are crude and inaccurate. As stated by Renfro,[16] "A characteristic relationship of sediment yields to erosion alone apparently does not exist."

Many factors and processes are significant contributors to the difference between upland erosion and sediment yield. Apart from possible measurement errors, these factors include redeposition of sediment in surface water storage, trapping of sediment by vegetation and plant residues, and local scour and redeposition in rills and channels. Roehl[17] listed sediment sources, climatic factors, texture of the material, environments of redeposition, and watershed characteristics as major factors affecting sediment delivery.

The sediment delivery estimation, and its accuracy and reliability, also is affected by the time scale. Estimation of delivery is most difficult for individual storms when the process is affected and depends on the dynamics of the storm. Long-term (for example, yearly) deliveries depend more on watershed morphological factors and are more reliable and easier to ascertain.

Renfro[16] listed the following factors controlling sediment delivery ratio:

1. In terms of sediment source, channel erosion produces sediment that is immediately available for transport by concentrated flows, while material from sheet erosion often moves only short distances and can be redeposited.

2. Magnitude and proximity of the sediment source is important. Large-magnitude sediment sources may have lesser impact if they are remotely located from the receiving water body. Overland flow has limited sediment transport capacity; therefore, a higher delivery ratio can be expected for clay and silt fractions than for sand and gravel because the larger particles require higher velocities and turbulence intensity for transport. Often the larger grain size fractions are made available for transport by channel erosion, whereas silt and clays generally are made available by sheet erosion.

3. Drainage basin characteristics such as the size of the basin, relief-length ratio, and channel density are correlated closely with annual sediment *DR*.[18] This relationship is attributed to the change in the geomorphic cycle stage which occurs with change in distance from the mouth of the drainage basin. Upstream, the drainage area decreases and factors promoting delivery—for example, steepness of the area and flood characteristics—intensify. In addition to drainage area, two other factors are correlated with delivery ratio parameters, namely, channel density and relief-length ratio. Channel density is defined as the ratio of total stream length within the system divided by the area of the basin. The apparent link between the channel density factor and *DR* is a close correlation between channel density and the average distance that sediment must travel overland before it reaches concentrated channel flow. Channel density also can be correlated with soil texture and other important geological characteristics of the area. The relief-length ratio is an index of drainage basin concavity as well as a determinant of cross-sectional slope characteristics of the basin. The relief-length ratio is defined as the ratio of elevation difference between watershed divide and outlet to watershed length. It can be correlated closely with stream gradients, valley side slopes, and other factors that affect sediment delivery. The watershed length characteristic is measured commonly along the main stream.

Simple Statistical Formulas for Delivery Ratio Estimation

A comprehensive analysis of the sediment delivery problem was first presented by Roehl.[17] The analysis was based on a comparison of sediment deposition in 15 reservoirs and upland sheet erosion in the Piedmont region of the Carolinas and Georgia. The statistical analysis of the most important watershed characteristics and the *DR* yielded the following log-log relationship:

$$\log DR = 3.59253 - 0.23043 \log W + 0.51022 \log R/L - 2.78594 \log BR \tag{5-12}$$

where:

DR = in percent
W = watershed area (km^2)

R/L = relief-length ratio
BR = bifurcation ratio

The *BR* is the ratio of number of streams of any given order to number in the next-higher order. The multiple regression coefficient for the equation was 0.961. The combined data from the Red Hills of Texas and Oklahoma, and Missouri Basin loess hills, sand-clay hills, and the southeastern Piedmont area revealed[17] a trend of decreasing *DR* delivery with watershed size (Fig. 5-14).

Mutchler and Bowie[19] noted that annual *DR* also is affected by runoff and precipitation.

The *DR*-watershed size relationship[17] has been used as a first-step guide, but it is apparent that under some circumstances other factors are more important in determining *DR*. Table 5-7 presents guidelines for the estimation of sediment *DC* for basins in Colorado.[20] A relationship for sediment delivery from construction sites has been proposed:[21]

$$DR = D^{-0.22} \tag{5-13}$$

or

$$DR = (\%DA)^{-0.51} \tag{5-14}$$

where D is the distance between the construction site and receptor stream in feet and $\%DA$ is the percent of the drainage basin exposed by construction.

An example of the correlation of *DR* with channel density is shown in Fig. 5-15.

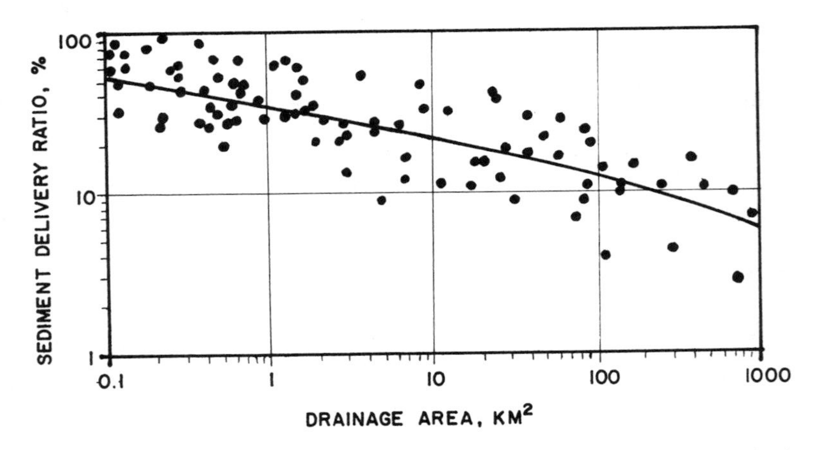

Fig. 5-14. Relation of the sediment delivery ratio factor to the watershed area. (Taken from reference 17.)

TABLE 5-7. Guidelines for Estimating *DR* in the Ryan Gulch Basin of Northern Colorado.[20]

DR	Channel Conditions
1.0	Unvegetated gullies with no deposits.
0.75	Unvegetated gullies with sediment deposits.
0.50	Gullies healed with vegetation indicating shallow flows.
0.3–0.5	Channels intermittently gullied.
0.0–0.4	Shallow vegetated, unregulated, or braided channels with evidence of deposition such as active alluvial fans or sediment deposits or bottomlands where flows spread naturally or are used for irrigation.

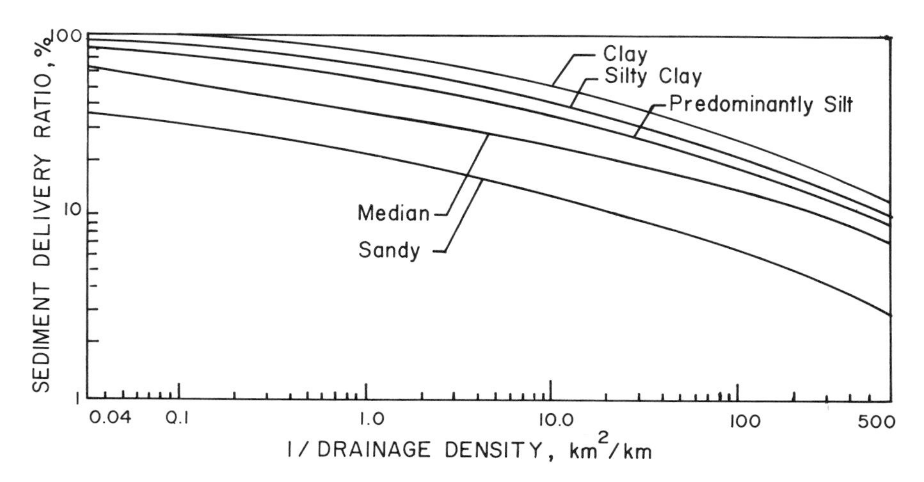

Fig. 5-15. Relation of the sediment delivery factor to the drainage density and soil texture. (Taken from reference 22.)

Example 5-2: Estimation of Sediment Delivery

A small lake has a drainage basin of 25 km^2. Most of the soils within the watershed are characterized as silt loams. The land use is mixed and includes agriculture, urban, and partially wooded areas. With use of the USLE, the upland potential erosion was estimated as 15 tons/ha/yr. The pertinent watershed characteristics are:

Length of the main stream is $L = 9$ km
Relief elevation difference is $R = 150$ m
Bifurcation ratio is $BR = 4.0$

Estimate the amount of sediment reaching the lake, and compare the results using Roehl's models.[17]

The total upland potential erosion is

$$A = 15 \text{ tons/ha/yr} \times 2500 \text{ ha} = 37{,}500 \text{ tons/yr}$$

From Fig. 5-15 DR is 17%. Then the sediment yield is approximately

$$Y = 0.17 \times 37{,}500 \text{ tons} = 6405 \text{ tons}$$

The DR based on Equation 5-12 is:

$$\log DR = 3.59253 - 0.23043 \log (25) + 0.51022 \log (0.150/9)$$
$$- 2.78594 \log (4) = 0.68585$$
$$DR = 4.85\%$$

and

$$Y = 0.0485 \times 37{,}500 = 1820 \text{ tons}$$

Hydrography Effect on Delivery of Sediments and Particulate Pollutants

In order to estimate the hydrograph effect on the sediment delivery, it is necessary to mention the basic considerations on which the USLE and most of the other sediment generation formulas are based.

The developers of the USLE utilized many years of data measured on about 10,000 small plots (average overland flow length 22 m). Under these conditions, the erosion particle load is proportional mostly to rainfall energy and attenuation is minimal and/or it is already included in the suggested values of the coefficients. Furthermore, since soil represents an infinite source of sediment particles, the amounts of sediment particles transported overland in rill and interrill portions of these experimental areas were close to the saturated sediment flow, with probable exceptions being surfaces covered by dense vegetation and mulches.

Foster and Meyer[23] then hypothesized that detachment and deposition rate of sediment flow is controlled by the sediment content of flow, hence,

$$\frac{D_F}{CT_c} + \frac{G_F}{T_c} = 1 \tag{5-15}$$

where:

D_F = detachment or deposition rate at a location (wt/unit area/time)
C = coefficient relating rate of deposition or detachment to the difference between transport capacity and sediment load
G_F = sediment load of flow at any location on the slope (wt/unit width/time)
T_c = flow transport capacity at a location on a given land profile (wt/unit width/time)

Equation 5-15 represents the basis for understanding the process of attenuation of particulate pollutants from nonpoint sources. The equation essentially states that when the sediment load reaches the transport capacity of flow, there is no detachment, and if G_F by some way becomes greater than T_c, D_F must become negative (deposition) in order to satisfy the relationship.

The transport capacity of overland flow can be estimated by the USLE (Equations 5-2 and 5-4), and it is believed that the USLE truly represents the transport capacity of flow from small eroded plots. However, with increased source area after the rain terminates, a significant portion of the overland flow still remains on the surface and moves to the nearest established channel. This portion is contained in the recession portion of the hydrograph. The recession portion of the flow hydrograph does not have the same energy as the runoff during the period of storm; therefore the excess sediment content must be deposited.

The relationship of the sediment-carrying capacity to the runoff flow rate can be expressed best by the Yalin equation.[24] The equation has been reported in the following form

$$p = 0.635s\left[1 - \frac{1}{as}\ln(1 + as)\right] \qquad (5\text{-}16)$$

where P is the dimensionless particle transport given by

$$P = \frac{p \times 10^{-6}}{(\rho_s - 1)Dv_*} \qquad (5\text{-}17)$$

The variables were defined as follows:

p = particle transport per unit width of flow (g/m-sec)
ρ_s = particle density (g/cm^3)
D = particle diameter (m)
$v_* = \sqrt{gHS}$ = shear velocity (m/sec)
g = gravity acceleration (m/sec^2)
H = depth of flow (m)
S = slope of the energy gradient (m/m)
$s = (Y/Y_{cr}) - 1$ (when $Y < Y_{cr}$, $s = 0$)
$a = 2.45\rho_s^{-0.4}\sqrt{Y_{cr}}$
$Y = v_*/[(\rho_s - 1)gD]$

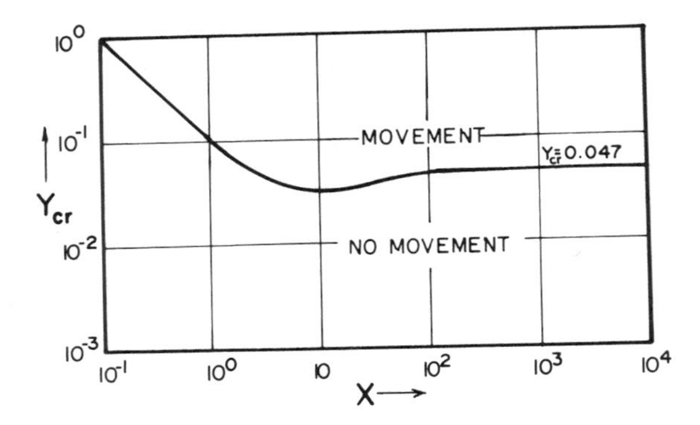

Fig. 5-16. Shields' diagram for particle bedload tractive force. (Taken from reference 25.)

The tractive force at which the sediment movement begins, Y_{cr}, can be found from the Shields diagram (Fig. 5-16), which is based on the particle Reynolds number

$$X = \frac{Dv_*}{\nu}$$

where ν is kinematic viscosity of the flow (m^2/sec).

By rearranging Equation 5-16 and introducing the Manning equation for overland flow rate q

$$q = \frac{1}{n} H^{5/3} S^{1/2}$$

where n is the Manning's roughness coefficient (Table 3-10). The saturated overland sediment load, p, or concentration, C_e, can be related to the flow rate as

$$p \sim D(qnS^{7/6})^{\beta} \tag{5-18}$$

or

$$C_e = \frac{p}{q} \sim D(nS^{7/6})^{\beta} q^{\beta-1} \tag{5-19}$$

where the exponent β could theoretically vary in ranges from $\frac{6}{5}$ to $\frac{3}{2}$.

Proportionalities 5-18 and 5-19 hence indicate that the sediment concentration decreases with $\beta - 1$ power as the flow decreases or with $7\beta/6$ power if slope decreases.

It must be realized that Yalin's equation was developed for fully turbulent channel flow conditions. Under shallow overland flow conditions, portions of the hydrograph tail can be under laminar conditions.[26] Under these circum-

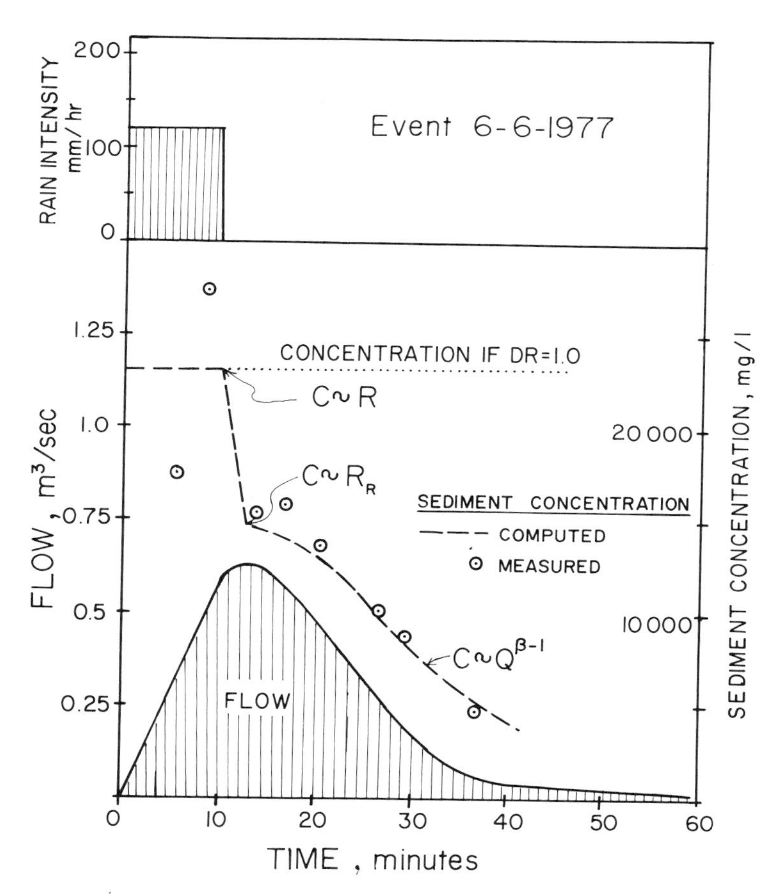

Fig. 5-17. Hydrograph effect on delivery of sediments and particulate pollutants during overland flow.

stances, the equilibrium sediment concentrations, C_e, may approach zero for all fractions. As a result, in a plot of C_e vs q, the exponent can have larger values than the theoretical estimate ($\beta - 1$).

Figure 5-17 shows the discussed concept. In the figure, note that the highest sediment concentrations that would result from both rain and runoff energy (Equation 5-4) can persist only as long as the rain exists. After the rain terminates, the excess sediment content settles out and the sediment concentrations then reflect the runoff energy. According to the principle of mass continuity it would follow that only when the sediment concentration remained constant would the delivery ratio be one.

The measured hydrograph and sediment concentrations were observed on an experimental 9.7-ha watershed located in Germantown, Wisconsin. The watershed is about 90% pervious and was disturbed by construction activities. No measures to control erosion were undertaken, so the effect of vegetation (cover factor *C*) and erosion control practices (factor *P*) on sediment generation were minimal.

Enrichment of Sediments

Enrichment ratio is defined as the concentration of a constituent in the eroded material divided by its concentration in oven-dried soil.[27] The enrichment ratio for various pollutants associated with sediment is discussed in Chapter 6. In this section, the analysis focuses on the difference in gradation of soils and the sediments originating from them. It should be noted that most of the sediment-associated pollutants are carried by or adsorbed on fine particle fractions, mainly clays and organics.

The erosion process is selective, that is, the fine fractions are transported more readily. Furthermore, during overland and channel flows, the coarser particles have a higher probability of redeposition, indicating that a relationship exists between *DR* and the enrichment factor as shown in Fig. 5-18. As *DR* decreases with increasing size of the watershed, the enrichment of the washload by clays and other fine fractions increases. The enrichment of sediments by clays is a two-step process, namely, enrichment during particle pickup and en-

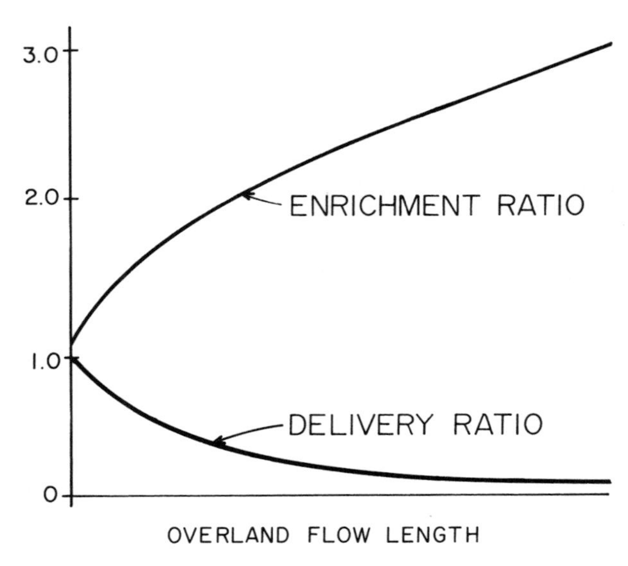

Fig. 5-18. Relation of the sediment delivery to enrichment of sediment by clays.

richment due to redeposition of coarser particles during overland and channel flows.

Pickup of Fines by Erosion. Two variables are necessary for determining the enrichment of sediment picked up by the runoff stream.[28] The variables are specific surface (SS) of soils or sediments and the clay ratio (CR) defined as

$$SS = [200(\%\text{clay}) + 40(\%\text{silt}) + 0.5(\%\text{sand})]\,100 \tag{5-20}$$

$$CR = \frac{(\%\text{clay})}{(\%\text{silt}) + (\%\text{sand})} \tag{5-21}$$

Free et al.[28] (ACTMO model) proposed the following relationships relating the above variables in eroded materials and in their soils of origin:

$$SS_e = 14.6 + 0.84 SS_m \tag{5-22}$$

and

$$CR_e = 0.021 + 1.08\,CR_m \tag{5-23}$$

The enrichment ratio then becomes

$$ER = \frac{(\%\text{clay})_e}{(\%\text{clay})_m} \tag{5-24}$$

The subscripts m and e refer to soil matrix and eroded material, respectively.

Example 5-3: Computation of Enrichment Factor

Determine approximate gradation and clay enrichment of the sediment eroded from a loam soil.

The texture of soils can be determined from maps with the aid of Fig. 5-19. In this case the loam soil has an average composition of: clay, 20%; silt, 40%; sand, 40%. The specific surface of the soil from Equation 5-20 is:

$$SS_m = [(200 \times 20) + (40 \times 40) + (0.5 \times 40)]/100 = 56$$

and the clay ratio from Equation 5-21 is:

$$CR_m = \frac{20}{40 + 40} = 0.25$$

From Equations 5-22 and 5-23 the specific surface and clay ratios of the eroded material are:

$$SS_e = 14.6 + (0.84 \times 56) = 62$$

$$CR_e = 0.021 + (1.08 \times 0.25) = 0.29$$

The textural composition of the eroded material can be obtained by solving the three equations

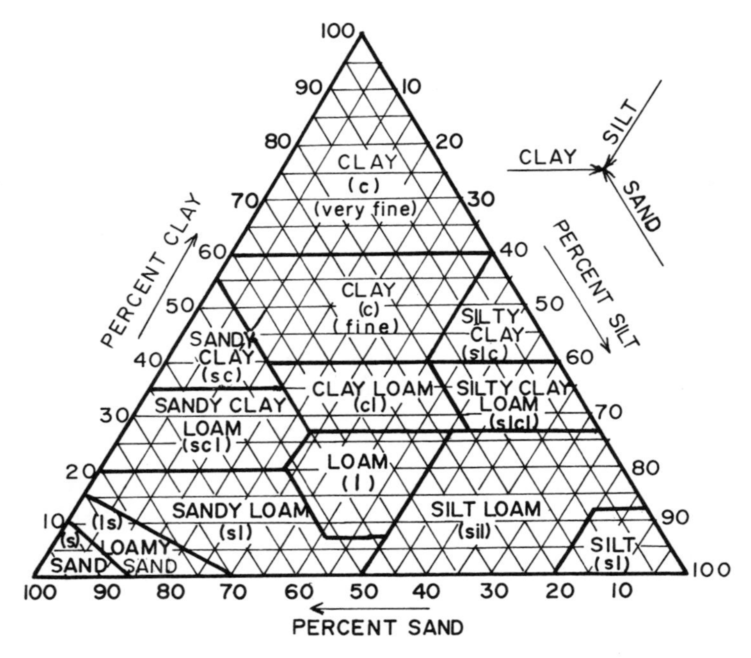

Fig. 5-19. Guide for USDA Soil Conservation Service textural classification.

$$\frac{1}{100}\,[200(\%\text{clay}_e) + 40(\%\text{silt}_e) + 0.5(\%\text{sand}_e)] = 62$$

$$\frac{(\%\text{clay}_e)}{(\%\text{silt}_e) + (\%\text{sand}_e)} = 0.29$$

$$(\%\text{clay}_e) + (\%\text{silt}_e) + (\%\text{sand}_e) = 100$$

which gives a textural composition for the eroded material of: clay, 22%; silt, 43%; sand, 35%. The clay enrichment ratio is then:

$$ER = \frac{(\%\text{clay}_e)}{(\%\text{clay}_m)} = \frac{22}{20} = 1.1$$

Change of the Sediment Texture and Its Enrichment During the Delivery Process. If the delivery ratio is less than one, the corresponding portion of the sediment equaling $(1 - DR)$ has been redeposited on the way. The redeposition process is selective, that is, according to the shear stress and the flow turbulence, gravel and sand fractions are redeposited first followed by silt, and only when the flow slows down to very small velocities will clays be deposited. Deposition of clays is enhanced if the clay particles move in aggregates.

The relation of the clay enrichment and the delivery ratios has been shown in Fig. 5-18. However, no accurate procedure or mathematical model exists that is capable of establishing an accurate functional relationship between the two factors. An approximate graphical procedure based on a grain size cumulative distribution curve can be used to determine changes in sediment texture during transport.

Example 5-4: Determination of Texture Changes in Sediment During Delivery

A watershed with a drainage area of 25 km^2 had a measured sediment yield of 10,000 Tonnes during a large spring storm. Estimated potential erosion from mostly loam agricultural soils was 50,000 Tonnes. Assume that the texture of the soils and the eroded materials is similar to Example 5-3. The eroded material has, therefore, the following texture: clay, 22%; silt, 43%; sand, 35%.

The delivery ratio is

$$DR = \frac{Y}{A} = \frac{10{,}000}{50{,}000} = 0.2$$

An approximate cumulative grain size distribution curve of the eroded soil is drawn in Fig. 5-20 as curve A. The delivery ratio of 20% indicates that most of the washload will be composited from clays. However, some coarser particles from areas closer to the receiving water body may still remain in the suspension. This can be represented by a tangent drawn to the curve A as shown on the figure. The triangular area on the left from the vertical line denoting the 20% delivery and the area between the tangent and the original curve on the right from the vertical delivery line are about the same. The shaded area represents the cumulative percentages of the remaining fractions in the washload with the tangent as the base line. To obtain the grain size distribution curve of the washload, the vertical coordinates of the remaining fractions must be transformed to cover the 100% scale (curve B).

From Fig. 5-20, the composition of the washload will be: clay, 91%; silt, 9%; sand, not detectable.

The ER for a soil originally containing 20% clay is:

$$ER = \frac{91}{20} = 4.5$$

Sediment Transport in Streams

The channel or lowland phase of sediment transport can be divided into the suspended fraction of the sediment carried by the stream, and the fraction of sedi-

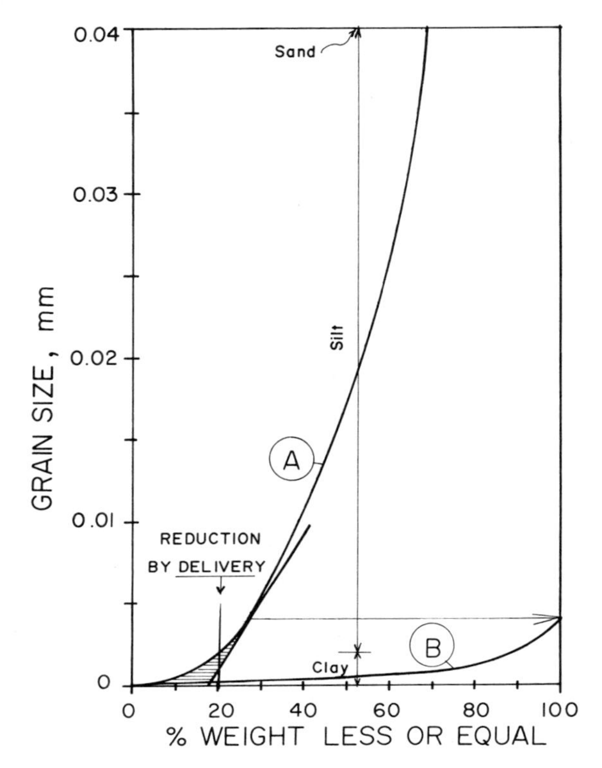

Fig. 5-20. Washload composition and enrichment ratio determination from grain size distribution curve.

ment contained in moving streambeds. The primary variables known to influence sediment transport in streams are: flow velocity, flow depth, slope of the energy gradient, density, viscosity of the water-sediment mixture, mean full diameter of the particles, gradation of the bed sediments, cohesiveness of the sediments, and seepage force on the streambed. Depending on channel hydraulic factors, the sediments may be deposited on or resuspended from the streambed into suspension or vice versa.

The suspended fraction of the sediment transport is termed washload and the portion that comprises the bed is called bedload. The sum of washload and bedload is total load. Measurements of sediment movement in lowland, largely agricultural areas indicate that washload may account for 90 to 95% of the total sediment load.[4,29] However, most of the research dealing with sediment movement has been devoted to bedload movement.

Sediments from upland sheet erosion are mostly cohesive. During the overland flow phase, these sediments move by saltation and rolling of the particles along the bottom of rills. Such sediments are detached from soils in the form of aggregates. When the flow slows, these aggregates readily deposit. Observations indicate that once aggregates are detached from cohesive agricultural soils, their transport is very similar to that of noncohesive particles.[30] During storm events, most of the sediment load in the rills nears saturation.

In channels, however, cohesive sediment flow differs from traditional models developed for noncohesive sediments, that is, for sand and gravel fractions. The main difference between the two classes of sediment transport is that the concentrations and transport rates of the coarser particles bear a distinct relationship to stream discharge whereas no such relationship exists for finer particles.[31] The generally accepted limit for washload is 0.06 mm, that is, clay and silt fractions.[29,31,32]

The behavior of washload in channels has been attributed to the nonsaturation of the sediment transport capacity of the flow.[31,32] The transport of the washload depends more on the availability of such sediments from upstream sources and not on flow characteristics. However, the interdependence of washload transport on flow conditions must have certain limits. Suspended sediments–including clays and silts–will settle in those stream zones of low velocity such as heads of reservoirs or sluggish, slow-moving streams. In such zones, the transport and depositional rates of fine sediments are indeed controlled by flow conditions. Therefore, it is evident that in studying sediment delivery one has to define the flow conditions where transport is controlled by the availiability of the sediment, from either an upstream source or a streambed. In the former case, *DR* is close to unity; in the latter case it will differ from unity. In areas where flow conditions are such that excess sediment is present, that is, the sediment transport capacity of the stream is exceeded, then sediment will eventually settle out and *DR* will be <1. Since the capacity of the flow to carry sediments depends on sediment particle mass and size, delivery must be determined separately for each fraction.

An empirical formula for channel sediment delivery was proposed by Williams.[33] In this formula, *DR* is estimated assuming that sediment deposition in channels depends upon the settling velocity of the sediment particles, the length of travel time, and the amount of sediment in suspension:

$$DR = e^{-BT_i\sqrt{D_{50_i}}} \tag{5.25}$$

where:

B = routing coefficient

T_i = travel time from source of sediment i to watershed outlet

D_{50_i} = the median particle diameter of the sediment (mm) for the source (subwatershed) i

The formula contains an empirical routing factor, B, which must be determined for each watershed from field data. The values reported for Texas watersheds ranged from 4.9 to 6.3 with an average value of 5.3.[33]

There are few other problems associated with the Williams formula and its concept. As will be seen later in Equations 5-31 and 5-32, the dependence of settling velocity on the square root of particle size is typical for large-grain particles. For small particles, settling velocity is proportional to the square of particle size. Also, the average particle size estimation causes a problem since the washload mean particle diameter is not the same as that of soil.

Saturated Sediment Flow. Many formulas and sediment transport models have been developed and published. Unfortunately, all the models are empirical or semiempirical in nature and are applicable—with varying degrees of success—mostly to sand and gravel bedload movement. The reader is referred to hydraulic and hydrologic sediment transport manuals[34,35] for detailed references on these models.

The sediment transport models for noncohesive sediments are based on the assumption that an equilibrium exists between the uplift turbulent momentum exchange forces and gravity forces acting upon the sediment particles. The sediment concentration profile under the saturated flow conditions can be described:[36]

$$C_y = C_B \left[\frac{\frac{H}{y} - 1}{\frac{H}{\Delta} - 1} \right]^r \tag{5-26}$$

where:

Δ = a small distance above the channel bottom, i.e., the roughness height
C_B = the near-bottom concentration of suspended sediment
C_y = the concentration of suspended sidement at the distance y above the bottom
H = depth

The factor r is determined from

$$r = \frac{w}{K\sqrt{gHS}} \tag{5-27}$$

where:

w = settling velocity of the particles
g = acceleration due to gravity
K = von Karman's constant
S = the energy slope

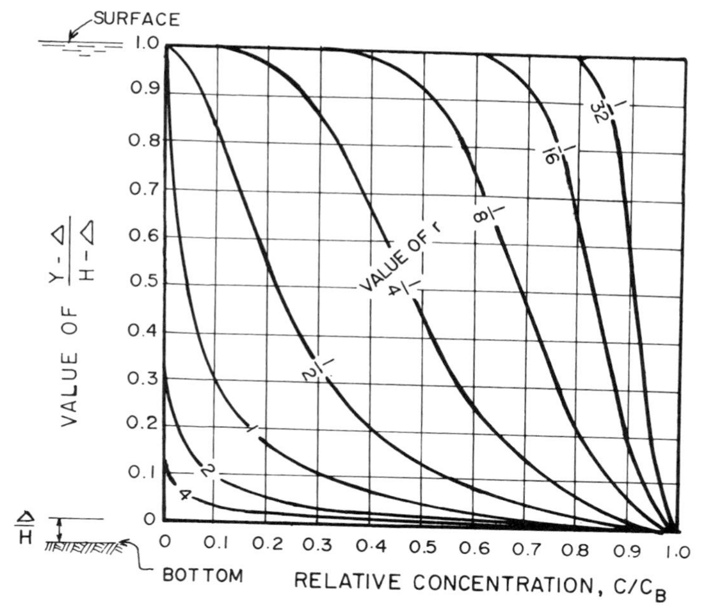

Fig. 5-21. Vertical distribution of suspended sediments in turbulent channel flow. (Taken from reference 35.)

Figure 5-21 shows the distribution of concentrations of suspended particles for different values of r.[31] The suspended sediment load (discharge) can be obtained by integrating the product of concentrations and velocities over the cross section. Then the flow of the suspended particles/unit width (g_{ss}) becomes

$$g_{ss} = \int_0^H uC_y \, dy \tag{5-28}$$

Substituting Equation 5-26 for C_y and assuming that the point velocity (u) can be related to maximum velocity on the vertical profile (U_{max}) as

$$\frac{u}{U_{max}} = \left(\frac{y}{H}\right)^{1/n} \tag{5-29}$$

Equation 5-28 can be solved as:

$$g_{ss} = \frac{U_{max}}{\left(\frac{H}{\Delta} - 1\right)^r} \cdot C_B H \xi \tag{5-30}$$

where

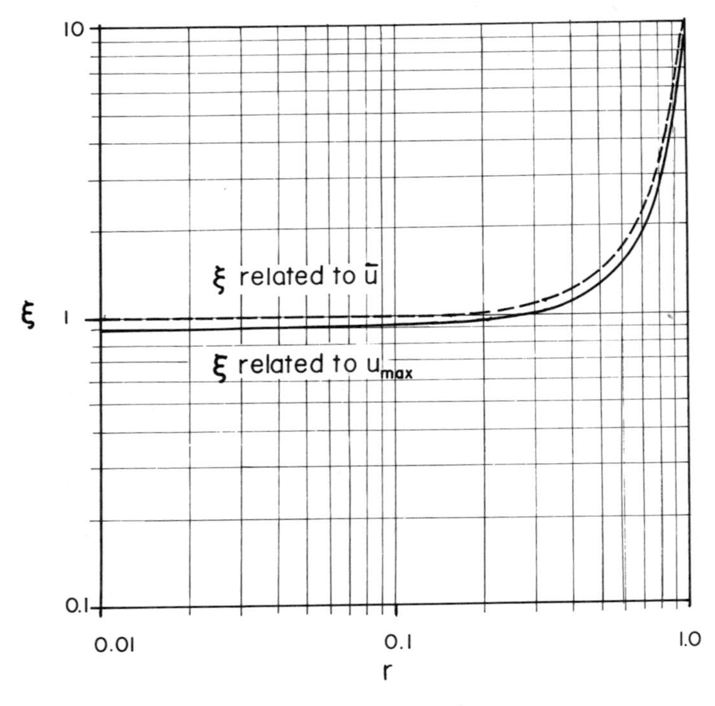

Fig. 5-22. Coefficient ξ vs r.

$$\xi = \frac{1}{1/n - r + 1} - \frac{r}{1!(1/n - r + 2)} + \frac{r(r - 1)}{2!(1/n - r + 3)} - \cdots$$

and the value of the coefficient ξ has been plotted against r in Fig. 5-22.

The exponent $1/n$ in the power expression of the vertical velocity distribution for sediment-carrying streams is a function of temperature as shown by Toffaleti[37]

$$1/n = 0.1352 + 0.00086(TC) \tag{5-31}$$

where TC is the stream temperature (°C).

Equations 5-26, 5-27, and 5-30 require a knowledge of the fall velocity of sediment particles. According to Stokes' law, the fall velocity of spheric particles in water is

$$w = \sqrt{\frac{4}{3} \frac{gD}{C_r} (\rho_s - 1)} \tag{5-32}$$

where:

w = settling or fall velocity (cm/sec)
ρ_s = the specific gravity of the grain with respect to fluid
D = grain diameter (cm)
g = acceleration due to gravity
C_r = the coefficient of resistance, which varies with the particle Reynolds number

For the laminar flow region

$$w = \frac{gD^2}{18\nu} (\rho_s - 1) \tag{5-33}$$

where ν is kinematic viscosity.

For most finer suspended particles originating from soils, the particle Reynolds number defined as

$$Re_s = \frac{wD}{\nu}$$

indicates laminar settling and Equation 5-33 applies.

The maximum velocity of flow (U_{max}) can be related to the average velocity ($\overline{U}$) as follows:

$$U_{max} = \overline{U}(1/n + 1) \tag{5-34}$$

and hydraulic roughness, Δ, can be approximately related to the Manning's coefficient n_M. According to Mostkow,[38] for Δ expressed in meters, the relationship is:

$$\Delta = 182 n_M^{6.0} \tag{5-35}$$

The value of the von Karman coefficient, K, has been suggested as 0.4 for clear water, but it may be quite variable for sediment-laden streams. In Vanoni's experiments,[36] K varied between 0.310 and 0.397 with an average value of 0.35.

The near-bottom concentration of sediments (C_B) necessary for estimating the saturated sediment flow is difficult to ascertain. For noncohesive sediments, this value could be related to bedload movement as shown by Einstein[39] or by Yalin.[24]

Partheniades[40,41] developed a model that may be applicable to the transport of cohesive sediments such as clays and other fine soil fractions. The models evolved from extensive laboratory investigations of behavior of suspended sediment flow. In the experiments, sediment mixtures (particles <0.06 mm) were introduced into a laboratory channel flow. It was observed that when deposition occurred, the concentration of the sediment reached a constant value termed the equilibrium concentration shortly after introduction. All experiments dealing with deposition indicated that C_{eq}, for constant flow conditions

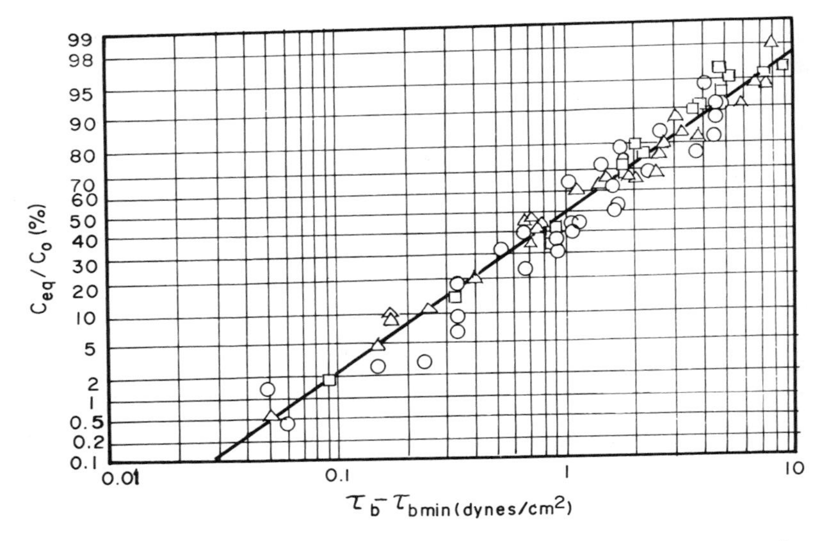

Fig. 5-23. Variation of C_{eq}/C_0 with bed shear stress. (Taken from reference 41.)

and channel geometry varied proportionally to the initial concentration, C_0, and the ratio C_{eq}/C_0 remains constant. This contradicts the theories developed for noncohesive sediment transport, which are based on simultaneous scour-deposition rates. If simultaneous scour and deposition occur, C_{eq} should be the same regardless of initial concentration of the suspension.

Furthermore, Partheniades found that the ratio of equilibrium and initial concentrations, C_{eq}/C_0, are related to bed shear stress, τ_b, by a logarithmic probability (error) function shown in Fig. 5-23. In the figure, $\tau_{b_{min}}$ is the maximum bed shear stress for 100% deposition. Bottom shear stress is a commonly used hydraulic quantity defined as:

$$\tau_b = \gamma HS = \rho gHS \tag{5-36}$$

where:

γ = specific weight of water
H = depth
S = energy slope
ρ = specific density of water
g = acceleration due to gravity

The observed behavior of clay and silt suspensions indicates that under equilibrium conditions ($C = C_{eq}$) neither simultaneous scour and deposition nor exchange between bed particles and suspended sediment takes place.

It has also been established that when erosion (scour) of the channel bed takes place, erosion rates are uniform for uniform bed conditions and suspended sediment concentrations increase linearly. As concluded by Partheniades, these fundamental properties do resemble the behavior of washload in streams. For the velocities and bed shear stresses found in natural free-flowing streams, C_{eq}/C_0 is close to unity and no deposition takes place. In slow-moving streams, reservoirs, or estuaries, when τ_b approaches the order of magnitude of $\tau_{b_{min}}$, sedimentation occurs, and when $\tau_b < \tau_{b_{min}}$, all particles settle provided that sufficient time is available for a particle to reach the bottom.

The critical shear stress ($\tau_{b_{min}}$) for deposition seems to have an order of magnitude of 0.01 psf $\simeq$ 5 dynes/cm^2. The critical shear stress for scouring (τ_c), that is, the shear stress above which bottom erosion takes place, has an order of magnitude of 10 to 25 dynes/cm^2. Its value depends on the physical-chemical characteristics of the settled sediments, their compactness, and their age.[42] The cohesiveness of the settled clays and fine soil particles makes them more resistant to scour.

Example 5-5: Determination of Sediment Deposition

A stream with an average depth of 1 m and a channel slope of 0.3 m/km (S_b = 0.0003) is carrying suspended cohesive sediments (washload). Determine deposition of the sediments and approximate magnitude of the delivery ratio.

Assume that steady-state flow conditions prevail. In this case, $S \approx S_b \approx 0.0003$. Then the bed shear stress becomes $\tau_b = \rho gHS = 1(g/cm^3) \times 981(cm/sec^2) \times 100(cm) \times 0.0003 = 29.4$ dynes/cm^2. The critical bottom shear stress for deposition, $\tau_{b_{min}}$, is approximately equal 1.6 dynes/cm^2. Then

$$\tau_b - \tau_{b_{min}} = 29.4 - 1.6 = 27.8 \text{ dynes/cm}^2$$

From Fig. 5-23 it can be seen that as C_{eq}/C_0 approaches 100%, no deposition of fine soil particles will occur. On the contrary, since τ_b has an order of magnitude of the critical shear stress, τ_c, for scour, erosion of bottom cohesive sediments can occur.

REFERENCES

1. Brant, G. H., Conyer, E. S., Lowes, F. J., Mighton, J. W. and Pollack, J. W. 1972. An economic analysis of erosion and sediment control analysis for watersheds undergoing urbanization. The Dow Chemical Co., Midland, Michigan.
2. Nordin, C. F. 1962. Study of channel erosion and sediment transport. *J. Hydraulics Div., Proc. ASCE*, **90**(Hy 4):172–192.
3. Stewart, B. A., Wollhiser, D. A., Wischmeier, J. H., Caro, and Frere, M. H. 1975. Control of pollution from cropland. U.S. EPA Report No. 600/2-75-026 or USDA Report No. ARS-H-5-1, Washington, D.C.

4. Gottschalk, L. C. 1964. Sedimentation. Part I–Reservoir sedimentation. In: V. T. Chow (ed.), "Handbook of Applied Hydrology." McGraw-Hill, New York.
5. Glymph, L. M. 1972. Evolving emphases on sediment yield predictions. Proc. Sediment-Yield Workshop, USDA Sed. Lab., Oxford, Mississippi.
6. Piest, R. F., Krammer, L. A., and Heineman, H. G. 1972. Sediment movement from loessial watersheds. Proc. Sediment-Yield Workshop, USDA Sed. Lab., Oxford, Mississippi.
7. Wischmeier, W. H., and Smith, D. D. 1965. Predicting rainfall-erosion losses from cropland east of the Rocky Mountains. USDA Agricultural Handbook No. 282, Washington, D.C.
8. Foster, G. R., Meyer, L. D., and Onstad, C. A. 1977. An erosion equation devised from basic erosion principles. *Trans. ASAE*, **20**:678–682.
9. Wischmeier, W. H., Johnson, C. B., and Cross, B. V. 1971. A soil erodibility nomograph for farmland and construction sites. *J. Soil Water Cons.*, **26**:189–193.
10. Wischmeier, W. H. 1972. Estimating the soil loss equation's cover and management factor for undisturbed areas. Proc. Sediment-Yield Workshop, USDA Sec. Lab., Oxford, Mississippi.
11. Ports, M. A. 1975. Urban sediment control design criteria and procedures. Paper presented at Winter Meeting, ASAE, Chicago, Illinois.
12. Ports, M. A. 1973. Use of the universal soil loss equation as a design standard. Water Resources Engineering Meeting, ASAE, Washington, D.C.
13. Negev, M. 1967. A sediment model on a digital computer. Tech. Report No. 62, Dept. of Civil Engineering, Stanford Univ., Palo Alto, California.
14. David, W. P., and Beer, C. E. 1975. Simulation of soil erosion–Part I, Development of a mathematical erosion model. *Trans. ASAE*, **18**(1):126.
15. Fleming, G., and Leytham, K. M. 1976. The hydrologic and sediment processes in natural watershed areas. Proc. 3rd Federal Interagency Sedimentation Conf., Denver, Colorado. Water Resources Council Sedimentation Committees, Washington, D.C.
16. Renfro, G. W. 1972. Use of erosion equation and sediment delivery ratio for predicting sediment yield. Proc. Sediment-Yield Workshop, USDA Sed. Lab., Oxford, Mississippi.
17. Roehl, J. W. 1962. Sediment source areas, delivery ratios and influencing morphological factors, Publ. No. 59, IASH Commission on Land Erosion, pp. 202–213.
18. Banner, S. B. 1962. Factors influencing sediment delivery ratios in the Blackland Prairie land resource area. USDA-SCS, Forth Worth, Texas.
19. Mutchler, C. K., and Bowie, A. J. 1976. Effect of land use on sediment delivery ratio. Proc. 3rd Federal Interagency Sedimentation Conf., Denver, Colorado. Water Resources Council Sedimentation Committees, Washington, D.C.
20. Hadley, R. F., and Shanon, L. M. 1976. Relation of erosion to sediment yield. Proc. 3rd Federal Interagency Sedimentation Conf., Denver, Colorado, Water Resources Council Sedimentation Committees, Washington, D.C.

21. Holberget, R. L., and Truelt, J. B. Sediment yield from construction sites. Proc. 3rd Federal Interagency Sedimentation Conf., Denver, Colorado. Water Resources Council Sedimentation Committees, Washington, D.C.
22. McElroy, A. D., Chiu, S. Y., Nebgen, J. W., Aleti, A., and Bennett, F. W. 1976. Loading functions for assessment of water pollution from nonpoint sources. U.S. EPA/600/2-76/151, Washington, D.C.
23. Foster, G. R., and Meyer, L. D. 1972. A closed form erosion equation for upland areas. In: H. W. Shen (ed.), "Sedimentation." Fort Collins, Colorado.
24. Yalin, M. S. 1963. An expression for bed load transportation. *J. Hydraulics Div., ASCE*, **89**:221–250.
25. Shields, A. 1936. "Änwendung der Ahnlichkeits-mechanik und der Turbulenzforschung auf die Geschiebebewegung." Preuss. Versuchanstalt für Wasserbaur und Schiffbau, Berlin.
26. Overton, D. E., and Meadows, M. E. 1976. "Stormwater Modeling." Academic Press, New York.
27. Massey, H. F., Jackson, M. L., and Hays, O. E. 1953. Fertility erosion on two Wisconsin soils. *J. Agron.*, **45**:543–547.
28. Free, M. H., Onstad, C. A., and Holtan, H. N. 1975. ACTMO–An agricultural chemical transport model. USDA Report No. ARS-H-3, Washington, D.C.
29. Shen, H. W. 1971. "River Mechanics," Vol. 1. Colorado State University, Fort Collins, Colorado.
30. Foster, G. R., and Meyer, L. D. 1972. Transport of soil particles by shallow flow. *Trans. ASAE*, **15**:99–102.
31. Einstein, H. A., Anderson, G. A., and Johnson, J. W. 1940. A distinction between bed load and suspended load in natural streams. *Trans. Amer. Geo. Union*, **21**(Part 2):628–633.
32. Partheniades, E. 1977. Unified view of wash load and bed material load. *J. Hydraulics Div., Proc. ASCE*, **103**(Hy 9):1037–1057.
33. Williams, J. R., and Berndt, H. D. 1972. Sediment yield computed with Universal Equation. *J. Hydraulics Div., ASCE*, **98**(HY 12):2087–2098.
34. Chow, V. T. 1964. "Handbook of Applied Hydrology." McGraw-Hill, New York.
35. American Society of Civil Engineers, V. A. Vanoni (ed.). 1976. "Sedimentation Engineering." ASCE Manuals and Reports on Engineering Practice, No. 54, 745 pp.
36. Vanoni, V. A. 1946. Transportation of suspended sediment by water. *Trans. ASCE*, **III**:67–102.
37. Toffaleti, F. B. 1969. Definitive computations on sand discharge in rivers. *J. Hydraulics Div., Proc. ASCE*, **95**(HY 1):225–248.
38. Mostkow, M. A. 1959. Basic theory of channel flow. (In Russian) Academy of Science of USSR, Moscow.
39. Einstein, H. A. 1950. The bed load function for sediment transportation in open channel flows. Tech. Bulletin No. 1026. USDA, Washington, D.C.
40. Partheniades, E. 1965. Erosion and deposition of cohesive soils. *J. Hydraulics Div., ASCE*, **91**(HY 1):105–139.

41. Partheniades, E. 1971. Results of recent investigations on erosion and deposition of cohesive sediments. In: H. W. Shen (ed.), "Sedimentation," Fort Collins, Colorado.

42. Ariathurai, R., and Arulanandan, K. 1977. Erosion rates of cohesive soils. *J. Hydraulics Div., ASCE,* **104**(HY 2):279–283.

6

Interaction of pollutants with soils

INTRODUCTION

Pollutant loadings to streams from pervious areas in urban and nonurban settings can be considerable. Enormous amounts of potential pollutants are part of topsoil per se or are deposited on the soil as a result of natural processes or man's activities.

Pollutants originating from topsoil losses include soil organic components, plant residues, nutrient elements, organic chemicals, toxic elements, and bacteria. In addition, sediment per se is considered a serious pollutant.

Soils can retain, modify, decompose, or sorb pollutants. Each year enormous amounts of organic materials, atmospheric pollutants, and liquid and solid wastes are deposited and incorporated into soils. These materials are decomposed largely into such safe bacterial decomposition products as CO_2, methane, nitrogen, and phosphorus. Due to the large numbers of microorganisms, the decompositon processes–under suitable environmental conditions–and degradation in soil probably constitute the best natural recycling process. For example, agricultural lands are being used increasingly as a means of disposal of municipal sewage and sludges. The nutrient content of sewage and sludges has been long recognized as a possible source of nutrients for increased crop production and as a source of irrigation water in arid and semiarid regions.

A properly balanced and managed soil system does not represent a great threat to water quality. In a balanced system, most of the nutrients added to soil, in amounts normally applied to increase crop production, will remain in the upper soil layers and/or will be taken up by crops. High pollutant loadings are associated with excessive nutrient applications, disturbed and unprotected lands, and unusual meteorological conditions. The excesses can be in the form of runoff from animal feedlots, and runoff and leachate from excessively fertilized and manured fields and urban lawns and gardens.

Large quantities of commercial fertilizers are applied to agricultural and grassed urban lands in the United States. Typical application rates range from 20 to 200 kg N/ha and from 10 to 50 kg P/ha. The application of fertilizers increases the amount of pollutants that potentially can be lost from these lands, but in some instances offsetting factors may compensate for the increased potential loss. Proper application, including matching the quantity and composition of fertilizers to crop needs and soil fertility, can reduce the amount of nutrient loss from croplands by increasing nutrient uptake by plants and by increasing crop density, thereby reducing surface runoff and erosion. Increased root density may improve soil permeability and reduce the hydrological activity of the land. Conversely, if the addition of the fertilizers creates nutrient imbalance, or if excessive rainfall occurs shortly after application, pollutional impact on receiving waters may be great.

It is difficult to estimate adequate loading figures for pervious lands. The process of pollutant-soil interaction is very complex. The affinity of soils to retain and decompose pollutants depends on many soil factors, such as texture (clay content), pH, exchangeable cation and anion content and capacity, organic matter content, permeability, and soil profile characteristics. Soils of high clay or organic matter content exhibit high retention capacity for many pollutants. It is known that phosphorus is largely immobile in such soils and remains near the point of application. The same phenomenon applies to most of toxic metals, pesticides, and to a certain degree ammonium. Due to the low permeability of such soils, erosion is the primary transporting vector for pollutants. The danger of groundwater contamination of clayey soils is low with the possible exception of such soluble components as nitrates. Under proper operating conditions, clayey soils can be used for wastewater disposal provided that permeability is sufficient to prevent excessive soil loss.

Sandy soils usually have low erosion potential, but their ability to retain pollutants is relatively low and pollutants may penetrate topsoil layers and reach the groundwater aquifer. Although the relatively high permeability of sandy soils makes them attractive for subsurface waste disposal systems (septic systems), the danger of groundwater contamination should always be considered carefully.

Description of the Soil Profile

The soil profile can be generally divided into three layers called horizons A, B, and C (Fig. 6-1). A typical profile may also contain a thin surface layer of decaying organic debris with little soil. The surface layer of organic debris is called horizon O.

The A horizon–usually several centimeters to a fraction of a meter thick–is the soil layer of greatest concern since roots, soil microorganisms, and organics occur there in greatest densities. It is also a layer of considerable leaching. The B horizon, underlying the A horizon, is a subsoil layer where most of the leached salts and clay may deposit. It usually has little organic matter and few plant roots (only larger plants–macrophytes–have root systems penetrating into subsoils). The C horizon extends from the bottom of the B horizon to the top of the parent bedrock from which the soil mostly evolved.

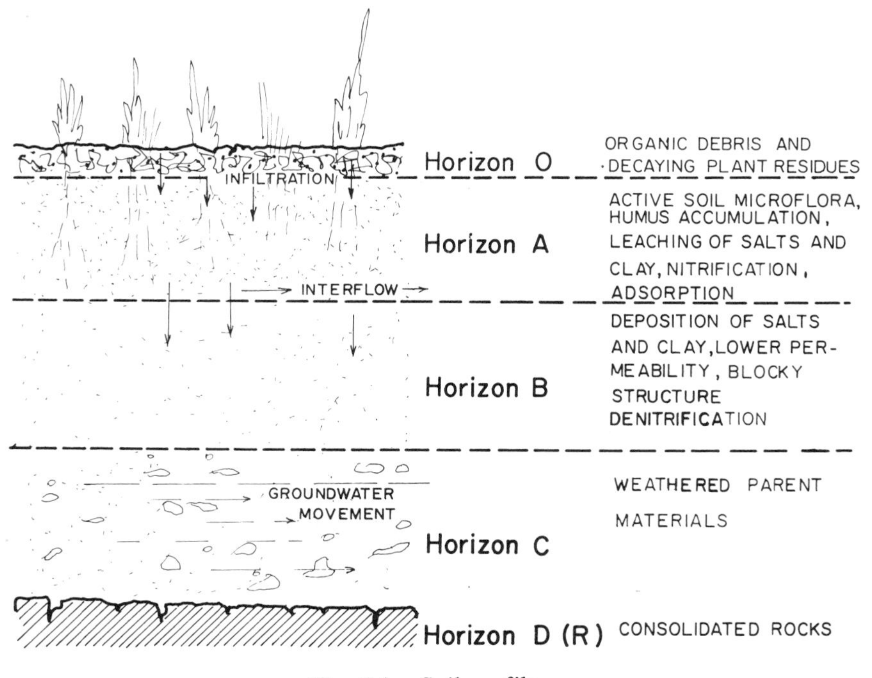

Fig. 6-1. Soil profile.

The A horizon is of considerable importance to nonpoint pollution studies since it is the soil layer where most of the adsorption of pollutants takes place. Also the microbiological processes by which pollutants and nutrients are decomposed or transformed are mostly confined to the A horizon. Only soluble—mobile—pollutants can penetrate into deeper soil zones and eventually pollute groundwater.

TRANSPORT OF POLLUTANTS AND LOADING FUNCTIONS

The transport of pollutants from topsoil is affected by its composition. Pollutants can be in a particulate form—i.e., sorbed by soil particles—or in a soluble form. Depending on the sorption capacity of soil particles for a particular pollutant, a sorption equilibrium can be established between the pollutants contained in the sorbed and solution phases of the soil.

Many pollutants (nutrients, organic chemicals, etc.) are subject to bacterial degradation, volatilization, and chemical breakdown. Those that persist are classified as *conservative* materials, while those that change their mass with time are *nonconservative.*

The transport of pollutants from pervious lands can take place along numerous pathways and can involve many transport mechanisms. However, for all practical purposes, water should be considered the primary transporting vector. Pollutants can be either transported in the particulate phase as a result of soil erosion by rain and overland runoff, or dissolved (Fig. 6-2). The particulate

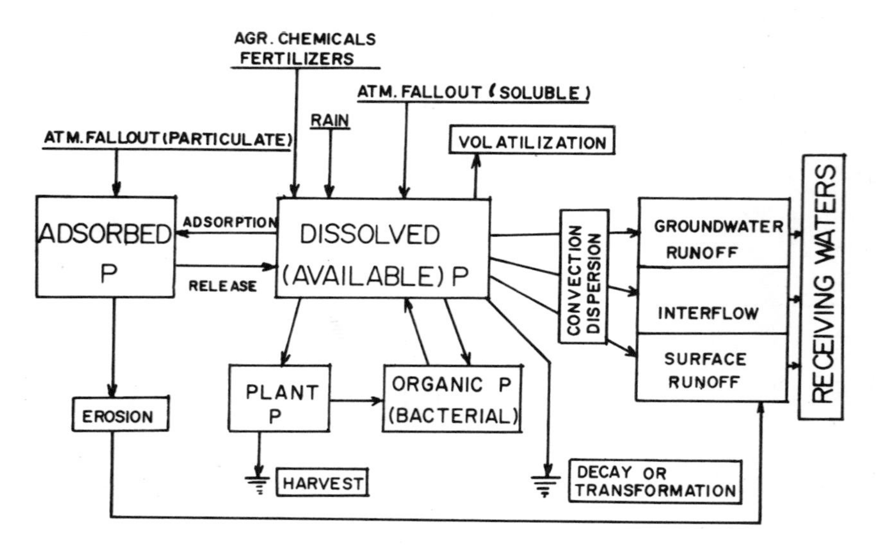

Fig. 6-2. Block diagram of the soil-water-pollutant interactions.

pollutants are subject to transport only by surface runoff, while dissolved pollutants can travel by surface and/or soil water and groundwater routes.

Since particulate pollutant transport is a part of sediment erosion and movement many studies use an arbitrary proportionality factor called the "potency factor" to equate sediment loading of other pollutants.[1,2] This relationship can be expressed by the equation:

$$Y_i = p_i \cdot Y_s \tag{6-1}$$

where:

Y_i = the loading or concentration of pollutant i
Y_s = the loading or concentration of sediment
p_i = the potency factor for the pollutant

The potency factor, p, can be related to the concentration of the pollutant in the parent soil and the enrichment factor for the pollutant, that is,

$$p_i = S_{is} \cdot ER_i \tag{6-2}$$

where:

S_{is} = the concentration of the pollutant in the soil (g/g of soil)
ER_i = the enrichment ratio for the pollutant between the source and the point of interest or watershed outlet

The enrichment ratio factor has been defined by Massey and Jackson[3] as the concentration of the constituent in the eroded material or stream sediment divided by its concentration in the parent soil expressed on an oven-dried basis.

The use of Equations 6-1 and 6-2 presumes a knowledge of the soil pollutant concentration, S_{is}, and the enrichment ratio, ER_i. For agricultural areas, S_{is} is often measured and may be available for organic matter, nutrients, and some other pollutants (Tables 6-1 and 6-2); however, the nature of the enrichment factor is not well understood and its estimation is—at best—a rough approximation.

Equations 6-1 and 6-2 are static, that is, they neither allow prediction of future conditions nor allow evaluation of the effect of future management practices. However, in some cases when a pollutant is conservative and immobile, approximate concentrations can be estimated from the pollutant-soil mass balance.

Since almost all particulate pollutants are usually associated with fine soil fractions and since erosion and sediment transport are selective for the fine materials, a simple relation or guidance for determining the potency factor apparently does not exist. Futhermore, this quantity is highly variable depending on soil characteristics, storm and overland flow characteristics, channel flow hydraulics, presence of sediment and pollutant sinks such as grassed areas or forest litter and mulch, and the nature of the pollutants. Also, the sorption

TABLE 6-1. Metal Concentrations (μg/g) in Surfical Materials in the U.S.A.[a]

Element[b]	Arithmetic Analysis		Geometric Means		
	Average	Range	Conterminous U.S.	W. of 97th Meridian	E. of 97th Meridian
Ba	554	15-5000	430	560	300
Ce	86	<150-300	75	74	78
Cr	53	1-1500	37	38	86
Co	10	<3-70	7	8	7
Cu	25	<1-300	18	21	14
Fe	25,000	100-100,000	18,000	20,500	15,000
Ga	19	<5-70	14	18	10
La	41	<30-100	34	35	33
Pb	20	<10-700	16	18	14
Mn	560	<1-7000	340	389	385
Mo	3	<3-7	–	–	–
Ni	20	<5-700	14	16	13
Nb	13	<10-100	12	11	13
Sc	10	<5-50	8	9	7
Sr	240	<5-3000	120	210	51
Ti	3000	300-15,000	2500	2100	3000

[a]After Shacklette et al.[4]
[b]Arithmetic range for As = <1000; Cd = <20; Ge = <10; Au = <20; Hf = <100; In = <10; Pd = <1; Re = <30; Ta = <200; Te = <2000; Tl = <50; Th = <200; U = <500.

TABLE 6-2. Composition of Soils in Menomonee River Watershed, Wisconsin.[a]

Soil[b]	Organic C (%)	Sand (%)	Silt (%)	Clay (%)	Total P (μg/g) in			Pb (μg/g) in		
					Sand	Silt	Clay	Sand	Silt	Clay
Ozaukee sil	1.6	24	57	19	119	241	2757	[c]	9.5	58
Mequon sil	–	35	36	29	186	443	2668	4.7	11	39
Hochheim sil	–	29	44	27	82	154	1821	5.5	9.8	56
Ashkum scl	4.3	21	44	35	491	397	2775	9.0	14	36
Pella sil	–	14	49	37	426	290	1570	9.8	10	39
Theresa sil	4.1	22	62	16	79	128	2336	[c]	6.0	55

[a]Data from the Water Resource Center, University of Wisconsin, Madison.
[b]Sil is slit loam and scl is silty clay loam.
[c]Not detected.

characteristics and capacity of soil particles to retain pollutants is different depending on whether the particles are in soil or are a part of the sediment wash-load and bedload in streams and other surface water bodies.[5,6] There are several processes that result in higher concentrations of pollutants in the eroded material and stream sediment than in the parent soil:

1. Selective removal of fine materials with higher pollutant concentrations than the remainder of the soil material, which is expressed as the clay enrichment ratio.
2. Diffusion of pollutants and salts from the topsoil layer into surface runoff.
3. Desorption of pollutants from soil particles due to low dissolved concentrations in the runoff water.
4. Flotation of low-density materials such as organic components from soil into surface runoff.
5. Deposition of coarse fractions containing little sorbed pollutant during overland and channel flow.

The transport and loading of dissolved pollutants is more complex because they can be transported by each of the three runoff components, namely, surface runoff, interflow, and groundwater.

It is evident that a simplistic approach to the determination of pollutant loadings from pervious surfaces may fail to provide adequate results, and prognoses using such approaches must be treated with caution and accepted only as rough estimates.

Example 6-1: Estimating Pollutant Concentration in Soils from Mass Balance

Application of 1 kg/ha of a conservative chemical that is immobile in soils was made to an area. Estimate the approximate concentration of the chemical on the soil particles if the pollutant is assumed to be uniformly distributed throughout the top 25 cm of the soil (approximate till depth). The specific density of the soil is 1.8 g/cm³.

From Equation 6-2:

$$S_{is} = \frac{1\ (\text{kg/ha}) \cdot 10^9\ (\mu\text{g/kg})}{25\ (\text{cm}) \cdot 10^8\ (\text{cm}^2/\text{ha}) \cdot 1.8\ (\text{g/cm}^3)} = 0.22\ \mu\text{g/g}$$

Biodegradable Organics

Organic matter is an integral part of soil. The organic matter content may vary from <1 to 40% in some organic soils and feedlots. Organic matter content is usually reported as organic matter (%) or organic carbon (%). Organic matter (%) $\cong$ 1.67 $\times$ organic carbon (%).

The organic content of soils, termed often as humus, is mostly a product of biodegradable processes by the soil microorganisms. It is rich in nutrients and remains for long periods as an important food supply for the soil microbial population.

Most of the soil organic matter is contained in the A and O horizons of the soil profile. As to their organic content, soils are divided into mineral soils that have low organic fraction and organic soils.

High organic content is not synonymous with the fertility of soils. Peat soils that usually have a high organic fraction exhibit lower organic activity due to poor drainage and aeration. When the peat soils are drained, the excess organic matter is converted, under aerobic conditions, to CO_2, and fertility of the soil is often restored.

Most of the organic matter in soils is in particulate form and, as such, can be related to erosion by applying Equations 6-1 and 6-2. However, the values of the enrichment factor are difficult to assess. The McElroy et al.[2] report noted that the values of enrichment ratio (ER) for organic matter can range from 1 to 5 with lower values typical for soils with fine texture and higher values for sandy soils. Massey and Jackson[3] found that ER values for organic matter increase with sediment concentration in runoff and the rate of erosion. The average enrichment ratio for three silt loam soils in Wisconsin measured by Massey and Jackson[3] was 2.1. Young and Onstad[7] proposed the following relationship for the enrichment ratio (ER_{or}) of organic matter based on regression analyses of Indiana and Minnesota soils:

$$ER_{or} = \frac{0.30}{(\%\ \text{soil organic matter})} + 1.08 \tag{6-3}$$

The organic content increases generally with finer textures, so Equation 6-3 indicates lower values of ER for clayey soils and higher values for sandy soils.

Since the specific density of organic particles is much less than mineral soil particles, ER may increase significantly during overland and channel flow.

Organic content of soils lost in agricultural erosion can be restored by application of organic fertilizers—manure and man-made organic mixtures. Increased organic content improves the permeability of soils and reduces their erodibility. As will be shown in a subsequent section of this chapter, a significant portion of the organic matter in fertile soils consists of soil microflora.

Sorption of Pollutants by Soils

Many conservative and nonconservative pollutants can be immobilized effectively in the soil environment. The immobilization is caused mainly by sorption on soil particles, precipitation of soluble components, and transformation by plants and bacteria into less mobile organic forms. It should be noted that soil-sorbed or precipitated pollutants, including nutrients, generally are considered unavailable to plants.

Adsorption is a physicochemical process by which molecules or ions are immobilized by soil particles. In the case of precipitation reactions, the amount of pollutant in the particulate fraction is governed by its solubility in the soil environment. If adsorption is the dominant process, the removal of pollutants from soil solution depends on the concentration of the pollutant in the solution, which, in turn, is in a dynamic equilibrium with the soil-adsorbed components. The preferred mathematical form for describing this distribution is to relate the concentration of pollutant adsorbed by the solid phase (S_e in μg/g) to the equilibrium solution concentration (C_e in mg/liter) at a fixed temperature. Mathematically, several descriptions of adsorption equilibria (isotherms) have been proposed, and the most commonly used are the Langmuir and Freundlich isotherms. The Langmuir adsorption model is valid for monolayer adsorption and is expressed in the form:

$$S_e = \frac{Q^\circ b C_e}{1 + bC_e} \tag{6-4}$$

where:

Q° = the adsorption maximum at the fixed temperature (μg/g)
b = a constant related to the energy of net enthalphy of adsorption (liters/mg or ml/μg)

Although the Langmuir isotherm is assumed to be valid for monolayer adsorption it adequately represents soil adsorption processes for such pollutants as phosphorus and some organics, metals, and ammonium.

The Freundlich isotherm is useful if the energy term, b, in the Langmuir isotherm varies as a function of extent of surface coverage, S. The Freundlich equation has the general form:

$$S_e = K \cdot C_e^{1/n} \tag{6-5}$$

where K and n are constants.

The parameters describing the isotherm equations for various pollutants are statistical quantities. In some special cases, laboratory adsorption studies may be available that can provide the magnitudes of the adsorption coefficients. However, for most modeling and pollution transport studies these variables are estimated only roughly from a few routinely measured soil parameters.

Sorption (and desorption) characteristics are related mostly to surface-active components of soils. Clay particles—due to their large specific area, active surface adsorption sites (mostly Al and Fe), cation exchange capacity, and negative surface charge–are active in retaining pollutants such as phosphates. Organic matter in soils also is known to have a high affinity for adsorbing various components. The soil adsorption activity, which is mostly physicochemical in nature, depends on the pH of the soil.

The kinetics of adsorption can be expressed as:

$$\text{Soluble Pollutant} \underset{1-K}{\overset{K}{\rightleftarrows}} \text{Adsorbed Pollutant} \tag{6-6}$$

where K is the adsorption coefficient. Very few examples are known that allow quantification of adsorption kinetics. Adsorption of some heavy metals and pesticides may be essentially complete within minutes or hours, while phosphates and ammonium do not adsorb as readily.

Decay, Sublimation, and Transformation

Although not important for phosphates and some metals, the processes of decay, sublimation, and transformation must be included in pollutant mass balance determination in soils. These processes are usually described by the first-order reaction:

$$\frac{dC}{dt} = -K_d C - \frac{K_s}{D_x} C \tag{6-7}$$

where:

K_d = the decay or transformation rate coefficient
K_s = the sublimation or stripping rate
Dx = the depth of the upper soil zone

Rates of decay for organic chemicals are related largely to soil organic matter content. High soil pH induces stripping of ammonia and temperature, and pH controls the rate of nitrification of ammonium soils.

Plant Uptake

Plant uptake is a form of immobilization of pollutants in soils. The nutrient and pollutant uptake process is a part of the overall transpiration process of plants. Nutrients are transported into plants only in the dissolved phase or as ions. In this case, pollutants and nutrients adsorbed on soil particles or precipitated in soils are not available to plants: thus, only the mobile components of the soil are of concern, and high uptake rates can be observed where the mobile component concentrations are high.

The solubility of almost all components of interest depends on soil pH, for example, most toxic metals have higher solubility in acid soils, resulting in higher rates of plant uptake. Total uptake rates and nutrient requirements depend also on crop type and yield.

Although promoting higher uptake rates by plants may appear to be a means of controlling pollution from nonpoint sources (e.g., by pH control or by selecting plants with higher uptake rates), the use of crops with high concentrations of accumulated contaminants is limited and benefits are suspect. The amounts of pollutants taken up from soils by plants must be subtracted in determination of the mass balance of pollutants from nonpoint sources. Table 6-3 contains N and P uptake for common crops.

Soil Phosphorus

In natural systems, phosphorus (P) occurs as the orthophosphate anion (PO_4^{3-}), which may exist in inorganic or organic forms. The source of all organic P is plant and organic biomass residues. The origin of all inorganic orthophosphate is the class of minerals known as apatites. These minerals are insoluble calcium phosphates existing in several forms, and the orthophosphate ions are liberated by chemical weathering processes.

Phosphorus is an important nutrient to aquatic ecosystems. Its excess causes accelerated eutrophication, a process during which surface water bodies age at increased rates due to excess productivity of organic matter by algae, macrophytes, and other chlorophyll-containing organisms. In fact, the rate of eutrophication in many water bodies can be related solely to P concentrations since other nutrients are not limiting growth (according to Liebig's law of minimum only the nutrient that is in shortest supply controls the growth rate of organisms).

The amounts of P pollution from nonpoint sources varies widely depending on environmental, pedological, and meteorological factors. Phosphate loadings from agricultural lands were summarized by Uttomark et al.[8] and ranged from 0.03 to 2.3 kg/ha-yr, with an average value of 0.4 kg/ha-yr. Somewhat higher values around 1 kg/ha-yr are typical for urban areas; however, most of the urban loadings include contributions from impervious surfaces.

Phosphorus, unlike nitrogen (N), is not particularly mobile in soils, and phosphate ions do not leach readily. Phosphorus is held tightly as a complex anion by clays and organic matter, and the amount of P in solution is small. Most of the P is removed from soils either by crop uptake or by soil erosion.[8]

Figure 6-3 shows P solubility in soils. At higher pH values–characteristic of calcareous soils–the P precipitates mostly in combination with calcium (Ca). Below a pH of 7, which is characteristic for soils with high clay and organic matter contents, Ca rapidly disappears from soils and the P reacts predominantly with Fe and Al ions in soils.

From Fig. 6-3 it can be seen that the maximum phosphate concentration in the soil solution is of the order 10^{-5} moles/liter, which corresponds to about

TABLE 6-3. Approximate Yields and Nutrient Contents of Selected Crops.[a, b]

Crop		Yield/hectare	Kg N/ha	Kg P/ha
Alfalfa		9 Tonnes	225	20
Barley	grain	100 bu	39	7
	straw	2.2 Tonnes	17	2
Beans[c]	(dry)	75 bu	84	11
Cabbage		45 Tonnes	165	18
Clover[c]	red	4.5 Tonnes	89	11
	white	4.5 Tonnes	145	11
Corn	grain	370 bu	151	27
	stover	10 Tonnes	111	18
	silage	56 Tonnes	225	34
Cotton	lint and seed	2.2 Tonnes	67	13
	stalks	2.2 Tonnes	50	7
Lettuce		45 Tonnes	100	13
Oats	grain	222 bu	62	11
	straw	4.5 Tonnes	28	9
Onions		17 Tonnes	50	9
Peanuts[c]	nuts	3.4 Tonnes	123	7
Potatoes	tubers	990 cwt	106	14
	vines	2.2 Tonnes	100	9
Rice	grain	225 bu	62	13
	straw	5.6 Tonnes	34	4
Rye	grain	75 bu	39	4
	straw	3.4 Tonnes	17	4
Sorghum	grain	150 bu	56	11
	stover	6.7 Tonnes	73	9
Soybean[c]	grain	111 bu	179	18
	straw	2.2 Tonnes	28	4
Sugarbeets	roots	45 Tonnes	95	16
	tops	27 Tonnes	123	11
Sugar cane	stalks	67 Tonnes	112	22
	tops	29 Tonnes	56	11
Tobacco		3.4 Tonnes	129	11
Tomatoes	fruit	56 Tonnes	162	22
	vines	3.4 Tonnes	78	11
Wheat	grain	123 bu	73	16
	straw	3.4 Tonnes	22	2

[a]Source: USDA.
[b]Values can vary by a factor of two across the country.
[c]Legumes that do not require fertilizer nitrogen.

Conversions: From Tonnes/ha to tons/acre multiply by 0.45.
From bu/ha to bu/acre divide by 2.47.
From kg/ha to lb/acre multiply by 0.89.

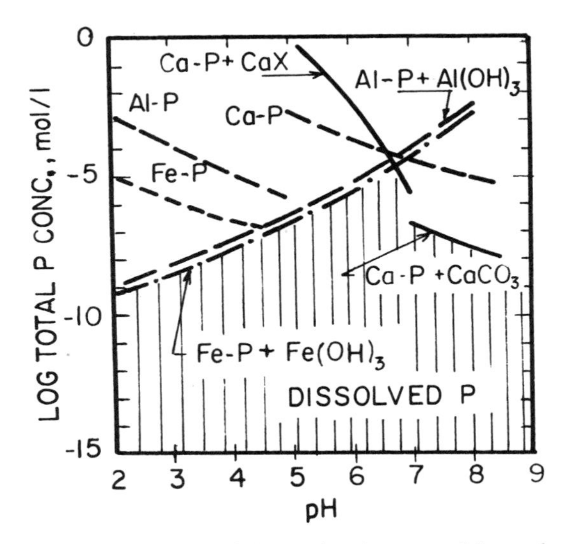

Fig. 6-3. Solubility diagram for calcium, aluminum, and iron phosphates in soil. (From reference 9. Copyright © 1960 The Williams & Wilkins Co., Baltimore.)

0.3 mg/liter. Depending on soil pH, the dissolved P concentrations may decrease to 0.01 mg/liter or less.

From the foregoing discussion it can be seen that the P concentration in the soil solution is very low and the excess P is fixed by the soil. Thus, almost all P arising from the weathering of minerals and/or from P application to soils by atmospheric fallout, commercial fertilizers, plant residues, or animal manure remains near the point of application.[10] The exception is in sandy or peat soils, which exhibit little tendency to react with phosphorus. Porcella et al.[10] have shown that after a normal growing season, fertilizer P applied in spring is confined to the surface 5 cm of the soil.

Some organic forms of P, for example, from farm manure, have a greater mobility in soils than inorganic P. This apparent increase in mobility is attributed to incorporation of P into soil microorganisms during the breakdown of soil organic matter.

The process of fixation of P is controlled by several factors:

1. Al and Fe oxides are responsible for P retention in acid soils.[11,13]
2. Calcium compounds control solubility of P in calcareous soils.[11]
3. Organic matter contributes to P adsorption.

Figure 6-4 shows typical adsorption isotherms for some selected soils. Several authors have attempted to correlate P sorptivity to various soil parameters. In

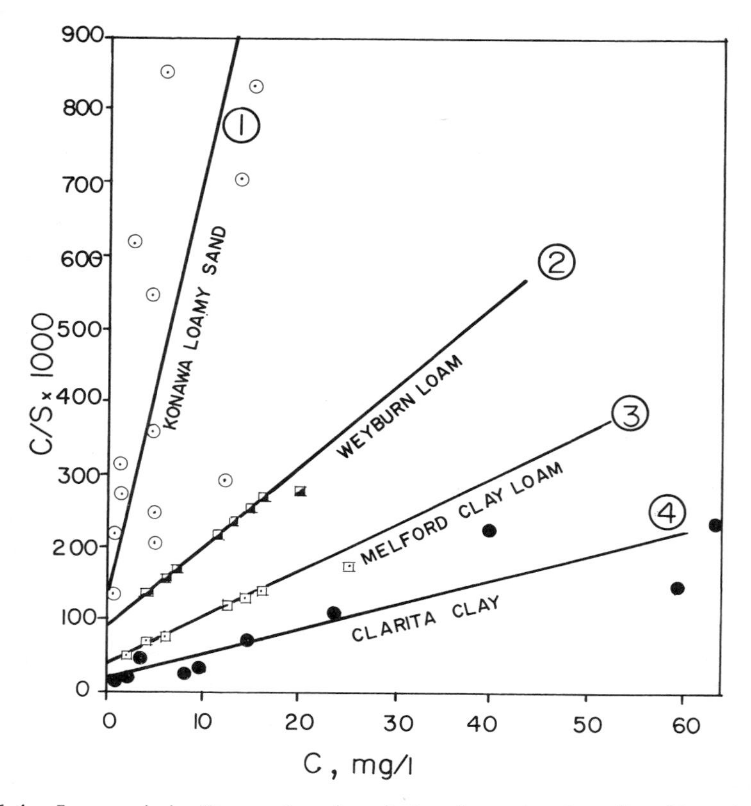

Fig. 6-4. Langmuir isotherms for phosphate adsorption in soils. (Data 1 and 4 taken from reference 18 and data 2 and 3 from reference 20.)

a summary of the literature findings, it was shown that parameters controlling P adsorption by soils are: pH; and Al, Fe, clay, and organic matter contents.

It must be noted that intercorrelations exist between the independent variables, for example, exchangeable Al is correlated inversely with soil pH.[14,15] The effect of Fe oxides and hydroxides is much less than that of Al,[15] and there may be a correlation between the Fe components, pH, and clay, Al and organic matter contents. The most satisfactory combination of variables found by the authors to assess the adsorption maxima ($Q°$) and the energy coefficient (b) was:[19]

$$Q° = -3.5 + 10.7\ (\%\ \text{clay}) + 49.5\ (\%\ \text{organic C}) \tag{6-8}$$

and

$$b = 0.061 + 169{,}832 \times 10^{-\text{pH}} + 0.027\ (\%\ \text{clay}) + 0.76\ (\%\ \text{organic C}) \tag{6-9}$$

The units for Q° are $\mu g/g$ and for b liters/mg. The coefficients of multiple correlation for Equations 6-8 and 6-9 were 0.83 and 0.54, respectively. It should be noted that the equations should be used for estimation of adsorption characteristic only for acid soils. For calcareous soils the distribution of particulate and dissolved P is governed by P solubility according to Fig. 6-3.

Example 6-2: Distribution of Sorbed and Dissolved Phosphate in Soils

A fertilizer was applied to a field at a rate of 100 kg P/ha. The fertilizer was plowed in and uniformly distributed to a depth of 30 cm. The soil was a silt loam and contained 20% clay, 55% silt, and 25% sand. The porosity of the soil was 40%, pH was 6.0, organic C content was 1%, and the specific density of the dry soil sample was 1.5 g/cm.³ Estimate the approximate groundwater contamination by P during a long rain period with an average intensity of 1 cm/hr. Assume that all water infiltrated into the soil and that the infiltration rate was slow enough that a full equilibrium between adsorbed and dissolved P was established. Assume also that the antecedent adsorbed P was negligible.

The P adsorption maximum, Q°, is then from Equation 6-8:

$$Q^\circ = -3.5 + 10.7 \times 20 + 49.5 \times 1 = 260 \ \mu g/g$$

and the adsorption energy coefficient from Equation 6-9 is:

$$b = 0.61 + 169{,}832 \times 10^{-6.0} + 0.027 \times 20 + 0.76 \times 1 = 1.53 \text{ liters/mg}$$

The total inorganic P content of the soil becomes

$$P = \frac{100 \text{ (kg/ha)} \times 10^9 \ (\mu g/kg)}{30 \text{ (cm)} \times 10^8 \text{ (cm}^2\text{/ha)}} = 33.3 \text{ g/cm}^3$$

This P concentration is distributed between adsorbed and dissolved fraction S. Hence:

$$P = S\rho + Cp$$

where:

ρ = the specific density of the dry soil
p = moisture content (assume that p = porosity)

Substituting Equation 5-4 for S

$$33.3 = \frac{260 \times 1.53 \times C}{1 + 1.53 \times C} \cdot 1.5 + C \times 0.4$$

or the dissolved P concentration C is 0.05 mg/liter and the adsorbed inorganic P concentration is:

$$S = \frac{260 \times 1.53 \times 0.057}{1 + 1.53 \times 0.057} = 20.85\ \mu g/g \text{ of soil}$$

The adsorbed P remains fixed, while the dissolved P can move downwards to the groundwater aquifer.

Adsorption of P by soils is not an instantaneous process. Adsorption studies of several days' duration revealed that an initial fast adsorption stage occurs that lasts for minutes or hours followed by a slow adsorption phase lasting days or weeks. If a first-order adsorption kinetic model is assumed to represent the adsorption process (a reasonable assumption) then

$$\frac{dS}{dt} = K(S_e - S) \tag{6-10}$$

where:

K = the adsorption coefficient
S = the amount of P adsorbed on the soil
S_e = the equilibrium adsorption concentration

Ryden et al.[19] estimated the adsorption kinetics coefficient for P to be about 0.12 hr^{-1}. The experimental data by Enfield[20] yielded a coefficient of about the same order of magnitude.

The enrichment ratio (*ER*) for P generally is higher than that for clay or organic matter due to desorption of P from soil into runoff water and possibly due to other causes. Data of Massey et al.[3,21] and Stoltenberg and White[22] indicate that the *ER* for P is about 1.5 to 2 times that for clay or soil organic components.

The adsorption characteristics for P, and evidently for other materials adsorbed to soil particles, are different for parent soils and suspended sediment in runoff.[5,6,19] This phenomenon seems to be logical since more adsorption surface is available on suspended soil particles than in partially compacted soils. An example of an adsorption isotherm of P on suspended solids in runoff is shown in Fig. 6-5. The ranges for suspended sediment adsorption maxima, Q°, in an Ohio watershed[5] were between 500 and 2000 μg/g (average 988 μg/g) while the ranges of Q° for parent soils were 199 to 287 μg/g (average 237 μg/g). Similarly, the adsorption energy coefficient, b, ranged from 0.11 to 0.45 liters/mg (average 0.32 liters/mg) for the suspended sediment as compared to 0.8 to 4.35 liters/mg (average 2.6 liters/mg) for the parent soil. Bottom sediments (bedload) showed adsorption characteristics between those of parent soil and suspended solids. More research is needed to develop a meaningful relationship for the enrichment ratios and adsorption characteristics in streams based on the characteristics of the parent soil.

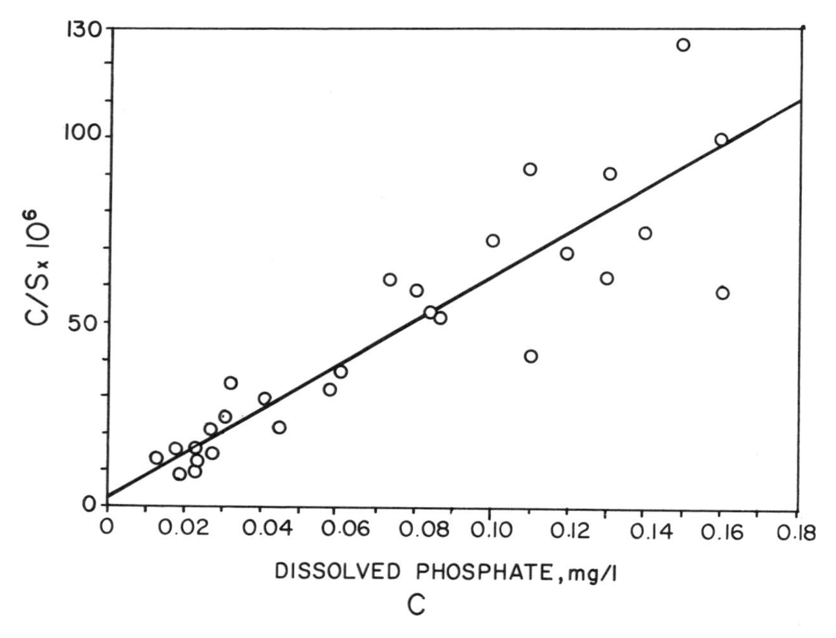

Fig. 6-5. Isotherm correlation of dissolved and adsorbed phosphate in stream-suspended solids. (From reference 17.)

Example 6-3: Phosphate Loss by Erosion

The distribution between dissolved and adsorbed fractions following application of 100 kg of P fertilizer was computed in Example 6-2. If a storm occurred after the fertilizer was incorporated, which resulted in soil loss by erosion of 1.5 Tonnes/ha, estimate P loss from the area.

The PO_4 loss can be approximated by Equations 6-1 and 6-2.

Let $S_{is} \approx S \approx 20.85$ µg/g or g/Tonnes and $ER_p \approx 2.0$. Then

$$Y_{PO_4} = S_{is} \times ER_p \times Y_s = 20.85 \text{ (g/Tonne)} \times 2.0 \times 1.5 \text{ (Tonnes/ha)}$$

$$= 62.55 \text{ g of P/ha}$$

Toxic Metals

The metallic content of soils generally is low (see Tables 6.1 and 6.2) and does not represent a major pollution hazard. However, localized exceptions can be found: (1) near urban and heavy transportation corridors where automobile exhausts can increase the environmental burden of Pb, (2) near ore deposits and mines, and/or (3) near ore and metal-processing industrial operations. In addition, sewage sludge applied to agricultural lands as a substitute for fertilizer or for disposal can contribute significant amounts of metals to the soil.

Metals can be divided into two groups according to their ecological effects:[23,24]

1. Biogenic metals are essential compounds of living organisms. The ecosystem is able to handle them without serious ecological consequences through its recycling and discharge mechanisms. Examples of these metals include Fe, Cu, Mg, and Al.
2. Nonbiogenic metals are part of the natural ecosystem. They act as poisons upon some step or parts of the ecosystem, and the ability of natural food chains to remove these materials is sometimes less than cumulation rates in tissues of organisms. Under these conditions, high concentrations of Hg, Pb, Cd, and Cr may accumulate in the tissue.

The uptake rates of metals by crops depends on their solubility in the soil solution. Most metals precipitate readily or can be adsorbed by soil particles. Those that are immobilized easily are considered less hazardous to plants than those that can be dissolved. The former group includes Mn, Fe, Al, Cr, As, Pb, and Hg. The metals most dangerous to crops due to the ease with which they are taken up by plants are Cd, Cu, Mo, Ni, and Zn.[25] Soil pH has an important influence on uptake rates of metals by plants. An almost tenfold reduction in Zn, Cd, and Mn content of plants may be achieved by liming acid soils to near neutrality.[25]

Lead. Lead (Pb) is an abundant naturally occurring metal that exists in four stable isotopes.[26] The accepted average value for the Pb content of the earth's crust is 15 μg/g, but ranges of 0.8 to 500 μg/g have been reported for arable soils.[26]

In addition to traffic and natural background concentrations, sources of Pb include atmospheric fallout, mining and smelting and insecticides (e.g., lead arsenate, which is now banned for use in the United States).

Lead is highly immobile in soils. The high degree of Pb immobilization may be related directly to the soil cation exchange capacity (CEC) and inversely with soil pH.[26] The sorption equilibrium for Pb is approached quite rapidly, and for all practical purposes, instantaneous adsorption occurs. It has also been shown[26] that soil pH, CEC, and organic matter and available P contents can affect soil adsorption capacity for Pb. A regression analysis yielded the following equation for adsorption maximum:

$$Pb_s = 34.3 + 0.774P + 5.358\,(pH) + 5.337\,(CEC) \qquad (6\text{-}11)$$

where:

Pb_s = the amount of Pb removed from soil solution in μ moles Pb/g of soil
P = the amount of dissolved phosphate

To convert Pb_s to Q° (μg/g), multiply Pb_s by 207. The multiple correlation coefficient for this regression equation was 0.997.

The transport of Pb from soil to receiving waters occurs almost solely by erosion with delivery ratios similar to those for clay particles. For all practical purposes, Pb remains only in the top few centimeters of the soil.[27]

Arsenic. Inorganic arsenic (As) compounds have been used in agriculture as pesticides and defoliants for many years, but due to serious pollution problems, their use has been banned in the United States since 1967. Other sources of As include sewage sludges applied to agricultural lands and industrial operations.

Once incorporated into soils, As reverts to arsenate, which is strongly held by the clay fraction of most soils.[25] As a consequence, As has a very low availability to plants and low mobility in soils of appreciable clay content.

Mercury. The background concentrations of mercury (Hg) in soils range from 0.01 to 0.05 μg/g.[25,28,29] In soils, Hg combines with the exchange complex forming ionic and covalent bonds.[25] Montmorillonite adsorbs Hg significantly and releases it only slightly on washing, while allophane and kaolinite exhibit very little adsorption affinity for Hg.[30] The rate of Hg fixation is influenced by the types of clay minerals, pH, and CEC. Evidence also exists that Hg can be chelated by organic matter.[25,28] Chemical and microbiological degradation of Hg-containing compounds can occur side by side in the soil, and the products are either retained by organic matter and clays or volatilized. Mercury may undergo biological methylation in sediments,[28,31] but no methylation has been observed in soils.

Because of the high affinity between Hg and soil surfaces, Hg persists in the upper soil surface layer; however, it can be leached from sandy or kaolinitic soils.

Cadmium. The chemistry of cadimum (Cd) in soils is not well understood, but Cd appears to be influenced by soil organic matter and clay contents, the types of hydrous oxides present, soil pH, and redox potential.[25] Increased soil pH leads to decreased plant uptake, indicating that more Cd is fixed at higher pH values.

In general, the mechanisms of retention of toxic metals by soils are numerous, complex, interrelated and poorly understood.[32] They can be grouped into: (1) ion exchange reactions with clays, (2) adsorption and precipitation reactions, and (3) complexation reactions. Toxic metal mobility decreases as pH increases, and is minimal above pH 6.5.

No literature or experimental data are available for estimating enrichment ratios for metals. Due to their tight bonding to clays and organic matter, it is expected that the enrichment ratios may be similar to those for clays and organic materials, but further research is necessary to substantiate this hypothesis.

Organic Chemicals

The use of organic chemicals is credited with substantially increasing yields of agricultural crops and assisting in the control of such tropical diseases as malaria. In a situation typical to most modern societies, the availability of agricultural land decreases in making way for urbanization. Thus the use of pesticides is inevitable and likely to increase in the future. As a matter of fact, increased crop yields are correlated closely to increased pesticide usage.[33]

In 1973, the total annual sales of pesticides in the United States reached 500,000 tons, and by the mid-1970s about 55% of corn fields were treated with pesticides and 35% with herbicides.[33]

Of the 21,700,000 ha of land used for crop production in the Great Lakes region, 62% received pesticide treatment at least once in 1971.[34] The increase of pesticide use is about 4% per year. Most pesticides used in the Great Lakes region are applied as preemergence herbicides, and only small portions of the soybean, small grain, and hay crops are treated for insects.

Although the use of pesticides dates back a hundred years, usage did not become substantial until about 1945 following the commercial manufacture of DDT. Rapid growth of the organic pesticide industry continued for about 25 yr. Pesticide usage revolutionized agricultural production to the point that most agricultural practices formerly used to control weeds, insects, and disease were shifted in favor of chemical control.

Approximately 34,500 pesticide products are registered currently with the U.S. EPA and the USDA.[35] The largest volume of pesticides is used in farming operations, but about one-half of the registered products are utilized for non-farm purposes.

Any pesticide application may represent a potential pollution hazard to aquatic ecosystems or drinking water. Many pesticides are considered serious pollutants, and their accumulation in the environment is limited by strict standards. The type and nature of pesticide residues reaching surface waters is controlled largely by: (1) the amounts, rates of application, and lengths of time pesticides have been applied: (2) the persistence and/or residence time of pesticides in the watershed; and (3) the mobility of pesticidal residues.[34]

Pesticides are divided into insecticides, herbicides, and fungicides.[34]

Insecticides (insect controlling) include organochlorine, organophosphorus, and carbamate chemicals. Organochlorine compounds such as DDT, dieldrin, aldrin, chlordane, heptachlor, and lindane are essentially conservative chemicals. Their persistence in soils and aquatic environments can be of several years' duration, and DDT is detected frequently 10 years following its application.

Use of organophosphorus and carbamate chemicals in the Great Lakes region increased rapidly following restrictions on the use of organochlorine insecticides. These chemicals are less persistent in soils than the "hard" organochlorine formulations. An extended discussion on the persistence of chemicals in soils and their removal from soils is presented largely in a subsequent section of this chapter.

Herbicides are the most commonly used agricultural chemicals for the controlling of weeds. They are less ubiquitous in the environment than organochlorine insecticides. However, such compounds as the s-triazines, picloram, monuron, and related substituted ureas, and 2,4,5-T often persist in soils for as much as a year following application. Atrazine, alone or in combination, and propachlor are the herbicides used most commonly for weed control on corn in the midwestern United States. The herbicides MCPA and 2,4-D amine are used most frequently on small grains. The carbamate herbicides and 2,4-D are short-lived in soils.

Fungicides are used to the greatest extent in orchards and on vegetable farms. The proportion of cropland treated with fungicides is small compared with the use of herbicides and insecticides. Mercurial fungicides have been of recent concern following the detection of mercurial contaminants in the Great Lakes and other surface waters. Subsequently, their usage has been drastically curtailed.

The fate and toxicity of pesticides have been studied extensively in the past two decades, and a substantial amount of information exists in the literature.[33-37] In the United States, pesticide levels in soils are monitored regularly, and the results of the national surveys are published in the *Pesticide Monitoring Journal.* The highest DDT levels (up to 80 μg/g) were detected in Michigan, while lower levels were detected in Alabama, California, Mississippi, and South Carolina. Average residue levels in these states were 1 to 2 μg/g in 1970.

Pesticide Mobility in Soil, Groundwater, and Surface Runoff Systems. Pesticides can be transported from the treated area through: (1) the atmosphere, (2) groundwater, and (3) surface runoff.

A general scheme of the distribution and fate of pesticides in the biosphere is shown in Fig. 6-6. Pesticidal contamination of the atmosphere can occur by drift during application, by volatilization, and by wind erosion. Drift is that portion of the application that does not reach the target area. The extent of drift losses and subsequent dispersion and transport of pesticides in air is governed primarily by meteorological conditions and method of application. Drift losses range from 25 to 75% of aerially applied pesticides.[34]

Volatilization has been recognized as a major pathway for the loss of pesticides from soils, plant, and water surfaces (see subsequent sections).

Pesticide contamination of groundwater can occur through leaching. Downward movement of agriculturally applied pesticides is controlled by soil type, pesticide composition, and climatic factors.[39-43] Leachability of a compound from soils depends primarily on the degree to which it is adsorbed. Pesticide adsorption is correlated more closely with organic matter content than clay content for nearly all pesticides, and pesticides are leached more readily from coarse-textured than from fine-textured soils. Furthermore, water infiltration is faster

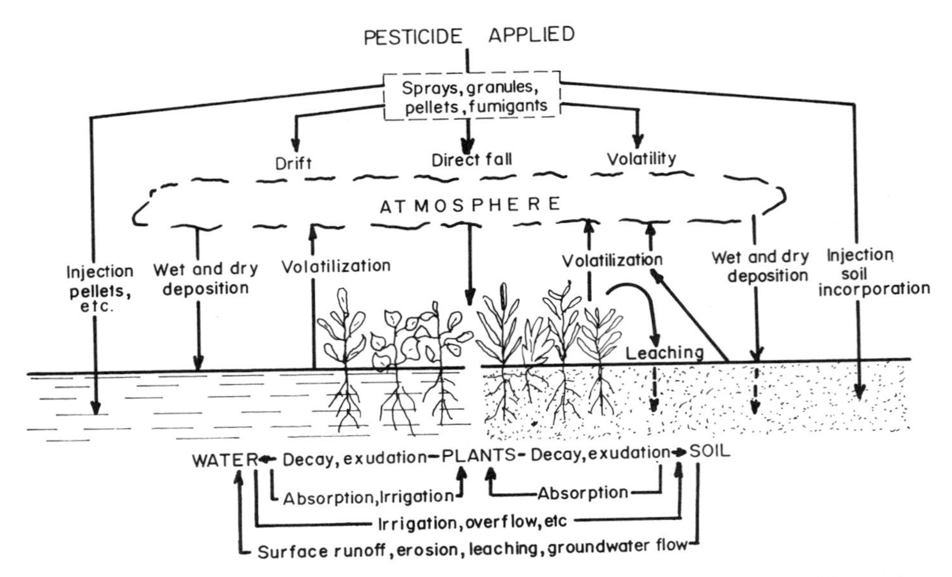

Fig. 6-6. Scheme showing the possible distribution and fate of pesticides and their degradation products in the biosphere. (Redrawn from Foy and Bingham.[38])

in coarse-textured soils. The solubility of pesticides plays an important role in their movement in the soil since solubility limits the concentration of the mobile portion of the compound in the soil solution. Solubility is correlated inversely with adsorptivity. Thus, pesticides of high water solubility are more subject to vertical movement than those of low solubility. The transport of pesticides through the soil also is dependent on the amount, intensity, and frequency of water infiltration. Water can facilitate desorption as well as dissolution of particulate or adsorbed compounds.

In summary, the entry of a particular pesticide to groundwater is defined by soil type, depth to water table, rainfall infiltration, and persistence of the compound in the soil. Available information indicates that pesticide contamination of groundwater under normal agricultural usage is minimal.[44] However, a need exists for more information on the extent of pesticide entering groundwater through soil cracks and sand layers.

Sorption of Pesticides by Soil. Sorption of nonionic pesticides–organchlorine and organophosphorous insecticides–in soils is correlated primarily with organic matter content[37,40] and to a lesser extent with clay content. Retention of acidic and basic compounds is affected markedly by soil pH.[37,45,46] Soil pH controls the overall charge of the molecule and hence its adsorptivity to clay and organic colloids. The organic cations–diquat and paraquat–are held strongly by clay minerals and often are adsorbed irreversibly. Weakly adsorbed, water-soluble compounds are desorbed readily by water and hence possess greater potential for leaching.

Numerous studies have been conducted to determine the relative mobilities of pesticides in soils (Table 6-4). Organochlorine insecticides–which have limited water solubility–are the least mobile, followed, in turn, by organophos-

TABLE 6-4. Relative Mobility of Pesticides in Soils.[a]

Mobility Class[b]				
5	4	3	2	1
TCA[c]	Picloram	Propachlor	Siduron	Neburon
Dalapon	Fenac	Fenuron	Bensulide	Chloroxuron
2,3,6-TBA	Pyrichlor	Prometone	Prometryne	DCPA
Tricamba	MCPA	Naptalam	Terbutryn	*Lindane*
Dicamba	Amitrole	2,4,5-T	Propanil	*Phorate*
Chloramben	2,4-D	Terbacil	Diuron	*Parathion*
	Dinoseb	Propham	Linuron	*Disulfoton*
	Bromacil	Fluometuron	Pyrazon	Diquat
		Norea	Molinate	*Chlorphenamidine*
		Diphenamid	EPTC	Dichlormate
		Thionazin	Chlorthiamid	*Ethion*
		Endothall	Dichlobenil	*Zineb*
		Monuron	Vernolate	Nitralin
		Atratone	Pebulate	C-6989
		WL 19805	Chlorpropham	*ACNQ*
		Atrazine	*Azinphosmethyl*	*Morestan*
		Simazine	*Diazinon*	*Isodrin*
		Ipazine		*Benomyl*
		Alachlor		*Dieldrin*
		Ametryne		*Chloroneb*
		Propazine		Paraquat
		Trietazine		Trifluralin
				Benefin
				Heptachlor
				Endrin
				Aldrin
				Chlordane
				Toxaphene
				DDT

[a]Adapted from Helling et al.[41] See reference 34 for more detailed literature sources.
[b]Class 5 compounds (very mobile) to class 1 compound (immobile); in each class pesticides are ranked in estimated decreasing order of mobility.
[c]Names of pesticides are set in normal type; insecticides, fungicides, and acaricides are set in italics.

phorous insecticides. The water-soluble, acidic herbicides are the most mobile. Most pesticides, including triazines, phenylureas, and carbamates, have intermediate degrees of mobility. For a diverse group of pesticides, relative mobility is related essentially to their solubility.

The effect of pH on adsorption of organic chemicals depends on their chemical composition. Six categories of chemicals have been distinguished based on their adsorption behavior as related to soil pH: strong acids, weak acids, strong bases, weak bases, polar compounds, and neutral compounds. The adsorption characteristics of such neutral materials as DDT and organochlorine compounds are virtually pH independent. Adsorption of other chemicals—including polar compounds—is pH dependent, but a simple correlation of the adsorption characteristics with pH may fail because of the strong effect of organic matter and clay contents on the relationship.[37]

For lower concentrations of the adsorbent in the soil solution, the Langmuir and Freundlich isotherms can be linearized to yield a relationship such as:

$$S_e = K_d C_e \tag{6-12}$$

where K_d = the distribution adsorption constant.

When the concentration of the adsorbed material is expressed per unit of organic C, the equation yields[37]

$$K'_{oc} = \frac{S'_e\ (\mu g/g \text{ of organic C})}{C_e} = \frac{K_d\ (\mu g/g \text{ of soil})}{C_e \times (\% \text{ organic C})} \times 100 \tag{6-13}$$

or

$$K_d = K'_{oc} \times (\% \text{ organic C})/100 \tag{6-14}$$

The value of K'_{oc} is more constant than K_d, and percent organic C can be calculated roughly from the content of soil organic matter with the use of a conversion factor ranging from 0.5 to 0.6. Values of the constant K'_{oc} may vary from soil to soil depending on the type of organic matter, its concentration, and the degree of compaction. Soils with low organic matter content tend to give higher values of K'_{oc} due to interference of clays and other factors. Figure 6-7 shows a relationship between the adsorption coefficient, K_d, and soil organic matter content for some pesticides. Table 6-5 shows the adsorption characteristics for a series of pesticides.

Adsorption characteristics of soils for organic chemicals also are affected by soil temperature and moisture content. Adsorption is an exothermic process, that is, adsorption decreases at higher temperatures. Furthermore as soil dries out, the degree of adsorption increases. From Equation 6-12 it follows that the total amount of chemical per unit soil volume is:

$$Q/m = C_w (\theta_{\text{eff}} + K_d) \tag{6-15}$$

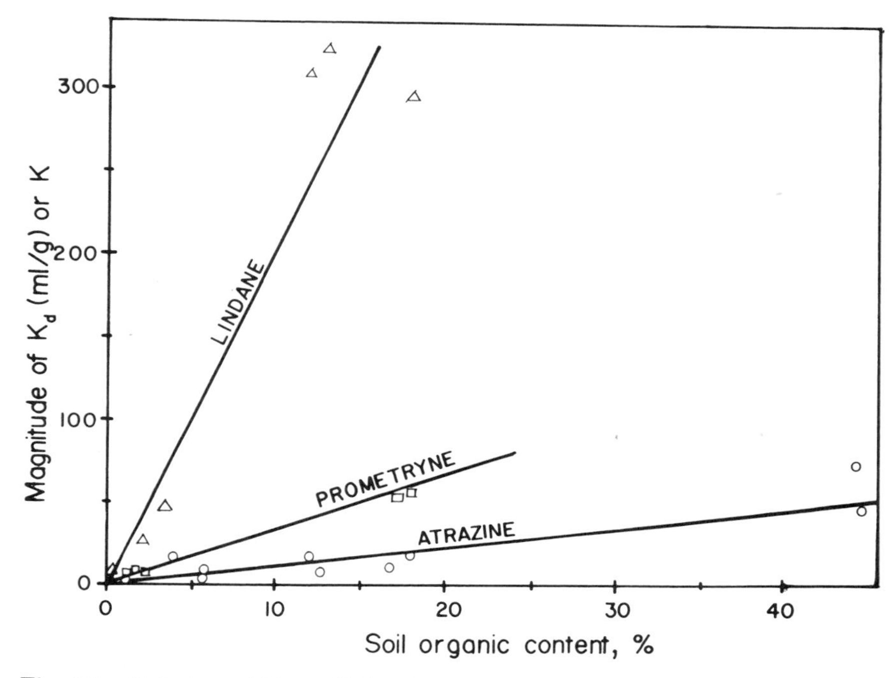

Fig. 6-7. Relation of Freundlich adsorption coefficients to organic carbon context of soils. (Data from references 37 and 47.)

where:

Q = the total amount of chemical added to the soil
m = the weight of the soil
C_w = the concentration of the chemical in the solution phase
θ_{eff} = the effective water content

The effective water content denotes that portion of the total soil moisture that is available for equilibrium with the applied chemical. It is identified as a positive difference between the actual water content, θ, and the soil wilting point (15-bar tension), $\theta_{15\ bar}$. Hence:

$$\theta_{eff} = \theta - \theta_{15 bar} \tag{6-16}$$

Adsorption of insecticides on lake sediments was studied in detail by Chesters et al.[47] Lindane (isomer of benzene hexachloride) adsorption was shown to be affected by sediment concentration, organic matter and clay contents, lindane concentration, and lindane : sediment ratio. The adsorption process followed the Freundlich equation.

Two empirical isothems were developed by multiple regression analysis of the

TABLE 6-5. Adsorption Characteristics of Some Pesticides in Soils.[a]

Pesticide	Mobility Class	Freundlich Isotherm				
		Adsorption Exponent $1/n \pm \sigma$	Adsorption Constant $K \pm \sigma$	Carbon[b] Related $K_{oc} \pm \sigma$	Linearized $K_d \pm \sigma$	Carbon[c] Related $K'_{oc} \pm \sigma$
Atrazine	3	0.77 ± 0.07	4.16 ± 2.2	120 ± 54		
Prometryne	2	0.85 ± 0.1	7.83 ± 2.4	395 ± 100	7.03 ± 0.62	513 ± 66
Fenuron	3	0.95 ± 0.1	0.55 ± 0.1	23.1 ± 2.9		
Monuron	3	0.76 ± 0.05	3.5 ± 2.4	190 ± 100		
Diuron	2				6.29 ± 0.6	351 ± 30
Linuron	2	0.71 ± 0.02	53 ± 12	1896 ± 10		
Aldrin	3	0.78	1.3 ± 0.08	253 ± 116		
Lindane	1				24.7 ± 0.67	1342 ± 201
2,4-D	4				1.6 ± 0.8	
2,4,5-T	3–4				1.05 ± 0.4	
DDT	1		>100,000		>100,000	
Dicamba	5		0		0	
Disulfoton	1	0.92 ± 0.02	20 ± 0.4	2132 ± 533		
Methyl parathion	1	1.04	18.4 ± 3.6	9799 ± 4017		
Parathion	1	1.03	21.9 ± 7.0	10,454 ± 3800		
Paraquat	1	0.56 ± 0.11	353 ± 330	20,152 ± 13,100		
Chloramben NH_4	5				0.32 ± 0.1	12.8 ± 4.0
Picloram	4				0.453 ± 0.14	12.7 ± 2.7

[a]Data reprinted from reference 37, by courtesy of Marcel Dekker, Inc.
[b]$K = K_{oc}$ (% org. C)/100.
[c]$K_d = K'_{oc}$ (% org. C)/100.
σ = standard deviation of the estimate.

lake sediment adsorption data. The parameters that were considered in the analysis were: X_1—lindane concentration; X_2—organic matter content; X_3—ratio lindane: sediment; X_4—organic matter content; and X_5—clay content as independent variables and adsorbed lindane concentration as dependent variable. The best fit equations were as follows:

$$Y = -0.063 + 0.093X_1 - 0.013X_2 + 0.034X_3 + 0.0034X_4 + 0.0058X_5$$

$$(\text{correlation coefficient } r = 0.92) \qquad (6\text{-}17)$$

and

$$Y = 0.203 + 0.095X_1 - 0.018X_2 + 0.042X_3$$

$$(\text{correlation coefficient } r = 0.94) \qquad (6\text{-}18)$$

where the adsorbed lindane, Y, is expressed in μg/mg of sediment.

Persistence of Pesticides in Soils. Pesticides remaining in soils are subjected continuously to dissipation processes, and their concentration decreases gradually after application. The mechanisms of dissipation include adsorption on soil particles and subsequent erosion, degradation (photochemical and microbial), volatilization, plant uptake, and leaching. Volatilization, leaching, and erosion are means by which pesticides are moved from one segment of the ecosystem (watershed or soil) to another (atmosphere, surface waters) and affect both nonconservative and relatively conservative pesticides. By definition, degradation affects only nonconservative pesticides. General magnitudes of overall persistence of pesticides in soils and aquatic environments are shown in Fig. 6-8.

Photodecomposition occurs only to pesticides located at the soil surface. The practical significance of photodegradation as a means of pesticide removal from soil and aquatic systems has not been determined quantitatively because of the difficulty of interpolating laboratory data to field conditions. In most cases, photodegradation reactions are comparatively slow and their rate depends on the physical state of the pesticide (vapor, dissolved, adsorbed), light intensity, photolytic efficiency, and other factors.

The rate of volatilization of pesticides depends on such factors as temperature, soil moisture content, pesticide vapor density, soil properties, water solubility, concentration of pesticides, ambient air humidity, and near surface wind velocity. Factors controlling and mechanisms of volatilization losses in soils have been discussed in a comprehensive review.[49]

After volatilization related to application has occurred, the remaining pesticides ultimately reach the soil. In soil, loss is evident for surface-applied and soil-incorporated pesticides. Although the vaporization rate is related to the vapor pressure of the compound, once a pesticide is in the soil its vapor pressure is modified by environmental variables.

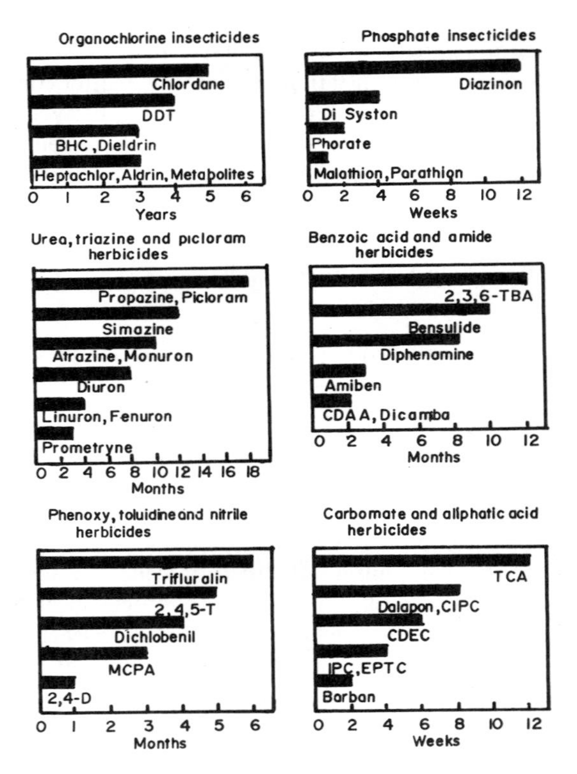

Fig. 6-8. Persistence of pesticides for soils. (Redrawn from Kearney et al.[48])

Field measurements indicate that significant volatilization loss may occur if pesticides are not incorporated in the soil. Losses have been observed for DDT,[50,51] TDE (DDD),[51] dieldrin,[52-54] endrin,[55] lindane,[53] heptachlor,[56] IPC,[57] and CIPC.[57]

In a study of pesticide losses from a watershed cropped to corn,[52,56] it was found that considerable portions of the soil-incorporated dieldrin and heptachlor were lost by volatilization. Losses in one growing season (from an application of 5.6 kg/ha) amounted to 2.8 to 2.9% for dieldrin and 3.9% for heptachlor. Under field conditions, the volatilization process is continuous, although its highest rate occurs immediately following pesticide application.

Many terrestrial and aquatic plants are capable of adsorbing and translocating pesticides,[58] followed by possible detoxification of the compound to less active components.

Several crops are known to adsorb chlorinated insecticides,[59-62] and evidence of metabolic breakdown was opined for DDT, heptachlor, endrin, γ-BHC, and aldrin. Corn—which is resistant to atrazine and simazine—was able to adsorb these herbicides from soils and metabolize them to nonphytotoxic compounds. Dissipation of atrazine from soils through uptake by corn, sorghum, and johnsongrass also has been studied.

It seems that the extent of pesticide detoxification by plants is small, as uptake is limited by the spatial availability of the pesticide and by the sorption capacity of the plant. However, a great need exists to better understand the mechanism of pesticide dissipation through plant uptake. Needed investigations include a search for terrestrial and aquatic plants that are efficient in detoxifying a wide spectrum of pesticides.

Nonbiological processes of pesticide breakdown in soils and sediments have long been recognized. Chemical reactions of pesticides may occur independently of the soil or they may be soil-catalyzed. Extensive coverage of this topic is provided in the comprehensive review by Helling et al.[41]

Investigations indicate that chemical breakdown may play a significant role in the dissipation of soil-adsorbed organochlorine insecticides during dry periods (aerobic conditions), while the dominant degradation mechanism for these pesticides in moist or ponded soils (anaerobic conditions) is microbial. However, much more information is required on the rates of conversion and on the environmental conditions that promote organochlorine insecticide dissipation from the environment.

Several organophosphorous insecticides and herbicides degrade rapidly by chemical hydrolysis. The process is catalyzed by adsorption at soil colloidal surfaces and follows first-order kinetics.[63,64] The longer-lived organophosphorous pesticides–including methyl parathion,[65,66] dimethoate, zinophos, and dursban[65]–are degraded primarily by microbial mechanisms.

Chemical hydrolysis of the 2-chloro-*s*-triazines in soils and sediments has been reported.[63,64] The hydrolysis of atrazine to nonphytotoxic hydroxyatrazine is enhanced by atrazine adsorption possibly to carboxyl groups present on the organic components of soils and sediments.[64]

Although certain pesticides are able to undergo rapid chemical degradation, the degradation products may be more persistent than the parent compound.

Microbial metabolism is considered to be the major pathway of degradation of many pesticides in soils and sediments. The efficiency of this mechanism depends on such environmental factors as temperature, moisture and organic matter content, aeration, pH, and pesticide concentration. Although intensive studies have been made on the mechanisms by which microorganisms degrade pesticides, the processes are not understood clearly.

In general, organochlorine insecticides are the most resistant to microbial attack. For example, partial degradation of DDT results in the formation of TDE (DDD) and DDE. Both components are stable in soils and aquatic systems, and their metabolic fate in these environments remains relatively unknown.

The oxygen status of soils and aquatic environments has a pronounced effect on the microbial breakdown of many organochlorine insecticides. In soils, DDT is rapidly converted to TDE (DDD) under anaerobic conditions and very slowly to DDE under aerobic conditions.[67-69] The conversion of p,p'-DDT to p,p'-DDD has been observed in flooded anaerobic soils and in oxygen-deficient lake water. The time required to convert 50% of the applied DDT to TDE (DDD) was 24 hr in the lake water and 8 weeks in flooded soils. However, TDE (DDD) formed in flooded soils seemed to resist further degradation.

Several organochlorine pesticides–heptachlor,[70] lindane,[71] and endrin[72]–have been shown to degrade in soils to compounds of lower toxicity and reduced insecticidal activity.

As stated earlier, many organophosphorous insecticides are hydrolyzed rapidly by nonmicrobial means. However, for the more persistent types, microbial breakdown may dominate.

Microbial degradation of organochlorine and organophosphorus pesticides in aquatic environments has been reviewed.[73,74] The newer vinyl phosphate insecticides, such as phosphamidon, chlorfenvinphos, and mevinphos, have a half-life of these soils ranging from 1 to 30 weeks.[75] Chlorfenvinphos appears to be the most resistant to bacterological decomposition.

Most agriculturally used herbicides are degraded primarily by microorganisms.[76] Many herbicides are quickly degraded by microorganisms in soils. Compared to organochlorine pesticides, phenoxyalkanoic acids–such as 2,4,5,-T and particularly 2,4-D–degrade rapidly in soil and sediment-water systems.[38] However, anaerobic conditions tend to retard 2,4-D metabolism, indicating the importance of oxygen in the metabolism of this compound.

The *s*-triazines are extensively degraded chemically in soils or sediments, but also are subject to microbiological metabolism.[76] Atrazine, the most widely used *s*-triazine, is quite stable compared with most herbicides and can persist in soil for more than one growing season.

Adsorption of pesticides by clay minerals may protect them from microbial attack. The dipyridyl herbicides–paraquat and diquat–are shown to be metabolized by soil microorganisms, but once adsorbed by clay minerals they become extremely resistant to microbial decomposition.[34]

A model for estimating herbicide persistence in soils with soil temperature and moisture as important variables[77] indicated that under controled conditions loss of some pesticides from soils follows first-order kinetics.

In spite of the enormous amounts of research on pesticide distribution and transport through the environment, no information is available to accurately estimate enrichment ratios for pesticides. It is evident that the enrichment ratio factor for tightly adsorbed pesticides (on the organic matter or clays) will approach that for clays and organic matter per se. Conversely, for pesticides that are mobile in soils and do not adsorb readily, the notion of enrichment ratio is meaningless. McElroy et al.[2] recommend that in such cases loadings should be computed from known stream concentrations of pesticides and from the volume of runoff. An approximate method of computation is shown in the following example.

Example 6-4: Pesticide Loss by Runoff

To control weeds, 5.55 kg/ha of atrazine was applied to a field. Estimate how much atrazine will be lost during a 5-cm storm resulting in 1.5 cm of surface runoff and 1.5 Tonnes/ha erosion soil loss, which occurred 4 weeks after application. Porosity of the soil p is 45%, 0.3 bar moisture content, $\theta_{0.3\ bar}$, *is 35%, 15 bar moisture content,* $\theta_{15\ bar}$, *is 15%, and the organic C content is 2.3%. The antecedent moisture,* θ, *was 31% and evapotranspiration was negligible. Specific density of the soil,* ρ *is* $1.5\ g/cm^3$.

From Fig. 6-7 it can be seen that atrazine persists in soils for approximately 10 months, which implies that during the 4 weeks between application and the storm, 10% of the pesticide was dissipated from the soil. Tables 6-4 and 6-5 show that the mobility of atrazine is intermediate with Freundlich adsorption coefficients of:

$$K = 120 \times (\%\ \text{organic C})/100 = 120 \times 2.3/100 = 2.76$$

and

$$1/n \approx 0.8$$

The depth of penetration of the pesticide in the soil can be approximated as 7.5 cm, with uniform distribution throughout the depth. Although this depth is arbitrary, it has been recommended[78] *as an average estimate of pesticide penetration in agricultural soils. Thus, the pesticide concentration per unit volume of soil is:*

$$P = 0.9 \frac{5.55\ (\text{kg/ha}) \times 10^9\ (\mu\text{g/kg})}{7.5\ (\text{cm}) \times 10^8\ (\text{cm}^2/\text{ha})} = 6.67\ \mu\text{g/cm}^3$$

The distribution between the adsorbed and dissolved phases can be computed from the volumetric mass balance equation:

$$P = S\rho + \theta_{\text{eff}} C = \rho K C^{1/n} + \theta_{\text{eff}} C$$

where:

S = *absorbed pesticide* ($\mu g/g$)
C = *dissolved pesticide* ($mg/liter = \mu g/cm^3$)
ρ = *specific density of the soil* (g/cm^3)
$\theta_{\text{eff}} = \theta - \theta_{15\ bar}$ = *effective moisture as a fraction*

Hence

$$6.67 = 1.5 \times 2.76 \times C^{0.8} + (0.31 - 0.15)C$$

and

$$C = 1.72 \text{ mg/liter}$$

and

$$S = 2.76 \times 1.72^{0.8} = 4.26\ \mu g/g\ (=g/Tonne)$$

Pesticide loss during the storm is estimated assuming that desorption from the adsorbed phase to solution is negligible. In this case, the adsorbed pesticide loss becomes (assume that the estimated ER for organic matter is ≈1.2)

$$Y_{PA} = S \times ER \times Y_{ss} = 4.26\ (g/Tonnes) \times 1.2 \times 1.5\ (Tonnes/ha)$$
$$= 7.67\ g/ha$$

If the pesticide was immobile and tightly bound to the soil organic matter, the above amount would represent the approximate pesticide loss during the storm. However, atrazine has intermediate mobility, and in this case the dissolved fraction must be estimated as well. As a first approximation assume that the storm water is completely mixed with the moisture in the upper soil layer (upper 7.5 cm). Then the mass balance equation becomes:

$$7.5\ (cm) \times \theta \times C = C'(5 + 7.5\theta)$$

from where the average pesticide concentration during and after the storm is:

$$C' = \frac{7.5 \times 0.31 \times 1.72}{5 + 7.5 \times 0.31} = 0.55\ mg/liter\ (=\mu g/cm^3)$$

and the dissolved pesticide loss in the runoff becomes

$$Y_{PD} = 1.5\ (cm) \times 10^8\ (cm^2/ha) \times 0.55\ (\mu g/cm^3) \times 10^{-6}\ (g/\mu g)$$
$$= 81.85\ g/ha$$

Relating pesticide loss only to erosion loss could lead to results that would grossly underestimate pesticide loss.

Since only gravitational water remains in the soil after the storm, the excess water will be leached into the lower soil layer. The excess infiltration amounts to

$$5 - 1.5 + 7.5 \times \theta - 7.5 \times \theta_{0.3\ bar} = 3.2\ cm$$

With this amount of excess water the amount of pesticide leached downward is:

$$3.2\ (cm) \times 10^8\ (cm^2/ha) \times 0.55\ (g/cm^3) \times 10^{-6}\ (g/\mu g) = 176\ g/ha$$

This loss has the potential to contaminate groundwater.

In the poststorm period, the adsorbed atrazine concentration is reduced and a new equilibrium with the dissolved component is established.

The computational method presented in this example is crude and should be considered only as a first approximation. More sophisticated modeling techniques will be discussed in Chapter 9.

Polychlorinated Biphenyls

As noted in the preceding sections, the polychlorinated biphenyls (PCBs) are alien to nature, and no natural (background) PCB concentrations in soils exist. Furthermore, a great majority of primary sources of PCBs are confined to urban and industrial areas. Therefore, concentrations of PCBs in rural soils are mostly below detectable levels in spite of measurable global atmospheric deposition of these substances. Of the 1556 soil samples collected in 1972 as a part of the National Soil Monitoring Program, only two samples from rural areas contained detectable levels of PCBs. On the othe hand, over 60% of urban soils sampled during the program showed detectable PCB contamination.[79]

The reactions of PCBs with soil are still not quite understood and literature data are sparse. PCBs are almost conservative components that are not readily decomposed by soil bacteria. Their removal from soil is primarily by volatilization or by biomodification of lower PCBs. Due to their hydrophobic nature they are not mobile in soils and are mostly tightly bound to soil particles. Their persistence in soils is expected to be of the same order of magnitude as for some of the most persistent organic chemicals such as organochlorine pesticides. When PCBs are absorbed on soil particles their loss by volatilization is greatly reduced.[80]

Soils with higher organic content exhibited the highest absorption of PCBs followed by clays. Sandy soils do not indicate significant adsorption capacity.[80] The values of K and $1/n$ of the Freundlich adsorption isotherm for illite clay and Woodburn silt loam were measured as 63.1 and 1.1, and 26.3 and 0.81, respectively.[80]

Soils containing detectable amounts of PCBs are often located near waste areas such as landfills, dumps, and areas with highest densities of items containing dissipative PCBs,[79] or near incinerators.

SOIL NITROGEN

Accumulation of Nitrogen in Soils

Nitrogen (N) is one of the four essential elements (carbon, oxygen, hydrogen, and nitrogen) that form the basic structure of proteins. Nitrogen also is the most abundant gas in the atmosphere, accounting for about 80%. Due to its several valence states it can exist in numerous forms and is abundant in the mineral and organic fractions of soil. However, not all N forms are related to water pollution or eutrophication of water bodies.

Soils contain 0.07 to 0.3% of total N or about 1500 to 6000 kg/ha in the top 15 cm. Nationally, almost 14 million Tonnes of N enter soils from various sources of which fertilizer applications account for 46%.[33,81] Other sources include N fixation from the atmosphere (20%), manure application (7%), plant residues (17%), and precipitation (10%). In suburban areas, significant amounts of N enter soils from the seepage fields of household septic systems.

Jenny[81] noticed that the nitrogen content of native soils depends on average meteorological factors of the region, namely moisture and temperature, soil nitrogen content declines as the temperature rises or moisture declines.

Although virgin lands rich with vegetation (forested areas, swamps, prairies) originally contained high amounts of N, on cultivation the natural supply of N is depleted and fertilization may be needed to supplement soil loss of N.

Most soil N ($>90\%$) is contained in soil organic matter, or in the case of ammonium ions, it can be sorbed by clays. In these forms, N may be considered immobile and not available to plants. Nitrogen is lost mostly by erosion and through the harvesting of crops. Only nitrates and to a certain degree ammonium ions are available to plants, and can be transported by soil water and infiltrate into groundwater. The time of migration of mobile N components in groundwater aquifers can extend to years and even decades.

Volumes of literature have been published on N behavior in soils and in the environment. Two particularly comprehensive reviews have been published.[82,83]

Nitrogen Cycle

The behavior and transformation of N in soils is complex, and the pathways from the soil to surface waters are numerous and not well defined. The processes of N transformations in soils can be schematically represented by the N cycle similar to that in Fig. 6-9.

In soils, N exists in four basic forms: ammonium, nitrate, organic phytonitrogen in plants and plant residues, and protein nitrogen in living and dead bacteria and small soil inhabitants. The following definitions are needed for an overall understanding of the N cycle:

Nitrogen fixation is a process by which soil microorganisms in symbiosis with some leguminous plants utilize atmospheric N and change it to an organic form.

Nitrogen accumulation (bacterial uptake) is the conversion of ammonium-N to protein and cell tissue by heterotrophic soil microorganisms.

Ammonification is a process by which protein and other organic forms of N are decomposed to ammonium.

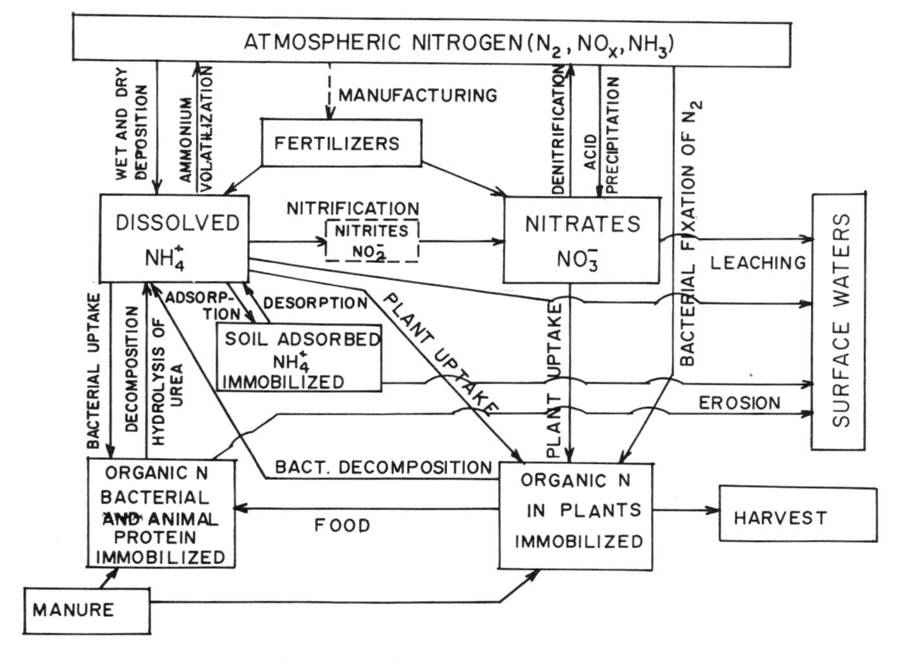

Fig. 6-9. Nitrogen cycle in soils.

Hydrolysis of urea involves conversion to ammonium ions in the presence of the enzyme urease, which is provided by many heterotrophic microorganisms.

Nitrification is a complex process by which ammonium-N (NH_4^+-N) is oxidized to nitrate (NO_3^-) with nitrite (NO_2^-) as an intermediate product. Nitrification is accomplished by two groups of bacteria: namely *Nitrosomonas* (oxidizing NH_4^+-N to NO_2^--N) and *Nitrobacter* (oxidizing NO_2^--N to NO_3^--N). Both groups of bacteria are autotrophic, utilizing the exothermic reaction as their source of energy and carbon dioxide from the atmosphere as their carbon source.

Denitrification is a process that occurs under reduced oxygen supply, that is, when water fills most of the available voids and pores in the soil. In this transformation, NO_3^--N serves as an electron acceptor and is reduced to gaseous nitrogenous forms including N_2, N_2O, NO, and NO_2.

Fixation of ammonium involves the sorption of NH_4^+-N in between the layers of expanding clay minerals such as montmorillonite. In this form, NH_4^+ is considered unavailable for plant growth or bacterial uptake.

Ammonia volatilization may occur at high soil pH values when the ammonium ion (NH_4^+) is converted to gaseous ammonia (NH_3), which volatilizes and is lost to the atmosphere.

Most of the reactions of the N cycle in soil are microbial and thus their rates are sensitive to temperature, moisture content, and aeration. Warm (32°C) and partially moist soils (80% of voids filled with water) are the optimum conditions for N cycling in soils.[33]

A tendency exists to associate suspected N contamination of surface waters with increased fertilizer use. However, as the foregoing discussion reveals, the behavior of N in soils is complex and all processes must be studied before conclusions on possible N pathways and sources from soils to surface waters can be drawn. As stated by Stanford et al.[84] the use of almost 7 million Tonnes of fertilizer N in 1969 was still insufficient to compensate for the large yearly drop in the capacity of soils to supply N for crop production. Nevertheless, there is no doubt that a portion of this N loss reached surface waters. Unlike phosphorus (P), most of which is adsorbed on soil particles and can be controlled by erosion control, the available N is mobile in soils and might leach to groundwater. Seepage represents the major pathway of nitrogen loss from agricultural areas and pervious urban and suburban lands.

Ammonium adsorption by soils has not been considered a major path of N in soil. However, in areas of high ammonia concentrations, such as downwind of industries, feedlots, or sewage sludge applications, soils may contain significant amounts of clay-fixed NH_4^+.[33]

Since the mobile components—NH_4^+ and NO_3^-—are carried by soil water, downward or lateral N movement occurs only if the moisture content is above that for gravitational water (0.3 bar tension). Maximum movement occurs when soil moisture content is near saturation and decreases rapidly with decreasing moisture content. Thus, between rains or irrigations, N movement is slow or nonexistent.

Nitrification and Denitrification. Over 90% of fertilizer used in the United States is in the form of ammonium salts.[84] Most organic N originating from manure application or from septic tanks can be quickly decomposed to ammonium. In addition, urea is readily hydrolyzed to ammonium. If the ammonium is applied to an aerated, microorganism-rich soil such as farm land, nitrification occurs resulting in conversion of NH_4^+ to NO_3^-. In aerated soils, the nitrification process proceeds as follows:

$$\text{Organic N} \rightarrow NH_4^+ \xrightarrow[O_2]{\textit{Nitrosomonas}} NO_2^- \xrightarrow[O_2]{\textit{Nitrobacter}} NO_3^-$$

The last reaction, that is, conversion of NO_2^- to NO_3^-, is faster than conversion of NH_4^+ to NO_2^-. Consequently, very little nitrite accumulates in soil.

Nitrates can readily move with the soil moisture front. The optimum temperature for nitrification is 22°C (Fig. 6-10), and the rate of nitrification decreases rapidly on both sides of the temperature curve. Stanford et al.[85] state

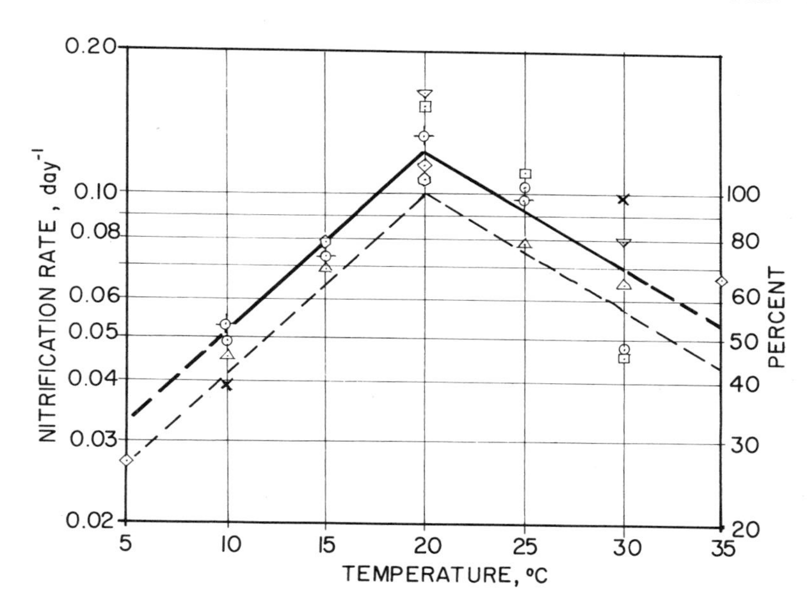

Fig. 6-10. Dependence of nitrification rate on temperature. (From Zanoni.[86])

that nitrification essentially ceases below 10°C and above 45°C. The nitrification rate depends also on the pH of the soil and its moisture content (see Figs. 6-11 and 6-12). Since the nitrifying bacteria depend on water as their living environment, nitrification rate decreases with decreasing moisture content (Fig. 6-12).

By contrast with many heterotrophic microorganisms, the growth of nitrifiers is slow and cell yield per unit of energy source oxidized is low. For practical

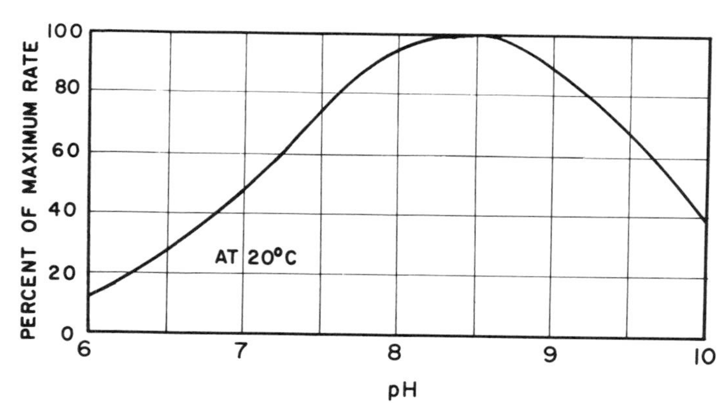

Fig. 6-11. Effect of pH on the rate of nitrification. (From Wilde et al.[87])

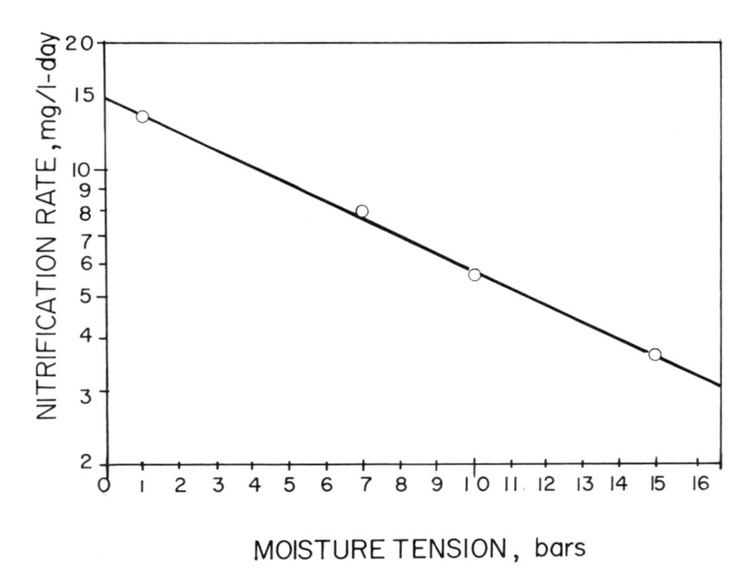

Fig. 6-12. Dependence of the soil nitrification rate on the moisture tension. (Data from Justice and Smith.[88])

purposes the NH_4^+ oxidation in soils can be simulated by an equation similar to the Michaelis-Menten equation:

$$-\frac{dC}{dt} = \frac{K_a C}{K_s + C} \tag{6-19}$$

where:

C = concentration of the substrate (NH_4^+)
K_a = maximum rate of nitrification
K_s = half-saturation concentration (50% saturation)

The values of K_a and K_s for soils were suggested to be 0.41 and 1.1, respectively[89,90] if rate changes are expressed in milligrams per liter in soil water per day at 20°C.

In the absence of oxygen or if the oxygen supply is depleted to a point below the oxygen demand, NO_3^- is reduced mostly to gaseous nitrogenous forms (largely N_2).

The process of denitrification occurs usually in subsoils with low permeability or in soils saturated with water for extended periods. Denitrification accounts for about 30% of average N losses in Illinois.[84] For soils with clayey horizons overlying sands, N losses by nitrification may be greater than 50%.[91]

Carter and Allison[92] concluded that very little, if any, N loss occurred under

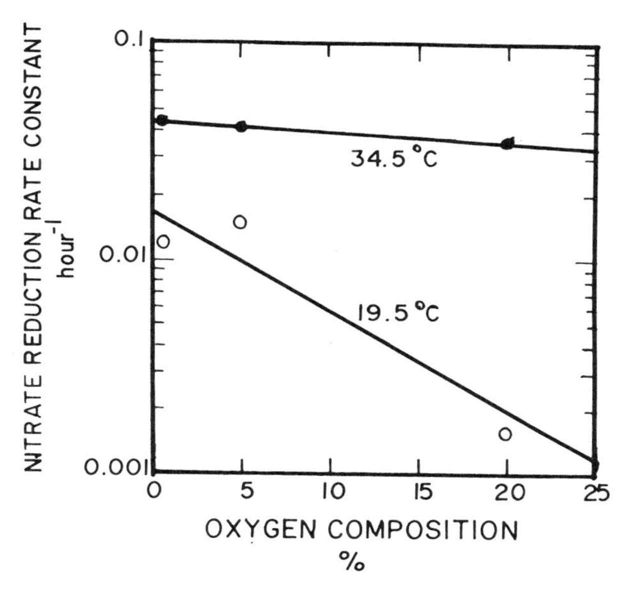

Fig. 6-13. Dependence of denitrification rate on oxygen composition and temperature. (From *Soil Science Society of America Proceedings*, 38:300–304, 1974.[94] Reprinted by permission of the publisher—Soil Science Society of America, 677 South Segoe Road, Madison, Wisconsin, USA 53711.)

aerobic conditions except where higher dosages of dextrose or other carbon-rich compounds were added in the presence of nitrates. They also stated that denitrification is of minor importance in soils that are maintained aerobic. However, it was emphasized that in heavy-textured soils, and in medium-textured soils following heavy rains, there were often periods of several hours or days when normal field soils were not well aerated.

The denitrification reactions are caused by heterotrophic soil and soil water bacteria and have been described by a first-order reaction[93,94]

$$\frac{d(NO_3^-)}{dt} = -K(NO_3^-) \tag{6-20}$$

The values of the denitrification coefficients were measured for sandy and silt loam soils.[94] No apparent effect of texture could be detected from the measured data. Figure 6-13 summarizes the results of the denitrification experiments.

The denitrification rate is a function of easily decomposable carbon and is expected to decrease with a decrease in the available energy source.

Nitrogen Fixation. Legumes (soybeans, peas, beans) fix nitrogen from the atmosphere by symbiotic microorganisms living on their roots. It has been assumed that the rate of nitrogen fixation can be related to root growth[95]

$$N_f = K_f r_g \tag{6-21}$$

where:

N_f = rate of fixation of N (mg N/day-cm^2)
K_f = a constant (0.011 mg N/cm^3)
r_g = the rate of root growth (cm/day)

Ammonification of Nitrogen. Three ways of producing NH_4^+ from organically bound N can be distinguished:[96]

1. From extracellular organic N compounds (e.g., urea) chemically or biochemically.
2. From living bacterial cells during endogenous respiration when cells are becoming smaller.
3. From dead and lysed cells.

Very little data are available on the breakdown of various N-containing components to NH_4^+. Complex organic compounds are (apparently) initially deaminated by appropriate exoenzymes or enzymes on the cells, and the ammonium is transported into the cell where it is used in synthetic reactions.

McLaren[89] reported that in a number of soils, urea decomposed at a maximum rate of 0.02 to 0.15 moles/g of soil-hour, which, on average, is equivalent to 34 mg/g of soil-day. This would lead to the following equation for the decomposition of urea:

$$-\frac{d(U)}{dt} = \frac{34(U)}{500 + (U)} \tag{6-22}$$

The rate of hydrolysis of urea varies between soils and is a temperature-dependent function. It proceeds at a much faster rate than nitrification, and due to the large half-saturation constant, K_s, it usually follows a first-order kinetics.

The process of hydrolysis of urea is important where manure or urea-containing sewage is applied to a field or land.

Soil organic N—other than urea—breaks down at a much slower rate than urea. Most of the organic N is not directly available to plants and must be initially converted to NH_4^+ and NO_3^-. Stanford et al.[85] concluded that in the mineralization of soil organic N other than urea, the rate-limiting process is ammonification. A first-order reaction fairly describes the process:

$$\frac{dN}{dt} = -KN \tag{6-23}$$

where:

N = the concentration of the organic N remaining in the soil
K = a rate coefficient

The coefficient, K, was found statistically uniform for all soils investigated[85] and had a value of (0.29 ± 0.007) weeks^{-1} at 25°C. As it is in all biochemical reactions, the coefficient, K, is temperature dependent.[85,97] An Arrhenius' plot yielded the following equation for the overall mineralization organic N:[85]

$$\log K = 6.16 - \frac{2299}{T} \tag{6-24}$$

where the temperature, T, is in degrees Kelvin (= 273 + °C).

Nitrogen Immobilization. Part of the soil N is dissolved and can move readily with soil moisture and groundwater. The dissolved fractions include NO_3^- and NH_4^+. But a significant portion of N components, especially NH_4^+, and most of the organic N can be immobilized. Nitrogen immobilization in soils results from physical-chemical attraction, chemical precipitation, and biochemical reactions. The known nitrogen fixation processes include:[84]

1. Ammonium fixation by clay minerals.
2. Ammonium fixation by lignin-derived substances contained in soil organic matter.
3. Reactions of amino acids derived from plant materials and microbial synthesis with quinones and subsequent polymerization.
4. Biological immobilization, which involves NH_4^+ uptake by heterotrophic bacteria participating in the decay of organic matter in soils.

The immobilized N is not readily available for mineralization and it appears that the deamination, that is, liberation of NH_4^+ from the fixed N, is much slower than the rate of nitrification.

Immobilization and deamination of soil N depend on the chemical composition of the material undergoing decomposition, primarily on its C:N ratio. Plant residues having large percentages of readily available C will stimulate the growth of microbial cells when incorporated in the soil under aerobic conditions.

Soil itself has an affinity to adsorb NH_4^+, and there is an impression that NO_3^- does not react with soil. Thomas,[98] however, states that most soils—with the exception of sands of low organic matter content—possess positively and negatively charged sites. An example is a soil in which Fe oxides thickly coat the negatively charged clay particles. The positively charged sites will react with NO_3^- and retard its progress through the soil as water moves. In most soils that have bright red colors, the movement of NO_3^- is substantially slowed.

Soils with a very dense negative charge on their surfaces do not only fail to attract NO_3^-, they actively repel it.

Shaffer et al.[99] statistically analyzed the rate of NH_4^+ immobilization by multiple regression technique. The proposed equation was:

$$R = \frac{d(NH_4^+)}{dt} = 0.892 - 0.00216T - 0.027\,(\text{organic N}) + 0.392\log_{10}(NH_4^+) \tag{6-25}$$

where T is temperature.

Preul and Schroepfer[100] investigated N adsorption by soils. The Freundlich adsorption isotherms for three soils are shown in Fig. 6-14.

Ammonium is adsorbed by clays in exchangeable form and by organic matter in calcareous soils.[101] Much of the NH_4^+ adsorbed by various kinds of organic matter is in a nonexchangeable form and is resistant to decomposition.

Volatilization of Ammonia. For soils of high pH and NH_4^+ content, ammonia (NH_3) can be lost by volatilization. Significant amounts of NH_4^+ (11 to 60%) can be lost from applied sludge, manure, or chemical fertilizers applied to calcareous soils.[32,102,103] Losses decrease as clay content increases and depend on the application rates, soil pH (can be correlated to clay content to which it is inversely proportional), moisture and NH_4^+ contents, and possibly other factors. Due to its effect on pH, liming of soils may enhance NH_3 volatilization.

Enrichment and Delivery of Nitrogen During Overland Flow. The enrichment and delivery of N from the soil of origin to the receiving surface waters by surface runoff depends on the form of N in the soil. Organic N as well as N adsorbed on organic matter will have enrichment ratios similar to that for organic matter. Ammonium can exist in dissolved and adsorbed forms, and higher enrichment ratios of this component can be expected. The notion of enrichment ratio is meaningless for NO_3^-.

Reported values of ER for N are 2 to 4.[4,22,104] The ratios of available N (including NH_4^+ and NO_3^-) rarely exceed 15%.[2] The suggested average value of available N : total N is about 8%.

Example 6-5: Distribution Between Mobile and Immobile N in Soil

A silt soil contains about 2.5% organic matter and 0.12% total N measured as TKN (total Kjeldahl N = organic N + NH_4^+-N) of which about 75% is organic N. Estimate the approximate magnitudes of mobile (available) and immobile N. Effective soil moisture $\theta_{eff} = 20\%$ and specific density of the soil $\rho = 1.5$ g/cm^3.

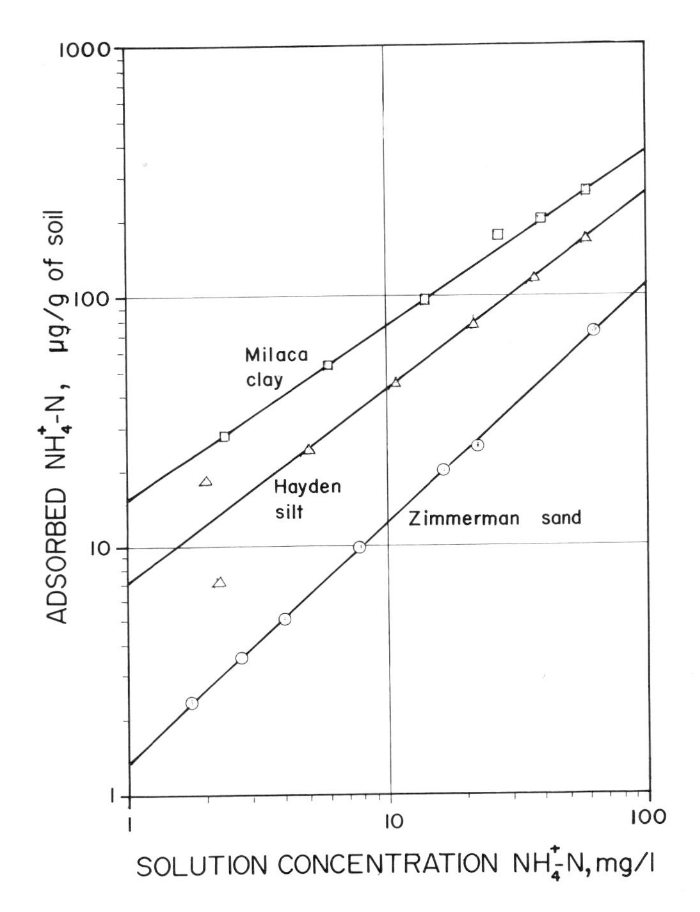

Fig. 6-14. Freundlich adsorption isotherm for ammonia on soils. (Source: H. C. Preul and G. J. Schroepfer, "Travel of Nitrogen in Soils," *J. Water Poll. Control Fed.*, **40**:30–48, 1968.[100])

From Fig. 6-14, the Freundlich adsorption isotherm for silt has the approximate form:

$$S = 7.0 \times C^{0.8}$$

Organic N constitutes part of the soil organic matter and, for the most part no equilibrium exists between dissolved and particulate fractions. The dissolved fraction of N is mostly NO_3^- and NH_4^+. Ammonium can exist in both dissolved and adsorbed phases. Since total NH_4^+ is 0.03% or 300 µg/g, the mass balance of exchangeable N per unit soil volume is:

$$300 = \rho S + \theta_{\text{eff}} C = 1.5 \times 7.0 \times C^{0.8} + 0.2 \times C$$

Solving these equations gives:

$$C = 62.63 \text{ mg/liter} \quad \text{and} \quad S = 7.0 \times 62.63^{0.8} = 191.66\ \mu\text{g/g}$$

The ratio of available NH_4^+ *to TKN is:*

$$100 \times [62.63 \times 0.2/(1200 \times 1.5)] = 0.7\%$$

Example 6-6: Nitrogen Loss Estimation

A 3-cm storm resulted in 0.5 cm of surface runoff and 1.3 Tonnes/ha sediment loss. Estimate N loss using the N concentrations computed in the previous example and evaluate the approximate enrichment ratio for N assuming that the enrichment ratio for organic matter has been estimated previously at $ER_{or} = 1.2$. *Porosity of the soil* $p = 45\%$ *and* $\theta_{15\ bar} = 10\%$.

The fixed N loss is

$$Y_{SN} = [(0.75 \times 1200 + 191.66)] \text{ (g/Tonne)} \times 1.3 \text{ (Tonnes/ha)} \times 1.2$$

$$= 1703 \text{ g/ha}$$

In order to determine the loss of dissolved N, the depth of mixing of the surface runoff with soil moisture containing the dissolved fraction must be known. In most cases, this depth and the extent of mixing depends on the intensity of the storm, surface storage characteristics, vegetative cover, and the stage of hydrograph when surface runoff occurs. For this example assume that the surface runoff penetration and mixing with the soil is approximately twice the depression storage (includes all interrill soil). If the surface depression storage for a typical field is 3 cm, then the dissolved N balance becomes

$$(\theta_{eff} + \theta_{15\ bar}) \times 2 \times 3 \text{ (cm)} \times C = [(\theta_{eff} + \theta_{15\ bar}) \times 2 \times 3 \text{ (cm)} + 3 \text{ (cm)}]\ C'$$

from where

$$\frac{C}{C'} = 1 + \frac{3}{(\theta_{eff} + \theta_{15\ bar}) \times 2 \times 3}$$

$$= 1 + \frac{3}{(0.2 + 0.1) \times 2 \times 3} = 2.67$$

and the approximate dissolved N concentration in the surface runoff is then:

$$C' = \frac{C}{2.67} = \frac{62.63}{2.67} = 23.46 \text{ mg/liter } (=\mu\text{g/cm}^3)$$

and the dissolved N loss in the surface runoff

$$Y_{DN} = 23.46\ (\mu\text{g/cm}^3) \times 10^{-6}\ (\text{g}/\mu\text{g}) \times 0.5 \text{ (cm)} \times 10^8 \text{ (cm}^2\text{/ha)}$$

$$= 1173 \text{ g/ha}$$

The total loss of N is:

$$Y_N = Y_{SN} + Y_{DN} = 1703 + 1173 = 2876 \text{ g/ha}$$

and the ratio of N loss: sediment loss gives the enrichment ratio when divided by the original N concentration in the soil. Hence

$$ER_N = \frac{2876 \text{ (g/ha)}}{1.3 \text{ (Tonnes/ha)}} \times \frac{1}{1200 \text{ (g/Tonne)}} = 1.84$$

Example 6-7: Mineralization of Ammonia

How much N will be mineralized during a 2-week dry period if the temperature is 22°C and the soil pH is 6.8? Assume average soil moisture tension close to 0.3 bar. Use the N values estimated in the previous two examples.

Only dissolved ammonia can be mineralized. Adsorbed or organic N first must be released into the solution before it can be mineralized.

From Equation 6-19 it can be seen that nitrification will apparently follow a zero-order reaction pattern and will—for most of the time—not depend on the NH_4^+ *concentration due to the high* NH_4^+ *concentration in the soil solution. Then*

$$\frac{dC}{dt} = \frac{0.41 \times 62.63}{1.1 + 62.63} = 0.41 \text{ mg/liter-day}$$

Since the temperature is near optimum, no thermal correction is necessary as indicated from Fig. 6-10. However, Fig. 6-11 shows that at pH 6.8 the reaction will proceed only at about 50% of the optimal rate. Since the moisture tension is near 0.3 bar, there is also no correction for moisture. Then

$$\frac{dC}{dt} = 0.2 \text{ mg/liter-day}$$

and in 14 days about 2.8 mg/liter of the dissolved NH_4^+ *will be mineralized to* NO_3. *This form of N can be taken up by plants, move with the soil water front, and/or be lost to the atmosphere as a result of denitrification during periods of high soil moisture or flooded conditions.*

SOIL MICROORGANISMS

As stated before, soils contain enormous densities of bacteria that decompose soil organic materials and/or are an integral part of the soil organic composition. Most of the soil microbial population is contained in the A horizon.

Soil contains five major groups of microorganisms:[105] bacteria, actinomycetes, fungi, algae, and protozoa. These organisms as well as their organic and inorganic food are a part of the soil ecosystem.

The bacteria are the most abundant group of microorganisms. They exist in the largest numbers and perform most of the decomposing processes in the soils. The bacteria in soils can be divided into two broad groups: autochthonous species that are true residents of soils and allochthonous species or invaders that have entered soils from precipitation, manure application, septic tank effluents, etc. The autochthonous species are the most important in the soil decomposition processes. The latter group of bacteria–allochthonous–may persist for some time but do not contribute significantly to any activity of the soil microflora.

The plate numbers indicate bacterial densities in soil in ranges from several thousands up to 200 million of bacteria per gram of soil. However, microscopic counts have revealed that the microbial densities can be higher, probably up to 10^{10} of microorganisms per gram of soil.[105] The bacterial population represents between 0.015 and 0.05% of the total mass of fertile soils, or some 300 to 3000 kg live weight of soil bacteria per hectare.

Soil bacteria are also divided according to taxonomic, morphological, and physiological categorizations. Taxonomy (classification of species) of soil bacteria is beyond the scope of this treatise. However, many bacterial and viral indicators of pollution are commonly found in soils both as allochthonous and autochthonous species.

Bacteria in soils can be divided into aerobic, anaerobic, and facultative groups. Aerobic organisms derive their energy from organic carbon sources and use soil O_2 as an electron acceptor. Some other organisms–strict anaerobes–grow only in the absence of oxygen. Most of the soil bacteria are facultative, which means that they can exist and grow in both aerobic and anaerobic environments.

Microorganisms can also be divided into heterotrophic and autotrophic forms according to their energy and carbon sources. Heterotrophic microorganisms require organic carbon as a source of energy while autotrophic microorganisms assimilate CO_2 as a sole carbon source. Autotrophs are of two types: photoautotrophs, which derive their energy from sunlight, and chemautotrophs, which use an inorganic exothermic reaction as their source of energy. *Nitrosomonas* and *Nitrobacter*, which derive their energy from the oxidation of ammonia to nitrate, are examples of soil chemoautotrophs. Similarly, *Thiobacillus* in soils derives energy from the conversion of inorganic sulfur to sulfate, or from the conversion of ferrous iron to ferric form.

Most of the bacteria in soils are facultative heterotrophs. During dry, well-aerated conditions these bacteria decompose organic matter as aerobes. During flooding or high moisture conditions, when O_2 is quickly depleted by the intensive decomposition processes, the facultative bacteria or strict anaerobes use other compounds as an electron acceptor, including NO_3^-, SO_4^{2-}, and Fe^{3+} ions.

The final product of bacterial decomposition is CO_2. Thus, the atmosphere in the soil during intensive decomposition processes has depeleted O_2 and increased CO_2 contents. This results in the increase of the partial pressure of CO_2 in the soil atmosphere which in the air above ground level is about 10^{-3} bar while partial pressure of CO_2 in soils can reach 10^{-1} bar.

Some groups of soil anaerobes, such as those that fix N from the atmosphere, decompose cellulose, form methane, or reduce nitrate to gaseous N, are of special interest to those interested in land disposal of wastes. In a few instances, such as DDT conversion to DDD, the reaction proceeds at a faster rate under anaerobic conditions. However, as a rule, aerobic decomposition is more rapid and complete for most other organic and inorganic biodegradable components.

Pathogenic Microorganisms in Soils

Various pathogenic (disease-causing) microorganisms—bacteria and viruses—can be found in soil. Their survival in soils depends on many environmental factors and, often, on the availability of an acceptable host organism. Because of the frequent contamination of soils with animal droppings, manure, sewage containing disease agents, wastewater effluents applied on land, and tissues of diseased plants, the survival of these microorganisms in soils has been investigated frequently.[105-108] References 109 and 110 contain comprehensive discussions on the survival rates of pathogenic microorganisms from sewage applied on land.

Bacteria. Bacteria may survive in soil for a period from a few hours to several months, depending on the type of organism, type of soil, moisture field capacity of soil, moisture and organic content of soil, pH, temperature, sunlight, rain, and predation of the resident microflora of the soil. In general, enteric bacteria persist in soil for 2 to 3 months; however, under certain favorable conditions, pathogenic microorganisms may actually multiply and increase in numbers.[109,110] Factors that influence the survival of bacteria in soil are listed in Table 6-6. Average survival times of selected pathogens are presented in Table 6-7.

The survival of *E. coli*, *S. typhi*, and *M. avium* is greatly enhanced in moist rather than in dry soil. Survival time is less in sandy, permeable soil than in soil with greater water holding capacity such as clayey soil and peat.[109,110]

Both pathogens and indicator organisms survive longer under lower temperatures. Figure 6-15 shows typical die-off curves for fecal coliforms and fecal streptococci under summer and winter conditions.

The competition and predation by resident soil bacteria is another important factor. Organisms invading sterilized soil survive longer than they would in unsterilized soil.

Bacteria move in the soil with the soil water. However, fine soil particles can effectively adsorb bacteria and soil, in general, is a very effective filtrating

TABLE 6-6. Factors that Affect Survival of Enteric Bacteria and Viruses in Soil.[a]

Factor		Comment
pH	Bacteria	Shorter survival in acid solids (pH 3 to 5) than in neutral and calcareous soils
	Viruses	Insufficient data
Predation by soil microflora	Bacteria	Increased survival in sterile soils
	Viruses	Insufficient data
Moisture content	Bacteria and viruses	Longer survival in moist soils and during periods of higher rainfall
Temperature	Bacteria and viruses	Longer survival at low temperatures
Sunlight	Bacteria and viruses	Shorter survival at the soil surface
Organic matter	Bacteria and viruses	Longer survival or regrowth of some bacteria when sufficient amounts of organics are present

[a]From reference 110.

TABLE 6-7. Survival of Selected Pathogens in Soils.[a]

Organism	Range of Survival Time
Salmonella	15 to more than 280 days
Salmonella typhi	1 to 120 days
Tubercle bacilli	More than 180 days
Entamoeba histolytica cysts	6 to 8 days
Enteroviruses	8 days
Ascaris ova	Up to 7 days
Hookworm larvae	42 days

[a]After reference 111.

medium. Bacteria are more mobile in sandy soils under high moisture conditions than in loamy or clayey soil. In fine-textured soil, bacteria can be filtered out by 1 to 2 m of soil. Soils containing clay remove most microorganisms through adsorption. Sandy soil removes them through filtration.[107]

Viruses. Viruses in soil are also considered invaders. There is little information on the survival of enteric viruses in the soil. Loehr et al.[109] summarized the available information on virus survival as follows:

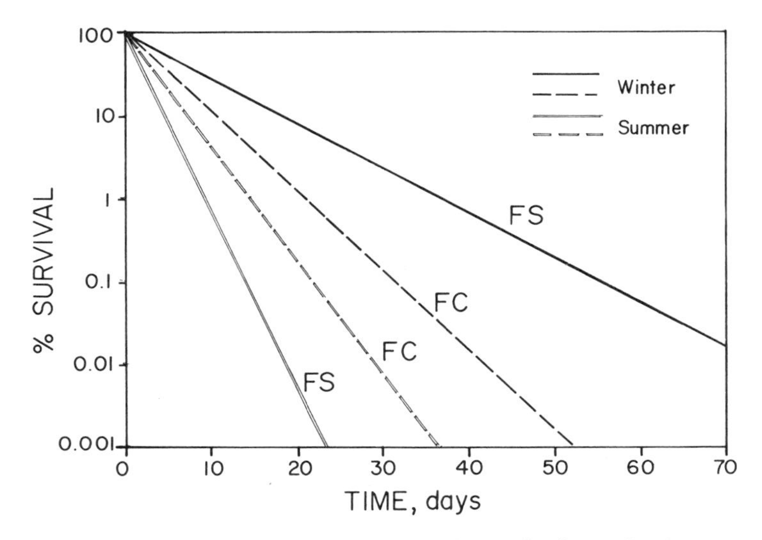

Fig. 6-15. Typical die-off curves for fecal coliforma (FC) and fecal streptococci (FS) under winter and summer conditions. (From Von Donsel; redrawn from reference 109.)

1. Virus survival in soil depends on the nature of the soil, temperature, pH, moisture, and possibly anatagonism from soil microflora.
2. Viruses readily adsorb to soil particles. Such viruses bound to solids are as infectious as free viruses.
3. Viruses survive for times as short as 7 days and as long as 6 months in soil. Climatic conditions, particularly temperature, have a major effect on survival time.
4. Enteric viruses can survive from 2 to more than 188 days in fresh water, temperature being the most important factor, with survival greater at lower temperatures.
5. Virus survival on crops is shorter than in soil, because viruses are more exposed to deleterious environmental effects.
6. Contamination of crops most commonly occurs when wastewater comes into contact with the surface of the crop.
7. In rare cases, the translocation of animal viruses from roots of the plants to the aerial parts can occur.
8. Sunlight is believed to be a major factor in killing viruses.
9. Viruses cannot reproduce at all in the soil, and they slowly die off.

Unlike bacteria, for which filtration appears to be the main factor limiting their movement through the soil, viruses are most effectively removed by

TABLE 6-8. Soil Adsorption Characteristics for Viruses-Freundlich Adsorption Isotherm.[a]

Soil	pH	Cation Exchange Capacity (meq/100 g)	Specific Surface Area (m^2/g)	Organic Carbon (%)	Freundlich Adsorption Coefficients[b]	
					K_F	$1/n$
Aastad clay loam	6.9	35.2	160	3.02	72.5	0.945
Kranzburg silt loam	6.2	31.4	154	2.38	161	0.908
Palouse silt loam	6.0	22.6	89.2	1.82	45.7	1.24
Parshall silt loam	6.8	13.7	67.6	1.32	4.6	0.916
Qunicy loamy sand	7.2	8.9	35.3	0.35	–	–

[a]From reference 114, by permission of ASA, CSSA, and SCS.
[b]S in number of viruses per gram of soil, C in number of viruses per milliliter.

adsorption on soil particles. The process of adsorption is strongly affected by soil pH. The best adsorption of viruses is achieved at pH 7, but it decreases on both acid and alkaline sides of the pH scale.[110] Adsorption of viruses by soil does not mean their complete immobilization since desorption of viruses can occur when pH or other environmental conditions change. Drewry and Eliassen[112] showed that virus adsorption follows the Freundlich adsorption isotherm (Table 6-8). Viruses can be desorbed from soil particles by heavy rains.

Scheuerman et al.[113] showed that in organic soil, water-soluble humic substances interfere with the sorptive capacity of soil towards viruses. Wetland soil such as peat and mulch were not effective in retaining viruses, and significant viral concentrations were observed in the leachate after land disposal of sewage. Sandy soils with little water-soluble humic substances retained most or all of the virus applied.

REFERENCES

1. Donigian, A. S., and Crawford, N. H. 1976. Modeling non-point pollution from the land surface. U.S. EPA/600/3-76/083, Washington, D.C.
2. McElroy, A. D., Chiu, S. Y., Nebgen, J. E., Aleti, A., and Bennett, F. W. 1976. Loading functions for assessment of water pollution from non-point sources. U.S. EPA/600/2-76/151, Washington, D.C.
3. Massey, H. F., and Jackson, M. L. 1952. Selective erosion of soil fertility constituents. *Soil Sci. Soc. Proc.*, **82**:353–356.
4. Shacklette, H. T., Hamilton, J. C., Boernagen, J. G. and Bowles, J. M. 1971. Elemental composition of surficial materials in the conterminous United States. U.S. Geological Survey Prof. Pap. 574-D, Washington, D.C.

5. Green, P. B., Logan, T. J., and Smeck, N. E. 1978. Phosphate adsorption-desorption characteristics of suspended sediments in the Maumee River Basin of Ohio. *J. Env. Quality*, **7**:208–212.
6. McCallister, D. L., and Logan, T. J. 1978. Phosphate adsorption-desorption characteristics of soils and bottom sediments in the Maumee River Basin of Ohio. *J. Env. Quality*, **7**:87–92.
7. Young, R. A., and Onstad, C. A. 1976. Predicting particle-size composition of eroded soil. Paper 76-2052 presented at the 1976 ASAE meeting, Lincoln, Nebraska.
8. Uttormark, P. D., Chapin, J. D., and Green, K. M. 1974. Estimating nutrient loadings of lakes from non-point sources. U.S. EPA/600/3-74/020, Washington, D.C.
9. Hsu, P. H., and Jackson, M. L. 1960. Inorganic phosphate transformation by chemical weathering in soils as influenced by pH. *Soil Sci.*, **90**:16–24.
10. Porcella, D. B., Bishop, A. B., Andersen, J. C., Asplund, O. W., Crawford, A. B., Greeney, W. J., Jenkins, D. I., Jurinek, J. J., Lewis, W. D., Middlebrooks, E. J., and Walkingshaw, R. W. 1973. Comprehensive management of phosphorous water pollution. U.S. EPA/600/5-74/010, Washington, D.C.
11. Hsu, P. H. 1965. Fixation of phosphate by aluminum and iron in acidic soils. *Soil Sci.*, **99**:398–400.
12. Tandon, H. L. S. 1970. Fluoride extractable aluminum in soils:2. As an index of phosphate retention by soil. *Soil Sci.*, **109**:13–17.
13. Vijayachandran, P. K., and Harter, R. D. 1975. Evaluation of phosphate adsorption by a cross section of soil types. *Soil Sci.*, **119**:119–125.
14. Syers, J. K., Brouman, M. G., Smillie, G. W., and Corey, R. B. 1973. Phosphate sorption by soils evaluated by the Langmuir adsorption equation. *Soil Sci. Soc. Amer. Proc.*, **37**:358.
15. Coleman, N. T., Weed, S. B., and McCracken, R. J. 1959. Cation-exchange capacity and exchangeable cations in Piedmont soils of North Carolina. *Soil Sci. Soc. Amer. Proc.*, **23**:146–149.
16. Franklin, W. T. and Reisenauer, H. M. 1960. Chemical characteristics of soils related to phosphorus fixation and availability. *Soil Sci.*, **90**:192.
17. Novotny, V., Tran, H., Simsiman, G. V., and Chesters, G. 1978. Mathematical modeling of land runoff contaminated by phosphorus. *J. WPCF*, **50**:101–112.
18. Rennie, D. A., and McKercher, R. B. 1959. Adsorption of phosphorus by four Saskatchewan soils. *Can. J. Soil Sci.*, **39**:64–75.
19. Ryden, J. C., Syers, J. K., and Harris, R. F. 1972. Potential of an eroding urban soil for the phosphorus enrichment of streams. *J. Env. Quality*, **1**:430.
20. Enfield, G. G. 1974. Rate of phosphorus sorption by five Oklahoma soils. *Soil Sci. Soc. Amer. Proc.*, **38**:404.
21. Massey, H. F., Jackson, M. L., and Hays, O. E. 1953. Fertility erosion of two Wisconsin soils. *Agron. J.*, **45**:543–547.

22. Stoltenberg, N. L., and White, J. L. 1953. Selective loss of plant nutrients by erosion. *Soil Sci. Soc. Amer. Proc.*, **27**:406–410.
23. Allaway, W. H. 1968. *Adv. Agron.*, **20**:235–271.
24. Viitasalo, I. 1978. Heavy metals in soil and cereals fertilized with sewage sludge. 9th conference IAWPR, Stockholm, *Prog. Water Tech.*, **10(5/6)**: 309–316, Pergamon Press, Oxford.
25. Anon. 1976. Application of sewage sludge to cropland. Appraisal of potential hazards of heavy metals to plants and animals. U.S. EPA/430/9-76/103, Washington, D.C.
26. Zimdahl, R. L., and Hasset, J. J. 1977. Lead in soil. In: W. R. Rogers and B. G. Wixson (eds.), "Lead in the Environment," NSF 770214, U.S. GPO, Washington, D.C.
27. Sanks, R. L., Plante, J. M., and Gloyna, E. F. 1975. Survey suitability of clay beds for storage of industrial solid wastes. Center for Research in Water Resources, The University of Texas, Austin, Texas.
28. Krenkel, P. A. 1973. Mercury: Environmental considerations, Part 1. CRC Critical Reviews in Environmental Control, pp. 303–373, Cleveland, Ohio.
29. Moraghan, V. T. 1973. Mercury in soil and plant systems, A review of literature. North Dakota State University, School of Agriculture, Fargo, North Dakota.
30. Aomine, S., and Inoue, K. 1967. Retention of mercury by soils, Part II. *Soil Sci. Plant Nutr.*, **13**:195.
31. Jernelov, A. 1972. Factors in the transformation of mercury to methyl mercury. In: R. Hartig and B. D. Dinman (eds.), "Environmental Mercury Contamination," Ann Arbor Science Publications, Inc., Ann Arbor, Michigan.
32. Keeney, D. R., Lee, K. W., and Walsh, L. M. 1975. Guidelines for the application of wastewater sludge to agricultural land in Wisconsin. Tech. Bull. No. 88, Wisc. Dept. Nat. Res., Madison, Wisconsin.
33. Steward, B. A., Woolheiser, D. A., Wischmeier, W. H., Caro, J. H., and Frere, M. H. 1975. Control of pollution from cropland. U.S. EPA/600/2-75/026, Washington, D.C.
34. Chesters, G., and Simsiman, G. V. 1975. Impact of agricultural use of pesticides on water quality of the Great Lakes. Task A-5, Water Resource Center, Univ. of Wisc., Madison, Wisconsin, and International Joint Commission, Windsor, Ontario.
35. Anon. 1973. Great Lakes water quality annual report to the International Joint Commission. Great Lakes Water Quality Board.
36. Guenzi, W. D. (ed.). 1974. "Pesticides in Soil and Water." *Soil Sci. Soc. Amer., Inc.*, Madison, Wisconsin.
37. Goring, G. A., and Hamaker, J. W. 1972. "Organic Chemicals in the Soil Environment." Marcel Dekker, Inc., New York.
38. Foy, C. L., and Bingham, S. W. 1969. Some research approaches toward minimizing herbicidal residues in the environment. *Residue Revs.*, **29**:105–135.

39. Adams, R. S. 1973. Factors influencing soil adsorption and activity of pesticides. *Residue Revs.*, **47**:1–54.
40. Bailey, G. W., and White, J. L. 1970. Factors influencing the adsorption and movement of pesticides in soil. *Residue Revs.*, **32**:29–92.
41. Helling, C. S., Kearney, P. C., and Alexander, M. 1971. Behavior of pesticides in soils. *Adv. Agron.*, **23**:147–240.
42. Stevenson, F. J. 1972. Organic matter reactions involving herbicides in soil. *J. Env. Quality*, **1**:133–343.
43. Weber, J. B. 1972. Interactions of organic pesticides with particulate matter and soil systems. In: R. F. Gould (ed.), "Fate of Organic Pesticides in the Aquatic Environment." Advanc. Chem. Ser. III, Amer. Chem. Soc., Washington, D.C., pp. 5–120.
44. Donaldson, T. W., and Foy, C. L. 1965. The phytotoxicity and persistence in soils of benzoic acid herbicides. *Weeds*, **13**:195–199.
45. Harris, C. R., and Lichtenstein, E. P. 1961. Factors affecting the volatilization of insecticidal residues from soils. *J. Econ. Entomol.*, **54**:1038–1045.
46. Scalf, M. R., Keely, J. W., and LaFevers, C. J. 1973. Groundwater pollution in the south-central states. U.S. EPA-R273-268, Washington, D.C.
47. Chesters, G. 1967. Terminal report on Phase I of insecticide adsorption by lake sediments as a factor controlling insecticide accumulation in lakes. Office of Wat. Res., Res. Grant No. B-008-Wis., University of Wisconsin, Madison.
48. Kearney, A. R. 1972. Decontamination of pesticides in soils. Residue Revs., **44**:73–113.
49. Spencer, W. F., and Cliath, M. M. 1973. Pesticide volatilization. Residue Revs., **49**:1–47.
50. Hindin, E., May, D. S., and Dunstan, G. H. 1966. Distribution of insecticides sprayed by airplane on an irrigated corn plot. In: R. F. Gould (ed.) "Organic Pesticides in the Environment," Advan. Chem. Series 60, Amer. Chem. Soc., Washington, D.C., pp. 132–145.
51. Willis, G. H., Parr, J. F., and Smith, S. 1971. Volatilization of soil-applied DDT and DDD from flooded and non-flooded plots. *Pestic. Monit. J.*, **4**:204–208.
52. Spencer, W. F., Farmer, W. J., and Cliath, M. M. 1973. Pesticide volatilization as related to water loss from soil. *J. Env. Quality*, **2**:284–289.
53. Willis, G. H., Parr, J. F., Smith, S., and Carrol, B. R. 1972. Volatilization of dieldrin from fallow soil as affected by different soil water regimes. *J. Env. Quality*, **1**:193–196.
54. Cars, J. H., and Taylor, A. W. 1971. Pathways of loss of dieldrin from soils under field conditions. *J. Agr. Food Chem.*, **19**:379–384.
55. Willis, G. H., Parr, J. F., Papendick, R. I., and Smith, S. 1969. A system for monitoring atmospheric concentrations of field-applied pesticides. *Pestic. Monit. J.*, **3**:172–176.
56. Cars, J., Taylor, A. W., and Lemon, E. R. 1971. Measurement of pesticide

concentrations in air overlying a treated field. Proc. Int. Symp.: Indentification and Measurement of Environmental Pollutants, Ottawa, Canada, pp. 72–77.

57. Parochetti, J. V., and Watten, G. F. 1966. Losses of IPC and CIPC. Weeds, **14**:181–185.
58. Edwards, C. A. 1970. "Persistence of Pesticides in the Environment." CRC Press, Cleveland, Ohio.
59. Lichtenstein, E. P., and Corbett, J. R. 1969. Enzymatic conversion of aldrin to dieldrin with subcellular components of pea plants. *J. Agr. Food Chem.*, **17**:589–594.
60. Lichtenstein, E. P., Schultz, K. R., Skrentny, R. F., and Tsukano, Y. 1966. Toxicity and fate of insecticide residues in water. *Arch. Env. Health*, **12**:199–212.
61. Nash, R. G., and Harris, W. G. 1973. Chlorinated hydrocarbon insecticide residues in crops and soil. *J. Env. Quality*, **2**:269–273.
62. Nash, R. G., Beal, M. L., Jr., and Woolson, E. A. 1970. Plant uptake of chlorinated insecticides from soils. *Agron. J.*, **62**:369–372.
63. Armstrong, D. E., Chesters, G., and Harris, R. F. 1967. Atrazine hydrolysis in soil. *Soil Sci. Soc. Amer. Proc.*, **31**:61–66.
64. Armstrong, D. E., and Chesters, G. 1968. Adsorption catalyzed chemical hydrolysis of atrazine. *Env. Sci. Technol.*, **2**:683–689.
65. Getzin, L. W., and Rosefield, I. 1968. Organophosphorus insecticide degradation by heat-labile substances in soil. *J. Agr. Food Chem.*, **16**:598–601.
66. Graetz, D. A., Chesters, G., Daniel, T. C., Newland, L. W., and Lee, G. B. 1970. Parathion degradation in lake sediments. *J. WPCF*, **42**:R76–R94.
67. Guenzi, W. D., and Beard, W. E. 1967. Anaerobic conversion of DDT and DDD in soil. *Science*, **156**:1–2.
68. Guenzi, W. D., and Beard, W. E. 1968. Anaerobic conversion of DDT to DDD and aerobic stability of DDT in soil. *Soil Sci. Soc. Amer. Proc.*, **32**:522–524.
69. Kearney, P. C., Woolson, E. A., Plimmer, J. R., and Isensee, A. R. 1969. Decontamination of pesticides in soils. *Residue Revs.*, **29**:137–149.
70. Miles, J. R. W., Tu, C. M., and Harris, C. R. 1969. Metabolism of heptachlor and its degradation products by soil microorganisms. *J. Econ. Entomol.*, **69**:1334–1338.
71. Yule, W. N., and Rosefield, I. 1964. Fate of insecticide residues. Decomposition of lindane in soil. *J. Agr. Food Chem.*, **15**:1000–1004.
72. Bouman, M. C., Schechter, M. S., and Carter, R. L. 1965. Behavior of chlorinated insecticides in a broad spectrum of soil types. *J. Agr. Food Chem.*, **13**:360–365.
73. Paris, D. F., and Lewis, D. L. 1973. Chemical and microbial degradation of ten selected pesticides in aquatic systems. *Residue Revs.*, **45**:95–124.
74. Sethunathan, N. 1973. Microbial degradation of insecticides in flooded soil and in anaerobic cultures. *Residue Revs.*, **47**:143–165.

75. Beynon, K. I., Hutson, D. H., and Wright, A. N. 1973. The metabolism and degradation of vinyl phosphate insecticides. *Residue Revs.*, **47**:55–142.
76. Kearney, P. C., and Kaufman, D.D. (eds.). 1964. "Degradation of Herbicides." Marcel Dekker, Inc., New York.
77. Walker, A. 1974. A simulation model for prediction of herbicide persistence. *J. Env. Quality*, **3**:396–401.
78. Bruce, R. R., Harper, L. A., Leonard, R. A., Snyder, W. M., and Thomas, A. W. 1975. A model for runoff of pesticides from small upland watersheds. *J. Env. Quality*, **4**:541–548.
79. Carey, A. E., and Gowen, J. A. 1976. PCBs in agricultural and urban soils. Nat. Conf. on Polychlorinated Biphenyls. U.S. EPA-QV-633N277c, Washington, D.C., pp. 195–198.
80. Hague, R., Schmedding, D. W., and Freed, V. H. 1974. Aqueous solubility, adsorption, and vapor behavior of polychlorinated biphenyl arochlor 1254. *Env. Sci. Technol.*, **8**:139–142.
81. Jenny, H. 1941. Factors of soil formation. McGraw-Hill Book Co., Inc., New York.
82. Barholomew, W. V., and Clark, F. E. (eds.). 1965. "Soil Nitrogen." Amer. Soc. Agron., Madison, Wisconsin.
83. Allison, F. E. 1973. Soil organic matter and its role in crop production. Elsevier, Amsterdam.
84. Stanford, G., England, C. B., and Taylor, A. W. 1970. Fertilizer use and water quality. Soil and Water Conserv. Research Div., ARS, USDA, ARS-41-168, Beltsville, Maryland.
85. Stanford, G., Free, M. H., and Schwaninger, D. H. 1973. Temperature coefficient of soil nitrogen mineralization. *Soil Sci.*, **115**(4):321–328.
86. Zanoni, A. E. 1969. Secondary effluent degradation at different temperatures. *J. WPCF*, **41**:640–659.
87. Wilde, H. E., Sawyer, C. N., and McMahon, T. C. 1971. Factors affecting nitrification kinetics. *J. WPCF*, **43**:1845–1854.
88. Justice, J. K., and Smith, R. L. 1962. Nitrification of ammonium sulfate in a calcareous soil as influenced by combination of moisture, temperature, and levels of added nitrogen. *Soil Sci. Soc. Amer. Proc.*, **26**:246–250.
89. McLaren, A. D. 1969. Steady state studies of nitrification in soil: Theoretical considerations. *Soil Sci. Soc. Amer. Proc.*, **33**:273–276.
90. McLaren, A. D. 1971. Kinetics of nitrification in soil: Growth of the nitrifiers. *Soil Sci. Soc. Amer. Proc.*, **35**:91–95.
91. Pratt, P. F., Jones, W. W., and Hunsaker, V. E. 1972. Nitrate in deep soil profiles in relation to fertilizer rates and leaching volume. *J. Env. Quality*, **1**:97–102.
92. Carter, J. N., and Allison, F. E. 1960. Investigation of denitrification in well-aerated soils. *Soil Sci.*, **90**:173–177.
93. Starr, J. L., Broadbent, F. E., and Nilson, D. R. 1974. Nitrogen transformation during continuous leaching, Parts I–III. *Soil Sci. Soc. Amer. Proc.*, **38**:283–304.

94. Misra, C., Nielsen, D. R., and Biggar, J. W. 1974. Nitrogen transformation in soil during leaching: III–Nitrate reduction in soil. *Soil Sci. Soc. Amer. Proc.*, **38**:300–304.
95. Duffy, J., Chung, C., Boast, C., and Franklin, M. 1975. A simulation model of bio-physico-chemical transformation of nitrogen in tile drain corn belt soil. *J. Env. Quality*, **4**:477–485.
96. Painter, H. A. 1970. A review of literature on inorganic nitrogen metabolism in microorganisms. *Water Res.*, **4**:393–450.
97. Larsen, V., Axley, J. H., and Miller, G. L. 1972. Agricultural waste water accommodation and utilization of various forages. Tech. Rep. No. 19, Water Res. Center, University of Maryland, College Park, Maryland.
98. Thomas, G. W. 1972. The relation between soil characteristics, water movement and nitrate contamination of ground water. Tech. Rep. No. 52, University of Kentucky WRI, Lexington, Kentucky.
99. Shaffer, M. J., Dutt, G. R., and Moore, W. J. 1969. Predicting changes in nitrogen compounds in soil-water systems. Collected papers: Nitrates in agricultural waste waters, FWQA, WPC Series, 13030 ELY 12/69, pp. 15–28.
100. Preul, H. C., and Schroepfer, G. J. 1968. Travel of nitrogen in soils. *J. WPCF*, **40**:30–48.
101. Bailey, G. W. 1968. Role of soils and sediment in water pollution control. Part I, Reactions of nitrogenous and phosphatic compounds with soils and geological strata. FWCA (present EPA, South-east Wat. Lab. Rep.) Athens, Georgia.
102. Ryan, J. A., and Keeney, D. R. 1975. Ammonia volatilization of surface-applied wastewater sludge. *J. WPCF*, **47**:386–393.
103. Terry, R. E., Nelson, D. W., Simons, L. E., and Meyer, G. J. 1978. Ammonia volatilization from wastewater sludge applied to soils. *J. WPCF*, **50**:2657–2665.
104. Viets, F. G. 1971. Fertilizer use in relation to surface and groundwater pollution. In: "Fertilizer Technology and Use," 2nd edition, Soil Sci. Soc. Amer., Madison, Wisconsin.
105. Alexander, M. 1977. "Introduction to Soil Microbiology." Wiley, New York.
106. Zibilske, L. M., and Weaver, R. W. 1978. Effect of environmental factors on survival of *Salmonella typhimurium* in soil. *J. Env. Quality*, **7**:593.
107. Lance, J. C. 1978. Fate of bacteria and viruses in sewage applied to soils. *Trans. Amer. Soc. Agr. Eng.*, **21**:114.
108. Schaub, S. A., and Sorber, C. A. 1977. Virus and bacteria removal from wastewater by rapid infiltration through soil. *Appl. Env. Microbiol.*, **33**:609.
109. Loehr, R. C., Jewell, W. J., Novak, H. D., Clarkson, W. W., and Friedman, G. S. 1979. "Land Application of Wastes." Van Nostrand Reinhold, New York.
110. Anon. 1977. Process design manual for land treatment of municipal wastewater. U.S. EPA/625/1-77/008, Washington, D.C.
111. Parson, D., Brownlee, D., Wetter, A., Mauer, A., Haugton, E., Kornder, L., and Slesak, M. 1975. Health aspects of sewage effluent irrigation. Pollut.

Cont. Branch, British Columbia Water Res. Service, Dept. of Lands, Forest, and Water Res., Parliament Bldg., Victoria, B. Columbia, 75 pp.

112. Drewry, W. A., and Eliassen, R. 1968. Virus movement in groundwater. *J. WPCF*, **40**:R257–R271.

113. Scheuerman, P. R., Bitton, G., Overman, A. R., and Gifford, G. E. 1979. Transport of viruses through organic soils and sediments. *J. Env. Eng. Div., ASCE*, **105**:629–640.

114. Burge, W. D., and Enkiri, N. N. 1978. Virus adsorption by five soils. *J. Env. Quality*, **7**:73–76.

7

Groundwater pollution

GROUNDWATER (BASE FLOW) AND NONPOINT POLLUTION

Extent of the Problem

Approximately one-half of the United States population depends on groundwater for its supply of potable water. Of the total population, about 29% use groundwater delivered by community systems and 19% use private wells.[1] As stated in Chapter 3, water recovered by wells is hydrologically considered as a part of groundwater runoff when it reaches surface drainage systems.

Ideally, groundwater should be characterized by clarity, bacterial purity, and constant temperature and chemical quality, and should require very little treatment prior to its use. However, increased usage of groundwater resources and general increase of inputs of surface pollution into groundwater zones have caused contamination and general deterioration of groundwater quality in many areas of the United States and throughout the world. Excessive mining of groundwater resources results in pumpimg from deepter geological zones, which yields water with high salt and ionic content.

With an increased emphasis on land disposal of sewage and industrial wastes instigated by the "zero discharge" legislation (PL-92-500 and PL-95-217), the trend in worsening groundwater quality may continue unless expensive measures are taken to protect groundwater resources.

While the basic composition of groundwater reflects its contact with soils, minerals, and rocks, the surface nonpoint pollution is often the primary source of groundwater contamination by many pollutants. For years, saltwater intrusion, septic tank seepage fields, irrigation leachate, and industrial waste injection (deep well disposal) have been considered by the regulatory agencies as primary causes of groundwater pollution. However, as shown in Table 7-1, many other potential sources with ranging degrees of influence can affect groundwater quality.

It should be realized that only a relatively small portion of groundwater problems have been discovered.[2] Groundwater pollution is not visible and is detected only when a water supply or spring is noticeably polluted or the pollutant is being discharged into surface waters. Monitoring of groundwater quality is still inadequate, using observation wells that are located great distances apart.

Groundwater movement is very slow. Practices and land-use activities that

TABLE 7-1. Principal Sources of Groundwater Contamination and their Relative Importance by Region.[2]

	Northeast	Northwest	South Central	Southwest
Septic tanks and cesspools	I[a]	I	I	I
Petroleum exploration and development	II	II	I	I
Landfills	I	II	II	II
Irrigation return flows	IV	I	I	I
Surface discharges	II	I	III	I
Surface impoundments	I	I	II	III
Spills	I	II	II	II
Mining activities	II	I	III	II
Agricultural activities				
Fertilizers	III	II	III	II
Feedlot and barnyard wastes	III	III	II	III
Highway deicing	I	III	IV	IV
Artificial recharge	III	IV	III	II
River infiltration	II	II	IV	IV
Land disposal of wastewater and sewage	III	IV	III	III

[a]I—high, II—moderate, III—low, IV—not significant.

may have an impact on groundwater quality have been occurring for the past 30 to 60 years. However, this time span is relatively short in geological terms, and most of the groundwater contamination is still confined near the source. As an example, most groundwater pumped from deeper aquifers infiltrated as rainfall hundreds or thousands of years ago.

In industrialized countries, groundwater pollution by toxic chemical has locally and regionally reached alarming levels. For years, industries have been disposing their toxic wastes on land or burying them without regard to possible contamination of groundwater. A few notable examples should be mentioned.

Love Canal Incident. Between 1947 and 1952, a chemical company in Niagara Falls, New York, used an abandoned canal as a toxic waste dump, burying thousands of drums under the water or in the canal banks. In 1953, the company sold the land, and after filling the canal, schools and residential houses were built in the area.

Chemicals from leaking and disintegrating drums contaminated groundwater, and in the late 1970s the chemicals began seeping through the basement walls into household drainage systems. Air pollution monitoring in the basements throughout the contamination area registered pollution that exceeded 100 to 5000 times the safe ambient standards. The toxic chemicals that created this extreme nonpoint pollution problem included chloroform, benzene, toluene, and many other dangerous and carcinogenic compounds.

New Jersey Groundwater Contamination. The Love Canal incident is an environmental nonpoint pollution disaster of extreme proportions but it is not an isolated case. There are many illegal and dangerous dumps of chemicals scattered throughout the United States. For example, in 1970 the State of New Jersey—the most industrialized state in the nation—sampled 1200 wells. The survey found toxic chemicals in groundwater in much higher concentrations than in surface waters. The chemicals most frequently found in the New Jersey groundwater were organic solvents such as trichloroethane, carbon tetrachloride, and tetrachlorethylene. Overall, 5% of the sampled wells had concentrations of synthetic organics greater than the safe drinking water standards.[3]

In many other cases, a leachate from a localized point source (a storage lagoon or dump) becomes a regional nonpoint pollution problem (Fig. 7-1).

The tasks facing those concerned with and responsible for groundwater quality are enormous. The solutions are expensive or nonexistent. There are basically two approaches that have been used to clean up contaminated groundwater:[2]

1. Containment of the source, for example, by lining the area of entry of the pollutant into the groundwater zones or by creating barriers against movement

Fig. 7-1. Illegal and unsanitary dumps and solid-waste disposal sites are the primary sources of groundwater contamination. (Photo Department of Agricultural Journalism—University of Wisconsin.)

of the pollutant by pumping, is effective if the degree of contamination has not reached dangerous levels.

2. Actual removal by drilled wells can be used when localized hazardous contamination has occurred.

Some problems, such as a widespread contamination of an aquifer by septic tank effluents or oil field brines, are uncontrollable at almost any cost.

Due to the slow movement and large volumes of water in groundwater systems, the recovery of a contaminated aquifer is slow, lasting many years.

There are a few engineering and scientific references[4-7] dealing specifically with groundwater problems to which readers are referred for more detailed information.

Groundwater Quality Standards

There are several pieces of legislation that can affect groundwater resources. The major legislative acts were introduced in Chapter 1. The three acts that have the greatest impact on groundwater pollution control are:

1. The Safe Drinking Water Act of 1974 (PL-93-523)
2. The Federal Environmental Pesticide Control Act of 1972 (PL-92-516)
3. The Toxic Substances Control Act of 1976 (PL-94-469)

The most important standards established by the federal legislation are those listed for drinking water. Table 7-2 presents the most important water quality standards applicable to groundwater resources.

The Federal Environmental Protection Agency (U.S.-EPA) issued primary

TABLE 7-2. National Interim Primary Drinking Water Standards.[8]

Effective as of June 24, 1977

Inorganic Chemicals		*Organic Chemicals*	
Arsenic	0.1 mg/liter	Chlordane	0.003 mg/liter
Barium	1.0 mg/liter	Endrin	0.0002 mg/liter
Cadmium	0.01 mg/liter	Heptachlor	0.0001 mg/liter
Chromium	0.05 mg/liter	Lindane	0.004 mg/liter
Cyanide	0.2 mg/liter	Methoxychlor	0.1 mg/liter
Lead	0.05 mg/liter	Toxaphene	0.005 mg/liter
Mercury	0.002 mg/liter	2,4-D	0.1 mg/liter
Nitrate (as N)	10.0 mg/liter	2,4,5-TP Silvex	0.01 mg/liter
Selenium	0.01 mg/liter	Azodrin	0.003 mg/liter
Silver	0.05 mg/liter	Dichlorros	0.01 mg/liter
		Dimethonoate	0.002 mg/liter
		Ethion	0.02 mg/liter

Bacteria in Raw Water Supply[9]

		Permissible	*Desirable*
Coliform organisms	N/100 ml	10,000	100
Fecal coliforms	N/100 ml	2,000	20

Microbiological limits are monthly averages based upon an adequate number of samples. Total coliform count may be relaxed if fecal coliform concentration does not exceed the specified limit.

Anticipated Secondary Contaminant Levels[10]

Chloride	250 mg/liter	Iron	0.3 mg/liter
Color	15 units	Manganese	0.05 mg/liter
Copper	1 mg/liter	Odor	3 Tresh. Od. Numbers
Corrosivity	none	Sulfate	250 mg/liter
Foaming agents	0.5 mg/liter	Zinc	5 mg/liter
Hydrogen sulfide	0.5 mg/liter		

drinking water regulations on December 10, 1975, as the first step in setting national standards for drinking water quality under the provision of the Safe Drinking Water Act of 1974.

The standards became effective July 1977. They apply to all public water systems. However, most of the drinking water standards are related to water quality at a point of discharge to the distribution system. It is assumed that intake waters will be sufficiently uncontaminated so that with the application of the most effective treatment method, a public water system would be able to protect public health.

GROUNDWATER MOVEMENT

Geological Formations of Aquifers and Aquitards

The basic chemical content of groundwater can be related to contact of water with rocks and soils. However, it should be noted that while mineralization (enrichment of water by salts and ions) occurs throughout the entire movement of water in the aquifer, groundwater contamination and pollution mostly occur in the recharge area (with the exception of deep wastewater disposal by wells).

The most common aquifer composition—about 90%—consists of unconsolidated materials.[4] Other materials capable of forming aquifers mainly include sedimentary rocks. Crystalline and metamorphic rocks are relatively impermeable and can form aquifers only when they are fractured and/or weathered near the surface.

Clay and clayey materials can retain large quantities of water, but they are considered generally impermeable.

Unconsolidated Deposits. The unconsolidated deposits are composed of gravel, sand, silt, or clay particles that are bound or hardened by mineral content, pressure, or thermal alteration. The three major types of unconsolidated deposits are:

1. Glacial deposits formed during the last glacial age, approximately 15,000 years ago.
2. Alluvial (fluvial) deposits that resulted from deposition of sediments by streams in valleys and flood plains.
3. Aeolian deposits that consist of finer soil materials transported by wind.

Glacial deposits are of particular hydrogeologic importance in the northern part of the United States, Canada, and Europe (Fig. 7-2).

Water in the unconsolidated materials can be divided into four categories: underground water courses, abandoned or buried valleys, plains, and intermontane valleys.

Fig. 7-2. Extent of glacial deposits in the United States.

Water courses consist of alluvium that forms and underlies stream channels as well as adjacent flood plains. The interconnecting water movement between stream channels depends on the general hydraulic grade. Stream water can be an influent into the groundwater zone or vice versa.

Abandoned or buried valleys are valleys no longer occupied by streams that formed them. In many areas of the United States—especially in the Great Plains and Florida—large regional plain aquifers are located near the surface in unconsolidated deposits. Intermontane aquifers are located in unconsolidated rock deposits, often of considerable thickness, formed by the geologic formation of mountains. These are mostly located in the western United States.

Water from unconsolidated deposits often has low mineral content; however, due to higher permeability of these materials and their near-surface location, these aquifers are most commonly affected by surface nonpoint pollution.

Sedimentary Rocks. These aquifers are formed by sandstone, limestone, or dolomite limestone. Limestone (mostly calcium carbonate) and dolomite (calcium and magnesium carbonates) are also called carbonate rocks.

Sandstone. Sandstone is a sedimentary rock with a porosity of 5 to 30%. Its potential as an aquifer is excellent. Sandstone is formed by the binding of sand or gravel by a cementing material such as calcite, dolomite, or clay. These minerals occur as a result of salt precipitation and filtration of finer clay particles from penetrating water during the geological times.

In fine sedimentary rocks such as shale, the pore space is virtually nonexistent, and in deeper formations these rocks (usually former bottom deposits of geological oceans) are considered aquicludes.

Carbonate Rocks. Carbonate rocks were formed by compaction and crystallization of shells and bones of aquatic animals living in geological oceans and lakes.

Although some of the porosity in carbonate rocks is retained, most of the groundwater movement occurs along joints, fractures, and channels formed mostly by the dissolving action of percolating water. The dissolution of limestone by groundwater can reach such proportions that underground streams and cavities occur. These limestone formations, called karst or karst limestone, are especially susceptible to contamination from surface sources. Notable groundwater pollution problems occurred in England[11] and elsewhere.

Basic Hydraulic Characteristics of Groundwater Movement

Permeability and Darçy's Law. The flow of water in an aquifer can be expressed by Darçy's law, which relates water velocity in a porous medium to the hydraulic gradient, S, as

$$v = kS = k\frac{\partial h}{\partial l} \tag{7-1}$$

where:

v = velocity
S = slope of groundwater or piezometric table
h = hydraulic head
l = distance in the direction of flow
k = constant of proportionality

In order to compute flow in the aquifer, Equation 7-1 should be multiplied by the cross-sectional area of pores of the media. Hence:

$$Q = A_p v = pAkS = KAS \tag{7-2}$$

where:

Q = the flow
A_p = the area of voids

TABLE 7-3. Typical Porosities.

	Porosity (Percent)
Soils	50–60
Clay	45–55
Silt	40–50
Medium to coarse mixed sand	35–40
Uniform sand	30–40
Fine to medium mixed sand	30–35
Gravel	30–40
Gravel and sand	20–35
Sandstone	10–20
Shale	1–10
Limestone	1–10

Source: Todd.[4] Courtesy J. Wiley & Sons, Inc. Publishers.

A = total cross-sectional area
p = cross-sectional porosity (Table 7-3)

The coefficient, K, is called the coefficient of permeability or hydraulic conductivity.[4,5] The coefficient of permeability as defined herein can be visualized as flow velocity in a porous medium under slope that equals unity. Table 7-4 presents coefficients of permeability for various geological materials.

Hydraulic conductivity, K, is a function depending on the porous media and also on the fluid itself. The media characteristics affecting hydraulic conductivity are grain diameter, porosity, packing, distribution, and slope. Fluid characteristics that also affect K are density, viscosity, and ionic nature. Since K depends on viscosity of water, permeability is affected by temperature.

Example 7-1: Flow Velocity in an Aquifer

An aquifer consisting of sandstone with average slope of the groundwater table of 1% has been contaminated. Estimate how fast the contaminated water will move through the aquifer.

This is a typical straightforward application of Darçy's equation. From Equations 7-1 and 7-2

$$v = \frac{KS}{p} \tag{7-3}$$

and substituting appropriate values for K and p from Tables 7-3 and 7-4, the velocity becomes

TABLE 7-4. Typical Hydraulic Conductivities.

	Hydraulic Conductivity (cm/sec)
Clay, sand, and gravel mixes (till)	10^{-6}–10^{-4}
Sand and gravel mixes	10^{-3}–10^{-1}
Gravel	10^{-1}–1
Coarse sand	10^{-2}–10^{-1}
Medium sand	10^{-2}
Fine sand	10^{-3}–10^{-2}
Loam soils (surface)	10^{-4}–10^{-3}
Deep clay beds	10^{-11}–10^{-5}
Clay soils (surface)	10^{-2}–10^{-1}
Volcanic rock	almost 0–1
Fractured or weathered rock (core samples)	almost 0–10^{-1}
Fractured or weathered rock (aquifers)	10^{-6}–10^{-2}
Dense, solid rock	$<10^{-8}$
Shale	10^{-10}
Carbonate rock with secondary porosity	10^{-5}–10^{-3}
Sandstone	10^{-6}–10^{-3}

Source: Bouwer.[7] Reprinted with the permission of McGraw-Hill Book Co.

$$v = \frac{10^{-4} \times 0.01}{0.15} = 6.67 \times 10^{-6} \text{ cm/sec} = 210 \text{ cm/yr}$$

Note the extremely slow advancement of the contaminated water parcel.

Typical groundwater velocities can range from less than 1 cm/yr in tight clays to more than 100 m/yr in permeable sand and gravel. Todd[4] indicated that the normal range for groundwater velocities is 1.5 m/yr to 1.5 m/day. However, highly permeable glacial outwash deposits, fractured basalts and granites, and cavernous limestone aquifers allow much greater velocities.

Transmissibility. Transmissibility of an aquifer is the hydraulic conductivity times the aquifer thickness.

Heterogeneity and Anisotropy of Hydraulic Conductivity.

Heterogeneity. A formation is homogenous if the hydraulic conductivity is uniform at all points within the aquifer. If the hydraulic conductivity varies with location, the formation is heterogenous. There are three types of heterogeneity that commonly exist in geologic formations–layered heterogeneity, discontinuous heterogeneity, and trending heterogeneity.

Layered heterogeneity is common in sedimentary deposits whereby layers of

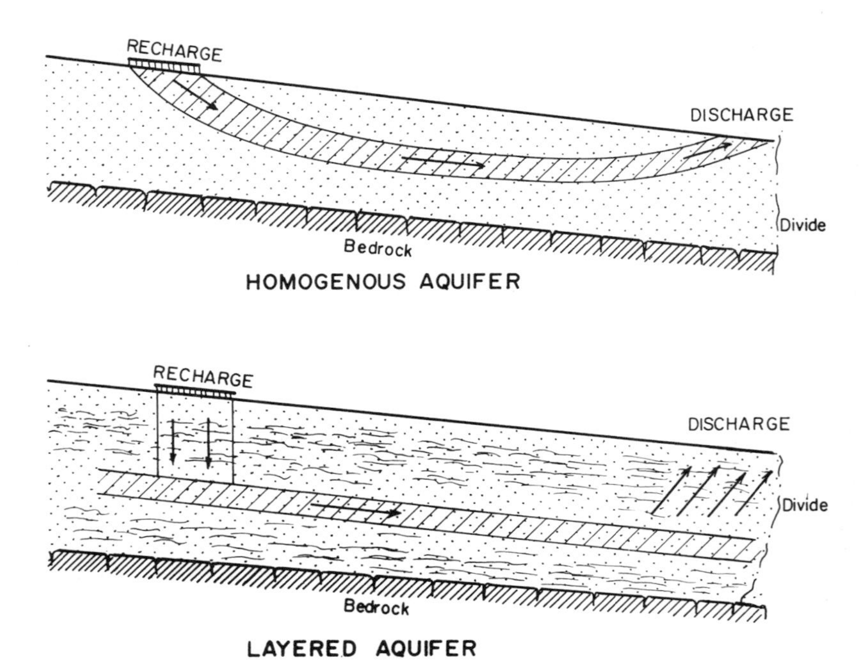

Fig. 7-3. Flow paths in homogenous and layered aquifers. (According to Freeze and Cherry.[5])

different materials have been deposited throughout the geological ages. Discontinuous heterogeneity is caused by the presence of faults or large-scale stratigraphic features. Trending heterogeneity results in the progressive change of hydraulic conductivity over a large aerial extent of a geological formation.

Anisotropy. A geologic formation is isotrophic at a given point if the hydraulic conductivity is the same in all directions. The formation is anisotrophic if the hydraulic conductivity varies with direction.

Heterogeneity and anisotropy can greatly influence flow patterns of contaminants in the aquifer (Fig. 7-3).

Dispersion. Average water movement alone does not fully explain the transport and spread of contaminants in groundwater aquifers. The processes that contribute to forming a concentration field of a contaminant in groundwater zones are advection, dispersion, and diffusion.

Advection is a process in which soluble contaminants are transported by the bulk of water. The rate of transport can be directly related to the average linear

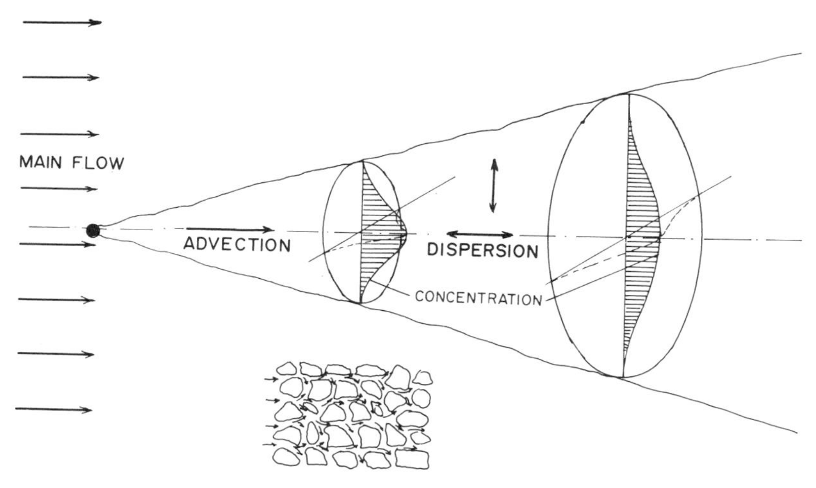

Fig. 7-4. Pollutant spread in groundwater zones from a continuous source by advection and dispersion.

groundwater velocity defined by Equations 7-1 and 7-3. However, the velocity field in groundwater zones is not uniform due to heterogeneity and anisotropy of the aquifer, to different pore sizes and branching of pores, and to many obstacles in the path of the groundwater movement. These factors cause a spread of mobile contaminants in all directions (longitudinal and transverse) as shown in Fig. 7-4. This process is called hydrodynamical dispersion. Molecular diffusion is a process by which mobile contaminants move as a result of kinetic activity of molecules and ions in the direction of their concentration gradient. Molecular diffusion can occur independently of groundwater movement. However, normal groundwater velocities and hydrodynamical dispersion are such that molecular diffusion can be neglected.[7]

The spread of contaminants in groundwater zones has a similar appearance to air pollution plumes or discharges of pollutants in surface water bodies. It is, basically, a mixing process. Nevertheless, it must be remembered that the dispersion of pollutants in air and water environments is caused primarily by turbulent mixing of air and water masses, while in groundwater zones, mixing occurs as a result of continuous splitting, slowing down, and deflecting of water particles in the pores.

The spreading of contaminants in the direction of the groundwater flow is called longitudinal dispersion. Transverse dispersion is perpendicular to the main direction of flow and is usually weaker than longitudinal dispersion.

In addition to the movement of water particles, spreading of semimobile contaminants can be caused by adsorption-desorption processes whereby a

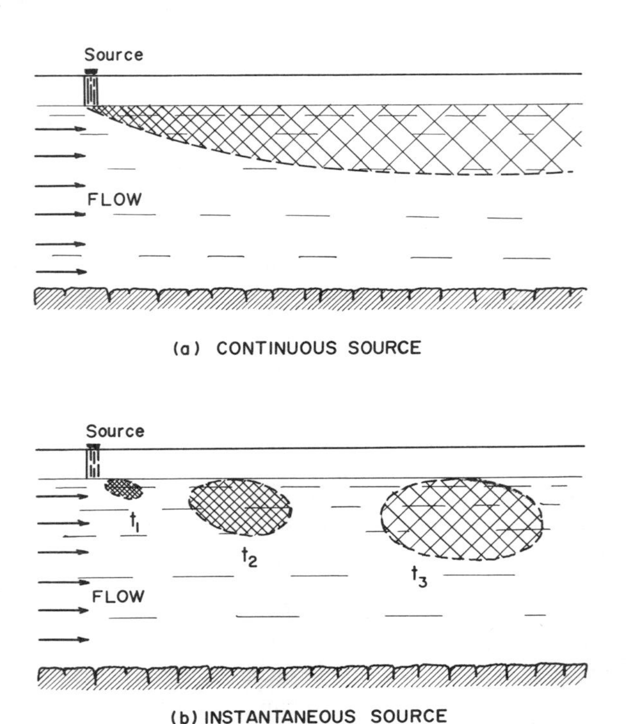

Fig. 7-5. Spreading of a pollutant in a two-dimensional uniform flow field in an isotrophic sand. (a) Continuous source–leaching, (b) instantaneous source–spill.

portion of the contaminant is adsorbed on soil particles at higher concentrations and released back to the solution when the concentration decreases.

Equations for Dispersive Movement. Most of the groundwater contamination problems can be analyzed assuming steady-state flow conditions, which implies that the flow velocity and dispersion characteristics remain constant with time. Furthermore, contamination problems can be limited to instantaneous injection of a pollutant such as a spill or a continuous release–leaching (Fig. 7-5). General equations of hydrodispersive movement have been published in many references. Reference 5 presents an adequate development of basic equations applicable to most common groundwater contamination problems.

The basic differential equation of the hydrodispersive movement can be written as[12]

$$\frac{\partial C}{\partial t} = \frac{\partial}{\partial X_i}\left(D_i \frac{\partial C}{\partial X_i} - V_i C\right) \tag{7-4}$$

where:

X_i = the coordinate
t = time
$C(X_i, t)$ = the concentration of the contaminant
V_i = the velocity in the ith direction
D_i = is the coefficient of dispersion in the ith direction

The solution of this equation for a continuous point source will yield an equation similar to the Gaussian plume dispersion model introduced in Chapter 4 (Equation 4-5). Due to slow movements of groundwater (a few meters per year or less), most of the groundwater contamination problems could be considered as an instantaneous release. There are two solutions available to the problem. One observes the concentration field from a fixed point, for example, the point of discharge. The other solution observes the movement of the cloud of a contamination as if an observer was located in the center of gravity of the cloud. The former concept will yield the following equation for a two-dimensional (longitudinal and transverse) movement of an inert dissolved contaminant.[13]

$$C(X, Y, t) = \frac{m}{4\pi p(D_L D_T)^{0.5}} \times \frac{1}{t} \times \exp\left(-\frac{(x - vt)^2}{4D_L t} - \frac{y^2}{4D_T t}\right) \qquad (7\text{-}5)$$

where m is the mass of the released contaminant per unit aquifer thickness.

The point of the coordinate origin for Equation 7-5 ($x = y = z = 0$) is the point of release of the contaminant in a two-dimensional aquifer.

Baetslé[14] published a solution for the latter case, when the coordinate origin is located at the center of gravity of the contaminant cloud. The cloud is carried away from the source with an average velocity, v, assumed to be equal to the linear velocity of the groundwater movement (Equation 7-1). The concentration distribution of the contaminant mass is then

$$C(X, Y, Z, t) = \frac{M}{8(\pi t)^{3/2}\sqrt{D_x D_y D_z}\, p} \times \exp\left(-\frac{X^2}{4D_x t} - \frac{Y^2}{4D_y t} - \frac{Z^2}{4D_z t}\right) \qquad (7\text{-}6)$$

where M is the total mass of the contaminant. In this equation the transformed coordinates are

$$X = x - vt$$
$$Y = y$$
$$Z = z$$

The maximum concentration occurs at the centerline of the cloud where $X = Y = Z = 0$). Hence

$$C_{\max} = \frac{M}{8(\pi t)^{3/2}\sqrt{D_x D_y D_z}\,p} \tag{7-7}$$

Very rarely can Equations 7-5 to 7-7 be used assuming a true dimensionless point discharge. For most cases, such as an areal application of a pesticide, the mass term, M, should be replaced by

$$\frac{M}{p} = C_0 V_0 \tag{7-8}$$

where V_0 is the volume of groundwater below the area of application and C_0 is the initial concentration. Then Equations 7-5 to 7-7 are applicable for distances where

$$8(\pi t)^{3/2}\sqrt{D_x D_y D_z} \gg V_0$$

Magnitude of the Dispersion Coefficients. The magnitude of the dispersion coefficient should be measured rather than estimated. The measured diffusion coefficients will have an advantage of accounting for heterogenity and anisotropy of the aquifer. Peaudecerf and Sauty[13] describe a method for determining the coefficients of dispersion from a dye pulse test by fitting an observed concentration curve to a dimensionless graphical form of Equation 7-5. Hydrodynamical dispersion in the transverse direction is a much weaker process than the dispersion in the longitudinal direction.[5] The magnitude of the dispersion coefficient can be related to the average linear groundwater velocity. The order of magnitude of the dispersion coefficient is approximately 10^{-8} to 10^{-6} cm^2/sec.[5]

ORIGIN OF GROUNDWATER QUALITY

Background (Natural) Groundwater Quality

The constituents that appear in groundwater can enter the aquifer with precipitation through the recharge area, as leachate from upper soil layers, or they can originate from dissolution of minerals during the groundwater passage through rocks and geological formations of the aquifer itself. Sometimes groundwater pollution can result from improperly designed deep well waste disposal systems.

Groundwater (Base Flow) Mineral Quality. Water entering groundwater zones from atmospheric precipitation is generally acidic, that is, its pH is below neutral. Under the normal conditions of unpolluted atmosphere, pH of precipitation is in an equilibrium with the saturated CO_2 concentrations in the atmosphere.

As reported in Chapter 4, the partial pressure of CO_2 in the earth's atmosphere is approximately 0.0003 bars, which will result in the normal pH of unpolluted precipitation of 5.6.

In soils, due to CO_2 production by soil bacteria from decomposing soil organic matter, the partial pressure of CO_2 can reach values up to 10^{-1} bars. This can result in pH values of soil and groundwater significantly below 5.

After entering into the soil and groundwater zones, the acidic water dissolves minerals until its dissolving capacity is exhausted. The salt and ionic content of groundwater depends on the type of minerals, their solubility, and time of contact.

Evolution of chemical groundwater quality begins in the upper (local) groundwater zone (defined in Chapter 3) by dissolution of soil and subsoil minerals. Due to its elevated acidity, water entering the aeration zone is a reducing agent with an ability to reduce various substances from their less soluble oxidized state. For example, ferric iron (less soluble) can be reduced to soluble ferrous iron.

The bicarbonate (HCO_3^-) content of water in the upper zone is a result of dissolution of limestone ($CaCO_3 \times nH_2O$) and dolomite ($Ca \times MgCO_3 \times nH_2O$) minerals and rocks by the acidic soil water and groundwater. A simple equation is often used to describe this process:

$$CaCO_3 + H_2CO_3 \rightleftarrows Ca^{2+} + 2HCO_3^- \qquad (7\text{-}9)$$

which indicates that carbonic acid is consumed in the reaction. Similarly stronger acids ($pH < 4.5$) in acid rain will also undergo a neutralization process in contact with limestone or dolomite rocks, for example,

$$CaCO_3 + 2H^+ \rightleftarrows Ca^{2+} + H_2O + CO_2 \qquad (7\text{-}10)$$

From the above equations it can be seen that the removal of acidity by limestone and dolomite minerals will increase the hardness of groundwater (hardness is defined as a content of divalent cations such as Ca^{2+}, Mg^{2+}, Fe^{2+}, and Sr^{2+} expressed commonly on a $CaCO_3$ equivalent basis).

The bicarbonate (HCO_3^-) and carbonate (CO_3^{2-}) content of groundwater represents the basic buffering system for neutralizing acids–alkalinity. From a chemical analysis, alkalinity can be determined as[15]

$$\{Alk\} = [HCO_3^-] + 2[CO_3^{2-}] + [OH^-]$$

and is expressed again as $CaCO_3$ equivalent.

Groundwater Quality Zones. Mineral groundwater quality is extremely variable. The shallow aquifer water has low mineral content but exhibits often significant seasonal or even day-to-day variations. Waters from deep underground zones have high mineral content but fairly constant quality and temperature. The longer water remains in the underground zones, the higher the measured salt content.

Freeze and Cherry[5] reported the evolution of mineral groundwater quality known as Chebotarev or Ignatovich and Souline sequence. According to this concept, groundwater evolution tends to be in the direction from atmospheric to seawater quality. For large sedimentary basins, the sequence can be described in three main zones that correlate well with the depth:

1. The upper zone is characterized by active groundwater flushing through relatively well leached rocks and soils. Water in this zone has HCO_3^- as the dominant anion and is relatively low in dissolved salts.
2. The intermediate zone has less active groundwater circulation and higher total dissolved solids. Sulfate is normally the dominant anion in this zone.
3. The lower zone has a very sluggish groundwater flow. Highly soluble minerals are commonly present. Very little flushing and leaching by groundwater has occurred due to the extremely slow movement of water. Dissolved solids are very high and chlorides are common anions.

These three water quality zones are similar to the groundwater flow zones by Todd, described in Chapter 3.

Estimating Base Flow Quality

The nature and concentrations of dissolved constituents in natural groundwater (base flow contribution) are dependent on the composition of the aquifer through which the groundwater flows. Water in calcareous sediments tends to be dominated by calcium, magnesium, and bicarbonate ions. Water in marine clay sediments tends to have higher proportions of sodium and chloride with lower concentrations of calcium and bicarbonate.

The background base flow water quality depends on many factors. The surface water is basically a mixture of groundwater (higher mineral content) and surface (lower mineral content) runoff. Their composition depends on the character of the drainage area, permeability of soils, elevation and slope, climatic conditions, percent of forest cover, general geology of the area, erosion rates, ion concentration in the precipitation, and surface waters natural biota. During low-flow nonevent conditions, the base flow quality is derived primarily from groundwater mineral contribution and pollution from point sources. In agricultural areas, base flow quality is affected by irrigation return flow.

Mineral loads in some cases can be estimated using an approach developed by Betson and McMaster.[16] The authors correlated the water quality of some undisturbed streams in the Tennessee Valley Authority area using the following logarithmic functional relationship between the quality and flow

$$C = a(Q/DA)^b \qquad (7\text{-}11)$$

where:

C = the concentration of a mineral constituent (mg/liter)
Q = the stream flow (ft^3/sec)
DA = the drainage area (sq mi) (If Q and DA are in m^3/s and km^2, respectively, multiply Q/DA by 27.83.)
a and b = empirically determined coefficients

The two coefficients in Equation 7-11 were related to land use, soils, and geological factors by a linear regression formula

$$a, b = N_1 F + N_2 CR + N_3 S + N_4 I + N_5 U \tag{7-12}$$

where:

a, b = the coefficients from Equation 7-11
F = the fraction of the watershed area that is forested
CR = the fraction of the watershed over carbonate rock
S = the drainage area fraction over shale-sandstone rock
I = the drainage area fraction over igneous rock
U = the drainage area fraction over unconsolidated rock
$N_1 \cdots N_5$ = regression coefficients

The four independent geological variables simply allocate the drainage area among the rock types present in the watershed and must sum one.

Table 7-5 shows the regression coefficients for 15 mineral constitutents obtained by analyzing 66 watersheds. As stated by the authors, the use of Equation 7-12 requires caution since constituent rating curves (Equation 7-11) have been found to display a hysteric effect with seasons and with rising and falling stages of the hydrograph. Variations among watersheds are also influenced by other factors, as previously mentioned.

Example 7-2: Base Flow Quality

Estimate base flow background nitrate content in a watershed with the following characteristics:

$$\text{Specific flow } \frac{Q}{DA} = 0.03 \text{ ft}^3\text{/sec-sq mi}$$

% forest	$F = 25\%$
% carbonate rock	$CR = 5\%$
% sandstone	$S = 35\%$
% igneous rock	$I = 2\%$
% unconsolidated rock	$U = 58\%$

TABLE 7-5. TVA Base Flow Mineral Quality Model–Regression with Forest and Geological Variables.[16]

Quality Parameter	Regression Coefficient		Regression Coefficient Value					Statistics		
		N[b]	N_1 F	N_2 C	N_3 S	N_4 I	N_5 U	R	Standard Deviation	F
SiO_2	*a*	64	-1.26	5.42	6.78	10.2	8.95	0.64	1.69	8.11
	b		-0.135	0.051	0.099	0	-0.029	0.40	0.114	2.26
Fe	*a*	29	0.035	0.020	0.009	-0.008	0.387	0.95	0.004	42.25
	b		-0.173	0.272	0.104	-0.125	0.397	0.38	0.482	0.80[a]
Ca	*a*	66	-8.52	53.9	13.4	8.32	8.41	0.85	9.0	31.42
	b		0.064	-0.116	-0.203	-0.229	-0.005	0.32	0.153	1.32[a]
Mg	*a*	66	-2.81	11.4	3.41	3.05	2.45	0.75	2.62	15.50
	b		-0.148	-0.145	-0.074	-0.104	0.513	0.67	0.197	9.99
Na	*a*	44	-1.79	2.23	2.84	3.00	3.74	0.74	0.50	9.45
	b		-0.318	0.079	0.122	0.110	-0.007	0.48	0.138	2.33
K	*a*	44	-1.08	2.51	1.94	1.58	1.80	0.72	0.47	8.12
	b		-0.152	-0.195	-0.061	0.033	-0.158	0.20	0.254	0.32[a]
HCO_3	*a*	66	-22.8	200	35.3	26.8	21.8	0.86	32.1	34.67
	b		0.110	-0.156	-0.294	-0.355	-0.139	0.35	0.132	1.71[a]
SO_4	*a*	66	-7.41	9.15	12.5	7.90	9.56	0.39	5.35	2.19
	b		-0.302	0.103	0.155	0.272	0.592	0.49	0.274	3.69
Cl	*a*	66	-1.86	3.21	2.95	2.58	3.81	0.60	0.93	6.68
	b		-0.171	0.010	0.088	0.099	0.067	0.24	0.14	0.72[a]
NO_3	*a*	63	-1.13	3.52	1.02	1.30	0.84	0.80	0.71	20.11
	b		-0.70	0.262	0.899	1.063	0.297	0.26	0.671	0.84[a]
TDS	*a*	66	-39.0	195.6	68.5	55.5	57.7	0.84	30.6	28.42
	b		0.016	-0.094	-0.146	-0.142	0	0.26	0.14	0.88[a]
$CaCO_3$	*a*	66	-33.4	182.8	48.1	34.3	31.5	0.86	29.3	33.06
	b		0.033	-0.150	-0.176	-0.222	0.131	0.49	0.151	3.78
Specific conductance	*a*	46	-145	357	180	142	128	0.88	54	26.74
	b		-0.015	-0.078	-0.134	-0.095	0.051	0.43	0.106	1.82[a]
pH	*a*	65	-0.573	8.37	7.32	7.33	6.86	0.73	0.42	13.53
	b		-0.003	-0.010	-0.003	-0.013	-0.003	0.16	0.021	0.32[a]
Color	*a*	63	-1.79	2.50	9.17	9.75	10.8	0.23	8.65	0.62[a]
	b		-0.376	0.211	0.339	0.204	0.448	0.32	0.364	1.26[a]

[a]Not significant at 0.9 level.
[b]N is number of watersheds analysed.

Using the Betson and McMaster concept, select the coefficients from Table 7-5. Then the coefficients for the rating equation become (Equation 7-12)

$$a = -1.13 \times 0.25 + 3.52 \times 0.25 + 1.02 \times 0.35 + 1.3 \times 0.02 + 0.84 \times 0.58 = 1.47$$

$$b = -0.70 \times 0.25 + 0.262 \times 0.25 + 0.899 \times 0.35 + 1.063 \times 0.02 + 0.297 \times 0.58 = 0.67$$

Hence, the nitrate concentration becomes (Equation 7-11)

$$C_{NO_3^-} = a\left(\frac{Q}{DA}\right)^b = 1.47(0.03)^{0.67} = 0.14 \text{ mg/liter}$$

Groundwater Contamination by Pollutants

Although mineral salts of low pollutional relevance are the most common constituents found in groundwater, some serious pollutants and pollution levels also can be detected. Nitrogen and some toxic metals are examples of pollutants that can seriously contaminate groundwater resources in quantities above acceptable levels. Table 7-6 shows natural and artificial sources of groundwater contamination.

Nutrients. Kreitler and Jones[18] reported nitrate levels in groundwater of Runnels County, Texas, that reached average values of 250 mg/liter, well above the drinking water standard of 10 mg/liter. It was found that almost 80% of the nitrate content leached from natural soil nitrogen as a result of cultivation during the last 50 years. Nitrate nitrogen is the most common contaminant identified in groundwater.

High levels of nitrate present a health problem and can cause infant methemoglobinemia (blue baby disease) and cancer.[19] Nitrate affects young babies less than three months old by depriving them of oxygen.

The primary source of NO_3^--N in groundwater is leaching from soils (see Chapter 6). However, geologic N, that is, N associated with certain geologic formations of sedimentary origin, is also known.[20,21] Chalk and Keeney[21] found widely varying concentrations of NH_4^+ and NO_3^- in Wisconsin limestones and suggest that many limestones are potential sources of NO_3^- to groundwater.

The presence of higher levels of NH_4^+ in groundwater usually indicates contamination from septic tank effluents, overfertilization, or leachate from barnyards and solid-waste disposal sites.

Toxic Metals. Natural levels of trace metals are commonly below the levels prescribed by the drinking water standards. Most of the metals precipitate and

TABLE 7-6. Natural Versus Artificial Sources of Various Chemical Constituents Found in Groundwater.[17]

Parameter	Sources in Natural Groundwater	Artificial Sources
Calcium	Dissolution of carbonates, evaporation and weathering of igneous rocks (anorthite).	Road salt, evaporation of irrigation water.
Magnesium	Dissolution of carbonates, weathering of ferromagnesium minerals in igneous rocks.	
Sodium	Dissolution of evaporites, weathering of igneous rocks (albite), cation exchange in clays.	Road salt, evaporation of irrigation water.
Potassium	Weathering of igneous rocks (orthoclase), cation exchange in clays.	Potash-fertilizer application.
Chlorides	Dissolution of evaporites, anion exchange in clays.	Road salt, evaporation of irrigation water, septic tanks, seepage of landfill leachates.
Carbonate/ bicarbonate	Dissolution of carbonates, precipitation, oxidation of organic materials.	
Sulfates	Dissolution of evaporites, oxidation of organic materials, oxidation of pyrite.	Oxidation of industrially produced sulfides, seepage from landfills.
Nitrogen	Oxidation of organic material.	Fertilizers, barnyard runoff, septic tanks.
Phosphorus	Oxidation of organic material.	Fertilizers, barnyard runoff, septic tanks.
Toxic metals	Concentration in natural waters usually low—sources are sulfide deposits and mineral veins.	Industrial discharges, automobile exhausts, seepage from landfill leachates.

have relatively low solubility. Mobility of metals and their solubility can be increased by lower pH that can be caused by the presence of some organic (humic) acids, by elevated CO_2 levels, or by acid precipitation.

Appreciable quantities of toxic metals can be detected in areas rich in mineral ore deposits. Klusman and Edwards[22] measured toxic metals in groundwater from the mineral belt of Colorado and noted that the drinking water standards were violated in 14% of the samples for Cd, 1% for Cu, 2% for Hg, and 9% for Zn.

Other Common Pollutants. Only a few pollutants have received some attention in groundwater investigations. Organic matter, phosphates, or pathogenic microorganisms should be absent unless considerable leaching from contaminated surfaces into shallow, highly permeable aquifers has occurred. These substances are considered immobile in most soil and groundwater systems.

Leachability of pesticides into groundwater depends primarily on the degree to which they are adsorbed to soil colloids. Pesticides are leached more readily in coarse-textured than in fine-textured soils.

From the pesticides presently in use, organochlorine insecticides are the least mobile, followed by the organophosphorous insecticides. The water-soluble acidic herbicides are most mobile. Most of the pesticides, such as triazines, phenylureas, and carbonates, have intermediate mobility.

Available information indicates that pesticide contamination of groundwater under normal agricultural usage is minimal.[23] However, in areas with a shallow and fluctuating water table, frequently found in coastal sandy soils and aquifers, appreciable amounts of pesticides may find their way to the underlying water; this is especially true for the soluble herbicides.[23]

SOURCES OF GROUNDWATER CONTAMINATION

There are many potential sources of groundwater contamination, including: septic tank systems, solid-waste disposal sites, land disposal of sewage, industrial wastes and sludges, irrigation and other agricultural practices, leaks and spills of chemicals and oils, and road deicing. Some of these sources can be viewed as point sources of pollution, resulting only in a localized aquifer contamination. Other sources, such as the use of septic systems in more densely populated suburban areas or irrigation, are typical nonpoint pollution problems.

Effect of Septic Tank Disposal Systems

The term septic system is commonly used to describe a subsurface, nonaerated sewage disposal system that uses soil filtration and adsorption for attenuating the effluent.

Approximately 20 million residences, or 29% of the U.S. population, dispose of their sewage by individual on-site systems.[24] Septic tanks represent the highest total volume of wastewater discharged directly to groundwater and are the most frequently recorded sources of contamination of groundwater and surface base flow. Sixty-one percent of all waterborne disease outbreaks in the United States between 1946 and 1960 could be attributed to groundwater contamination.[22] Considering the quantity of sewage discharged by septic systems, it is quite likely that septic tank effluents could cause a significant portion of these outbreaks.

The amount of discharge from septic tank systems is commonly estimated as 280 liters/cap-day (75 gpcd). Septic tank effluents contain 40 to 80 mg N/liter, 11 to 31 mg P/liter, and 20 to 450 mg/liter of BOD_5.[25] There have been many failures of septic systems caused by inadequate subsoil conditions or, more commonly, by overloading. Failures can originate from one or two causes:[26]

Fig. 7-6. Effluent surfacing from septic systems caused by overloading. (Photo USDA Soil Conservation Service.)

1. Inadequate infiltration of effluent into the soil due to soil clogging, which results in surfacing of septic tank effluents (Fig. 7-6).

2. Inadequate purification in the soil during percolation because of short travel times, which can be due to the presence of permeable, shallow soils or to local overloading. This type of failure is not visible; however, inadequate purification may result in pathogenic pollution of groundwater and contamination of groundwater and base flow by organics and nutrients.

Surfaced effluent excess commonly contributes to the pollution of the base flow and often results in faulty septic conditions and high bacterial counts of small local streams.

Local ordinances do not often pay enough attention to the preservation of groundwater resources from pollution by septic tank effluents. The standard percolation test simply favors highly permeable, sandy soils. Such soils do postpone the failure of the system by hydraulic overloading; nevertheless, adsorption and the purifying capacity of such soils is greatly reduced and pollutants can move downwards and contaminate groundwater.

Problems of groundwater contamination by on-site systems can be classified as individual, local, or regional. An individual problem occurs when one disposal system contaminates one or more wells in the immediate vicinity. A problem is considered local when systems in a small high-density residential area contaminate the aquifer used by that area. A regional problem exists when a large number of on-site systems contaminate extensive aquifers that serve a broad area such as one or more counties.

The state of Illinois is using a pollution potential index to assess hazards to water quality caused by septic tank effluents. The index is defined by the following equation:[27]

$$PI = \frac{200\sqrt{D}}{\sqrt{I}\,(T - 5)} \tag{7-13}$$

where:

PI = pollution index
D = housing density in numbers of residences with septic systems within a circle having an area of one-fourth of a square mile
I = time in minutes for water to fall 6 in. (15 cm) as measured in a standard percolation test
T = thickness, in feet, of soil between the discharge level and the underlying aquifer to be protected; if $T < 5$, assume $(T - 5) = 0$.

Values of PI greater than 10 suggest that some potential for pollution exists. If the PI value is less than 10 the potential for significant groundwater pollution is assumed to be low.

Nitrogen Contribution from Septic Systems. While organic matter (BOD_5, pathogenic microorganisms) and phosphates are usually effectively removed by most systems and usually do not penetrate more than 1.5 m below the level of discharge or beyond the immediate vicinity of the seepage field,[28-30] nitrogen migration into groundwater systems is a serious problem.

Nitrogen, which in the septic tank effluent exists as a mixture of approximately 75% of NH_4^+ and 25% of organic N, can undergo nitrification in aerated soils.

Field investigations were performed using 24 shallow observation wells that monitored effluent migration in soil and groundwater aquifers adjacent to a septic tank-soil adsorption system in a rural Wisconsin high school.[30] The soil at this site is a deep silt loam underlain by glacial outwash sand and gravel at a depth from 1.8 to 2.7 m. The investigation showed that nitrification is quite complete in well-drained soils.

The product of nitrification—nitrate N—is highly mobile in soils and aquifers. Nitrate contamination of groundwater occurs most often with septic systems installed in sandy soils. The appearance of ammonia in groundwater indicates failing and exhausted soil adsorption field and/or shortcutting.

An investigation of nitrogen mass balance in Long Island, New York, revealed

TABLE 7-7. Major Groundwater Recharge Operations in California That Use Reclaimed Water From Wastewater.[32]

Wastewater Reclamation Plant	Treatment Processes[a]	Recharge Method	Type of Use
San Jose Creek (Whittier)	PS, AS, C, F, Ch	Surface spreading	Groundwater replenishment
Whittier Narrow	PS, AS, C, F, Ch	Surface spreading	Groundwater replenishment
Water Factory 21[b] (Orange County)	PS^+, AS^+, C, F, NS	Direct injection	Seawater intrusion barrier and groundwater replenishment.
Chino Basin (Ontario)	PS, TF	Surface spreading	Groundwater replenishment
Palo Alto	PS, AS, C, F, Ch, ozonation	Direct injection	Seawater intrusion barrier

[a] AS–activated sludge
C–coagulation
CA–carbon adsorption
Ch–chlorination
F–filtration
NS–ammonia stripping
PS–primary sedimentation
RO–reverse osmosis
TF–trickling filters

[b] Treated at Orange County Sanitation District.

that more than 20% of the total nitrogen contribution from subsurface disposal systems leached as nitrate N into groundwater zones.[31]

Land Application of Water and Sewage

Sewage and Wastewater Disposal. The "no pollution discharge" policy of the Federal Water Pollution Control Act cause an increased interest in land disposal of liquid wastes such as conventionally treated sewage effluents, processing-plant wastes, animal wastes and feedlot runoff, and sewage sludge. In many areas (Southern California, desert states) land application of sewage helps replenish the groundwater supply for many communities (Table 7-7).

Although the soil has a great capacity for attenuation of pollutants, the "renovated" water is obviously not of the same quality as the native groundwater.[33]

The problems associated with land application of wastewater are similar to septic systems discussed previously; however, much greater volumes of wastewater are concentrated in a relatively small area. Mobile pollutants such as nitrates are of greatest concern since evidence indicates that several other common pollutants (BOD, pathogenic microorganisms, and phosphates) remain near the area of application. Bacteria and viruses die off quite rapidly as wastewater passes through the soil material.[34]

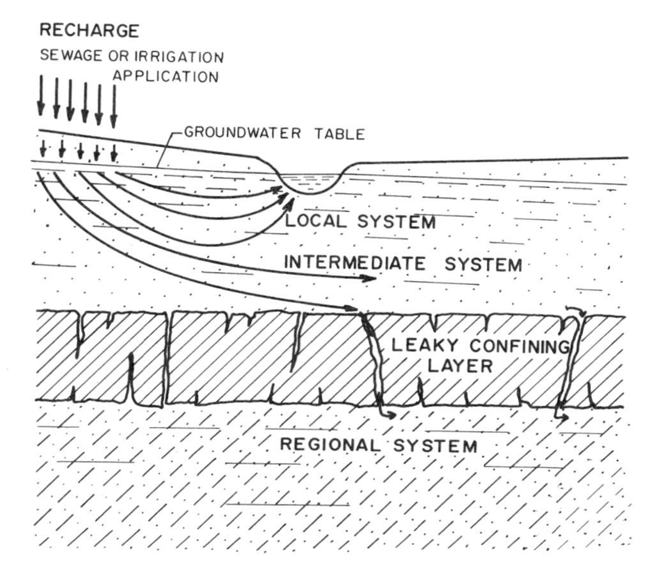

Fig. 7-7. Possible routes of pollutants in groundwater systems.

Freeze and Cherry[5] state that groundwater contamination by mobile organics may become a serious problem. Treated wastewater contains many dissolved organic compounds, and some of the potentially dangerous components—chlorinated hydrocarbons—may be created by the treatment process itself. Little is known about the toxicity and mobility of such components. Since many of these components are not biodegradable and can be mobile, their impact should always be evaluated.

As shown in Fig. 7-7, land application systems may contaminate all three groundwater systems, that is, local, intermediate, and regional aquifers.

Sludge generated by wastewater treatment facilities is commonly applied to agricultural lands as a fertilizer and soil conditioner. The effect of land application of sludges on groundwater quality depends on transformations that occur within the top soil horizon and unsaturated subsoils. Most heavy metals and phosphates will normally be retained by the soil.

Application Systems. Sewage can be applied at a low rate—in centimeters per week—or at a high rate—in meters per week.[33] The low application rates are based on matching the nutrients from the sewage with their uptake by crops, which reduces nitrate cumulation and pollutant buildup in soils. These systems represent the minimal danger to groundwater resources. The disadvantage of of low-rate application systems is large area requirements (approximately 30 ha of land per 1000 m^3/day of sewage).

The high-rate systems require only a fraction of the land area needed for low-rate systems. Since water in these systems is applied in excess of the evapotranspiration rate a portion of the applied volume will leach into groundwater zones. To minimize the impact on groundwater resources, the land application systems should be designed and operated (a) to obtain recharge water of the best possible quality (particularly with regard to nitrogen), and (b) to restrict the spread of recharged water into the native groundwater.[33]

Maximizing Nitrogen Removal. The danger of nitrogen migration into groundwater from high-rate application systems requires operation of the system that would maximize the nitrogen loss from the soil. There are only two processes by which nitrogen can be effectively removed: plant uptake and denitrification.

In high-rate systems, plant uptake accounts for only a small portion of the total nitrogen input (see Table 6-3 for values of N uptake by crops). To maximize denitrification, Bouwer[33] proposed alternate dry and wet periods. Before nitrogen can be lost by denitrification, ammonia and organic nitrogen in the applied wastewater must be converted to nitrate under aerated conditions when the soil moisture is below saturation. Denitrification, that is, conversion of nitrate to N_2 gas, occurs when soil is flooded and oxygen is depleted so that anaerobic conditions prevail. Bouwer suggests that flooding periods of 2 to 3 weeks followed by dry periods of the same length will result in approximately 30% nitrogen removal. Denitrification is almost impossible in sandy soils where saturated conditions are difficult to achieve.[35]

Restricted Use of the Aquifer. The aquifer that is recharged by sewage should not be used as a source of drinking water. The portion of the aquifer that receives sewage or other liquid wastes should be restricted, and water should be taken out at some distance from the recharge area (Fig. 7-8). This may occur naturally as a base flow into a nearby stream or lake, or artificially by drains (shallow aquifer) or wells (deep aquifer). After the collection, water can be reused for irrigation, recreation, and other nonpotable uses, or discharged into receiving streams for purposes of flow augmentation.[33] With such aquifer zoning, the portion of the aquifer between the land application site and the point of discharge is used as a natural filtration system.

Increased Salinity and Pollution of Irrigation Return Flows. Water applied to land either in the form of a treated effluent or as water withdrawn from a nearby surface water body contains certain amounts of salt and ions. A portion of the irrigation water, after application on an irrigated area, is returned to the atmosphere by evapotranspiration. Since evaporated or transpired water has negligible salt content, there is a subsequent salt and pollution buildup in soils. The portion returned to the atmosphere may range from less than 20% in high-rate

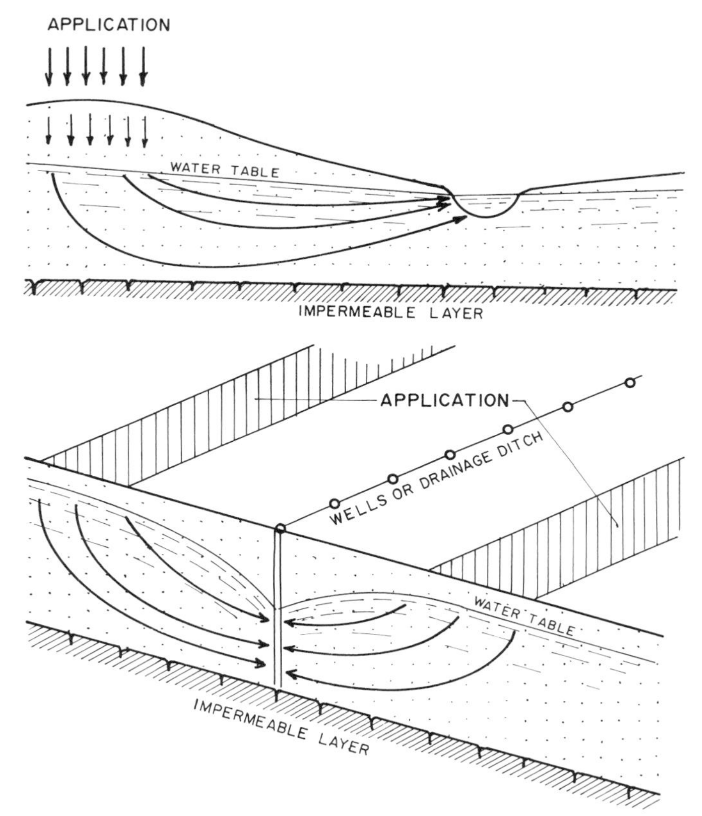

Fig. 7-8. Restriction of aquifer contamination by sewage application, natural drainage (upper portion), and artificial drainage (lower portion). (According to Bouwer.[33])

application systems in humid climatic conditions to almost 100% in low-rate application in arid or semiarid climatic zones.

In order to maintain an acceptable salt content of soils to sustain crop growth, excess irrigation water must be applied if natural precipitation is not sufficient to control the salt buildup in soils. The excess water containing increased salinity and leachate from soils is collected by drainage (natural or man-made) and/or will percolate into groundwater zones. The irrigation tail water collected by drainage systems (Figs. 7-9 and 7-10) is called irrigation return flow, and it represents one of the serious problems associated with pollution from agriculture.

The concentration of salts in water percolating through the soil root level zone into the irrigation return flow or to the groundwater can be computed from the following mass balance:[33]

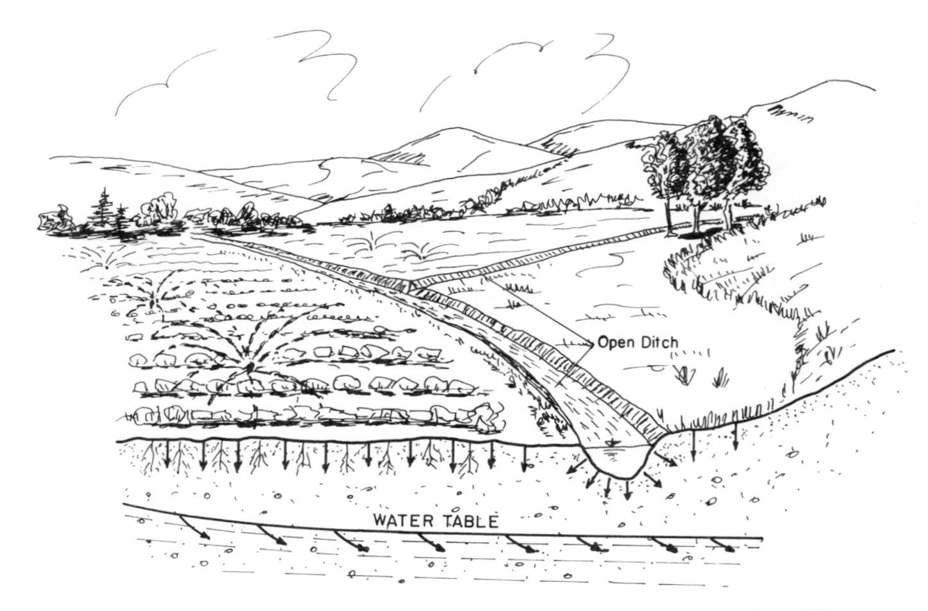

Fig. 7-9. Irrigation return flow systems. (Redrawn from Walker.[36])

$$C_i D_i = C_d (D_i - D_e) \tag{7-14}$$

where:

C_i = salt concentration of water or wastewater used for irrigation
D_i = amount of irrigation water also including effective precipitation
C_d = salt concentration of water percolating from the root zone downward
D_e = amount of water used by evapotranspiration

The amount of excess irrigation water that has to be applied to control salt buildup in soils depends on the salt tolerance of crops, the salinity of irrigation water, and the evapotranspiration rate plus other losses from the system. The leaching ratio is then computed from Equation 7-14 as

$$\frac{D_i}{D_e} = \frac{C_d}{C_d - C_i} \tag{7-15}$$

Salinity of irrigation water is usually expressed as conductivity in millimhos per centimeter or in micromhos per centimeter (1000 μmhos/cm ≃ 640 mg/liter of total dissolved solids–TDS). The salt tolerance of crops ranges from less than 500 μmhos/cm for salt-sensitive crops, such as most fruit trees and some vegetables (beans, celery, or strawberries), to more than 1500 μmhos/cm for salt-

Fig. 7-10. Drainage ditch in agricultural fields. Drainage water contains often significant amounts of pollutants leached from soils by percolated water and irrigation return flows. (Photo USDA Soil Conservation Service.)

tolerant crops such as cotton, beets, barley, and asparagus. Most common grain crops and vegetables have medium tolerance (500 to 1500 μmhos/cm) to salts. The leaching requirement is then defined as

$$LR = \frac{EC_i}{EC_d} = \frac{C_i}{C_d} \tag{7-16}$$

where:

EC_i = electric conductivity of irrigation water
EC_d = salt tolerance of crops or drainage water

Combining Equations 7-15 and 7-16, the leaching ratio becomes

$$\frac{D_i}{D_e} = \frac{1}{1 - LR} \tag{7-17}$$

The quality of excess irrigation water (irrigation return flow) remained for long unnoticed, and no control measures were undertaken to control its water quality. However, the Water Pollution Control Act of 1972 recognized irrigation water reaching groundwater or discharged as irrigation return flow into surface waters as nonpoint pollution. Investigations in central Wisconsin[35] and elsewhere[37] revealed that nitrate concentrations in the subsurface water in agricultural areas receiving irrigation are well above their background levels. In the

Wisconsin study,[35] nitrate N concentrations ranged up to 56 mg/liter. Chapter 10 contains typical loading values of some pollutants from irrigated agricultural lands.

The salinity problem is especially troublesome in arid zones of the western United States. It must be realized that water in some streams may be reused for irrigation several times as downstream users irrigate with mostly irrigation return flows from upstream farms, with increased salinity at each reuse.

Example 7-3: Amount and Quality of Irrigation Discharge Water

An agricultural field growing crops was irrigated by a treated effluent with the following quality characteristics:

TDS–300 mg/liter (EC_i = 470 μmhos/cm)

nitrogen–10 mg/liter

The effluent was applied at a rate of 10 cm/week. The evapotranspiration rate during the irrigation period (lasting 2 months) was 5 cm/week. The crop yield per hectare was approximately 5 tonnes/ha containing approximately 20 kg/tonne of nitrogen. Estimate the amount and concentration of nitrogen leached into the groundwater and the salinity of the soil water. No significant precipitation occurred during the irrigation period.

Nitrogen uptake assuming 4 months growing period is

$$UP_N = \text{crop yield} \times \text{nitrogen content/growing period}$$

$$= 5\ (\text{tonnes/ha}) \times 20\ (\text{kg/tonnes})/16\ \text{week}$$

$$= 6.25\ \text{kg/ha-week} = 6250\ \text{g/ha-week}$$

Nitrogen input from the effluent is

$$0.1\ (\text{m/week}) \times 10{,}000\ (\text{m}^2/\text{ha}) \times 10\ \text{g/m}^3 = 10{,}000\ \text{g/week-ha}$$

The amount of nitrogen leached can be obtained by subtracting the plant uptake from the nitrogen input. Hence:

$$\text{nitrogen leached} = 10{,}000 - 6250 = 3750\ \text{g/week}$$

Nitrogen concentration in the leachate is

$$C_{d_N} = \frac{\text{nitrogen leached}}{\text{volume of water leached}} = \frac{3750\ (\text{g/week})}{(0.1 - 0.05) \times 10{,}000} = 7.5\ \text{mg/liter}$$

Estimate the salinity of the leachate from Equation 7-17. The leaching requirement factor is:

$$LR = 1 - D_e/D_i = 1 - \tfrac{5}{10} = 0.5$$

and by combining Equations 7-16 and 7-17

$$C_d = \frac{C_i}{LR} = \frac{300}{0.5} = 600 \text{ mg/liter} = 940\ \mu\text{mhos/cm}$$

Groundwater Pollution from Solid-Waste Disposal Sites

In 1970 there were about 20,000 solid-waste disposal sites in the United States; however, only 6% were classified as sanitary landfills (landfills that do not cause environmental problems and are properly operated). Figure 7-11 shows an example of a solid-waste disposal operation.

Although solid-waste disposal sites are considered mostly as point sources of pollution, leachate from landfills may severely pollute large portions of adjacent aquifers and appear as contaminated base flow in a diffuse manner. Futhermore,

Fig. 7-11. Solid-waste disposal sites. Note the prevention of runoff input from surrounding areas. The refuse is covered daily by clayey soils. (Photo A. Zanoni.)

TABLE 7-8. Leachate Characteristics from Municipal Solid Waste.[39]

Components	Median Value (mg/liter)[a]	Ranges of All Values (mg/liter)[a]
Alkalinity ($CaCO_3$)	3,050	0-20,850
Biochemical oxygen demand (5 days)	5,700	81-33,360
Calcium (Ca)	438	60-7,200
Chemical oxygen demand (COD)	8,100	40-89,520
Copper (Cu)	0.5	0-9.9
Chloride (Cl)	700	4.7-2,500
Hardness ($CaCO_3$)	2,750	0-22,800
Iron, total (Fe)	94	0-2,820
Lead (Pb)	0.75	<0.1-2.0
Magnesium (Mg)	230	17-15,600
Manganese (Mn)	0.22	0.06-125
Nitrogen (NH_4)	218	0-1,106
Potassium (K)	371	28-3,770
Sodium (Na)	767	0-7,700
Sulfate (SO_4)	47	1-1,558
Total dissolved solids (TDS)	8,955	584-44,900
Total suspended solids (TSS)	220	10-26,500
Total phosphate (PO_4)	10.1	0-130
Zinc (Zn)	3.5	0-370
pH	5.8	3.7-8.5

[a]Where applicable.

some dangerously toxic compounds are commonly a part of the overall composition of the landfill leachates, especially when the landfill is used for disposal of toxic chemicals.

During the decade of 1970-1980, a large portion of the land fills and clumps that represented a danger to the environment were discontinued, including some receiving radioactive wastes. However, stored decomposing waste and leaching disintegrating drums left on these sites will represent a serious problem for decades. As shown in the Love Canal incident, improperly operated and maintained toxic waste dispoal sites can give a rise to most serious nonpoint pollution problems.

Leachate Characteristics and Management. Leachate from solid-waste disposal sites is a highly mineralized liquid containing constitutents such as chloride, iron, lead, copper, sodium, nitrogen, and various organic chemicals. Manufacturing wastes can add hazardous constitutents such as cyanide, cadmium, chromium, chlorinated hydrocarbons, and PCBs. Table 7-8 shows the ranges in concentration for various chemical constitutents and physical parameters of typical leachate from municipal solid waste. It should be noted that in countries that still use coal for household heating, the composition of leachate may be quite different from that typical for U.S. conditions.[40]

There are several general methods for managing leachate: natural attenuation by soils, prevention of leachage formation, collection and treatment, pretreatment to reduce volume or solubility, and detoxification of hazardous wastes prior to landfilling. Leachate undergoes natural attenuation by various chemical, physical, and biological processes as it migrates through soil. Whether natural attenuation will be adequate to prevent groundwater pollution should be evaluated for each site. The generation of leachate can be minimized by restricting water from infiltrating the waste. This is accomplished by providing appropriate

TABLE 7-9. Leachate Control Methods.[39]

Method	Effectiveness	Degree of Use	Cost (Examples)
Natural Attenuation			
Clay	Promising research	Unknown	Natural
Silt	Unknown	Unknown	Natural
Sand	Unknown	Unknown	Natural
Preventing Leachate Generation	Ranges from complete to partial control	Limited	Not available
Collection and Treatment			
Liners	Promising research	Limited	\$1.50 to \$4.00/yd^2
Biological treatment	Promising research	Very limited	Not available
Physical-chemical	Promising research	Very limited	Not available
Recirculation	Promising research	Very limited	Not available
Spray irrigation	Promising research	Very limited	Not available
Immobilization			
Chemical stabilization	Research progressing looks promising	Limited but growing	\$10 to \$20/ton
Encapsulation	Research progressing looks promising	Very limited	\$16/ton
Fixation and encapsulation	Research progressing looks promising	Not in use	\$40/ton
Volume Reduction			
Dewatering	Effective	Widely practiced in water pollution	\$5 to \$20/ton
Incineration	Effective for organics	Moderate	\$20 to \$100/ton
Detoxification	Varies widely by process and waste	Limited to specific wastes	Varies widely

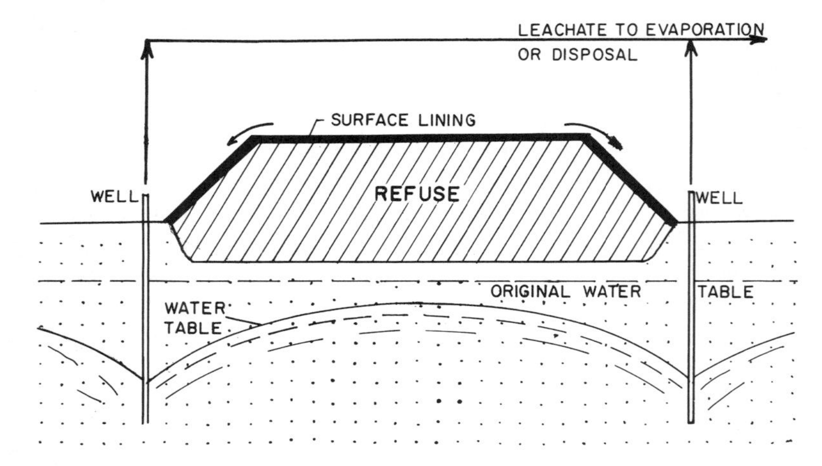

Fig. 7-12. Control of groundwater pollution from landfills by creation of a groundwater depression barrier.

surface drainage and/or placement of an impermeable liner over the landfill (preferably daily). Another method of controlling leachate is to collect it at the bottom of the landfill and treat if before discharging it to the surface water or land. Table 7-9 summarizes the various methods for leachate control and the relative effectiveness, degree of use, and cost for each method. Figure 7-12 shows a well barrier, which prevents leachate from reaching groundwater.

Most rules and regulations recommend or require that landfill sites be developed on uplands rather than floodplains and in low-permeability rather than high-permeability soils. Geologically ideal sites are seldom found; however, the sites must also be socially and politically acceptable. Therefore, it is usually necessary to settle for sites that do not meet all the recommended criteria.

Tracing of contaminant plumes from land disposal sites requires testing the levels of certain water quality parameters. Although the most appropriate parameters may vary somewhat, depending on the types of solid waste and geologic conditions, certain parameters have been found to be generally suitable. Key indicators of leachate presence that have been suggested by the U.S. Environmental Protection Agency[41] are specific conductance, pH, temperature, chloride, iron, color, turbidity, and COD. Specific conductance can be measured quickly and easily with inexpensive, accurate, and reliable portable equipment.

GROUNDWATER QUALITY MODELS

In spite of the fact that many groundwater flow problems can be solved by electric analogs, almost all of the vast number of models reported in the literature are numerical digital computer models. Among the numerical groundwater models, four different categories can be distinguished:[42]

1. Predictive models, which simulate the behavior of the groundwater system and its response to stress.
2. Resource management models, which integrate prediction with explicit management decision procedures.
3. Identification models, which determine input parameters for both of the above.
4. Data manipulation and storage procedures, which process and manage input for all above.

Predictive Models. Predictive models represent the vast majority of models developed for groundwater-related problems.[42] Bachmat et al.[42] surveyed over 250 groundwater models developed in 14 countries. The majority of the models address various types of water supply (quantity) problems. A smaller but sizable number of models exists for predicting contaminant levels and the temperature of groundwater. Only a few models have been developed for coupled groundwater–surface water systems that have the best applicability to nonpoint pollution problems.

Walker[36] surveyed 43 models that would be applicable to the problem of irrigation return flow. These models deal mostly with water and contaminant movement in unsaturated zones as opposed to the aquifer models surveyed by Bachmat et al. Chapter 9 presents a discussion on the types of watershed models and reviews some nonpoint models, including those containing a groundwater component.

The groundwater movement models (quantity and quality) can be either distributed parameter models or lumped parameter models. A distributed parameter model is one in which the variables are determined at several discrete points or nodes in the groundwater system (aquifer), while in a lumped parameter model, the groundwater system is treated as a point in space and only spatially averaged values of the variables are considered.

Gelhar-Wilson Model.[43] This model is a simple lumped parameter aquifer-stream model treating an aquifer as a point. As pointed out by Gelhar and Wilson,[43] the justification for using a lumped parameter model is that when long-term basin-wide changes in groundwater quality are desired, spatial variation becomes less important than temporal variation. The fluctuation of the water table is simulated mathematically by the following equation:

$$p \frac{dh}{dt} = -q + \epsilon + q_r - q_p \qquad (7\text{-}18)$$

where:

h = average thickness of the saturated zone
p = average effective porosity
ϵ = natural discharge rate
q = natural outflow from the aquifer
q_r = artificial recharge/unit area
q_p = pumping rate/unit area
t = time

It can be demonstrated that $q = a(h - h_0)$ where h_0 is the elevation of the river and $a = 3T/L^2$ where:

$T = h_0 K$ transmissivity
K = hydraulic conductivity
L = length of the aquifer

The change in concentration is represented by an equation of the form:

$$ph\frac{dc}{dt} + (\epsilon + q_r + \alpha ph)c = \epsilon c_L + q_r c_r \qquad (7\text{-}19)$$

where:

c = concentration
c_L = concentration of the natural recharge
c_r = concentration of the artificial recharge
α = a first-order rate constant that accounts for degradation of the contaminant

Dispersion is assumed to be negligible, which is a reasonable assumption if only regional average concentrations are sought.

The hydraulic response time and the solute response time are measures of the lag observed in the response of the system to a given input. Hydraulic response time (t_h) is defined as follows:

$$t_h = p/a \qquad (7\text{-}20)$$

where $a = 3T/L^2$ and solute response time (t_c) is defined as:

$$t_c = ph_0/\epsilon_0 \qquad (7\text{-}21)$$

where ϵ_0 is the initial recharge rate. In general, t_c is time dependent but can be estimated from Equation 7-21. A representation of the model is shown in Fig. 7-13.

Gelhar and Wilson[43] base their model on the concept of a well-mixed linear

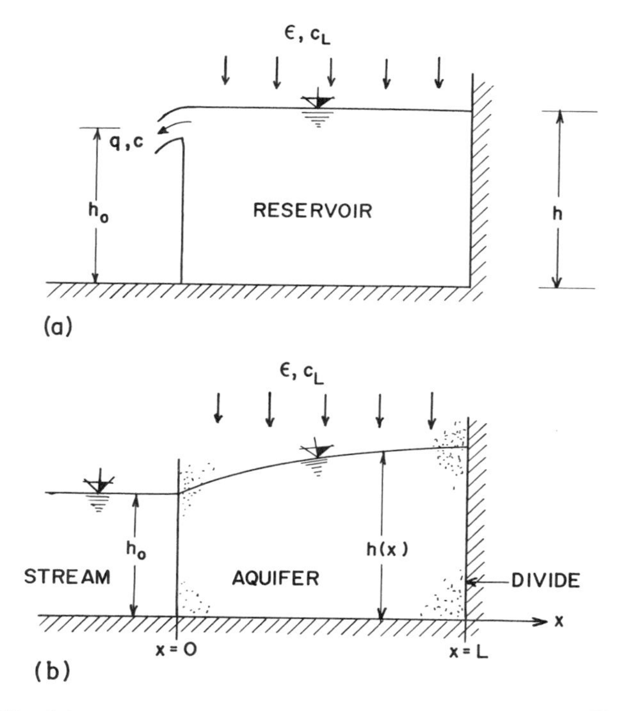

Fig. 7-13. Representation of the Gelhar-Wilson model.[43]

reservoir. They postulate that aquifer response to a given input will be similar to the response of a well-mixed linear reservoir. They show that the concentration of water leaving the aquifer is representative of the average concentration within the aquifer. Therefore, such a model is ideally suited for determining the quality of groundwater discharging to surface waters.

Example 7-4: Application of the Gelhar-Wilson Model

The model was applied to modeling of chloride contamination of groundwater in the Menomonee River watershed caused by road salting.[17] *Figure 7-14 shows modeling areas of the watershed. The hydraulic characteristics of the areas are given in Table 7-10.*

The upper part of the watershed is mostly rural, the middle is developing, and the lower part of the watershed is a densely populated urban center–Milwaukee, Wisconsin. Since the urban areas are highly impervious, it was estimated that only 10% of the road salt reached the water table in heavily urban areas, 75% in the lightly urban and suburban areas, and 100% in rural areas.

Chloride loading rates were estimated from road salt use during the winter of

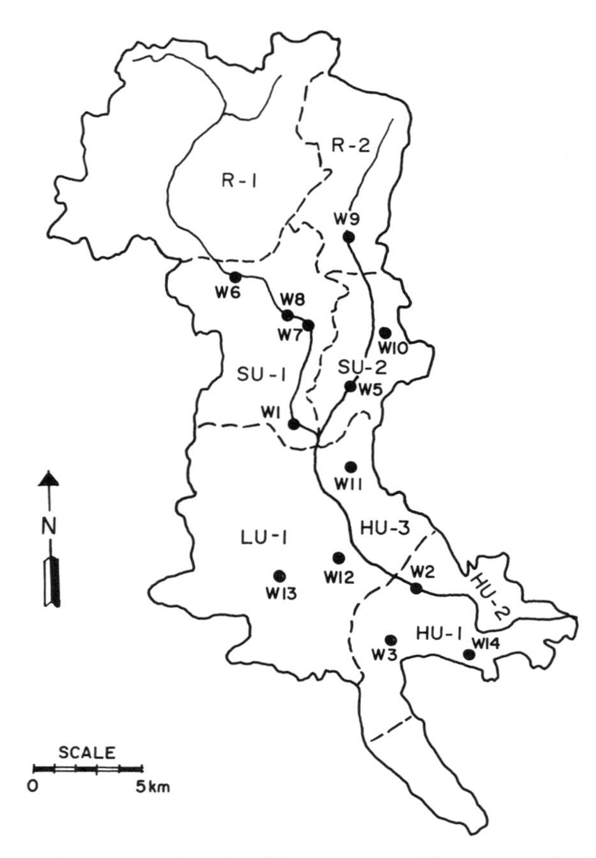

Fig. 7-14. Modeling areas in the Menomonee River watershed used in the Gelhar-Wilson model. (Source: Menomonee River Pilot Watershed Study.[17])

TABLE 7-10. Aquifer Parameters.[17]

Area	Hydraulic Response Time (yr)	Solute Response Time (yr)	Aquifer Length (m)	Hydraulic Conductivity (m/day)	Winter Loading Rate of Chloride (g/m^2/month)
HU-1	191	1000	6100	2.4	57
HU-2	29	1000	2400	2.4	57
HU-3	75	1000	2700	1.2	35–57
LU-1	146	23	7300	4.5	12
SU-1 West	67	100	3600	2.4	3.5–5
SU-1 East	74	100	1200	2.4	3.5–5
SU-2	25	100	1800	1.6	7.7
R-1	284	100	4800	3.0	1.4
R-2	50	100	1800	0.8	1.4

TABLE 7-11. Simulated Chloride Concentrations Compared to Field Data.[a, b]

Area	(1) Yearly Average Cl Concentrations at Upland Well Sites (mg/liter)		(2) Yearly Average Cl Concentrations at Well Sites Near the River (mg/liter)	(3) Average for the Area (mg/liter)	(4) Chloride Loading Rate (g/m^2/month)	(5) Simulated Cl Concentrations (mg/liter)
HU-1	Site W3	Site W14	Site W2 West			
	246	20	235	167	57	187
HU-2	–		Site W2 East			
			147	–	57	214
HU-3	Site W11		–	–	57	205
	145				35	150
LU-1	Site W13		Site W12			
	71		58	64	12	52
SU-1 West	–		Site: W1 W6 W7	89	3.5	64
			39 106 121		5.0	88
SU-1 East	–		Site: W1 W6 W7 W8	130	3.5	90
			270 113 115 20		5.0	125
SU-2	Site W10		Site W5			
	9		338	174	7.7	165
R-1	–		–	–	1.4	23
R-2	–		Site W9	–		
			34		1.4	42

[a]Compare columns 2 and 3 with column 5.
[b]From Anderson et al.[17]

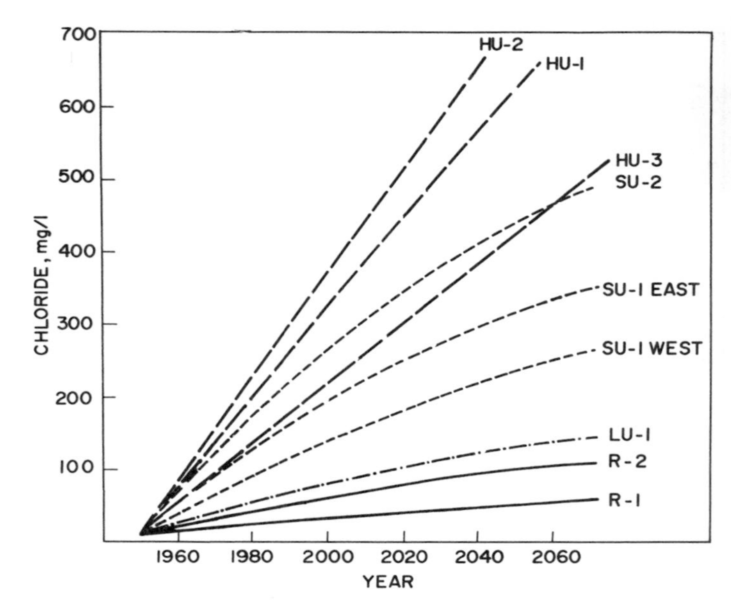

Fig. 7-15. Response of the Menomonee River aquifer to salt inputs computed by the Gelhar-Wilson model. (Source: Menomonee River Pilot Watershed Study.[17])

1968. It was assumed that road salt was applied at this rate continuously since 1950, the year that marked a nation-wide increase in the use of salt for ice control on highways.

Since the Gelhar-Wilson model is a lumped parameter model, it must be calibrated. A historical record of chloride concentration in wells or base flow is needed for calibration of the model parameters.

A comparison of computed chloride concentrations and observed concentrations in the walls is presented in Table 7-11. Based on the comparison and considering the limitation of the model, it was concluded that the model does simulate the probable response of the system.

Figure 7-15 shows long-term simulation-prediction results, indicating a marked effect of road salt application on the salinity of the shallow aquifer in the Menomonee River watershed. Once the model is calibrated for chlorides, it can be used to predict the impact of other conservative and possibly nonconservative pollutants on the groundwater and base flow quality.

REFERENCES

1. Anon. 1977. The report to Congress: Waste disposal practices and their effect on groundwater. U.S. EPA, Washington, D.C.

2. Miller, D. W., and Scalf, M. R. 1974. New priorities for groundwater quality protection. *Groundwater*, **12**:335–347.
3. Dallaire, G. 1979. Toxics in the N.J. environment: Microcosm of the U.S. ills. *Civil Engineering*, **49**:74–79.
4. Todd, D. K. 1959. "Groundwater Hydrology." Wiley, New York.
5. Freeze, R. A., and Cherry, J. A. 1979. "Groundwater." Prentice-Hall, Englewood Cliffs, New Jersey.
6. Davis, S. N., and DeWiest, R. J. M. 1966. "Hydrogeology." Wiley, New York.
7. Bouwer, H. 1978. "Groundwater Hydrology." McGraw-Hill, New York.
8. Anon. 1975. National Interim Primary Drinking Water Standards. U.S. EPA, Washington, D.C.
9. Anon. 1973. Water quality criteria—1972. National Academy of Sciences, National Academy of Engineering, EPA-R3-73-033.
10. Hernandez, J. W., and Barkley, W. A. 1976. Training seminar on the National Safe Drinking Water Act. New Mexico State University, College of Engineering, Las Cruces, New Mexico.
11. Edworthy, K. J., Wilkinson, W. B., and Young, C. P. 1978. The effects of the disposal of effluents and sewage sludge on groundwater quality in the Chalk of the United Kingdom. *Prog. Water Tech.*, **10**:479–493.
12. Bear, J. 1972. "Dynamics of Fluids in Porous Media." American Elsevier, New York.
13. Peaudecerf, P., and Sauty, J. P. 1978. Application of a mathematical model to the characterization of dispersion effects on groundwater quality. *Prog. Water Tech.*, **10**:443–455.
14. Baetslé, L. H. 1969. Migration of radionuclides in porous media." In: A. M. F. Duhamel (ed.). "Progress in Nuclear Energy Series XII, Health Physics." Pergamon Press, Elmsford, New York, pp. 707–730.
15. Sawyer, C. N., and McCarty, P. L. 1978. "Chemistry for Environmental Engineering." McGraw-Hill, New York.
16. Betson, R. P., and McMaster, W. M. 1975. Nonpoint source mineral water quality model. *J. WPCF*, **47**:2461–2473.
17. Anderson, M. P., Eisen, C. E., and Hoffer, R. N. 1978. Vol. 5. Groundwater hydrology. Menomonee River pilot watershed study. International Joint Commission, Windsor, Ontario.
18. Kreitler, C. W., and Jones, D. C. 1975. Natural soil nitrate: The cause of the nitrate contamination of groundwater in Runnels County, Texas. *Groundwater*, **13**:53–61.
19. Cleary, R. W., and Miller, D. W. 1978. Introduction to groundwater pollution and hydrology. Seminar at Princeton University, Princeton, New Jersey.
20. Boyce, J. S., Muir, J., Edwards, A. P., Seim, E. C., and Olson, R. A. 1976. Geologic nitrogen in Pleistocene Loess of Nebraska. *J. Env. Quality*, **5**:93–96.
21. Chalk, P. M, and Keeney, D. R. 1971. Nitrate and ammonium contents of Wisconsin limestone. *Nature*, **229**:42–47.

22. Klusman, R. W., and Edwards, K. W. 1977. Toxic metals in groundwater of the Front Range, Colorado. *Groundwater*, **15**:160–169.
23. Chesters, G., and Simsiman, G. V. 1974. Impact of agricultural use of pesticides on the water quality of the Great Lakes. Task A-5, Water Resources Center, University of Wisconsin, Madison. International Joint Commission, Windsor, Ontario.
24. Anon. 1977. Environmental effects of septic tank systems. U.S.-EPA-600/3-77/096, Washington, D.C.
25. Sikora, L. J., Bent, M. G., Corey, R. B., and Keeney, D. R. 1976. Septic nitrogen and phosphorus removal test system. *Groundwater*, **14**:304–314.
26. Otis, R. J., Bouma, J., and Walker, W. G. 1974. Uniform distribution in soil adsorption fields. *Groundwater*, **12**:409–416.
27. Cartwright, K., and Sherman, F. B. 1974. Assessing potential for pollution from septic systems. *Groundwater*, **12**:239–240.
28. Brown, K. W., Wolf, H. W., Donnelly, K. C., and Slowey, J. F. 1979. The movement of fecal coliform and coliphages below septic lines. *J. Env. Quality*, **8**:121–125.
29. Reneau, R. B., Jr., and Pettry, D. E. 1976. Phosphorus distribution from septic tank effluent in coastal plain soils. *J. Env. Quality*, **5**:34–39.
30. Dudley, J. G., and Stephenson, D. A. 1973. Nutrient enrichment of groundwater from septic disposal systems. Upper Great Lakes Regional Commission Report.
31. Andreoli, A., Bartilucci, N., Forgione, R., and Reynolds, R. 1979. Nitrogen removal in a subsurface disposal system. *J. WPCF*, **51**:841–854.
32. Asano, T. 1979. State-of-the art review of groundwater recharge operations in California. Groundwater Symposium, Pomona, Cal., Summary published in *Municipal Wastewater Reuse News*, No. 26, U.S. Dept of Interior, Washington, D.C.
33. Bouwer, H. 1974. Design and operation of land treatment systems for minimum contamination of groundwater. *Groundwater*, **12**:140–147.
34. Bell, R. G., and Bole, J. B. 1978. Elimination of fecal coliforms from soil irrigated with municipal sewage lagoon effluent. *J. Env. Quality*, **7**:193–196.
35. Saffigna, P. G., and Keeney, D. R. 1977. Nitrate and chloride in groundwater under irrigated agriculture in Central Wisconsin. *Groundwater*, **15**:170–177.
36. Walker, W. R. 1978. Identification and ititial evaluation or irrigation return flow models. U.S.-EPA-600/2-78/144, Ade, California.
37. Burwell, R. E., Schuman, G. E., Saxton, K. E., and Heineman, H. G. 1976. Nitrogen in subsurface discharge from agricultural watersheds. *J. Env. Quality*, **5**:325–329.
38. Zanoni, A. 1972. Groundwater pollution and sanitary landfills—A critical review. *Groundwater*, **10**:3–13.
39. Anon. 1977. The report to the Congress: Waste disposal practices and their effect to groundwater. U.S. EPA, Washington, D.C.

40. Johansen, O. J., and Carlson, D. A. 1976. Characterization of sanitary landfill leachates. *Water Res.*, **10**:1129–1134.
41. Anon. 1977. Procedures manual for groundwater monitoring at solid waste disposal facilities. U.S.-EPA 530/SW-611, Washington, D.C.
42. Bachmat, Y., Andrews, B., Holtz, D., and Sebastian, S. 1978. Utilization of numerical groundwater models for water resources management. U.S.-EPA-600/8-78/012, Ada, Oklahoma.
43. Gelhar, L. W., and Wilson, J. L. 1974. Groundwater quality modeling. *Groundwater*, **12**:399–408.

8

Pollution from impervious urban areas

INTRODUCTION

The nonpoint pollution generation in urban areas is quite different from that in nonurban or suburban lands. Several factors cause this difference:

1. Large portions or urban areas are impervious, resulting in their much higher hydrological activity (Fig. 8-1).
2. Except for construction sites, most of the pervious surfaces in residential or city areas are well protected by lawns, and as a consequence, erosion is reduced.
3. The pollution loadings in urban areas are affected mainly by litter cumulation, wet and dry atmospheric fallout, and traffic, while in nonurban or suburban areas soil erosion and soil-adsorbed pollutants cause most of the pollution.
4. Over a longer time period (e.g., a season or year) almost all of the pollution deposited on impervious surfaces

Fig. 8-1. Highly impervious urban areas are one of the largest sources of non-point pollution. (Photo University of Wisconsin.)

that has not been removed by street cleaning, wind, or decay will eventually end up in surface runoff. On the other hand, soil represents an infinite pool of sediments and potential pollutants associated with soils in non-urban and suburban areas. Their removal rate depends on surface protection and on the energy of rain and surface runoff that liberate the soil particles.

The amount of refuse accumulated on impervious surfaces depends on various factors and inputs. Atmospheric fallout, street litter deposition, animal and bird fecal waste, dead animals and vegetation, fallen leaves and grass residues, and road traffic impact are the major inputs. The factors that affect the quality of the street refuse include land use, protection of surrounding pervious areas from wind erosion, population, traffic flow and frequency, effectiveness of street cleaning, season of the year, meterological conditions, and street surface type and condition. The most significant refuse component deposited on impervious surfaces in terms of producing runoff pollution is the "dust and dirt" fraction, or that fraction of street refuse passing a 3.2-mm ($\frac{1}{8}$-in.) hardware cloth.[1]

The atmospheric input (which was discussed in Chapter 4) represents only a portion of the total pollution loading from urban nonpoint sources. Randall et al.[2] concluded that the atmospheric fallout in urban areas is sufficient to have a significant impact on runoff quality in a large metropolitan area. Based on the study in the greater Washington, D.C., area, the wet fallout is more significant than dustfall during dry periods. A study in Sweden[3] revealed that 20% of the organic matter, 25% of the phosphorus, and 70% of the total nitrogen in storm water can be attributed to atmospheric fallout. In addition, urban snow was found to have significantly higher pollutant concentrations than average storm water from the same area. The study recommended that the urban snow from congested areas should be treated as sanitary sewage.

The magnitude of pollution generated from impervious urban areas is of the same order as the raw sewage contribution. In Durham, North Carolina, the annual pollution from urban nonpoint sources equaled 91% of the COD load of raw sewage generated in the areas, BOD_5 yields were 67%, and the suspended solids yield was 20 times higher than that contained in the raw municipal sewage.[4] Akerlinch[5] in Scandanavia, Shigorin[6] in the Soviet Union, and Palmer[7] in the United States found BOD_5 values in storm runoff from urban areas that approached 10 times the concentrations in sewage after secondary treatment. Pravoshinsky[8] analyzed the quality of storm sewer discharges in Minsk, USSR. The storm sewers received rainfall runoff, snow-melt, water from street washings, and some cleaner industrial wastewaters. The discharge from storm sewers into receiving water is permitted without treatment. He found a distinct relationship between the concentrations of BOD_5 and suspended solids and rainfall intensity. The BOD_5 ranges were 85 to 95 mg/liter for snow-melt, 25 to 30 mg/liter for washwater, and 11 to 100 mg/liter for rainfall. Suspended solids concentration ranged from 300 to 3500 mg/liter, with the snow-melt water exhibiting the highest pollution potential.

The temporal pattern of storm runoff and its effect on receiving waters is also important. While sewage "dry-weather" flow exhibits some diurnal variations, the effect of treatment plant effluents on the receiving waters is more or less uniform. The storm water, on the other hand, represents shock loadings that can be 100 to 1000 times greater than that from the dry-weather sewage. These loadings can have a significant and often devastating effect on receiving waters.

Combined sewer systems in older urban areas represent a specific and very serious problem. The overflows from these systems, in addition to the storm-water pollution, contain diluted sewage, foul organic deposits and bacterial growths from sewers, and catchbasin deposits and debris. In the Milwaukee, Wisconsin, area, over 40% of the pollution reaching receiving waters on an annual basis is from a relatively small urban area served by combined sewers.

This chapter will explain and describe the mechanisms and processes that contribute to pollutant accumulation and wash-off from impervious areas. The general loading magnitudes related to the land use and other causative factors will be discussed in Chapter 10. General magnitudes and statistical U.S. averages of loadings from urban areas were presented in Ref. 9.

DEPOSITION AND ACCUMULATION OF POLLUTANTS ON IMPERVIOUS SURFACES

Street Refuse Characteristics and Deposition

Figure 8-2 is a schematic representation of the street surface pollutant accumulation process. The primary sources of pollution are wet and dry atmospheric deposition, litter cumulation, and traffic. The pollutants deposited on the surface during a dry period can be carried by wind and traffic and cumulate near the curb or median barrier. Thus, many urban pollution studies report the street pollution cumulation rates related to the unit length of the curb instead of an apparently more logical area loading. This seems to be justified since observations indicate that nearly 80% of the street refuse can be found within 15 cm from the curb and over 95% within 1 m from the curb[10] (Fig. 8-3). Table 8-1 depicts some pollutant loading magnitudes on impervious urban areas.

A strong correlation between curb length density and percent imperviousness of residential areas (Fig. 8-4) can be used to convert curb loadings to areal pollution generation or to estimate total curb lengths. The American Public Works

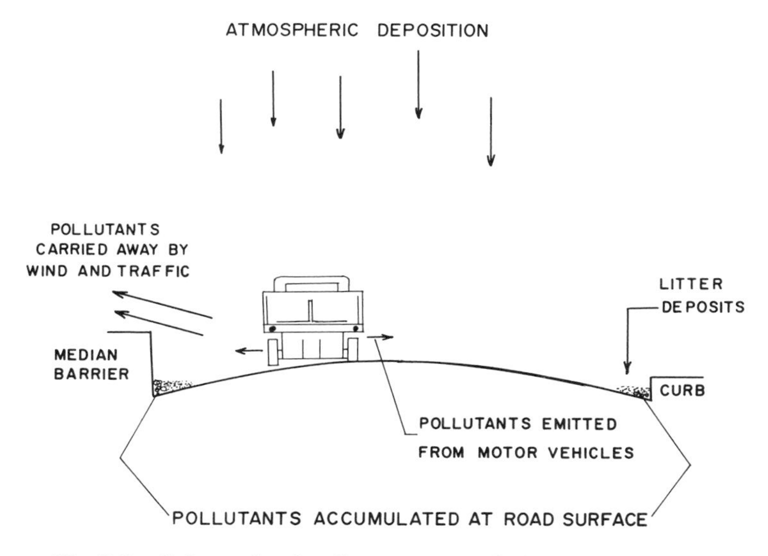

Fig. 8-2. Schematic of pollutant accumulation in urban areas.

Fig. 8-3. Most of the street refuse remains within 1 m from the curb. (Photo authors.)

TABLE 8-1. Street Refuse Accumulation.

	Solids Accumulation (g/curb m/day)	
Land Use	Chicago[a] Dust and Dirt	Eight American Cities[b] Total Solids
Single family	10.4	48
Multiple family	34.2	66
Commercial	49.1	69
Industrial	68.4	127
Average of above (weighted)	22.3	–

[a]American Public Works Association.[1]
[b]Sartor et al.[10,11]

Approximate dust and dirt content of total solids = 75%.

Association developed a regression formula between the curb length of urban areas and population density based on an analysis of many American cities. The resulting regression equation (converted to metric units) was:

$$CL = 311.67 - 266.07 \times 0.839^{(2.48PD)} \tag{8-1}$$

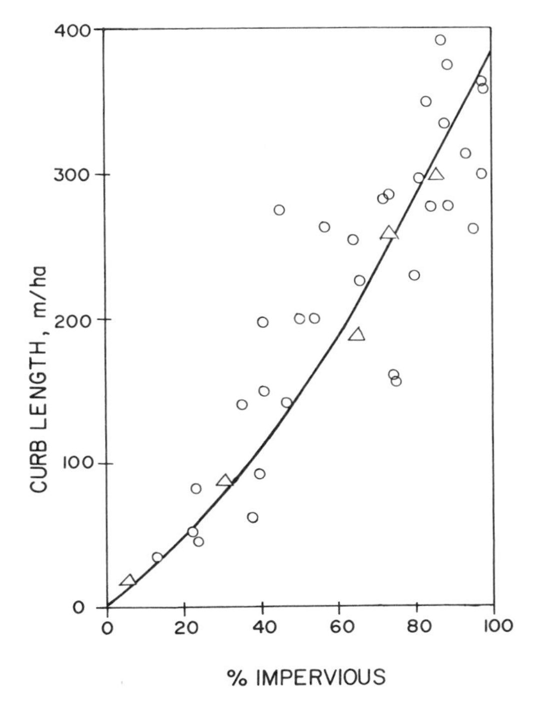

Fig. 8-4. Relationship of curb length density to total imperviousness of urban areas.

where:

CL = curb length (m/ha)
PD = population density (persons/ha)

The street refuse washed by runoff to surface waters contains many contaminants. Significant amounts of organics, heavy metals, pesticides, and bacteria are commonly associated with street refuse, especially its dust and dirt fraction. Tables 8-2 and 8-3 show typical contamination values. Although these values are thought to be typical, they are not uniform, but represent averages from a wide range of street refuse deposition and contamination for a limited number of investigated municipalities. Figure 8-5 shows that PCBs can be readily adsorbed on fine fractions of the street dust and carried in suspension by the street runoff.

Atmospheric Fallout. As shown in Chapter 4 (Table 4-6), in most larger cities the deposition rates of atmospheric particulates in wet or dry fallout range from 7.0 tonnes/km^2-month (=g/m^2-month) to more than 30 tonnes/km^2-month. As

TABLE 8-2. Pollutants Associated with Street Refuse (μg/g of Total Solids).

Constituent	Land Use			
	Residential	Industrial	Commercial	Transportation
BOD_5	9,166[a]	7,500[a]	8,333[a]	2,300[b]
COD	20,822[a]	35,714[a]	19,444[a]	54,000[b]
Volatile solids	71,666[a]	53,571[a]	77,000[a]	51,000[b]
Total Kjeldahl nitrogen	1,666[a]	1,392[a]	1,111[a]	156[b]
PO_4^- P	916[a]	1,214[a]	833[a]	610[b]
NO_3^- N	50[a]	64[a]	500[a]	79[b]
Pb	1,468[c]	1,339[c]	3,924[c]	12,000[b]
Cr	186[c]	208[c]	241[c]	80[b]
Cu	95[c]	55[c]	126[c]	120[b]
Ni	22[c]	59[c]	59[c]	190[b]
Zn	397[c]	283[c]	506[c]	1,500[b]
Total coliforms (No./g)	160,000[a]	82,000[a]	110,000[a]	NR
Fecal coliforms (No./g)	16,000[a]	4,000[a]	5,900[a]	925[b]

[a] Sartor et al.[10,11]—average of several American cities.
[b] Shaheen[16]—Washington, D.C., area.
[c] Amy et al.[13]
NR—not reported.

TABLE 8-3. Mean Concentrations of Organic Chemicals in Urban Dust and Dirt (μg/g). Averages of Several United States Cities.

Constituent	Concentration	Standard Deviation
Endrin[a]	0.00028	0.00078
Dieldrin[a]	0.028	0.028
PCBs (overall)[a]	0.78	0.76
Methoxychlor[a]	0.50	1.1
Lindane[a]	0.0022	0.0063
Methylparathion[a]	0.0024	0.0073
p, p-DDD[a]	0.082	0.080
p, p-DDT[a]	0.075	0.12
Asbestos[b]	160,000 fibers/g	–

[a] References 11, 14.
[b] Shaheen[16]—transportation zones, Washington, D.C.

expected, higher deposition rates occur in congested downtown and industrial areas and lower rates are typical for residential and rural suburban areas.

In addition to particulates, the wet and dry atmospheric fallout contain many other pollutants. As Table 4-8 shows, dustfall in urban areas may contain 30 to

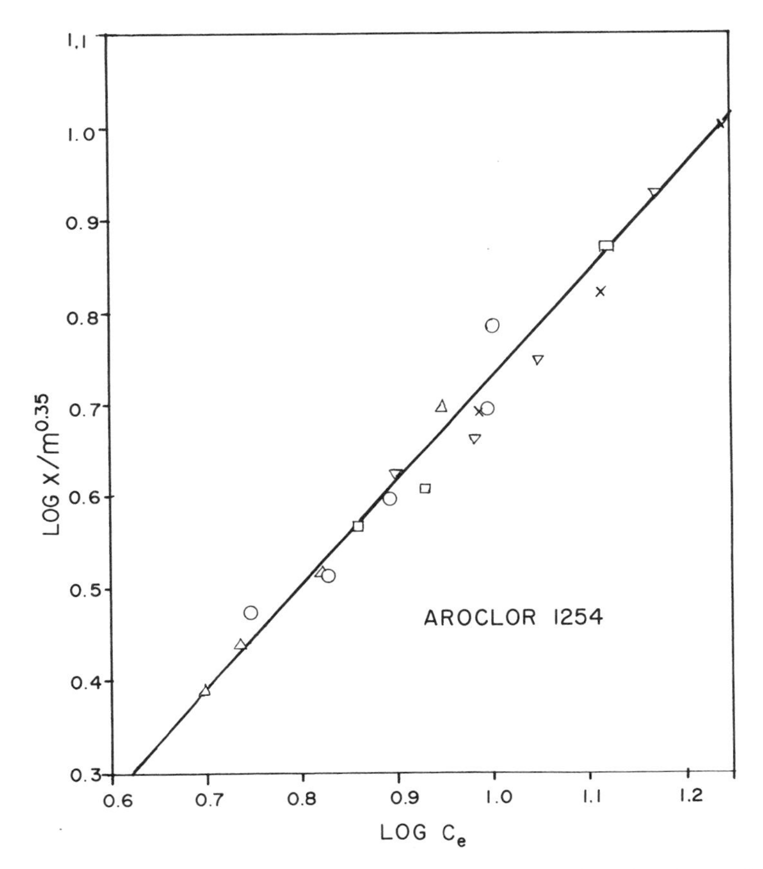

Fig. 8-5. Adsorption isotherm of PCB Aroclor 1254 on fine fractions of urban dust.

40% organic matter, 0.05 to 0.2% lead, 0.01 to 0.05% zinc and other pollutants. The mean concentrations of suspended solids in rainwater (wet fallout) can range from 1.2 to 10 mg/liter, nitrate content from 0.6 to 3.6 mg/liter, and total phosphorus from 0.016 to 0.135 mg/liter.[15]

Pitt,[14] after an extensive study in the San Francisco Bay area, concluded that a majority of street surface contaminants depend on local geological conditions with added fractions from motor vehicle emissions and road wear. Most of the street refuse particles originate from local erosion of soils and are transported by air.

Lead, which is associated with combusiton of leaded gasoline, is transported with atmospheric particulate matter. The influence of traffic on lead concentrations in the atmosphere and its deposition rates is profound. For example, in Munich, Federal Republic of Germany, traffic was prohibited for some days, and deposition rates dropped to 10 to 15% of normal ranges after 1 day. Also, cadmium, strontium, zinc, nickel, and many organic chemicals including PCBs are transported with atmospheric aerosols and can be found in the atmospheric fallout.

Effect of Wind. The following factors are important in assessing the impact of wind erosion: climate, soil characteristics, surface roughness, vegetation cover, and length of the eroding surface. The great dust storms that devastated portions of the southwestern United States during the first part of this century resulted from a combination of the most adverse factors. Wind erosion will cause the greatest losses on loose, dry, finely granulated soils. A roughened soil surface or a surface covered by vegetation and tree windbreaks can significantly reduce wind velocity near the surface and, therefore, the extent of erosion. In urban areas, the primary source of materials eroded by wind are open, ungrassed areas and construction sites.

Wind—natural or traffic-induced air movement—can translocate deposited particles on the surface and move them towards the curbs, median barriers, and other obstacles that reduce air velocity. These obstacles and the pockets of calmer air they form represent depositional traps for airborne particulates. This is why almost all surface pollution by particulate matter is found near these obstacles and not on the road surfaces. Since air movement translocates the street surface particles, one would suspect that the curb or obstacle height might influence the deposition rates. Figure 8-6 shows such a relationship measured in the Washington, D.C., area. This graph indicates that the curb may have a profound effect on cumulation rates of finer dust and dirt particles (<3.2 mm), while larger litter particles were not affected.

The amount of particulates that wind removes from the impervious surfaces and curb storage increases with the amount of particulates present or remaining. This leads to a classical "first-order removal concept," which also means that the amount of pollutants deposited on the surface during a dry period will have a decreasing rate of increase and not a uniform linear increase. Figure 8-7 shows that the deposition rates seem to be higher following a rain or street cleaning. Wind and traffic can remove more pollutants as they cumulate on adjacent pervious areas, and the deposition curve tends to level off after a few dry days.

The leveling-off phenomenon is less profound in areas where little or no adjacent pollution traps (pervious areas) are available, that is, when the area's imperviousness approaches 100%.

Effect of Traffic. Traffic can significantly contribute either directly or indirectly to the deposition rates of pollutants in urban areas and near traffic corridors. High amounts of toxic metals, especially lead, are often attributed to motor vehicle emissions and the breakdown of vehicle parts and road surfaces.

Table 8-4 lists major traffic-related sources of various pollutants. It should be

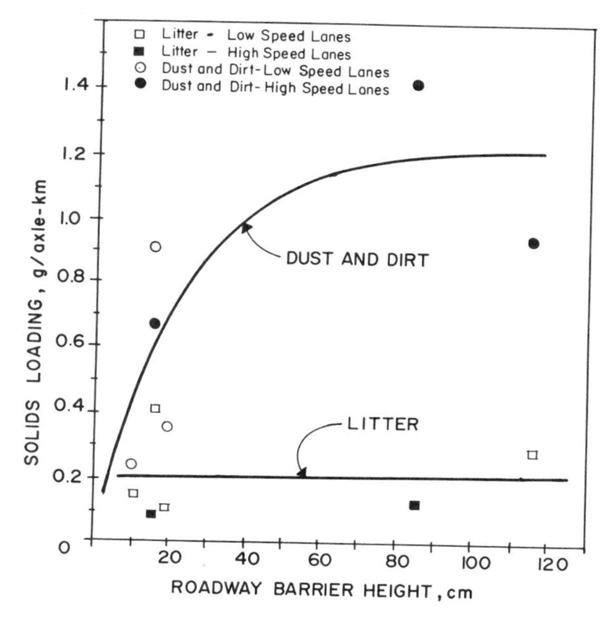

Fig. 8-6. Effect of curb height on particle loading of street refuse. (Replotted from reference 16.)

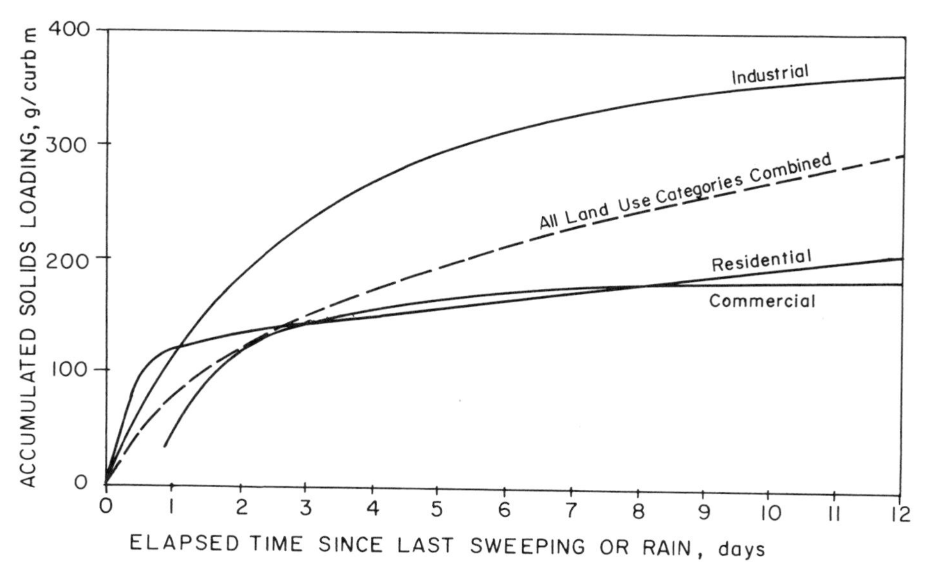

Fig. 8-7. Particle accumulation rate decreases with increased period of accumulation. (Replotted from reference 11.)

TABLE 8-4. Traffic-Related Sources of Roadway Pollution.[a]

Pollutant	Traffic Source
Asbestos	Clutch plates, brake linings
Copper	Thrust bearing, bushing, and brake lining
Chromium	Metal plating, rocker arms, crankshafts, rings, brake linings, and pavement materials
Lead	Leaded gasoline, motor oil, transmission babbitt metal bearings
Nickel	Brake linings and pavement material
Phosphorus	Motor oil additive
Zinc	Motor oil and tires
Grease and hydrocarbons	Spills and leaks of oil and *n*-paraffins lubricants, antifreeze, hydraulic fluids
Rubber	Tire wear

[a]Adapted from Shaheen.[16]

realized that only a small portion (<5%) of traffic-related pollution can be directly traced to motor vehicles.[16] However, the pollutants that motor vehicles emit are among the most important because of their potential toxicity.

In addition to traffic density, the pavement condition and compactation are significant in determining the traffic impact on pollution accumulation. Streets paved entirely with asphalt have loadings about 80% higher than all concrete streets.[10,11] Streets whose conditions were rated "fair-to-poor" were found to have total solids loadings 2.5 times greater than those rated "good-to-excellent." Similar conclusions were also reported by Pitt,[14] who showed that pollutant loadings from "good" asphalt in downtown San Jose, California, were less than half of those from "poor" asphalt. General composition of traffic-related street dust and dirt, along with the dust and dirt emission rates, are given in Table 8-5.

Lead, which is the most recognized traffic-related component on the surface runoff pollution, is a gasoline additive. The quantities of lead addition may vary according to local conditions. In the United States, lead use is curtailed and most new cars are required to use unleaded gasoline. In Europe, the permissible lead concentrations vary from 0.1 g/liter in the Federal Republic of Germany to 0.45 g/liter in England.[17] Many nations still do not restrict lead content in the fuel. Some industrial air pollution aerosols also contain significant amounts of lead.[18]

The deposition rates of lead may vary according to traffic density. They can range from 2.35 mg/m^2-day on a busy highway to 0.046 to 0.186 mg/m^2-day for background deposition rates. The measured levels of lead associated with street dust and dirt ranged from 1.0 to 20.0 mg/g of solids, with average values from 2 to 4 mg/g.[17] These values are considerably higher than the lead levels found in soils (see Tables 6-1 and 6-2).

TABLE 8-5. Composition of Traffic-Related Depositions from the Washington, D.C. Area.[a]

	% of Total Solids by Weight
Volatile solids	5.1
BOD_5	0.23
COD	5.4
Grease	0.64
Total P	0.06
TKN	0.016
Nitrate	0.008
Asbestos	3.6×10^5 fibers/g
Lead	1.2
Chromium	0.008
Copper	0.012
Nickel	0.019
Zinc	0.15
Emission rates of total solids	0.671 g/km-axle

[a]After Shaheen.[16]

Litter Deposition. Litter deposits in urban areas include solid wastes deposited on surfaces by careless public, private, and municipal waste deposition and collection operations, animal and bird fecal droppings, fallen tree leaves, grass clippings, and other deposits. The dust and dirt component is regarded as having the greatest pollution potential.

Although most of the litter deposit occurs originally in sizes greater than 3.2 mm, it is possible that a significant portion of the street dust and dirt originates from mechanical fracture of litter. The Public Works Association[1] reported that residential areas had greater amounts of street surface dust and dirt as population density increased, reflecting increased pedestrian and roadway traffic. Greater street surface deposition rates by public and private solid-waste storage and collection practices would also be expected to be greater in high-density urban areas. The general attitude of the public towards the cleanliness of the area is also an important factor.

Effect of Vegetation. Fallout of leaves and grass clippings in urban areas may contribute significantly to dust and dirt accumulation on adjacent impervious areas. The rates of litter deposition from vegetation increase substantially during the fall.

From the work by Carlisle et al.[19] Heaney and Huber[20] estimated an average leaf fallout of 14.5 to 26 kg/tree-yr. The wooded area investigated was stocked with uneven-aged (about 40 to 120 yr) trees with a 90 to 96% closed canopy and 155 trees/ha. The tree species were mainly oak (*Quercus* sp.) and birch

(*Betula* sp.). A typical value of leaf fallout in Minnesota in a forested area with about 420 trees/ha is about 380 ton/km^2-yr. Of these yearly values, about 65% occurs during the fall. The fallen leaves are about 90 to 95% organic and contain about 0.04 to 0.28 phosphorus.

Only the portion of the vegetation residues that cumulates on impervious areas represents a great pollution hazard to surface waters. The vegetation residues on soils become an integral part of the soil composition, and in most cases may even improve soil permeability and erosion resistance.

Particle Size Distribution of Street Refuse. Table 8-6 and Fig. 8-8 indicate that most of the street refuse is in coarser fractions, roughly in sand and gravel equivalent sizes. However, most of the pollutants, as they are in soils, are associated with fine fractions. For example, 6% of solids with sizes less than 0.43 μ (equivalent to clay and silt fractions in soils) contain more than 50% phosphorus.

Street cleaning practices are selective for certain fractions, and the removal efficiency expressed as solids removal rates may not be the same as pollutant removal rates. Street sweeping is selective for coarser particles, while street flushing and surface runoff are selective for finer particles.

Pollution Accumulation Model

Pollution accumulation in urban areas is a highly random process and no sophisticated mathematical formulations will yield highly reliable results. In such cases, simplified semiempirical models may be as good as or even better than detailed analytical concepts. On the other hand, oversimplified concepts, such

TABLE 8-6. Percent of Street Pollutants in Various Particle Size Ranges.[a]

	Particle Size (μm)					
Pollutant	>2000	840–2000	240–840	104–246	43–104	<43
Total solids	24.4	7.6	24.6	27.8	9.7	5.9
Volatile solids	11.0	17.4	12.0	16.1	17.9	25.6
COD	2.4	4.5	13.0	12.4	45.0	22.7
BOD_5	7.4	20.1	15.7	15.2	17.3	24.3
TKN	9.9	11.6	20.0	20.2	19.6	18.7
Phosphates	0	0.9	6.9	6.4	29.6	56.2
All toxic metals	16.3	17.5	14.9	23.5	–	27.8
All pesticides		27.0			73.0	
PCBs		66.0			34.0	

[a]After Sartor et al.[11]

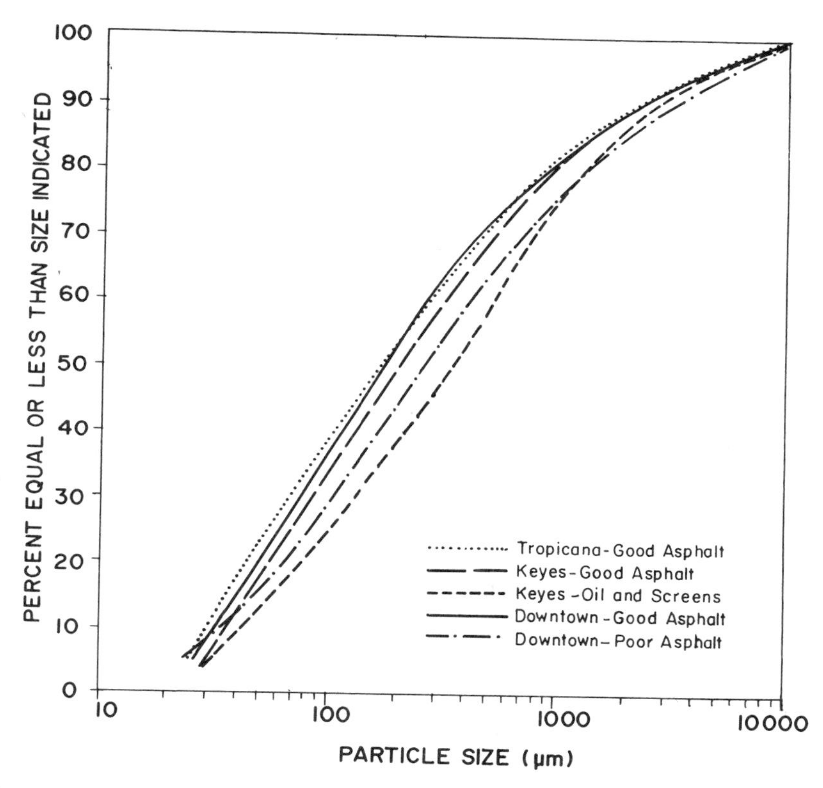

Fig. 8-8. Particle size distribution of street refuse in San Francisco Bay area before sweeping. (Reprinted from reference 14.)

as simple surface or curb loadings, may not be adequate because the kinetics of the pollution accumulation process is apparently non-linear.

As in an storage-input-output balance schematics, the amount of accumulated pollutant on impervious surfaces could be described by a simple mass balance formula:

$$\frac{dP}{dt} = \Sigma I - \Sigma L \tag{8-2}$$

where:

P = the amount of street refuse or dust and dirt present on the surface
ΣI = sum of all inputs
ΣL = sum of losses

This simple mass balance equation can be expanded by identifying the significant factors that affect the deposition and removal of pollutants on and from street surfaces and the curb storage. The inputs which have been discussed in the previous sections include atmospheric fallout, litter deposits, vegetation residues, and traffic impact. The losses include mainly translocation of the accumulated dust from the surface towards the curb and on adjacent pervious areas. Wind, both natural and traffic-induced, curb height, presence or absence of adjacent pervious areas that act as pollutant traps, and general airflow patterns in the street "canyon" affect the losses. The basic concepts of the pollutant cumulation model are shown in Fig. 8-2.

Figure 8-6 demonstrates that the effect of the curb on dust and dirt losses can be described approximately by an exponential equation of the type:

$$\zeta = af(TS, WS)e^{-bH} \tag{8-3}$$

where:

ζ = the fraction of the pollutant lost from the curb storage due to a combined effect of traffic and wind
TS = traffic speed (km/hr)
WS = wind speed (km/hr)
a, b = coefficients
H = curb height (cm)

If the curb height is zero, it will cause no obstruction and the particles will be removed by wind. When the curb becomes higher, particles are trapped in the curb storage. If Equation 8-3 is introduced into Equation 8-2, the following mass balance formula is obtained

$$\frac{dP}{dt} = \Sigma I - af(TS, WS)e^{-bH}P = A - \zeta P \tag{8-4}$$

where

$$A = \Sigma I \quad \text{and} \quad \zeta = af(TS, WS)e^{-bH}$$

After integration, Equation 8-4 becomes

$$P = \frac{A}{\zeta}(1 - e^{-\zeta t}) + C \tag{8-5}$$

The inputs can be combined and expressed per unit length of curb to yield

$$A = \Sigma I = \left(\frac{SW}{2}\right)(ATMFL + LIT) + C_1(TD)(RCC)$$

where:

A = the pollutant deposition rate per unit curb length (g/m-day)
SW = street width

$ATMFL$ = atmospheric fallout rate (g/m^2-day) = tonnes/km^2-day
LIT = litter deposition rate (g/m^2-day)
C_1 = emission rate of pollutants from vehicle traffic (g/km-axle)
TD = traffic density (thousands axles/day)
RCC = a function expressing road conditions

Using the data from Washington, D.C.,[16] Novotny and Goodrich-Mahoney[21] obtained the best-fit equation for A and ζ in the form:

$$A = (ATMFL + LIT)\left(\frac{SW}{2}\right) + 1.15(TD)$$

$$\zeta = 0.0116e^{-0.088H}(TS + WS)$$

$$C = 0.0$$

which is applicable to the dust and dirt cumulation.

Sutherland and McCuen[22] analyzed the data of Sartor et al.[11] and many other sources. They developed an accumulation component of the street refuse accumulation and removal model using the pattern search technique. The independent variables used in analyzing the curb solids cumulation were annual average daily traffic volume (AADT) and pavement conditions expressed by the present serviceability index (PSI). The PSI is an established standard used throughout the United States to classify pavement conditions. The result of the search and the final form of the statistical equations are given in Table 8-7.

Most of the pollution cumulation equations are suggested types of formulas with limited regional applicability. An equation needed to fit local conditions should be selected from and tested by surveys in the particular area of study.

TABLE 8-7. Street Refuse Accumulation Equations.[a]

	Land Use			
	Multifamily[b]	Single Family[b]	Industrial[c]	Commercial
Equation	$P_R = Pt/(1 + Kt)$	$P_R = Pt/(1 + Kt)$	$P_I = P(1 - e^{-Kt})$	$P_C = P(1 - e^{Kt})$
Parameter	$P = e^{(6.29 - 0.18C_p)}$	$P = e^{(5.96 - 0.17C_p)}$	$P = 204 - 13.6C_p + 0.0057T^{1.1}$	$P = 130 - 20C_p + 0.013T^{0.85}$
	$K = 1.25 + 0.016C_p^{1.27}$	$K = 1.173 + 0.017C_p^{1.18}$	$K = 0.175T^{0.035} - 0.03C_p$	$K = 0.335$

P = total accumulation (g/m)
T = traffic volume (AADT)
C_p = present serviceability index (PSI)
t = time in days since last rainfall or sweeping

[a] After Sutherland and McCuen.[22] Used with permission of American Water Resources Association.
[b] For residential land use and all concrete pavements, decrease P by 8%.
[c] For industrial land use and concrete pavement, decrease P by 10%.

Example 8-1: Refuse Accumulation

Estimate the amount of refuse that accumulated in the curb storage in a medium-density residential area. The area has a mixture of single and multiple-family houses. The period of cumulation started after a large storm and lasted 10 days. Average atmospheric fallout measured in the areas was 15 tonnes/km^2-month, litter accumulation amounted to about 1.2 tonnes/km of curb per month, and street widths are about 15 m. Wind speed was about 20 km/hr and posted traffic speed was 50 km/hr. Pavement conditions were rated $C_p = 2.0$ and average traffic flow in the street was 2000 cars (4000 axles)/day. Curb height is 20 cm.

Use Equation 8-5 and the Sutherland and McCuen[22] models to estimate the solids cumulation and compare the results.

Solids Cumulation Estimate by Equation 8-5. *Select solids emission rate 1.0 g/axle-curb km and RCC = 1.0, which reflects approximately average conditions. Then*

$$A = ATMFL \times \frac{SW}{2} + LIT + C_1 \times TD \times RCC$$

$$= \frac{15}{30} \times \frac{15}{2} + \frac{1200}{30} + 1.15 \frac{4000}{1000} 1.0 = 47.75 \text{ g/m-day}$$

If there was no possibility of translocating the accumulated solids on adjacent pervious areas, the amount of solids accumulated during the 10-day period would be $10 \times A = 477.5$ g/m. However, due to the wind and traffic effects, part of the solids will be carried away from the curb storage. Therefore

$$\zeta = 0.0116 e^{-0.088H}(TS + WS) = 0.0166 e^{-0.088 \times 20}(20 + 50) = 0.14 \text{ day}^{-1}$$

and

$$P = \frac{47.75}{0.14}(1 - e^{-0.14 \times 10}) = 256.96 \text{ g/m}$$

Solids Cumulation by Sutherland and McCuen's Models. *For residential multi-family area*

$$P_R = Pt/(1 + Kt)$$

and when

$$C_p = 2.0$$

$$P = e^{(6.29 - 0.18C_p)} = e^{(6.29 - 0.18 \times 2)} = 376.15$$

$$K = 1.125 + 0.016 \times 2^{1.27} = 1.16$$

then

$$P_R = 376.15 \times 10/(1 + 1.16 \times 10) = 298.53 \text{ g/m}$$

Residential single-family area

$$P = e^{(5.96-0.17\times 2.0)} = 275.89$$

$$K = 1.173 + 0.017 \times 2.0^{1.18} = 1.21$$

$$P_R = 275.89 \times 10/(1 + 10 \times 1.21) = 210.6 \text{ g/m}$$

Average loading for the area

$$P_R = \frac{210.6 + 298.53}{2} = 255 \text{ g/m}$$

From Fig. 8-7, the average curb loading for a 10-day period would be around 200 g/m.

Example 8-2: Effect of Curb Height

If the curb height was lowered to 5 cm what would the approximate effect on the amount of pollutants accumulated in the curb storage be?

Use the information from Example 8-1.

$$\zeta_{5\,\text{cm}} = 0.0116 e^{-0.88\times 5} = 0.52$$

Then the curb loading becomes

$$P_{5\,\text{cm}} = \frac{47.75}{0.52}(1 - e^{-0.52\times 10}) = 92.32 \text{ g/m}$$

which would indicate about 64% reduction in loadings.

Example 8-3: Accumulation of Phosphates and Lead

Estimate curb accumulation loadings for phosphates and lead for conditions of Example 8-1.

No loading formulation for phosphates is available. Therefore, the phosphate loading must be estimated using the "potency factor" given in Table 8-2. For residential areas, the average phosphate content of the street refuse is about 900 μg/g.

Then the phosphate content of 257 g/m of curb solids is

$$P_{PO_4} = 900 \times 10^{-6} \times 257 = 0.22 \text{ g/m}$$

The lead content of street refuse in residential areas with lower traffic volumes is about 1000 μg/g. Hence

$$P_{Pb} = 1000 \times 10^{-6} \times 257 = 0.257 \text{ g/m}$$

The lead accumulation can also be computed using Equation 8-5 if appropriate values for lead content of the atmospheric fallout and litter deposits and ranges of the removal functions are known.

The deposition rates of lead from the atmosphere in lower traffic density areas is about 1.6 mg/m²-day (see Laxen and Harrison).[17] *The emission rate of lead from cars can be estimated as follows: Assume the average lead content of the gasoline is 0.4 g/liter (depending on the proportion of cars using unleaded fuel) and average fuel consumption is 0.138 liter/km (corresponding to 17 miles/gal).*

Then if all lead is emitted the emission rate becomes

$$C_1 = 0.4 \text{ (g/liter)} \times 0.138 \text{ (liter/km)} = 0.055 \text{ g/road km-car}$$

$$= 0.027 \text{ g/road km-axle} = 0.014 \text{ g/curb km-axle}$$

From that

$$A = ATMFL \times \frac{SW}{2} + C_1 \times TD \times RCC$$

$$= 0.0016 \times \frac{15}{2} + 0.014 \times \frac{4000}{1000} \times 1.0 = 0.068 \text{ g/m-day}$$

The removal coefficient, ζ, for lead is probably not the same as it was estimated for solids since lead is more associated with finer dust fractions. As a rough estimate, increase the removal coefficient for solids by 100%.

Then

$$\zeta_{Pb} = 2.0 \times 0.14 = 0.28 \text{ day}^{-1}$$

and the accumulated lead during the 10-day period is

$$P_{Pb} = \frac{0.068}{0.28}(1 - e^{-0.28 \times 10}) = 0.23 \text{ g/m}$$

REMOVAL OF SOLIDS FROM STREET SURFACES

Street Cleaning Practices

Street cleaning practices include sweeping of street and parking lot surfaces by mechanical vehicles or flushing from tanker trucks (Figs. 8-9 and 8-10). Sweeping is more common in the United States, while street flushing is mainly practiced in Europe. Both practices are selective for certain particle size fractions. Sweeping is more selective for coarser dust and litter particles, while flushing is more selective for finer fractions. Flushing does not remove the particles from the

Fig. 8-9. Street sweeping vehicle. (Photo courtesy of Wisconsin DNR.)

Fig. 8-10. Street flushing vehicle. (Photo courtesy of University of Wisconsin.)

system, it only translocates the street refuse towards the drainage system. The effect of flushing on pollution reduction is marginal or negligible in areas served by separate sewers since most of the flush is collected by storm sewers and conveyed to receiving waters. In many cases the volume of water used for flushing is insufficient to transport the accumulated refuse to the nearest drain.[14]

As Pravoshinsky[8] has shown, the quality of runoff from street washing is very poor and is apparently inferior to storm runoff. However, street flushing may be advantageous in areas with combined sewer systems. Flushing cleans a larger street area and is more effective for picking up fine particles. The street runoff this practice generates during a dry period is collected by the combined sewer system and conveyed to the dry-weather sewage treatment plant. The capacity of the flow separators, as well as the treatment plant, is designed for flows that exceed the dry-weather flow by four to six times. This unused capacity can be effectively used to remove street contaminants. In addition, frequent flushing can help to clean deposited solids and organics in sewers which otherwise would be discharged untreated into receiving waters during storm overflow.

Street Sweeping. Current street sweeping practices are mostly for aesthetic purposes, and the removal efficiency of sweepers for finer dust and dirt fractions is low.[3,10] Two types of sweepers are presently used to remove solids from impervious urban surfaces. The most common design (mechanical street cleaners) uses a rotating gutter broom to remove the particles from the gutter area and place them in the path of a large cylindrical broom which rotates to carry the material onto a conveyor belt and into the hopper. Vacuum-assisted street cleaners use gutter and main pickup brooms to loosen and move street refuse into the path of a vacuum intake. The vacuum places the debris in the hopper.

Both types of sweepers are relatively ineffective for removing fine particles.[1,3,10,11,22] Broom mechanical cleaners are ineffective for particles in the dust and dirt range ($<$3.2 mm), and their overall efficiency is only about 50%. Vaccuum sweeper efficiency is higher, but is still ineffective for silt and clay-size particles.[1,10,11]

Cars parked by the roadside of clogged congested urban streets also reduce the effectiveness of street sweeping practices—often to zero. Street flushing is not affected by parked cars. An overnight ban on street parking and nightly sweeping may partially alleviate the problem.

In spite of relatively low sweeper efficiencies, Malmquist[3] reported significant improvements in water quality from areas with regular sweeping practices. Water flushed from unswept streets contained on the average 2.3 times more suspended solids and heavy metals than that from swept and cleaned streets. It seems that although a single pass of the sweeper may remove relatively low percentages of solids and pollutants, more pollution can be removed by repeated passes between the rains.

TABLE 8-8. Mechanical Street Cleaner Efficiencies for Various Equipment Passes (%).[a]

Size Range	50–500 g/curb m		
	1 pass	2 passes	3 passes
43	15	28	39
43 104	20	36	49
104 → 246 μ	50	75	88
245 → 840 μ	60	84	94
840 → 2000 μ	65	88	96
2000 μ → 6370 μ	80	96	99

[a]From Sartor et al.[10] and Pitt.[14]

Tables 8-8 and 8-9 show the street sweeper efficiency for removal of particles of various sizes. If the efficiency is combined with the pollutant distribution on particles of various sizes, the sweeper removal efficiency for the pollutants can be estimated (Table 8-10).

Sweeper efficiency can be related to the sweeper effort, a variable expressing the amount of time devoted to cleaning an area. Normal sweeping effort was defined by Sartor et al.[10,11] as 2.56 equipment min/1000 m^2 of cleaned area, which can be translated to the average sweeper vehicle velocity of about 10 km/hr. The information on street sweeper efficiencies contained in Table 8-8 refers to the normal operating effort.

TABLE 8-9. Removal Efficiencies for Vacuumized Street Cleaner at Different Initial Particulate Loadings and for Various Equipment Passes (%).[a,b]

Size Range	Street Surface Loading and Number of Passes								
	5–50 g/curb m			50–280 g/curb m			280–2800 g/curb m		
	1	2	3	1	2	3	1	2	3
44 → 74 μ	3	6	9	20	36	49	70	91	97
74 → 177 μ	50	75	88	60	84	94	75	94	99
177 → 300 μ	50	75	88	60	84	94	80	96	99
300 → 500 μ	60	84	94	65	88	96	70	91	94
750 → 1000 μ	50	75	88	60	84	94	70	91	97

[a]From Clark and Cobbins[25] and Pitt.[14]
[b]From cleaner path (0 to 2.5 m from curb), not total street loading.

TABLE 8-10. Removal Efficiencies from Cleaner Path for Various Street Cleaning Programs (%).[a, b]

Street Cleaning Program and Street Surface Loading Conditions	Total Solids	BOD_5	COD	KN	PO_4	Pesticides	Cd	Sr	Cu	Ni	Cr	Zn	Mn	Pb	Fe
Vacuum street cleaner															
1 pass: 5–50 g/curb meter total solids	31	24	16	26	8	33	23	27	30	37	34	34	37	40	40
2 passes	45	35	22	37	12	50	34	35	45	54	53	52	56	59	59
3 passes	53	41	27	45	14	59	40	48	52	63	60	59	65	70	68
Vacuum Street Cleaner															
1 pass: 50–280 g/curb meter total solids	37	29	21	31	12	40	30	34	36	43	42	41	45	49	59
2 passes	51	42	29	46	17	59	43	48	49	59	60	59	63	68	68
3 passes	58	47	35	51	20	67	50	53	59	68	66	67	70	76	75
Vacuum street cleaner															
1 pass: 50–500 g/curb meter total solids	48	38	33	43	20	57	45	44	49	55	53	55	58	62	63
2 passes	60	50	42	54	25	72	57	55	63	70	68	69	72	79	77
3 passes	63	52	44	57	26	75	60	58	66	73	72	73	76	83	82
Mechanical street cleaner															
1 pass: 50–500 g/curb meter total solids	54	40	31	40	20	40	28	40	38	45	44	43	47	44	49
2 passes	75	58	48	58	35	60	45	59	58	65	64	64	64	65	71
3 passes	85	69	59	69	46	72	57	70	69	76	75	75	79	77	82
Flusher	30	c	c	c	c	c	c	c	c	c	c	c	c	c	c
Mechanical street cleaner followed by a flusher	80	d	d	d	d	d	d	d	d	d	d	d	d	d	d

[a] Sources: Calculated from Clark and Cobbin,[25] Sartor et al.[10] and Pitt.[14]
[b] These removal values assume all the pollutants would lie within the cleaner path (0 to 2.5 m from the curb).
[c] 15 → 40% estimated.
[d] 35 → 100% estimated.

Increased street sweeping efficiency can be achieved by operating the vehicle at a slower speed or by conducting multiple passes. From the work by Sartor et al.[10,11] it follows that the removal of particles by street sweepers can be approximated by the equation

$$P = P^* + (P_0 - P^*)e^{-kE} \tag{8-6}$$

where:

P = the amount of street surface particulates in a given size range remaining after sweeping

P_0 = the initial amount of particulates in the size range

E = the amount of sweeping effort involving (min/1000 m^2) or as a relative effort (i.e., actual effort divided by the standard effort)

P^*, k = empirical constants depending on sweeper characteristic and design, particle size or particulates, and street surface characteristics.

If E is normalized by expressing it as a relative effort, the constant k becomes dimensionless. From the Sartor et al.[10,11] data it seems that k can be approximated best by the following equation

$$k = \alpha d^{\beta} \tag{8-7}$$

where α and β are empirical coefficients and d is particle size in microns. The approximate magnitudes of α and β for street dust and dirt and common mechanical sweepers are around 0.027 and 0.55, respectively, if d is in micrometers. Actual magnitudes of α and β should be determined for each type of equipment by the manufacturer or by a testing laboratory.

Example 8-4: Removal of Pollutants by Sweeping

A street dust sample had the following particle size distribution:

Particle size (μm)	*% distribution*
<43	*5*
43-104	*10*
104-246	*20*
246-840	*25*
840-2000	*25*
>2000	*15*

The magnitudes of the sweeper efficiency coefficients, α and β, were given by the manufacturer as $\alpha = 0.03$ and $\beta = 0.55$.

Determine the overall removal efficiency of the sweeper if it moves at a sweeping speed of 5 km/hr.

The removal efficiency for each particle size can be computed using Equation 8-6. Then

$$\text{Eff}(\%) = 1 - e^{-kE}$$

where $k = 0.03d^{0.55}$.

The normalized sweeper effort variable is

$$E = \frac{\text{actual effort (hr/km)}}{\text{standardized effort (hr/km)}}$$

$$= \frac{\dfrac{1}{5\ \text{km/hr}}}{\dfrac{1}{10\ \text{km/hr}}} = 2.0$$

The solution can be obtained either graphically or numerically as shown in the following table:

Particle Size (μm)	*Initial Distribution (%)*	*k*	e^{-kE}	*Average Removal* $1 - e^{-kE}$	*% Removed*	*% Remaining*
0		0	1.0			
	5			0.19	0.95	4.05
43		0.24	0.62			
	10			0.46	4.6	5.4
104		0.39	0.46			
	20			0.625	12.5	7.5
246		0.62	0.29			
	25			0.81	20.25	5.75
840		1.22	0.09			
	25			0.945	23.62	1.38
2000		1.96	0.02			
	15			0.98	14.7	0.3
Total	100				76.62	23.38

$$\%\ \text{removed} = (\text{initial}\ \%)\,(1 - e^{-kE})$$

The overall removal efficiency is 76.6%.

Example 8-5: Lead Removal by Sweeping

By analyzing the lead distribution with particle sizes of the street dust sample from the previous example, it was found that most of lead is associated with fine fractions as follows:

Particle Size (μ)	*% Lead*
0–43	*48*
43–104	*25*
104–246	*20*
246–840	*16*
840–2000	*1*

Estimate how much lead will be removed by the sweeper.

The computation can be arranged in the same fashion as in the previous example:

Particle Size Range (μ)	*Initial Distribution*	*Average Removal*	*% Removed*
0–43	*48*	*0.19*	*9.12*
43–104	*25*	*0.46*	*11.50*
104–246	*20*	*0.625*	*13.5*
246–840	*16*	*0.81*	*12.96*
840–2000	*1*	*0.94*	*0.94*
2000	*0*	–	*0*
Total	*100*		*48.02*

Only 48% of lead will be removed as compared to 76% of solids.

Washoff of Pollutants by Surface Runoff

When surface runoff occurs on impervious surfaces (Fig. 8-11), as a result of either natural storms or street flushing practices, the splashing effect of rain droplets and drug forces of the flow put particles in motion. Many hydraulic models have appeared in the literature on sedimentation which potentially could be applicable to the problem of particle pickup and transport.

From the numerous equations published in the literature, the Yalin equation[26] (see Chapter 5 for a detailed description of the model) is viewed as the one that best describes pickup and transport of particles by shallow flow typical for rills[27] and street gutters.[22]

Fig. 8-11. Wash-off of street refuse during a rainstorm.

Example 8-6: Wash-off Estimation by Yalin Equation

Estimate wash-off of dust and fractured debris particles (50% dust and 50% debris) from the curb storage following a 1-hr rain that resulted in a gutter flow 1.5 cm deep. The slope of the gutter is 2% and its width is 1 m. The debris particles, mostly organic, had a particle size of 2 mm and specific density ρ_s = *1.5 g/cm³, while dust, mostly soil, had an average particle size of 0.1 mm and specific density* $\rho_s = 2.5$ *g/cm³.*

For shallow flow conditions (width ≫ 12 × depth) the hydraulic radius approximately equals the depth of flow. Then the shear velocity becomes (see Equations 5-16 and 5-17 in Chapter 5):

$$v_* = \sqrt{gHS} = \sqrt{9.81 \times 0.015 \times 0.02} = 0.054 \text{ m/sec}$$

The particle Reynolds number for debris:

$$X = \frac{v_* D}{\mu} = \frac{0.054 \times 0.002}{10^{-6}} = 108$$

From the Shields diagram (Fig. 5-16) $Y_{cr} = 0.043$
for dust:

$$X = \frac{0.054 \times 0.0001}{10^{-6}} = 5.4$$

and

$$Y_{cr} = 0.034$$

The actual tractive force for debris:

$$Y = \frac{v_*^2}{(\rho_s - 1)gD} = \frac{0.054^2}{(1.5 - 1) \times 9.81 \times 0.002} = 0.297$$

for dust:

$$Y = \frac{0.054^2}{(2.4 - 1) \times 9.81 \times 0.0001} = 1.982$$

The coefficients are then for debris:

$$s = \frac{Y}{Y_{cr}} - 1 = \frac{0.297}{0.043} - 1 = 5.90$$

$$a = 2.45 \rho_s^{-0.4} \sqrt{Y_{cr}} = 2.45 \times 1.5^{-0.4} \times \sqrt{0.043} = 0.427$$

for dust:

$$s = \frac{1.982}{0.034} - 1 = 57.29$$

$$a = 2.45 \times 2.5^{-0.4} \times \sqrt{0.034} = 0.313$$

and the particle transport becomes for debris:

$$p = 10^6 (\rho_s - 1) D v_* 0.635 s \left[1 - \frac{1}{as} \ln (1 + as)\right]$$

$$= 10^6 (1.5 - 1) \times 0.002 \times 0.54 \times 0.635 \times 5.90$$

$$\times \left[1 - \frac{1}{0.427\ 5.9} \ln (1 + 0.427 \times 5.9)\right] = 101 \text{ g/m-sec}$$

for dust:

$$p = 10^6 (2.5 - 1) \times 0.0001 \times 0.054 \times 0.635 \times 57.3$$

$$\alpha \left[1 - \frac{1}{0.313\ 57.3} \ln (1 + 0.313 \times 57.3)\right] = 246 \text{ g/m-sec}$$

The above values of maximum (saturated particle transport) are per 1-m width. After multiplication by the width of the gutter (=1 m) the particle transport of debris is 101 g/sec and that of dust is 246 g/sec. If the total mass of particles in the 100-m long curb storage is greater than 101 g of debris and 246 g of dust, the rate of debris and dust removal will equal the computed values until the storage is exhausted.

Sutherland and McCuen Wash-off Model. Foster and Mayer[27] realized that the Yalin equation is extremely sensitive to local flow parameters. It is therefore often necessary to use a hydraulic model to accurately predict the hydraulic radius of the gutter flow throughout the duration of a given rainfall. Using a computer model developed for this purpose, Sutherland and McCuen[22] statistically analyzed the Yalin equation and simulated results of particle removal based on rainfall volume and curb flow characteristics. The statistical equations were formulated first for a standard 12.7-mm ($\frac{1}{2}$-in.) rainfall lasting 1 hr. The equations obtained from the analysis are given in Table 8-11. With the exception of the largest particle range, the effect of rainfall intensity on particle removal was negligible. For a given rainfall volume the total particle removal is then computed from

$$TS_j = K_j(TS_i) \tag{8-8}$$

where:

TS_j = the percentage removal of total solids in a particle range due to total rainfall volume j measured in mm
TS_i = the percentage removal of total solids in the particle range due to a total rainfall volume of 12.7 mm ($\frac{1}{2}$ in.)
K_j = a factor relating to TS_j and T_i taken from Table 8-12

TABLE 8-11. Total Solids Removed by a Uniform Rainfall of 12.7 mm ($\frac{1}{2}$ in.) Lasting 1 Hr.[a]

Particle Size Range (μ)	Equation
<43	$T_1 = 91.4 + 0.76(S^{1.92}) + 0.1I - 0.035P_1^0 - 0.01L$
43–104	$T_2 = 95.6 + 0.65(S^{1.55}) + 0.061I - 0.02P_2^0 - 0.09L$
104–246	$T_3 = 83.6 + S^{2.14} + 0.2I - 0.067P_3^0 - 0.21L$
246–840	$T_4 = 64.2 + 1.35(S^{2.45}) + 0.39I - 0.127P_4^0 - 0.024L$
840–2000	$T_5 = 33.6 + 1.58(S^{2.7}) + I^{0.9} - 0.22P_5^0$
>2000	$T_6 = (S - 1.44)(-3.7 + 0.5(I^{0.95}) - 0.071P_6^0 + 0.085L)$

[a]After Sutherland and McCuen.[22] Used with permission of American Water Resources Association.
T_i = removal percentage for particle range i.
S = slope of gutter (%).
I_0 = impervious area (%).
P_1^0 = initial total solids loading (g/curb m).
L = length of gutter (m).
Note: If T_i exceeds 100, it is assumed equal to 100.

TABLE 8-12. K_j **Values to be Used in the Equation** $TS_j = K_j(TS_i)$.[a]

	Total Volume j											
TS_i	$\frac{1}{8}$ 3.1	$\frac{1}{4}$ 6.3	$\frac{1}{2}$ 12.7	$\frac{3}{4}$ 19	1 25.4	$1\frac{1}{4}$ 31.7	$1\frac{1}{2}$ 38.1	$1\frac{3}{4}$ 44.4	2 50.1	$1\frac{1}{2}$ 65.5	3 76.2	in. mm
100	0.84	1.0	1.0	1.0	1.0	1.0	1.0	1.0	1.0	1.0	1.0	–
97	0.835	1.0	1.021	1.031	1.031	1.031	1.031	1.031	1.031	1.031	1.031	–
94	0.808	0.915	1.0	1.032	1.053	1.064	1.064	1.064	1.064	0.064	1.064	–
93	0.731	0.882	1.0	1.048	1.075	1.075	1.075	1.075	1.075	1.075	1.075	–
90	0.778	0.889	1.0	1.055	1.089	1.1	1.111	1.111	1.111	1.111	1.111	–
88	0.670	0.852	1.0	1.063	1.091	1.108	1.119	1.131	1.136	1.136	1.136	–
84	0.417	0.798	1.0	1.083	1.119	1.143	1.161	1.178	1.190	1.190	1.190	–
76	0.303	0.658	1.0	1.125	1.184	1.224	1.25	1.263	1.270	1.289	1.316	–
72	0.139	0.542	1.0	1.125	1.208	1.264	1.35	1.333	1.347	1.369	1.389	–
64	0.156	0.375	1.0	1.219	1.344	1.406	1.469	1.515	1.563	1.563	1.563	–
61	0.082	0.295	1.0	1.230	1.352	1.426	1.492	1.533	1.582	1.606	1.639	–
45	0.044	0.178	1.0	1.489	1.689	1.811	1.911	1.978	2.044	2.149	2.222	–
44	0.057	0.159	1.0	1.477	1.704	1.841	1.954	2.045	2.114	2.182	2.232	–
15	0.0	0.133	1.0	2.6	3.933	4.733	5.233	5.6	5.9	6.233	6.333	–
2	0.0	0.0	1.0	4.0	11.0	20.0	26.5	30.5	33.75	38.0	41.0	–

[a]After Sutherland and McCuen.[22] Used with permission of American Water Resources Association.
TS_j = the percentage removal of total solids in a particle size range to a total rainfall volume measured in inches.
TS_i = the percentage removal of total solids in a particle size range due to a total rainfall volume of 12.7 mm ($\frac{1}{2}$ in.)

Example 8-7: Wash-off Estimation by Sutherland and McCuen Model

Compute particle removal from the curb storage for a 20-mm rain volume using the Sutherland-McCuen model if the curb loading is 200 g/m and other information is the same as in the previous example. The imperviousness of the gutter is 100%.

For the debris component (D = 2 mm = 2000 μ) the removal for the standard rain is (Equation 5 in Table 8-11):

$$TS_5 = 33.6 + 1.58(S^{2.7}) + I^{0.9} - 0.22P_5^0$$
$$= 33.6 + 1.58(2^{2.7}) + 100^{0.9} - 0.22\,(100) = 84.96\%$$

From Table 8-12 for a 20-mm rain volume and TS_5 = 85% K_j = 1.08. Then from Equation 8-9: TS = 1.08 × 84.9 = 91.7%.

For the dust component (D = 0.1 mm = 100 μ):

$$TS_2 = 95.6 + 0.65(S^{1.55}) + 0.061I - 0.02P_2^0 - 0.09L$$
$$= 95.6 + 0.65(2^{1.55}) + 0.061(100) - 0.02(100) - 0.09(100)$$
$$= 94.63\%$$
$$TS = 1.08 \times 93.63 > 100\%$$

Total removal of all dust particles can be expected.

Sartor et al.[10,11] Wash-off Function

Sartor et al.[10,11] analyzed their data using a simple first-order removal concept

$$\frac{dP}{dt} = -k_U rP \tag{8-9}$$

where:

- r = rainfall intensity
- k_U = a constant depending on street surface characteristics called "urban wash-off coefficient"
- P = amount of solids remaining
- t = time

The constant, k_U, was found to be almost independent of particle size within the studied range of 10 to 1000 μm.

Equation 8-9 integrates to

$$P_t = P_0(1 - e^{-k_U rt}) \tag{8-10}$$

where:

P_0 = is the initial mass (weight) of solids in the curb storage
P_t = mass (weight) of material removed by rain with duration t

The value of the urban wash-off coefficient was almost arbitrarily chosen[28] as being 0.19 if the rain intensity is in millimeters per hour. Interestingly, this value has been recommended by all subsequent urban runoff models that utilize this concept. The value was derived from an assumption that 90% of the solids are removed by a 12.7-mm/hr ($\frac{1}{2}$-in./hr) rain.

In spite of the high empirical nature and arbitrarily chosen constants, the Sartor et al.[10,11] concept has been incorporated in most urban runoff models.

The authors of STORM[29] modified Equation 8-10 by assuming that not all the solids are available for transport. Then

$$P_t = AP_0(1 - e^{-k_{ur}rt}) \quad (8\text{-}11)$$

where

$$A = 0.057 + 0.04(r^{1.1}) \quad (8\text{-}12)$$

This accounts for the heterogeneous makeup and for the variability in travel distance of the dust and dirt particles. The maximum value for A is 1.0.

Example 8-8: Wash-off Estimation by Sartor et al. Model

Estimate solids removal from the curb storage by the Sartor et al.[10,11] concept for a 20-mm/hr rain with P_0 = 100 g/m.

From Equation 8-12 the availability factor is

$$A = 0.057 + 0.04(r^{1.1}) = 0.057 + 0.04(20^{1.1}) = 1.13$$

therefore

$$A = 1.0$$

Then the amount removed by the rain is

$$P_t = 1.0 \times P_0(1 - e^{-k_{ur}rt}) = 1.0 \times 100.0 \times (1 - e^{-0.19 \times 20 \times 1})$$

$$= 97.76 \text{ g/m}$$

The particle removal is therefore almost 98%.

POROUS PAVEMENT

Porous pavement presents a new idea to the storm-water management problem. The use of porous pavements for storm-water management and nonpoint pollution control was initiated by the Franklin Institute in Philadelphia, Pennsylvania,

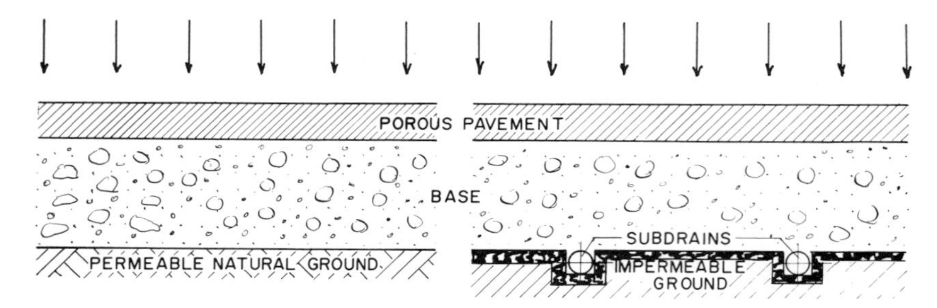

Fig. 8-12. Porous pavement installed on permeable and impermeable ground.

under the sponsorship of the U.S. Environmental Protection Agency.[30] However, porous pavements have been installed in other parts of the United States (Woodland, Texas) and in some European countries (Denmark and Germany).

The primary benefit of porous pavements is a significant reduction in runoff rate and volume from otherwise impervious areas. If the pavements are designed properly, all or most of the runoff can be stored and subsequently allowed to infiltrate into the natural ground.

Physical Considerations

Most porous pavements are made from asphalt in which the fine filling particles are missing. Absence of these particles may require increased amounts of asphalt, but otherwise there is not much difference between the porous and regular impervious pavements. The asphalt mixture is installed on a gravel base (Fig. 8-12). In areas where permeability of underlying natural soil is not adequate or porous pavement is installed over an impervious base, a drainage system can be installed. If the drains are not installed, subgrade softening may occur.

Porous pavements can also be installed in established urban zones over existing impervious pavements. This may especially be advantageous in areas with combined sewers (reduces frequency of overflow) or in areas with inadequate storm drainage.

Hydraulic Characteristics of Infiltration

The Franklin Institute study[30,31] indicated that porous pavements can be designed with hydraulic conductivities much greater than typical rainfall rates. The typical hydraulic conductivity quoted by Jackson and Ragan[31] is 250 cm/hr. This means that the infiltration into the base of the pavement should occur without ponding. Jackson and Ragan[31] and Diniz and Espey[32] discuss modeling techniques for the hydraulic design of porous pavements.

Benefits of Porous Pavements

Simply stated, the pollution loading by surface runoff from a porous pavement should be zero if all water infiltrates. However, this may not happen, especially if the porous pavement is installed on an impervious surface. In this case the porous pavement and the base act as a filter. If the accumulated dust and dirt on the top of the pavement is not periodically and frequently removed (e.g., by sweeping) clogging may occur. Therefore, the maximum benefits of porous pavements are achieved when combined with proper maintenance practices.

Even when the ground is impervious, the porous base and pavement are beneficial. The gravel base serves as a storage area, and if the storm water requires treatment, it may be stored in the porous media until the treatment capacity becomes available.

Porous pavements may sometimes cost slightly more than conventional surfaces for parking lots, roads, and other urban surfaces. However, their benefits and savings on sewer and drainage capacities, as well as savings on treatment, will offset any additional cost for the pavement itself.

REFERENCES

1. American Public Works Association. 1969. Water pollution aspects of urban runoff. U.S. Dept. of Interior, FWPCA (present EPA), Washington, D.C. WP-20-15. 200 pp.
2. Randall, C. W., Helsel, D. R., Grizzard, T. J., and Hoehn, R. C. 1978. The impact of atmospheric contaminants on storm water quality in an urban area. *Prog. Water. Tech.*, **10**:417–431.
3. Malmquist, Per-Arne. 1978. Atmospheric fallout and street cleaning–Effect on urban storm water and snow. *Prog. Water Tech.*, **10**:495–505.
4. Colson, N. V., and Tafuri, A. N. 1975. Urban land runoff considerations. pp. 120–128. In: W. Whipple (ed.). Urbanization and water quality control. Proc. No. 20, AWRA, Minneapolis, Minnesota.
5. Akerlinch, G. 1950. The quality of stormwater flow. *Nord. Hyg. Tidskr.* (Norway).
6. Shigorin, G. C. 1956. The problem of city surface runoff water. *Vodosnabzh. Sanit. Tekh.* (USSR).
7. Palmer, C. L. 1950. The pollutional effects of stormwater overflows from combined sewers. *Sewage Ind. Wastes*, **22**:154–165.
8. Pravoshinsky, N. A. 1975. Basic principles for determining regulating structure parameters to prevent rain sewer receivers from contamination. *Prog. Water Tech.*, **7**(2):301–307.
9. Bradford, W. L., 1977. Urban stormwater pollutant loading–a statistical summary through 1972. *J. WPCF*, **49**:613–622.
10. Sartor, J. D. and Boyd, G. B. 1972. Water pollution aspects of street surface contaminants. U.S. EPA Report No. R2-72-081, Washington, D.C.

11. Sartor, J. D., Boyd, G. B., and Agardy, F. J. 1974. Water pollution aspects of street surface contaminants. *J. WPCF*, **46**:458–667, Washington, D.C.
12. Pitt, R., and Amy, G. 1973. Toxic material analysis of street surface contamination. U.S. EPA Report No. R2-73-283, Washington, D.C.
13. Amy, G., Pitt, R., Singh, R., Bradford, W. L., and LaGraff, M. B. 1975. Water quality management planning for urban runoff. U.S. EPA Report No. 44019-75-004, Washington, D.C.
14. Pitt, R. 1979. Demonstration of nonpoint pollution abatement through improved street cleaning practices. U.S. EPA Report No. 600/2-79/161, Cincinnati, Ohio.
15. Goettle, A. 1978. Atmospheric contaminants, fallout, and their effect on storm water quality. *Prog. Water Tech.*, **10**:455–467.
16. Shaheen, D. G. 1975. Contribution of urban roadway usage to water pollution. U.S. EPA Report No. 600/2-75-004, Washington, D.C.
17. Laxen, D. P. H., and Harrison, R. M. 1977. Review paper—The highway as a source of water pollution: An appraisal with the heavy metal lead. *Water Res.*, **11**:1–11.
18. Corrin, M. L., and Nausch, D. F. S. 1977. Physical and chemical characteristics of environmental lead. In: W. R. Boggers and B. G. Wixson (eds.). "Lead in the Environment." NSF-RA-770214, U.S. Government Printing Office, Washington, D.C.
19. Carlisle, A. A., Brown, H. F., and White, E. J. 1966. Litter fall leaf production and the effect of defoliation by *Tortrix varidiana* in a sissle oak (*Quercus petraea*) woodland. *J. Ecol.*, **54**:65–86.
20. Heaney, J. P., and Huber, W. C. 1973. Stormwater management model. Refinements, testing and decision-making. Department of Environmental Engineering and Sciences, University of Florida, Gainesville.
21. Novotny, V., and Goodrich-Mahoney, J. 1978. Comparative assessment of pollution loadings from nonpoint sources in urban land use. *Prog. Water Tech.*, **10**:775–785.
22. Sutherland, R. C., and McCuen, R. H. 1978. Simulation of urban nonpoint source pollution. *Water Res. Bull.*, **14**:409–428.
23. Oberts, G. L. 1977. Water quality effects of potential urban best management practices. A literature review. Tech. Bull. 97, Wisconsin Department of Natural Resources, Madison, Wisconsin.
24. Pitt, R. 1978. The potential of street cleaning in reducing nonpoint pollution. International Symposium on Urban Water Management. University of Kentucky, Lexington, 24–27 July.
25. Clark, D. E., and Cobbins, W. C. 1963. Removal effectiveness of simulated dry fallout from paved areas by motorized and vacuumized street sweepers. U.S. Naval Radiological Defense Laboratory, San Francisco, California.
26. Yalin, M. S. 1963. An expression for bed load transportation. *J. Hydraulics Div., ASCE*, **89**(3):221–250.
27. Foster, G. R., and Meyer, L. D. 1972. Transport of soil particles by shallow flow. *Trans. ASAE*, **15**:99–102.

28. Roesner, I. A. 1974. Quality aspects of urban runoff. Short course on application of SWMM, University of Massachusetts, Amherst.
29. U.S. Army Corps of Engineers. 1974. Urban stormwater runoff—STORM. The Hydrologic Engineering Center, Davis, California.
30. Thelen, E., Grover, W. C., Hoiberg, A. J., and Haigh, T. I. 1972. Investigation of porous pavements for urban runoff control. U.S. EPA Report No. 11034 DUY 03/72.
31. Jackson, T. J., and Ragan, R. M. 1974. Hydrology of porous pavement parking lots. *J. Hydraulics Div., ASCE*, **12**:1739–1752.
32. Diniz, E. V., and Espey, W. H. 1979. Maximum utilization of water resources in a planned community. U.S. EPA Report No. 600/2-79-050, Cincinnati, Ohio.

9

Nonpoint pollution simulation models

BASIC CONCEPTS

Types of Models

Unlike the pollution from point sources, which enters the hydrologic transport during the channel or estuary flow phase, nonpoint sources of pollution are always associated with the early phases of the hydrological cycle–rainfall formation, detachment of particles by rainfall impact, and early phases of overland flow. Thus, modeling nonpoint pollution generally means modeling a larger portion if not the entire hydrological cycle.

Nonpoint pollution simulation models are part of a category of "loading models," which represent the inputs and movement of materials from the point of origin to watercourses. These models interface with water quality models and usually provide input concentrations and flow rates.[1]

Water quality models simulate the movement of materials through streams, rivers, impoundments, estuaries, or near-shore ocean dispersion.

Only very few models have attempted to integrate loading components with the receiving waters component. Most of the stream models are designed to simulate steady-state or quasi steady-state low-flow conditions, while nonpoint loading models simulate highly dynamic storm events that often result in high flows and flooding.

This section briefly describes the concepts and basic design approaches of loading models. The reader is referred to other extensive literature sources covering the water quality modeling of receiving water bodies (e.g., Biswas, "System Approach to Water Quality Management," McGraw-Hill, New York, 1976; Krenkel, Novotny, "Water Quality Management," Academic Press, New York, 1980; or Shen, "River Modeling," Wiley Interscience, New York, 1979).

Every model can be represented by a "black box" concept (Fig. 9-1). The model, as the real system, produces output to various inputs. The structure of the model is always a very simplified version of the interactions and reactions taking place in the real system. The variables describing the physical state of the system are called *system parameters*; the variables affecting the state of the system are *state variables.* Some input variables may be considered state variables and vice versa. Watershed size, slope and roughness characteristics, erodibility and texture of soils are examples of system parameters, while temperature, radiation, and vegetation cover are considered state variables. Rain, atmospheric fallout, and daily litter contributions on impervious areas can be considered

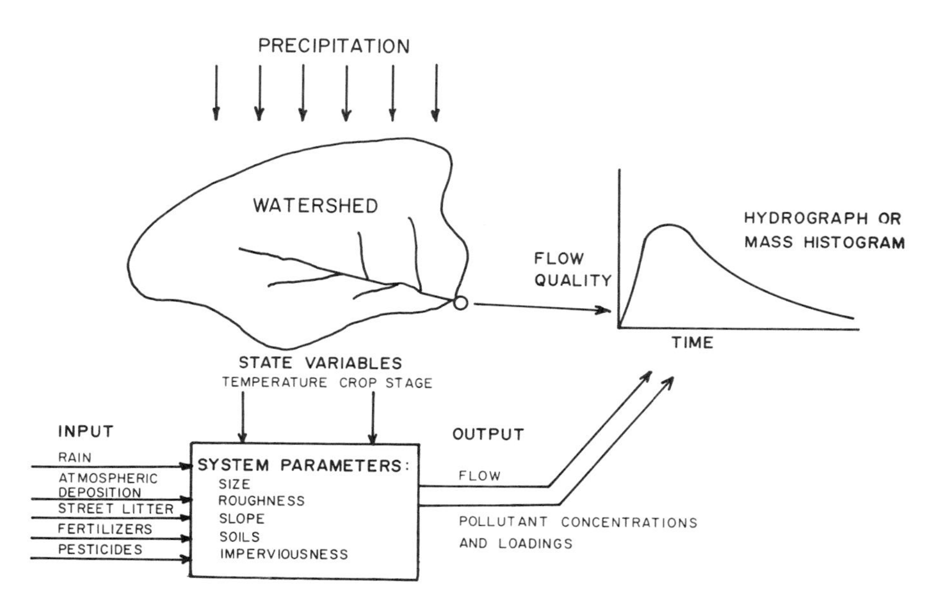

Fig. 9-1. "Black box" concept in watershed modeling.

inputs to most nonpoint water quality models. The model (watershed) response to these inputs and state variables is expressed often as water loading or pollutant loading and time relationships (hydrograph and pollutograph).

There are basically two approaches to modeling nonpoint pollution. The more widely used are the *lumped parameter* models, while some more complex models developed recently are based on the *distributed parameter* concept.

The lumped parameter models treat the watershed or a significant portion of it as one unit. The various characteristics of the watershed are lumped together, often with the use of an empirical equation, and the final form and magnitude of the parameters are simplified to represent the modeled unit as a uniform system. The coefficients and system parameters for each unit are determined mostly by calibrating the response of the model against extensive field data. The models–once they have been calibrated and verified–can produce long-time series of outputs reflecting different hydrologic and meteorologic conditions (input variables). However, a change of system parameters, such as a change in land use, requires that the coefficients and variables describing the system be recalculated. Considering that many of the relationships used by these models are of an empirical nature, the use of such models for simulating changed conditions is less reliable.

The distributed parameter approach involves dividing the watershed into smaller homogeneous units with uniform characteristics (soil, imperviousness, crop, slope, etc.). Each areal unit is modeled separately, and the total output is obtained by summing all individual outputs from the homogeneous units. The input to each unit consists of the distributed inputs such as rain or atmospheric fallout plus output from upstream or uphill adjacent units.

Theoretically, the lumped parameter model can provide only one output location, while outputs can be obtained throughout the system from distributed parameter models, that is, from each modeled subunit. Distributed parameter models require large computer storage and extensive description of system parameters, which must be provided for each unit. However, changes in the watershed and their effect on the output can be modeled easily and more effectively.

The U.S. Environmental Protection Agency has divided loading models into three groups according to their geographic scale, namely, basin scale (>500 km^2), area-wide (>55 km^2) and small watersheds or fields.[1] The complexity and detail of the process decreases with increasing size of the modeled unit. Lumped parameter models are more flexible and usable for large-scale systems, while distributed parameter models may provide a better understanding of smaller systems. A trade-off always exists between the size of the modeled area and the detail of the model.

Models can be designed or run on an event or continuous basis. Discrete event modeling simulates the response of a watershed to a major rainfall or snowfall snow-melt. The principal advantage of event modeling over continuous simulation is that it requires relatively little meteorological data and can be operated with shorter computer run times. The principal disadvantage of event modeling is that it requires specification of the design storm and antecedent moisture conditions, thereby assuming equivalence between the recurrence interval of a storm and the recurrence interval of the runoff. This disadvantage can be partially eliminated in modeling urban (mostly impervious) areas.

Continuous process modeling sequentially simulates processes such as precipitation, available surface storage, snow accumulation and melt, evapotranspiration, soil moisture, surface runoff, infiltration, soil water movement, pollutant accumulation, and erosion. Such models typically operate on a time interval ranging from a day to a fraction of an hour, and continuously balance water and pollutant mass in the system.

The principal advantage of continuous modeling is that it provides long-time series of water and pollutant loadings that can be analyzed statistically as to their frequency and occurrences. A principal disadvantage of continuous modeling is that it requires long simulation runs, thus imposing restrictions on the amount of alternatives that can be investigated. It also requires historical data on precipitation often in less than hourly intervals, which is not always available.

Structure of the Models

Models of nonpoint pollution loadings can be in the form of either simple statistical unit loading equations (see Chapter 10 for more detailed discussion of the concept) that express long-term loading related to land use and other areal characteristics, or deterministic time variable models allowing computations for each individual event. Only deterministic time variable models are discussed in this chapter. These models are basically a description of the hydrologic rainfall-runoff transformation process with attached quality components. Most of the models have the following basic components (Fig. 9-2):

1. The surface runoff generation component describes the transformation of rainfall into runoff and its overland surface flow component. Most of nonpoint pollution originates from hydrologically active areas, that is, areas from which the surface runoff is generated. The purpose of this component is to locate these areas and determine the magnitude of the overland flow. Modeling surface runoff includes the following processes:
   a. Exhaustion of surface storage.
   b. Evaporation.
   c. Snow accumulation and melt.
2. The soil and groundwater component (not common to all models) describes movement of water through the unsaturated soil zone and into saturated groundwater zones (aquifers). This component balances current soil moisture

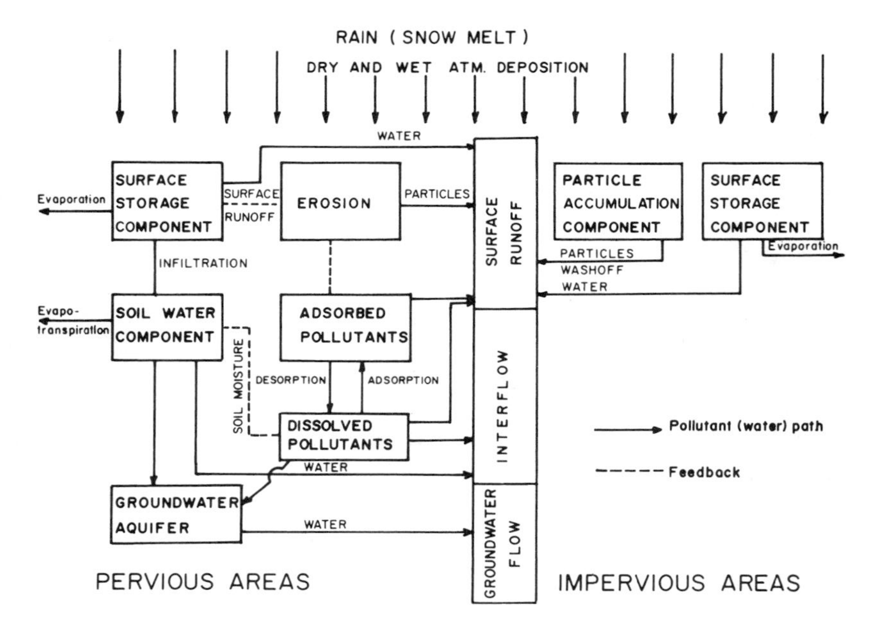

Fig. 9-2. Components of watershed nonpoint pollution models.

with infiltration rate, evapotranspiration, and water loss into deep groundwater zones. Since infiltration is a function of soil moisture content, an iterative procedure usually is employed. If the soil component is not included in the model, infiltration rate is estimated by an empirical equation (e.g., Horton's equation—Equation 3-4).

A schematic representation of processes involved in the preceding two hydrological components is presented in Figs. 3-2 and 3-3.

3. Erosion component estimates soil loss from pervious areas. The most common models used to represent the erosion process are the Universal Soil Loss Equation and Negev's model (see Chapter 5).

4. Particle accumulation and wash-off from impervious areas accounts for the particles that accumulate near the curb in urban areas and their removal by street cleaning practices and storm-water wash-off.

5. The soil adsorption/desorption component (not common to all models) determines the distribution of adsorbed and dissolved fractions of pollutants in soils. This component also may include volatilization and decay equations and reactions in order to simulate the history of such materials as pesticides or nitrogen forms. In the absence of the soil adsorption segment, modeling of pollutants is accomplished by the use of factors that relate pollutant concentration to that of sediment.

Most common nonpoint pollution loading models handle pollutants as sorbed components on particulate matter. In this case, estimation of the enrichment of surface runoff by soil and dust and dirt particles is necessary because of significant differences in pollutant concentrations in soils and washload. Reviews of some available models already have appeared in the literature.[1,2,3,4]

Reliability and Usefulness of the Models

It must be understood that as with any simulation of "real world" systems, mathematical models are only a rough approximation. The accuracy and reliability of models is limited. Although many models represent the best available technology for analysis of environmental systems, a common error made by many planners is that they accept simulation results as true and absolute results for unknown conditions. In order to avoid disappointments and court challenges, users should be aware of the limited accuracy of the models.

The most accurate models (±a few percent) are hydrologic models simulating runoff from small, uniform, impervious areas, the least reliable (an order of magnitude or more) are water quality models for large watersheds. Figure 9-3 shows the approximate accuracy limits for modeling water and pollutant loadings from nonpoint sources. It should be noted that in determining pollutant transport, hydrology must be calibrated and determined first, followed by sediment, and finally pollutant transport. Any errors that appear in the hydrologic or erosion components will be transferred and magnified in all dependent components.

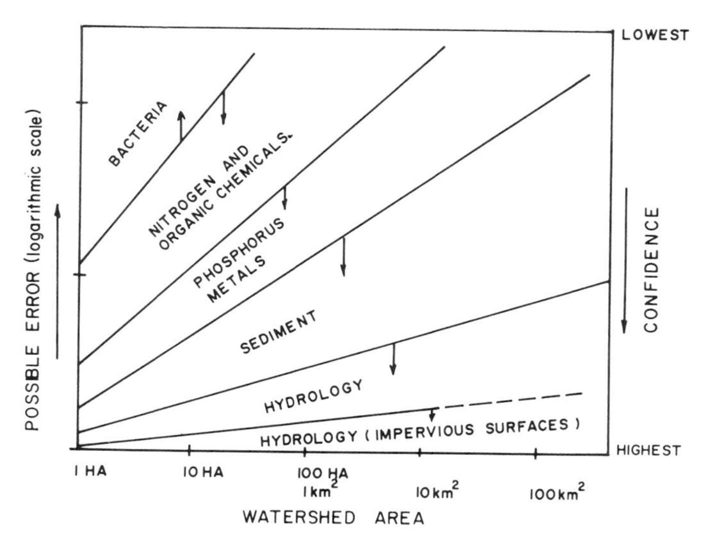

Fig. 9-3. Reliability and possible errors for nonpoint pollution models.

The highest error associated with bacterial estimations is caused primarily by analytical techniques and to a lesser degree by modeling reliability.

In spite of the errors involved in modeling complex environmental systems such as nonpoint pollution, the model as a planning tool cannot be replaced by any "rule-of-thumb" approach which some planners unfamiliar with the capability of models might suggest. The use of models is beneficial and greatly enhances the planning process for the following reasons:

1. Models can provide a forecast of the impact (although only approximate) of planned actions on water quality and pollution loadings.
2. Models provide an understanding of the processes involved in pollution generation from nonpoint sources.
3. The data base necessary to construct and calibrate the model is useful for other planning activities. Many problems will be answered or become clearer just by evaluating the data and compiling them into an appropriate input format.
4. Critical processes and areas of concern can be delineated and detected by modeling.
5. Models can be updated continuously according to the state-of-the-art of modeling technology and understanding of the modeled processes.
6. Models can generate numerous alternatives according to the specification of the users. Various strategies can be investigated, and the impact of remedial measures can be evaluated.
7. Although the absolute accuracy of the outputs from the model is limited and sometimes even small, a comparison and ranking of outputs for various alternative remedial measures often are reliable and in most cases more than adequate.
8. Models can estimate and analyze trade-offs between planning objectives. A system providing the lowest pollutant load may not be optional for other objectives such as agricultural production (e.g., limiting or eliminating pesticide use may result in significant yield reduction; urban development–no future development, etc.). If the environmental objective is known, the alternatives to achieve it can be measured in terms of economic efficiency by considering the willingness of those involved by the measures to pay for the consequences. If there is a financial limit, it must be treated as a constraint.

Data Requirements

Although models may vary in their scope and capabilities, some generalizations of the input data requirements can be made. The data can be divided into the following groups with an approximate list of variables:

1. System parameters
   a. watershed size
   b. subdivision of the watershed into homogenous subareas

c. imperviousness of each subarea
d. slopes
e. fraction of impervious areas directly connected to a channel
f. maximum surface storage (depression plus interception storage)
g. soil characteristics including texture, permeability, erodibility, and composition
h. crop and vegetation cover
i. curb density or street gutter length
k. sewer system or natural drainage characterisitcs

2. State variables
   a. ambient temperature
   b. reaction rate coefficients
   c. adsorption/desorption coefficients
   d. growth stage of crops
   e. daily accumulation rates of litter
   f. traffic density and speed
   g. potency factors for pollutants (pollutant strength on sediment)
   h. solar radiation (for some models)
3. Input variables
   a. precipitation
   b. atmospheric fallout
   c. evaporation rates

Data related to land and land use can be obtained from maps and/or aerial photographs and remote sensing. The data obtainable by high plane or satellite (LANDSAT) imagery can be digitized in order to provide more specific information on the degree of imperviousness, vegetation cover, surface roughness etc.

Soil data are available from the U.S. Soil Conservation Service (U.S. SCS) maps. These maps provide information on soil type distribution and include data on slope, texture, and other important soil characteristics (Fig. 9-4).

Land-use data can be obtained from the U.S. Bureau of Land Management and/or the U.S. Census Bureau.

Hydrologic and basic water quality information is collected and published regularly by the U.S. Geological Survey (USGS) and the U.S. Environmental Protection Agency (U.S. EPA). Both agencies have computerized systems for storing and retrieving water quality and quantity information.

Information on dust and dirt accumulation and traffic densities in urban areas usually is available from local city engineer offices, sanitation departments, and planning agencies. Meteorological data is routinely measured by the U.S. National Oceanic and Atmospheric Administration (NOAA).

It must be emphasized that the format and frequency of data obtainable from various agencies usually do not conform to modeling requirements, and often

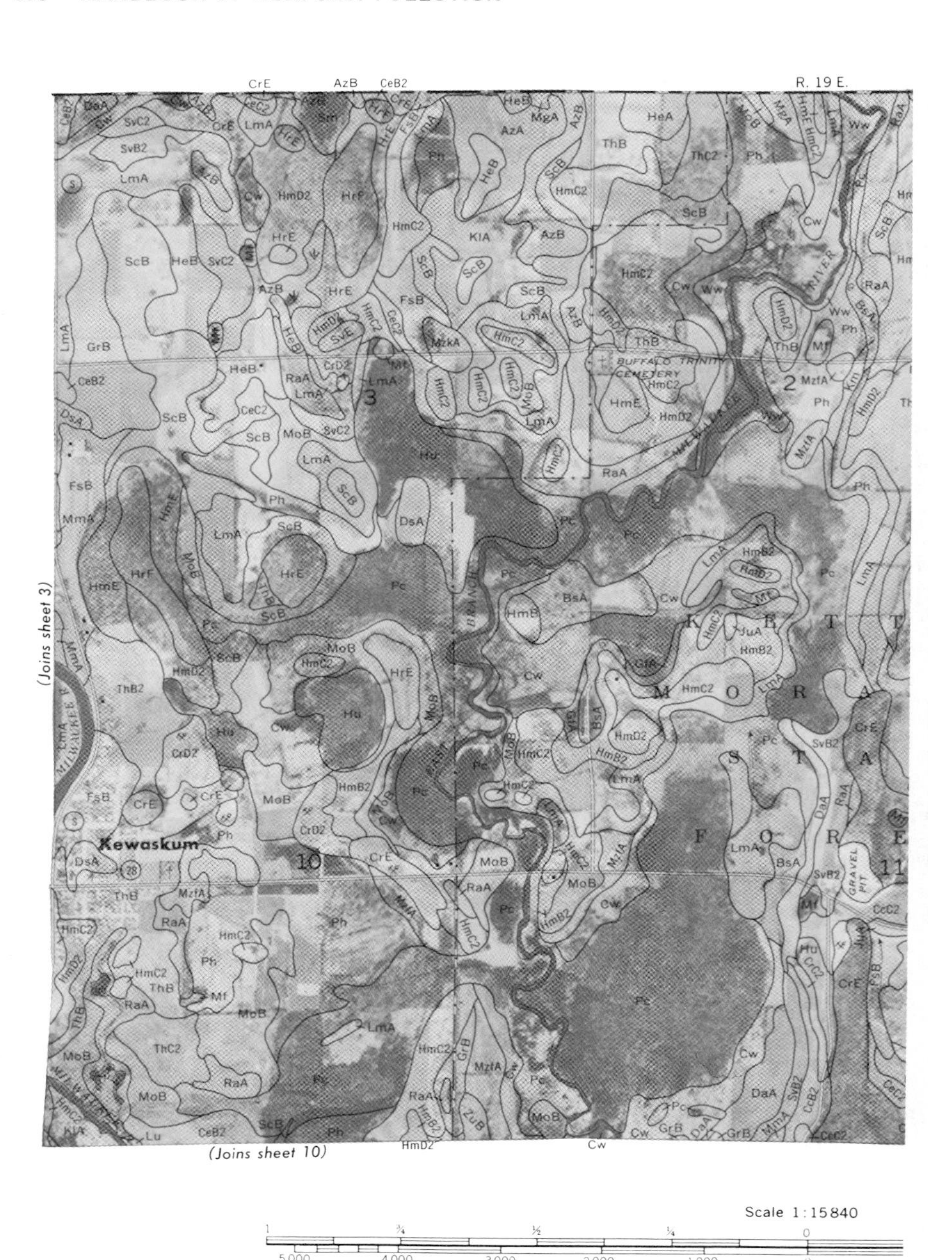

Fig. 9-4. Example of soil map by the U.S. Soil Conservation Service.

Fig. 9-5. Typical monitoring station for small watershed used for gathering calibration and verification data. (Photo authors.)

significant time and financial resources must be spent to put available information into conformity with the requirements of the models. Additional in situ surveys and monitoring are necessary to gather missing information (Fig. 9-5). Expenses associated with data collection activities often exceed the expenses for setting up and running the model.

Model Calibration and Verification

Many mathematical equations and formulas used in hydrological and water quality models are of an empirical or semiempirical nature requiring a knowledge of many coefficients and reaction rates. On average, twenty or more coefficients must be input for each subunit of the modeled system, and often only approximate ranges or orders of magnitude of these coefficients are known.

As an example, consider the soil permeability and surface storage. Although the magnitudes of surface runoff and pollutant loadings are sensitive to these variables, their exact measurement is tedious if not impossible even for relatively small uniform areas. The permeability ranges reported in the U.S. SCS soil maps are broad, for example, silt loam soils are listed as having permeability between

1.6 and 5 cm/hr. In addition, no adequate physical method is available to measure surface storage.

Under these circumstances, the only way to arrive at a set of coefficients adequately describing the watershed is by calibrating the model against measured data. Calibration means varying the coefficients within acceptable limits until an agreement between measured and computed values is obtained. The best method involves a selection of three or more representative storms (low, medium, and high intensity or volume storms) and calibration of the coefficients by comparing outputs from the model with the measured data. For nonpoint pollution modeling, the hydrologic components must be calibrated first, followed by erosion, and finally the pollutant component.

Trying to adjust all twenty or more coëfficients for each land use at the same time is tedious and often impossible. Fortunately, all models are more sensitive to a few important variables and less to others. As a rule-of-thumb, the variables to which the model is most sensitive should be calibrated first, with the other coefficients kept at their optimal or average levels. The hydrologic components of nonpoint pollution models are sensitive to the magnitude of surface storage and soil permeability, which determine surface runoff volume. Surface roughness affects the magnitude and time location of the runoff peaks. Other variables—slope, soil moisture characteristics, etc.—are not as important.

In the erosion component, the factor to which soil loss is most responsive is the vegetative cover factor, *C*. Adsorption characteristics and attenuation rates are most important for the pollutant submodel.

Once the model has been calibrated, that is once a satisfactory fit of computed and simulated data has been achieved for one calibration storm, the model must be verified. Verification is accomplished by running the model with the coefficients established during calibration and with inputs corresponding to another (verification) storm. If a satisfactory fit of computed and measured data is obtained, calibration and verification are accomplished. Very often this is not the case, and calibration and verification must be repeated until a satisfactory fit is found.

BRIEF DESCRIPTION OF AVAILABLE NONPOINT POLLUTION SIMULATION MODELS

During the past decade numerous models have been developed that can be used to simulate nonpoint pollution loadings to surface waters:

Urban Pollution Simulation Models:

- Storm Water Management Model (SWMM)[5]
- Storage, Treatment, and Overflow Model (STORM)[6]
- Battelle Urban Runoff Management Model[7]

Models for Agricultural and Rural Areas
- Agricultural Chemical Transport Model (ACTMO)[8]
- Agricultural Runoff Management Model (ARM)[9]
- Unified Transport Model (UTM)[10]
- Chemicals, Runoff, and Erosion from Agricultural Management Systems Model (CREAMS)*
- Areal, Nonpoint Source Watershed Environment Response Simulation Models (ANSWERS)**

Urban and Nonurban Distributed Parameter Models
- Nonpoint Simulation Model (NPS)[11]
- Land Runoff Model (LANDRUN)[12]

Complex Watershed Models
- Hydrocomp Simulation Program (HSP)[13]

Almost all of the above models can be used only for modeling relatively small watersheds. Most of the processes and the mathematical formulations used by the models have been described in previous chapters. The following section includes descriptions of some models in current use and some case studies.

The Stormwater Management Model (SWMM)[5]

This model was developed for the U.S. EPA by Metcalf & Eddy, Inc., the University of Florida, and Water Resources Engineers, Inc. The initial and most common version of the model developed in the early 1970s was a single runoff event model that was rather complicated and took several years to "debug." A simplified version of SWMM that provides for continuous simulation has become available.

The basic structure of the model is shown in Fig. 9-6. The first computational block, RUNOFF, simulates the quantity and quality of the overland flow caused by a storm. The design of the block is based on the distributed parameter concept whereby the designated area is divided into small homogenous subcatchments (maximal number of subcatchments is 200). Each subcatchment has a single outlet, usually a manhole on the storm or combined sewer system. This manhole becomes an inlet into the following block–TRANSPORT.

The TRANSPORT block takes the storm-water runoff generated by RUNOFF and routes it through the sewer system. During the flow, infiltrated water is

*Model documentation not available at the time of the manuscript preparation. The model is developed by the U.S. Dept. of Agriculture and consists of three components: Hydrology, erosion/sedimentation, and chemistry.

**Model documentation not available at the time of the manuscript preparation. The model is developed by the Dept. of Agricultural Engineering, Purdue University, West Lafayette, Ind. The model is based on the distributed parameter concept and simulates water, sediment, and nutrient transport during and immediately following a rainfall event.

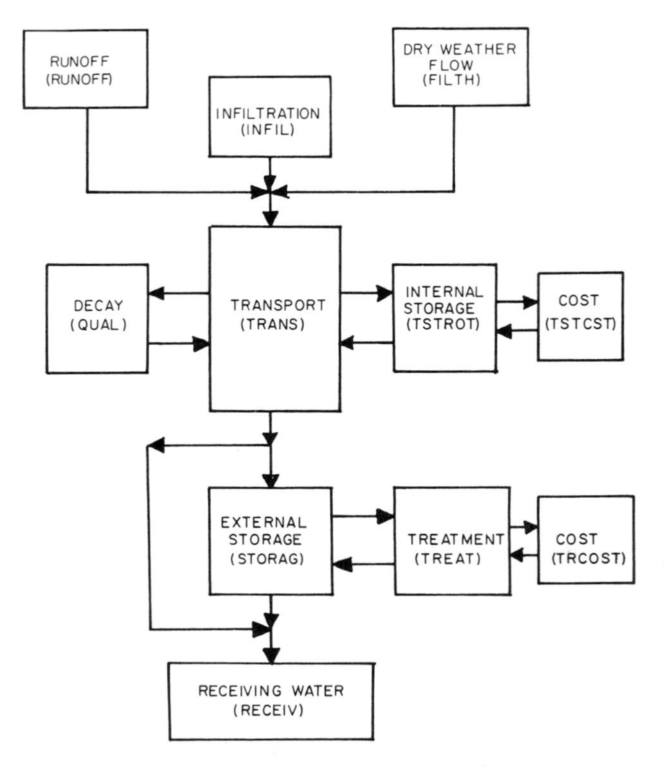

Fig. 9-6. Overview of the Storm Water Management Model (SWMM). Subroutine names are shown in parentheses. (Replotted from reference 5.)

added to the routed flows from the block INFIL, and dry-weather flow and quality generated from each subarea is entered by the FILTH block in areas served by combined sewers. The TRANSPORT block can route flows in 13 different conduit shapes and seven nonconduit elements (flow dividers, lift stations, storage units, etc.). The mass of degradable pollutants in the sewers can be reduced by decay accomplished by the QUAL block.

The STORAGE TREATMENT blocks estimate the effect of external natural or man-made storages on dampening the peaks of storm-water runoff and the effects of treatment practices on reducing pollutant loadings caused by sewer systems.

The last block of the program, RECEIV, simulates the response of the receiving water body to the sewer discharge or overflow. The types of surface waters that can be simulated include rivers and estuaries (one-dimensional representation) or lakes and bays (two-dimensional representation). For the simulation, the receiving water system is divided into a set of one- or two-dimensional homogenous elements (nodes) connected with neighboring nodes. The RECEIV blocks simulates quantity and water movement first (SWFLOW) followed by quality and pollutant movement (SWQUAL). Quantity (flow) simulation utilizes the equation of continuity and motion for channel flow, and quality simulation is accomplished by application of the mass continuity equation within the element. Since the computational stability of the SWQUAL subroutine is better than that for SWFLOW, the quantity subroutine requires shorter simulation intervals (fraction of an hour), while the time step for quality may be several hours.

A detailed description of the model and instructions for its use are presented.[5]

Although SWMM is designed for simulation and design of combined or stormwater sewer systems, nonpoint pollution from areas with no sewers also can be estimated. In this case, the TRANSPORT and STORAGE-TREATMENT blocks are not run.

The net rain computation in the RUNOFF block utilizes the concept of surface water storage balance, which includes infiltration and surface storage losses. Infiltration is estimated by Horton's formula (Equation 3-4).

The original version of the SWMM has been improved further by its authors and users. The University of Florida included in the model the Universal Soil Loss Equation that estimates soil loss from pervious areas.[14] The particle accumulation on impervious surfaces utilizes the concept of linear daily (hourly) increase of pollutants in the curb storage during a dry period with the Sartor et al. wash-off model (see Chapter 8, Equation 8-10). The amount of particles on impervious surfaces can be reduced by street sweeping practices. Loading of pollutants other than sediment is estimated by employing various potency factors; some of the factors as well as the daily accumulation rates are incorporated in the model as default values.

The SWMM model is capable of representing drainage areas that range from 5 to 2000 ha. The program itself is rather large, requiring an extensive amount of input data. Its use for simulating nonpoint pollution processes and problems is limited.

Example 9-1: Use of the SWMM

There are many reported applications of SWMM.[4,5,14] *The following excerpt describes an application of the model to a 335-ha combined sewer area in Racine, Wisconsin.*[15]

The drainage boundaries, main and branch sewers were located according to the city sewer maps. The total drainage area was subdivided into a number of smaller areas based on the layout of the sewerage system. After the runoff patterns were determined, the land use and distribution within each common runoff area were used for further subdividing into 56 subareas as follows:

Single family residential	*20*
Multifamily residential	*13*
Commercial	*14*
Industrial	*6*
Open land	*3*
Total	*56*

The above five land-use categories are the only ones recognized by SWMM.

The next step involved determination of the drainage system for each subarea. After each subarea was defined, the degree of imperviousness (%) and curb length density were estimated from aerial photographs.

Since the task of the project was to model combined sewer systems, the dry-weather sewage flow characteristics had to be determined. The quantity and quality of the dry-weather flow was based on data from the local sewage treatment plant distributed according to housing density and location of industries throughout the modeled area.

The amount of rainfall was measured by three rain gages. The contributing area for each rain gage was determined graphically by constructing perpendicular bisectors of the lines joining the location of each rain gage on the map.

The model was calibrated and verified on three storms with rainfalls of 6.6, 22, and 41 mm. Figs. 9-7 and 9-8 show comparsions of measured and computed quantity and quality from the combined sewer overflow before and after treat-

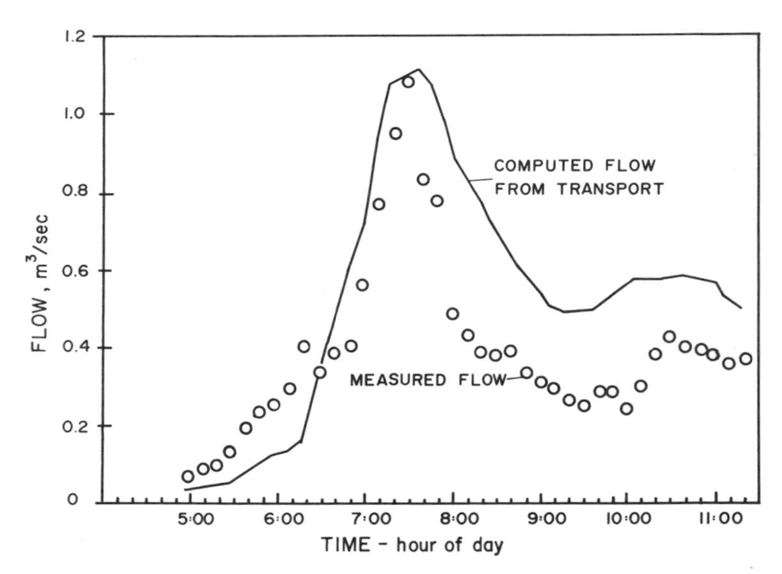

Fig. 9-7. Comparison of observed runoff and simulation results by SWMM for a combined sewer overflow, in Example 9-1. (From Meinholz et al.[15])

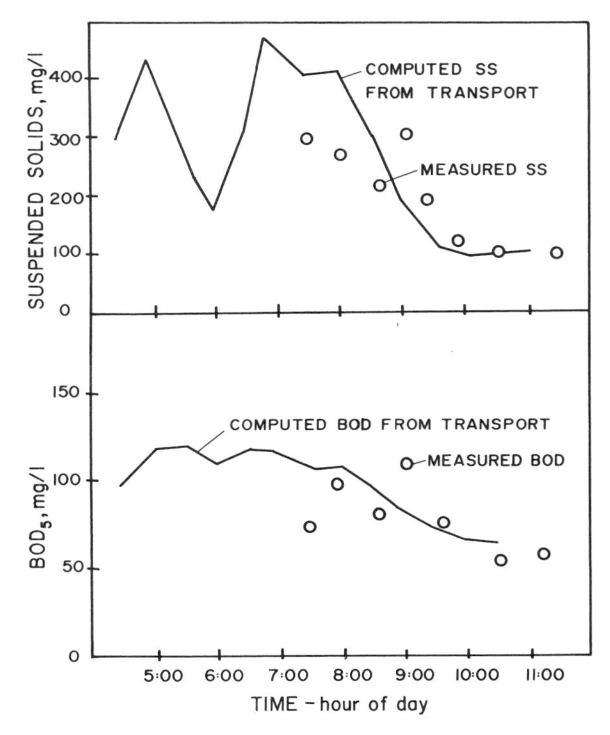

Fig. 9-8. Comparison of measured and simulated suspended solids and BOD_5 for Example 9-1. (From Meinholz et al.[15])

ment. The experimental treatment unit consisted of bar racks and dissolved air flotation units.

Storage, Treatment, and Overflow Model, STORM[6]

This model was developed for the U.S. Army Corps of Engineers by Water Resources Engineers, Inc. The program is capable of simulating quantity and quality of runoff from small–primarily urban–watersheds, however, nonurban areas also can be included. The modeled water quality parameters include total and volatile particulates, biochemical oxygen demand (BOD_5), total nitrogen (N), and orthophosphate. These water quality parameters are assumed to be a part of or related to the particulate matter. The model considers the interactions of seven storm-water transport processes:

1. Precipitation and air temperature to determine rainfall or snowfall–snow-melt.

2. Surface runoff.
3. Pollutant accumulation.
4. Land surface erosion.
5. Treatment rates.
6. Storage of storm-water runoff.
7. Overflows from the storage/treatment system.

Although the program is discrete, it can be used for simulating longer (up to several years) time series of runoff quality and quantity. The basic concept of the program is shown in Fig. 9-9.

In the first portion of the program computation, a decision is made whether precipitation is in the form of rain or snow according to the ambient air temperature. The snow-melt from the accumulated snow storage is computed by the degree-day method (Equation 3-10), and runoff is computed by a modified rational formula (Equation 3-16). No provision is made for flow routing in the program and—due to the questionable nature of the runoff coefficient in the rational formula—runoff quantities can be highly inaccurate even for calibrated models.

Before the runoff from urban areas is combined with the nonurban runoff, it is subjected to diversion into a storage unit.

The pollutant accumulation component for impervious surfaces assumes a uniform daily accumulation rate of pollutants during a dry period. Accumulated

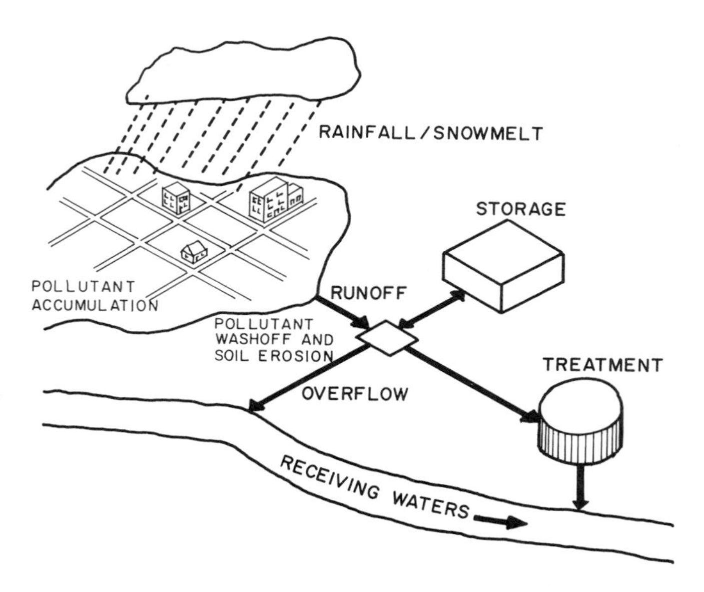

Fig. 9-9. Conceptualized view of urban system represented by STORM. (Replotted from reference 6.)

pollutants can be removed by sweepers (Equation 8-6). Wash-off rate is estimated by the modified Sartor et al. wash-off function (Equation 8-11) using the availability factors defined in Equation 8-12.

Pollutant wash-off from pervious areas is computed by an equation similar to that for impervious surfaces, but no analysis of the availability of pollutants is made for nonurban areas.

Sediment generation from pervious surfaces is computed by the Universal Soil Loss Equation (USLE). In order to determine how much of the eroded soil reaches a given point on the watershed, the amount is multiplied by the sediment delivery ratio that must be input.

The STORM model is a medium-sized program but still requires a substantial amount of input data. The inputs should include information on:

1. Meteorological inputs (precipitation, temperature, potential evaporation). The precipitation data are input hourly, temperatures daily (average, minimum, and maximum), and evaporation monthly.
2. Street lengths, depression storage for impervious and pervious areas, and gutter density.
3. Land-use distribution for urban areas.
4. Degree of imperviousness for each land use.
5. Street sweeping efficiency and interval.
6. Pollutant accumulation rates.
7. Overall runoff coefficient for nonurban areas.
8. Soil distribution and coefficients for the USLE determination for each soil in nonurban areas.
9. Trap efficiency for sediment detention reservoirs.
10. Treatment efficiencies for runoff.

References 4 and 6 contain detailed examples of the use of the model.

Calibration of the model parameters is accomplished through the use of the zero storage-zero treatment alternative. In this case, rainfall excess is not altered by storage and treatment, and a detailed comparison of computed and observed hydrographs and pollutographs can be made for selected storms. It should be remembered that no routing procedure is included; therefore only total daily or weekly runoff volumes and pollutant masses can be calibrated.

The zero storage-zero treatment alternative also is used to predict nonpoint pollution from uncontrolled areas.

The Battelle Wastewater Management Model[7]

This is a flow-economical model for determining the required control operations of sewer systems during real-time rainstorm events. The model considers the flow rate and quality of the sewage and the optimal allocation of available storage and treatment capacities and determines the best distribution of overflows to receiving waters.

The model computes surface runoff by considering the losses from precipitation by surface storage and infiltration. The excess rain is routed overland by use of the unit hydrograph concept.

The quality of the storm runoff is computed by simple linear multiple-regression formulas that relate concentrations of the runoff quality constituents to the storm runoff rate, cumulative runoff during the storm, and antecedent conditions. The equations were developed from and verified by observed data from Cincinnati, Ohio.

After the runoff enters the sewer system, the dry-weather sewage flow and quality are added and routed through the sewer system.

The use of this model for nonpoint simulation studies is limited.

An Agricultural Chemical Transport Model—ACTMO[8]

This model was developed by the Agricultural Research Service of the U.S. Department of Agriculture. The model consists of three submodels: hydrological, erosion, and chemical transport. The primary purpose of the model is to simulate transport of organic chemicals from agricultural lands.

The hydrological component is a slightly modified version of the USDAHL-70 Model of watershed hydrology.[16] The basic areal subunit for the hydrologic model is called a "zone" and is constructed by grouping together fields of the same crops, or major soil area, or land-use classes, or whatever physical feature is recognized as being important. The percent of each zone overflowing onto lower zones is estimated from soil and topographical maps. Figure 9-10 represents a schematic breakdown of a watershed into computational units-zones.

The model continuously accounts for soil moisture by balancing infiltration, evapotranspiration, and seepage into lower soil layers. The evapotranspiration rate is estimated from crop stage, pan evaporation, and soil moisture characteristics (Equation 3-12).

Infiltration capacity is expressed by the Holtan model (Equation 3-5a). Infiltration and rainfall excess are computed for each zone by comparing rainfall to infiltration capacity and available surface storage.

Rainfall in excess of infiltration and surface storage is routed across each zone and cascaded if it overflows on adjacent soil segments. A modified kinematic wave concept is used for routing excess rainfall.

Groundwater recharge rate is input in the model and can be estimated on a regional basis from average annual rainfall, average annual evapotranspiration, and average annual stream flow.

The Erosion-Deposition submodel predicts soil loss by the modified Universal Soil Loss Equation (Equations 5-2 and 5-4). The model includes rill and interrill

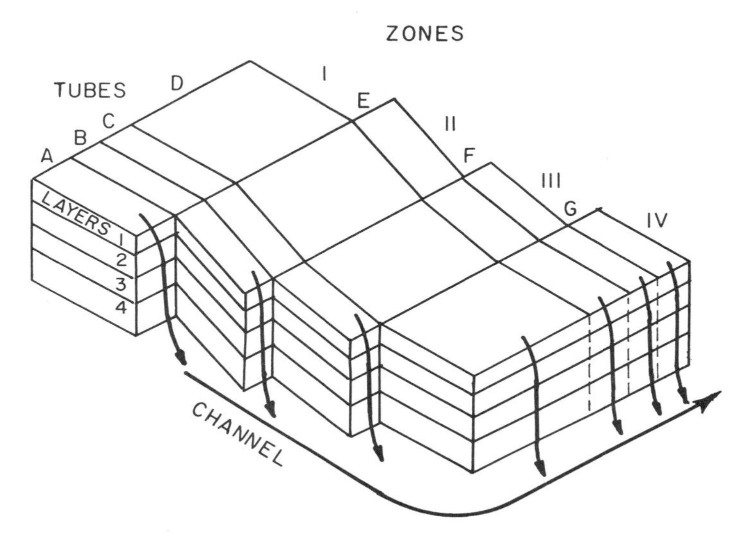

Fig. 9-10. Three-dimensional transformation of the watershed into computational units of ACTMO model. (From Free et al.[8])

contributions and is capable of estimating particle size distribution by calculating clay enrichment ratios (Equations 5-22 to 5-24).

The Chemical submodel traces the movement of a single application of a chemical through and over the watershed. Cultivation is assumed to redistribute the chemical. The model simulates the sorption-desorption process by a linear isotherm and the process is assumed to be instantaneous. Other processes included in the chemical submodel are decomposition of the chemical and its dispersion. The sorbed pollutant fraction moves with the detached soil particles, while surface runoff during ponding mixes with the soil water containing the dissolved phase.

The model also is capable of simulating N movement and transformations through a special option. Organic N is mineralized to NH_4^+ and then to NO_3^- according to a first-order reaction scheme similar to Equation 5-20. The rate coefficients depend on soil moisture and temperature.

The size of the watershed is limited by the requirement that only one rain gage input is permitted. A farm-size watershed is recommended as the most convenient simulation unit.

Example 9-2: Use of ACTMO Model

Two test examples of the use of the model are included in reference 8. The only complete example reported by the authors of the model involves a 2-yr experi-

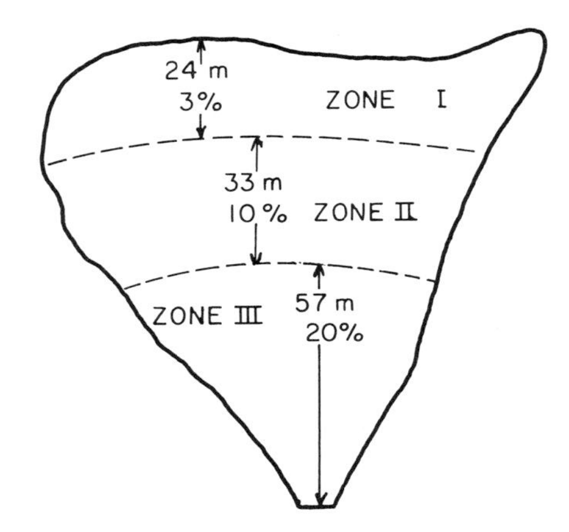

Fig. 9-11. Watershed 113 divided into three hydrological zones for Example 9-2. (Replotted from Free et al.[8])

ment with the pesticide carbofuran on watershed No. 113 at the North Appalachian Experimental Watershed, Coshocton, Ohio.

The 0.6-ha watershed was divided into three equal zones as shown in Fig. 9-11; corn was grown on the watershed for 2 yr.

The input data required for the hydrological submodel included a physical description of the three zones and the amount of flow (%) overflowing from the upper zone onto the adjacent lower zone, temperature and pan evaporation, groundwater recharge, soil moisture characteristics and permeability, crop stage, and rainfall inputs.

The erosion submodel requires information on the parameters of the USLE (soil erodibility, slope, length of overland flow, texture, crop factor, and management factors), and from the hydrological submodel, the rainfall factor computed from the storm intensity and volume and peak runoff rate.

The chemical submodel input data include information on application rate, date of application of the chemical, depth of mixing, adsorption coefficient, decomposition coefficient and preference for size fractions, soil moisture and texture characteristics, hydrological characteristics from the hydrological submodel, and erosional characteristics from the erosion submodel. Table 9-1 contains a comparison of predicted and measured values. Some problems with the performance of the model can be detected in this case study, and more research has been recommended.

Hydrocomp Models

Several models applicable to nonpoint pollution problems have been developed by Hydrocomp, Inc., Palo Alto, California. The Hydrocomp models evolved

TABLE 9-1. Measured and Predicted Water, Sediment, and Carbofuran Losses From Coshocton Watershed 113 for Five Storms.[8]

Item	Storm Date				
	5-6-71	6-13-72	6-15-72	7-10-72	11-7-72
Runoff (in.)					
Measured	0.21	0.24	0.41	0.20	0.22
Predicted	0.21	0.46	0.27	0.001	0.15
Peak flow (cfs)					
Measured	0.28	0.85	3.54	4.24	...
Predicted	0.23	1.15	0.83	0.01	0.27
Erosion (tons/acre)					
Measured	1.73	2.12	4.69	0.69	0.38
Predicted	4.27	9.53	4.80	0.80	3.60
Carbofuran (lb)					
Measured	0.036	0.021	0.046	0.004	0.001
Predicted	0.071	0.15	0.104	0.04	0.013
Concentration (mg/liter)					
Measured	0.47	0.19	0.22	0.05	0.02
Predicted	0.78	0.56	0.70	4.24	0.14

Conversions: 1 in. = 2.54 cm, 1 lb = 0.454 kg, 1 ton/acre = 2.24 tonnes/hectare, 1 cfs = 0.028 m^3/sec.

from the Stanford Watershed Model IV (SWM).[17] The latest and most completely tested model derived from the SWM is the Hydrocomp Simulation Program (HSP).[13,18]

The HSP system is a proprietary model; however, several simplified versions of the HSP have been ordered by the U.S. Environmental Protection Agency and are in the public domain. These models are the Nonpoint Simulation Model (NPS)[11] and the Agricultural Runoff Model (ARM)[9] and HSPF.

The HSP modeling system requires division of the watershed into overland flow segments and stream channels. The system includes three basic components: LIBRARY, LANDS, and CHANNELS. The water quality model (QUALITY) is a separate unit used with LIBRARY, LANDS, or CHANNELS.

LIBRARY is a data-file-handling component and, basically, the master program.

LANDS is a flow-generating submodel that computes overland inputs into channels from precipitation, meteorological factors, and overland flow characteristics. Essentially, LANDS describes the first part of the hydrological cycle: precipitation-runoff transformations.

The basic structure of LANDS is identical to SWM (Fig. 9-12). The rainfall data can be input in time intervals from 5 min to 6 hr. In addition to rainfall,

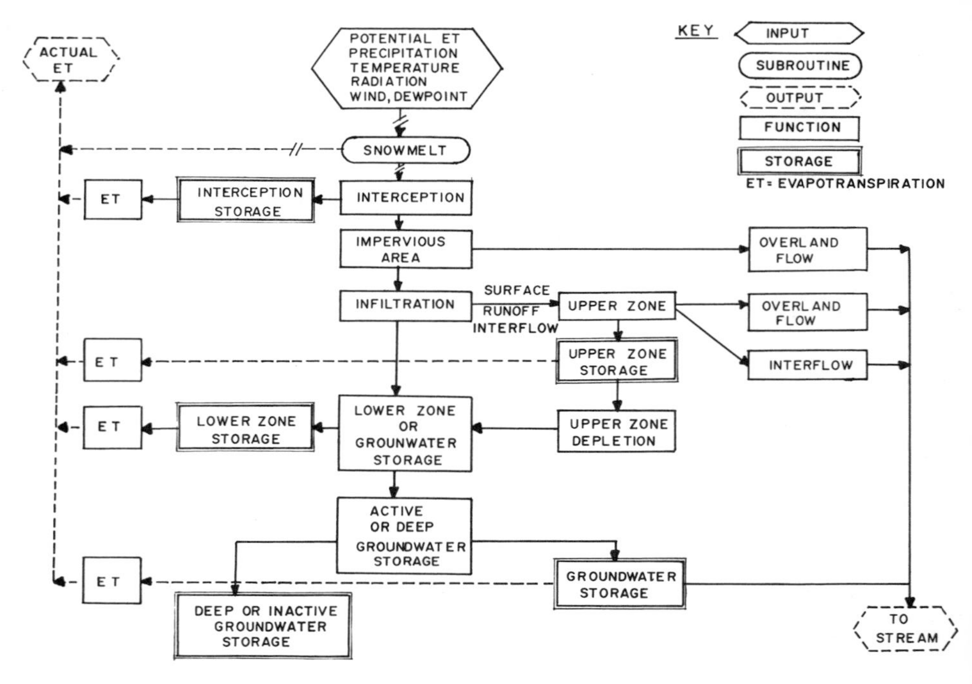

Fig. 9-12. LANDS submodel of the Hydrocomp Watershed Model. (Replotted from Donigian and Crawford.[11])

LANDS requires daily information on potential evapotranspiration, interception storage, soil moisture and saturation permeability characteristics, slope of the overland flow, imperviousness of the watershed, plus other physical and meteorological variables. The magnitude of many system parameters, especially those related to soil and groundwater flow, is determined by calibrating the model against measured hydrological flow data.

Rainfall excess is estimated by subtracting the hydrologic losses, infiltration, surface storage, etc., from precipitation. The snow-melt component balances heat inputs and losses from the accumulated snow pack.

Runoff from impervious areas equals precipitation minus surface storage. However, an impervious area that overflows onto an adjacent pervious area or into soil is not counted, and the runoff volume is added to the water balance of the pervious area.

On pervious lands, water infiltrates into soil horizons. The present version of the LANDS model is using a modified and simplified Philip's infiltration formula (Equation 3-9) in which the soil storage parameter, S, is determined from calibration of the model, and the wetting front conductivity, A, is neglected.

As soil moisture storage is increased by infiltration, it is depleted by evapotranspiration and by percolation into deeper horizons. Soil storage is divided into two components: upper soil storage, which represents soil water in the top soil zone that can be removed by evaporation and transpiration; and lower soil zone storage, which is assumed to represent soil moisture in the zone of aeration to the bottom of the root zone.

Interflow is simulated by an empirical equation relating lateral movement of soil water to the exhaustion of the lower zone storage (Equations 3-27 and 3-28).

The excess precipitation that escapes the upper zone and surface storage is routed overland to form the surface runoff component of flow input to the receiving channel. The difference between seepage from the lower soil zone and deep percolation enters groundwater storage. A portion of groundwater that is above the elevation of the receiving channels reappears as groundwater or base flow. The amount diverted to deep storage (below the channel elevation) is lost from the system; however, it may serve as useful information for studies concerned with groundwater recharge.

Routing in the CHANNELS segment is accomplished–in earlier versions of the HSP and SWM models–by a modified Muskingum routing procedure and by kinematic wave approximation in the most recent versions (HSP II).

The QUALITY model in the newer version of the HSP program is able to simulate erosion from pervious areas by the Negev model (Equations 5-7 and 5-8) plus street dust and dirt accumulation and wash-off from impervious surfaces by the Sartor et al. concept (See Chapter 8). Loading of pollutants other than sediment is estimated by application of potency factors by which loadings of sediment are multiplied.

The HSP modeling system is large, requiring great computer hardware, and input data matrices are extensive. After calibration the model adequately simulates flow and quality time series for any hydrological and meteorological condition. Its lumped parameter character makes it somewhat less reliable for simulations when some physical features of the watershed are changed from the situation existing during calibration.

The Nonpoint Simulation Model, NPS[11]

This model is one of the packages developed by Hydrocomp, Inc., which is available from the U.S. EPA.

The NPS model simulates nonpoint source pollution from a maximum of five different land-use categories in a single operation. In addition to runoff, water temperature, DO, and sediments the NPS model allows for up to five user-specified pollutants from each land-use category.

The NPS model is composed of three major components: MAIN, LANDS, and QUAL. The model operates sequentially, reading parameter values and meteorological data, performing computations in LANDS and QUAL. The LANDS segment is similar to the HSP model (Fig. 9-12). This component simulates the hydrologic response of the watershed to the precipitation input and the process of snow-melt and accumulation.

The QUAL component simulates the erosion process, street refuse accumulation, and sediment and pollutant wash-off from the land surface. During storm events, LANDS and QUAL operate in 15-min intervals. For nonstorm periods, LANDS uses a combination of 15-min, 1-hr, and 24-hr time intervals to simulate evapotranspiration and soil water percolation processes that determine soil moisture content. Since nonpoint pollution from the land surface occurs only during storms, QUAL operates on a daily interval between storm events to estimate accumulations of pollutants on the land surface that will be available for transport at the next storm event. MAIN is the master executive routine.

Erosion in the QUAL submodel is computed by the Negev model. The mass of accumulated particles on impervious surfaces is estimated by a first-order accumulation function similar to Equation 8-4 whereby the street refuse removal rate, ξ, is input. The washoff function is similar to that used by SWMM and STORM.

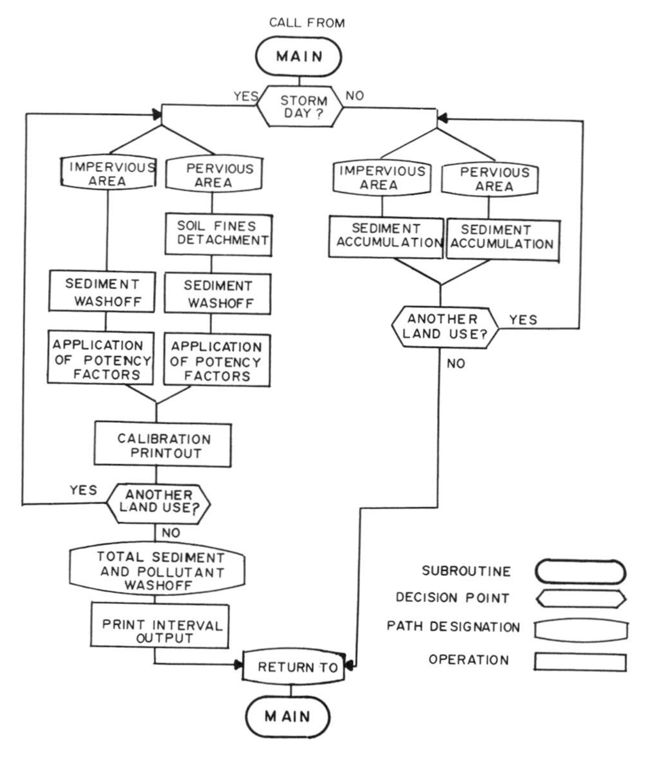

Fig. 9-13. Flow chart of the QUAL subroutine in the NPS model. (Replotted from Donigian and Crawford.[11])

Pollutants other than sediment are estimated by multiplying sediment loss from soils and impervious surfaces by specified appropriate potency factors. Figure 9-13 shows a block diagram of the QUAL subroutine.

Example 9-3: Application of the NPS Model

The NPS model was used for simulation of flow, sediment, and nutrient transport from the Third Fork Creek in Durham, North Carolina.[11, 19]

The upper Third Fork Creek watershed is a typical urbanized area in the Piedmont region of the southeastern United States (Fig. 9-14). The area of the basin is 433 ha and encompasses a variety of land uses including: high and low density residential areas, commercial, industrial, and open lands. A continuous record of 15 min precipitation was developed from data supplied by an in situ rain gage and augmented for missing periods with recorded hourly data from the regular U.S. Weather Bureau station. Daily pan evaporation data were obtained at the Chapel Hill station, which is located about 16 km from the test site. Continuous daily stream flow and monthly runoff volumes were obtained

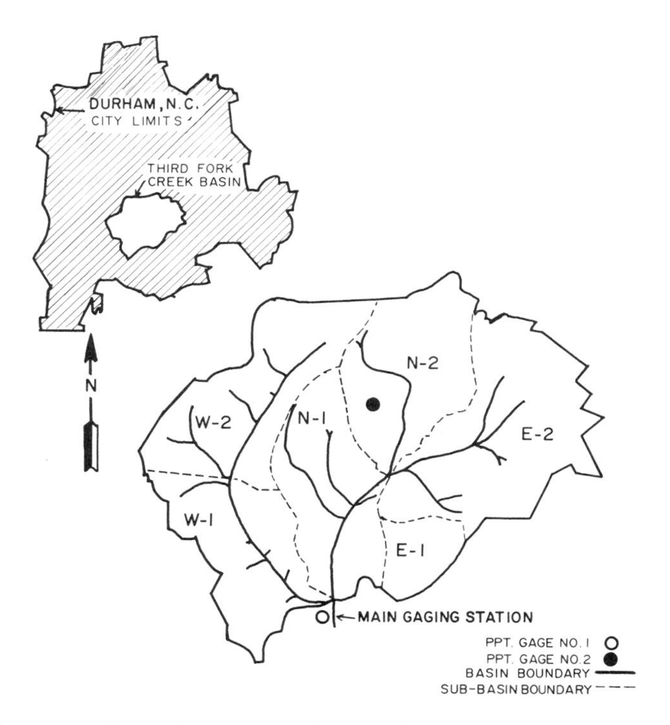

Fig. 9-14. Third Fork Creek watershed in Durham, North Carolina, for Example 9-3. (Replotted from Donigian and Crawford.[11])

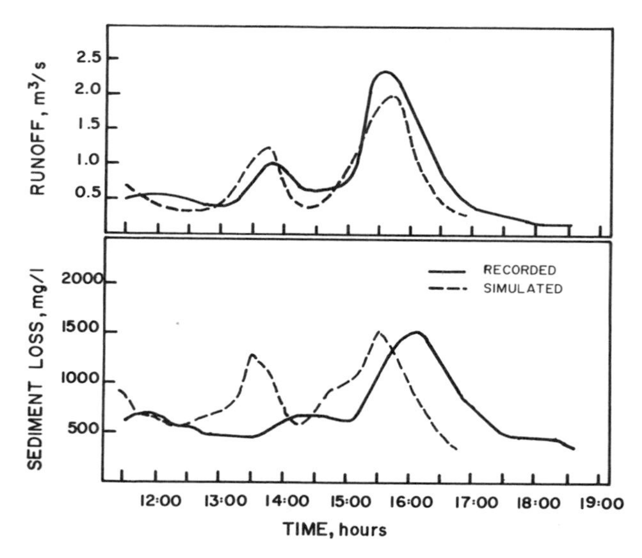

Fig. 9-15. Comparison of measured and simulated runoff and sediment loss by the NPS model for Example 9-3. (Replotted from Donigian and Crawford.[11])

from USGS records. Physical description of the watershed was taken from previous research reports.

Suspended solids and BOD concentrations are shown for a selected storm in Figs. 9-15 and 9-16. Although the simulation results indicate adequate agreement with measured data, an apparent effect of channel processes is evident. With an area of 433 ha (4.33 km^2), the Third Fork Creek watershed approaches the upper limit of applicability for the NPS model.[11]

Nutrient modeling by the NPS for the same watershed was attempted later.[19]

The Agricultural Runoff Management Model, ARM[19]

This is another overland flow version of the HSP model based on the Stanford Watershed model. The ARM model is a continuous model that simulates runoff (including snow accumulation and melt), sediment, pesticide, and nutrient loadings to surface waters from surface and subsurface sources.

The major components of the model (Fig. 9-17) are: the hydrological component (LANDS), sediment production (SEDT), pollutant adsorption/desorption (ADSRB), pesticide degradation (DEGRAD), and nutrient transportation (NUTRNT). The executive subroutine, MAIN, is the controlling master program.

The LANDS submodel, except for a few modifications, is identical to the HSP and NPS models. The SEDT subroutine utilizes the Negev sediment model introduced previously.

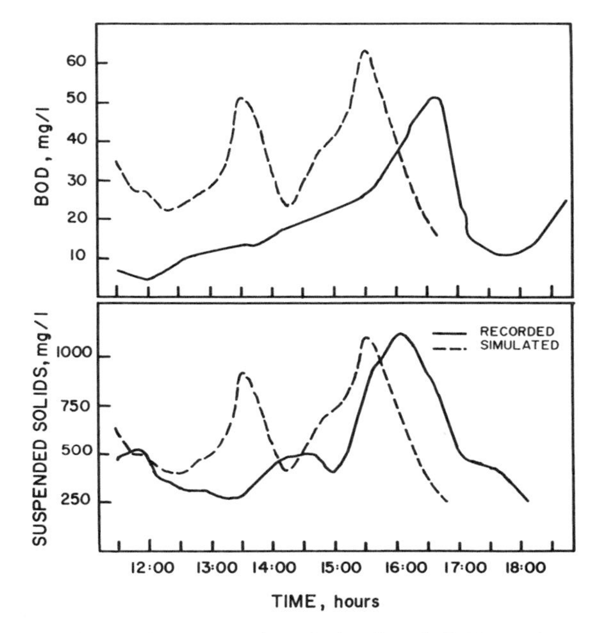

Fig. 9-16. Comparison of measured and simulated BOD and suspended solids by the NPS model for Example 9-3. (Replotted from Donigian and Crawford.[11])

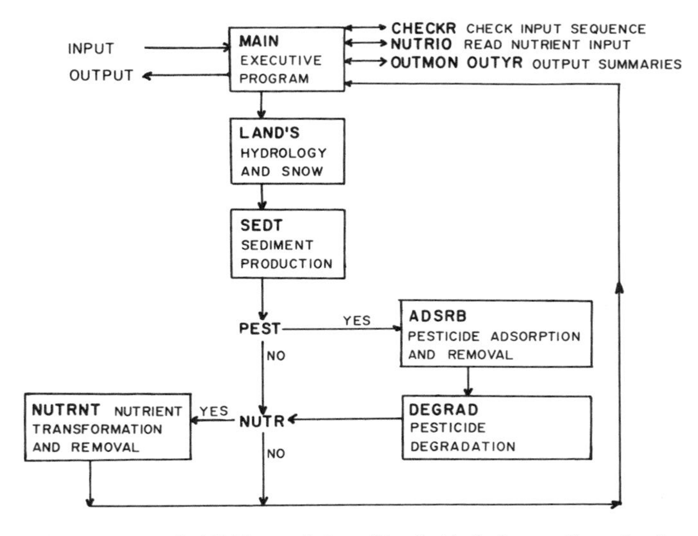

Fig. 9-17. Structure of ARM model. (Replotted from Crawford and Donigian.[20])

The adsorption/desorption model assumes an instantaneous adsorption or desorption described by the following modified Freundlich adsorption isotherm:

$$S = KC^{1/N} + F/M \tag{9-1}$$

where:

S = pesticide adsorbed, μg/g of soil
F/M = pesticide adsorbed in permanent fixed state, and F/M is $\leqslant FP/M$, where FP/M is the permanent maximal fixed capacity of the soil in μg/g for each pesticide
C = the equilibrium pesticide concentration in solution (mg/liter)
N = exponent
K = coefficient

Volatilization of pesticides was modeled by an equation adopted from basic heat transfer processes. Diffusion is assumed to be the rate-limiting mechanisms for pesticide flux occurring from the land surface for pesticides incorporated in soils. Volatilization of surface-applied pesticide is simulated by a function that relates flux of pesticides to ambient wind velocity and vapor pressure.

A first-order degradation model has been assumed to represent the removal of pesticides by biological and chemical decomposition and transformation. In the most recent version of the model,[22] a first-order attenuation function that calculates combined degradation by volatilization, microbial decomposition, and other attenuation mechanisms has been used instead of more sophisticated degradation models.[21]

The nutrient transformation model is shown in Fig. 9-18. The model assumes first order reaction rates. The processes simulated include immobilization, mineralization, nitrification/denitrification, plant uptake, and adsorption/desorption. The ARM model simulates nutrient movement in the watershed by water or sediment. Reaction rates are input on a per-day basis for each soil zone. Plant uptake rates are modified monthly by an input parameter that depends on the stage of crop growth.

The model can be used for event and continuous simulations. Only small, relatively uniform watersheds can be simulated successfully.

Example 9-4: Application of ARM Model

References 9, 20, and 22 contain several examples of calibration runs for small (about 1 ha) controlled watersheds. The results show fair agreement of simulated and measured data for some pesticides (paraquat, diphenamid), but some problems are still unresolved.[23]

A paper by Beyerlein and Donigian[21] *contains an interesting example of the use of the ARM model to evaluate the effect of various soil and water conserva-*

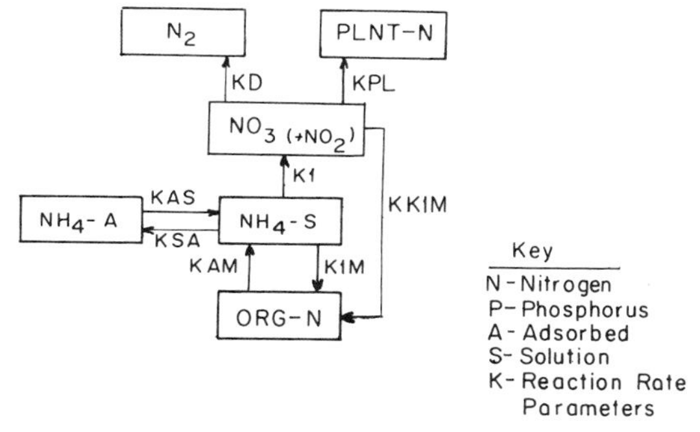

A. Nitrogen transformation in ARM model

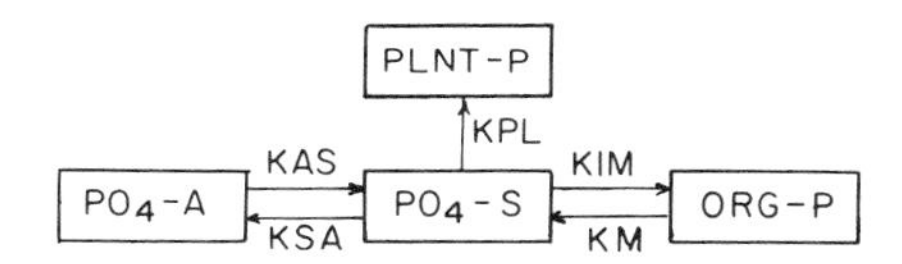

B. Phosphorus transformation in ARM model

Fig. 9-18. Nitrogen and phosphorus transformation diagrams for ARM model. (Replotted from Crawford and Donigian.[20])

tion practices on loadings of pollutants. Three selected practices were tested by modeling the response of an experimental watershed, ARS-EPA-P2, located near Watkinsville, Georgia (Fig. 9-19).

The three practices are the use of contours, terraces, and minimum tillage. The base references practice for the watershed was planting row crops in straight rows up and down the slope with disk tillage in spring.

In the model, a total of eight parameters were identified as related to the soil and water conservation practices. These parameters included upper zone moisture storage capacity (increases for contour and terrace practices), Manning roughness factor (increases for all soil and water conservation practices except no till), length of overland flow (decreases for terraces), slope of overland flow (decreases for terraces), fraction of land covered by vegetation (decreases for terraces), erosion control practice factors (decreases for terraces), sediment wash-off coefficient in the Negev model (decreases for all soil and water conservation practices), and sediment fine particles from tillage (decreases for no-till practice). The magnitude of the change of these factors was based on experience and scientific guesses rather than on direct measurements.

The size of the P2 watershed is 1.3 ha. From 1973 through 1975, U.S. EPA and USDA ARS extensively monitored the watershed, and the data were used for calibration of the model.

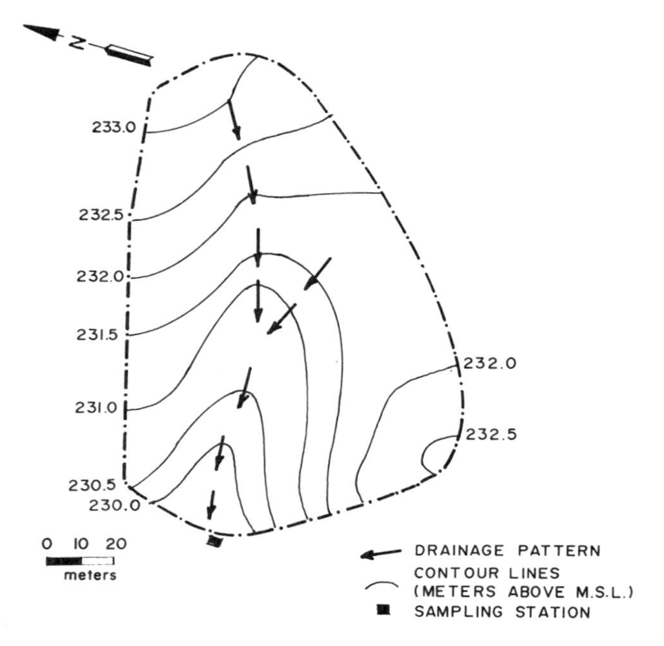

Fig. 9-19. P2 watershed, Watkinsville, Georgia, for Example 9-4. (Replotted from Crawford and Donigian.[20])

To study the effect of the soil and water conservation practices on runoff, sediment loss, and pesticide and nutrient wash-off, the P2 watershed was simulated for the 10-yr period 1966 to 1975. The continuous information produced by the 10-yr ARM model simulations was analyzed in two ways: (1) mean annual runoff and pollutant losses, and (2) frequency of occurrence of runoff, pollutant concentrations, and loadings. The mean annual values are summarized in Table 9-2. The results in the table indeed show that loadings of pollutants decrease with the suggested conservation practices, while contours and terracing provide the best results.

The Overland Flow and Pollution Generation Model, LANDRUN[12]

This model was developed during PLUARG (Pollution from Land Use Activities Reference Group) studies sponsored by the U.S.–Canada International Joint Commission.[12] The model was used primarily for modeling effects of various land uses on the pollutant loading to the Great Lakes; however, other uses also have been reported.[24]

LANDRUN is similar to the majority of models discussed in previous sec-

TABLE 9-2. Mean Annual Runoff and Pollutant Losses Simulated with the ARM Model.[a,b]

	Base Conditions	Minimum Tillage	Contours	Contours and Terraces	Percent Change from Base: Minimum Tillage	Percent Change from Base: Contours	Percent Change from Base: Contours and Terraces
Total runoff (in.)	8.03	7.74	7.16	5.21	-3.6	-10.9	-35.2
Overland flow (in.)	7.55	7.23	6.66	4.58	-4.2	-11.9	-39.4
Interflow (in.)	0.48	0.51	0.50	0.63	+5.8	+4.8	+31.8
Sediment loss (ton/ac)	2.84	1.99	1.80	1.29	-30.0	-36.8	-54.6
Total atrazine loss (lb)	0.163	0.150	0.130	0.063	-7.7	-20.1	-61.4
In solution (lb)	0.162	0.150	0.130	0.063	-7.5	-19.9	-61.3
On sediment (lb)	0.0013	0.0009	0.0008	0.0004	-30.8	-38.5	-69.2
Paraquat on sediment (lb)	0.976	0.738	0.647	0.475	-24.4	-33.7	-51.3
Nitrogen (lb/ac)							
Organic	6.30	8.36	4.91	3.84	+32.7	-22.1	-39.0
NH_4^+–Solution	1.96	2.02	1.89	1.88	+3.0	-3.9	-4.2
NH_4^+–Adsorbed	0.680	0.508	0.460	0.339	-25.3	-32.4	-50.1
$NO_3^- + NO_2$	2.67	2.80	2.89	3.62	+4.8	+8.4	+35.8
Phosphorus (lb/ac)							
Organic	0.68	0.56	0.52	0.40	-17.6	-23.5	-41.2
PO_4^{2-}–Solution	0.376	0.408	0.372	0.367	+8.5	-1.1	-2.4
PO_4^{2-}–Adsorbed	2.21	1.81	1.66	0.45	-18.2	-24.9	-79.7

[a] From Beyerlein and Donigian.[21] Courtesy Ann Arbor Science Publishers Inc.
[b] Obtained from 10-yr simulations (1966–1976) on the P2 watershed in Watkinsville, Georgia.

tions. It has three major components: net rain estimation, sediment and pollutant generation, and overland routing of sediments and pollutants.

LANDRUN is a deterministic watershed model capable of simulating the following processes:

1. Snow accumulation and melt by the degree-day method.
2. Infiltration by the Holtan or Philip models.
3. Excess rain as the difference between precipitation, infiltration, and surface storage.
4. Soil water movement and moisture storage.
5. Routing of excess rain by an Instantaneous Unit Hydrograph method.
6. Dust and dirt accumulation on urban impervious surfaces and wash-off.
7. Removal of accumulated pollutants on impervious surfaces by cleaning practices.
8. Surface erosion by a modified Universal Soil Loss Equation that takes into account rainfall energy and runoff.
9. Routing of sediment and sediment-adsorbed pollutants.
10. A dynamic soil adsorption segment is an optional feature of the model that enables detailed studies of pollutant-soil interreactions.[25]

LANDRUN is a medium-sized program. It can be run on a storm event basis or continuously. The input data requirements are similar to other overland flow and quality models (e.g., NPS). The time interval at which the rain data are input and computations are performed is specified by the user. Daily information is entered during nonevent times. Up to 25 land-use segments can be simulated at the same time, and it can be used successfully for watersheds up to 10 km^2 (i.e., 1000 ha) in size provided that channel effects are not predominant.

Example 9-5: LANDRUN Application

The LANDRUN model was tested on the Menomonee River watershed in southeastern Wisconsin. Three test sites were selected for model testing (Fig. 9-20), and extensive monitoring was conducted during the 1975 to 1978 period along with a detailed land-use inventory.

Figures 9-21 and 9-22 show examples of calibration and verification runs of the model on the Donges Bay Road subwatershed. This rather large watershed (21 km^2) was divided into 25 land segments according to soil type, slope, and land use. Although many soil types can be found throughout the watershed, their hydrologic response and the texture of the soils can be divided into about five primary soil types and three slope categories. This categorization limited the number of segments to a reasonable range. Information on soil characteristics and distribution are given in Table 6-2.

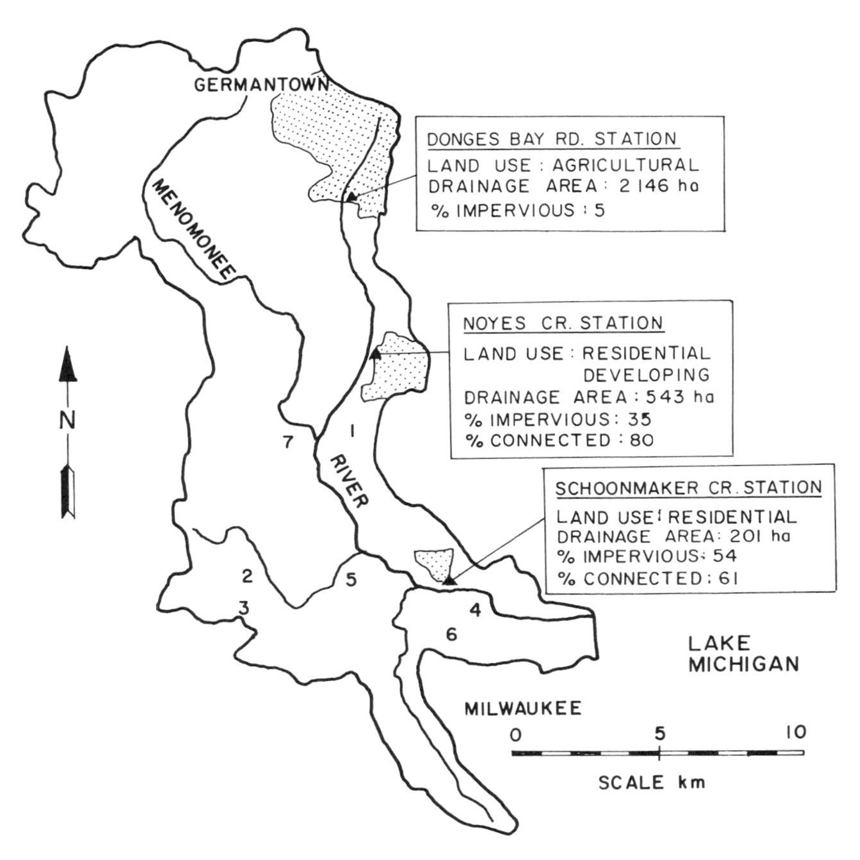

Fig. 9-20. Menomonee River watershed and location of experimental watersheds. Numbered points indicate locations of small homogeneous watersheds used for establishing loads of pollutants. (Source: Menomonee River Pilot Watershed project–International Joint Commission.[12])

The watershed has gentle hills, and the soils as well as the terrain configurations were formed during the last glacial period.

The simulation results indicate fairly low hydrologic activity of the majority of the agricultural and other pervious lands in the watershed. The impervious areas (about 5%) were responsible for almost all of the surface runoff contributions during medium storm events. Only very large storms–especially during high soil moisture in spring or frozen ground conditions–yielded surface runoff from the pervious lands.

The April 24, 1976, storm was an example of a storm during saturated soil conditions. This storm was preceded by a wet period, soil moisture was high,

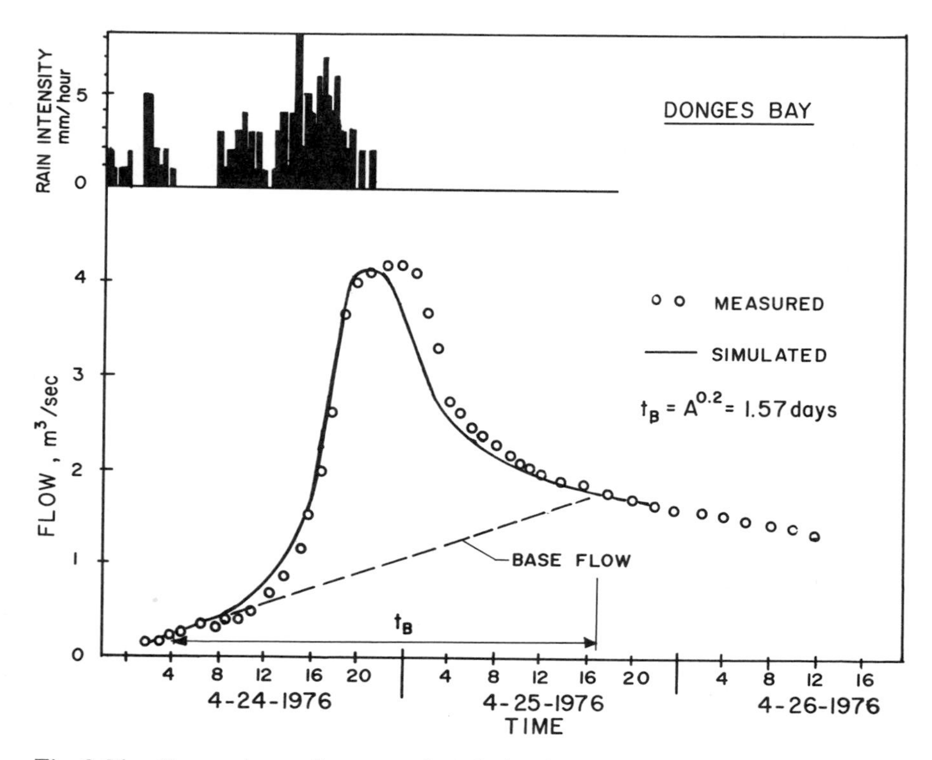

Fig. 9-21. Comparison of measured and simulated runoff by the LANDRUN model. (Source: Menomonee River Pilot Watershed project.[25])

and probably some subsoils remained frozen. Under these circumstances, high runoff and sediment volumes originated from pervious lands.

The simulated sediment loads from pervious surfaces were reduced by a delivery ratio factor ($DR \simeq 0.2$) estimated according to the methodology proposed by Roehl (see Example 5-2).

The Unified Transport Model, UTM[10,26]

The basic block of the UTM is the Wisconsin Hydrologic Transport Model (WHTM)—a Fortran version of the Stanford Watershed model. The Oak Ridge National Laboratory environmental team added new submodels to the WHTM, primarily determined for simulation of transport of toxic metals throughout a watershed.

The Wisconsin Hydrologic Transport block (WHTM) simulates the flow of toxic metals and water from their deposition on the surface of a watershed,

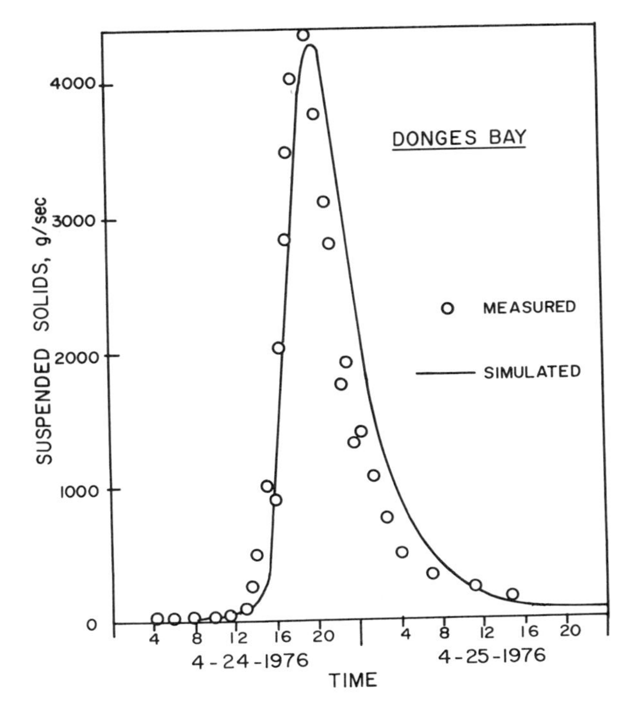

Fig. 9-22. Comparison of measured and simulated suspended solids by the LANDRUN model. (From Menomonee River Pilot Watershed project.[25])

through infiltration into the soil profile, to runoff into the stream network due to overland flow, interflow, and groundwater flow. The toxic substances transport was built into the SWM for the purpose of studying transport of radioaerosols resulting from atmospheric nuclear weapon tests.

The Atmospheric Transport Model (ATM) simulates deposition onto the land surface of toxic materials transported through air. The model is based on Gaussian plume formulations of the movement of dust or gas from an emission source to the point of deposition. The basic principles of the model are described in Chapter 4.

The Terrestial Ecology and Hydrology Model (TEHM) describes soil-plant water fluxes, interception, infiltration, and storm and groundwater flow, and replaces the parametric equations for these processes incorporated in the original SWM.

The Soil Chemical Exchange Model (SCEHM) simulates the transport of toxic metals through the soil system. The processes considered in SCEHM include deposition, infiltration, adsorption and desorption (ion exchange), and the flow of contaminants. The model assumes an instantaneous adsorption-desorption process in several soil layers.

The Forest Stand Biomass Dynamics Model (CERES) predicts both short-term effects and long-term accumulation of trace contaminants in forest plants. It estimates growth rates of plants, and in combination with the Plant Uptake and Recycle of Trace Metal Models (DIFMAS and DRYADS), it estimates contamination and the recycling of plant toxic metals incorporated in leaves, stems, and roots.

Example 9-6: UTM Application to Lead Transport

The model has been applied to the Crooked Creek Watershed (Fig. 9-23) located in the "New Lead Belt" of southeastern Missouri.[26] *The watershed is contaminated by toxic metals, primarily lead, from lead smelting operations in the south portion of the watershed. Metal particulates are transported into the air from a smelter stack and by fugitive dust generated from the smelter yard and roads by vehicular traffic, by heavy equipment operations, and to a lesser degree, by wind.*

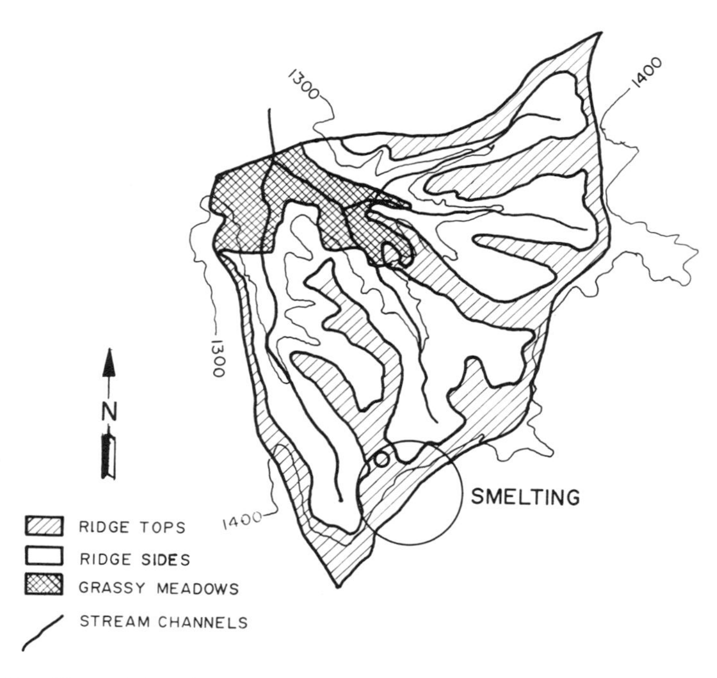

Fig. 9-23. Map of Crooked Creek Watershed for Example 9-6. (Replotted from Munro et al.[26] Courtesy Oak Ridge National Laboratory, Union Carbide Corporation.)

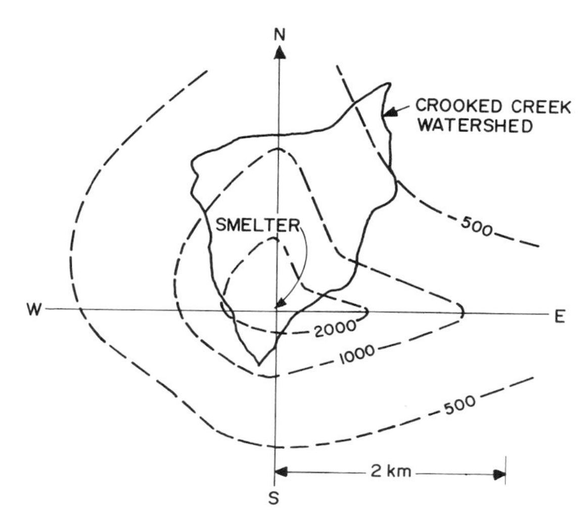

Fig. 9-24. Simulated average monthly deposition rates (mg/m^2-month) for lead near Crooked Creek Watershed using a deposition velocity of 10 cm/sec for Example 9-6. (Replotted from Munro et al.[26] Courtesy Oak Ridge National Laboratory, Union Carbide Corporation.)

The first segment of the model–ATM–generated deposition isopleths (Fig. 9-24) of the metals. The measured deposition rates deviated from the simulated isopleths by a factor of five during calibration and verification runs.

Figure 9-23 shows segmentation of the watershed into three zones:

1. *Ridge tops that include most of the bare areas of the watershed and most activities associated with mining and smelting operations. This segment receives most of the dust associated with movement of ore, ore concentrate, and slag.*
2. *Hillsides that are almost entirely covered by forest.*
3. *Grassy meadows used mainly for agricultural purposes.*

The segmentation was input in the WHTM model. Other inputs necessary for running the WTHM are similar to the SWM and Hydrocomp models and include meteorological data, soil characteristics, etc.

The deposition rates were simulated only for Pb. The ATM was not run for Cd, Zn, and Cu since the deposition rates were deemed similar to those of lead.

Table 9-3 shows the comparisons of average daily flows and average metal concentrations simulated by the UTM and corresponding measured values in the stream. Table 9-4 presents a comparison of simulated and measured values in the vegetation of the watershed. Due to the complexity of the processes involved

TABLE 9-3. Comparison of Simulated and Measured Stream Flows and Toxic Metal Concentrations for Several Storms at Crooked Creek Watershed, New Lead Belt, Missouri.[a]

Heavy Metal	Storm Dates	Daily Flows (cfs)		Simulated Total Metal Transported Per Day (kg)	Daily Average Metal Concentration (Unfiltered Samples) (mg/liter)	
		Measured	Simulated		Measured	Simulated
Pb	8-28-74	1.69	3.15	58.3	3.7	7.56
	8-29-74	6.75	10.7	22.4	9.0	0.86
	9-11-74	12.7	11.8	110.0	9.9	3.81
Cd	8-28-74	1.69	3.15	2.01	1.8	0.26
	8-29-74	6.75	10.7	3.01	2.8	0.11
	9-11-74	12.7	11.8	6.72	0.57	0.23
Zn	8-28-74	1.69	3.15	14.1	4.57	1.83
	8-29-74	6.75	10.7	10.4	10.01	0.40
	9-11-74	12.7	11.8	62.2	5.43	2.15
Cu	8-28-74	1.69	3.15	0.47	0.18	0.06
	8-29-74	6.75	10.7	1.08	0.50	0.04
	9-11-74	12.7	11.8	6.64	0.55	0.23

[a]From Munro et al.[26] Courtesy Oak Ridge National Laboratory, Union Carbide Corporation.

and the lack or quality of adequate input, differences between the simulated and observed data often greater than an order of magnitude have been noticed, indicating a necessity for further research.

Watershed hydrological models such as those discussed in the previous sections of this chapter are adequate for modeling transport of pollutants from a terrestrial source (soil or impervious urban areas) to a nearest receiving water body. However, watershed water transport models may not be sufficient if regional pollution problems involving air transport, soil storage and transformation, biological modification, and water transport are involved.

The multimedia models, in which the history of a pollutant released at a location into the environment is traced from the point of its origin to its final sink, are on the drawing boards of several institutions and consulting companies. It is interesting to note that these efforts may not necessarily lead to the development of a giant "universal" computer model but, at least in one instance, the modules of the overall model can be programed on a desk or pocket calculator.[28,28]

The UTM is one example of the multimedia modeling concept that encompasses air transport, soil interaction, and hydrologic transport. Several efforts

TABLE 9-4. A Comparison of Simulated and Experimental (Measured) Results for Crooked Creek Watershed.[a]

	Simulated	Experimental
Total Litter Mass (g/m^2)		
Litter Conc. (ppm)	2,448.0	3,600.0
Lead	51,000.0	72,000.0
Cadmium	5.3	150.0
Zinc	121.0	2,350.0
Copper	375.0	1,400.0
Lead Tissue Conc. (ppm)		
Lead	170.0	467.0
Cadmium	10.0	3.9
Zinc	50.0	47.0
Copper	10.0	10.0
Root Tissue Conc. (ppm)		
Lead	1,150.0	2,400.0
Cadmium	52.0	46.0
Zinc	52.0	48.0
Copper	10.0	6.0
Stem Tissue Conc. (ppm)		
Lead	27.0	680.0
Cadmium	9.0	3.9
Zinc	35.0	30.0
Copper	18.0	13.0
Soil Conc. Al Layer (ppm)		
Lead	271.0	334.0
Cadmium	4.4	3.55
Zinc	52.0	25.0
Copper	46.0	18.0

[a]From Munro et al.[26] Courtesy Oak Ridge National Laboratory, Union Carbide Corporation.

have been undertaken by the developers and users of the SWMM to broaden the versatility of the model and to expand its capability. The work of Perez et al.[29] at the University of Florida is of particular interest since it represents an effort to develop a set of interfacing modules that could be used in sequence (Fig. 9-25) to model pollutant input from complex watersheds.

Large and complex models may be very difficult to calibrate and, therefore, they must rely on tested and verified subroutines.

Two most recent models that unify some of the discussed concepts should be also mentioned in the conclusion to this chapter:

The ANSWERS (Areal, Nonpoint Source Watershed Environment Response Simulation) model is developed by the Purdue University[30]. This model is based on the distributed parameter concept that divides the watershed into small

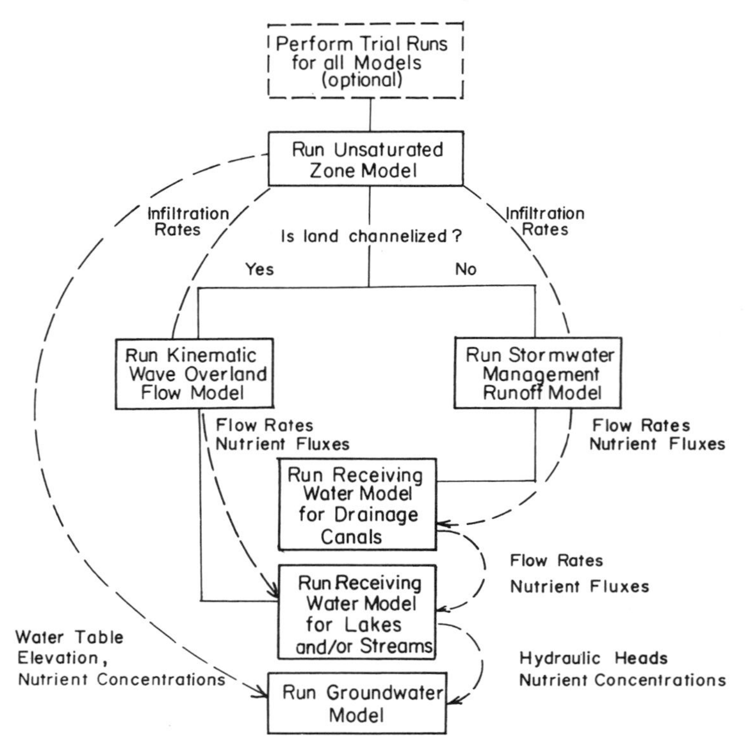

Fig. 9-25. Sequential interfacing in Perez et al.[29] complex watershed modeling concept.

square (1 to 4 ha) areal elements which interact, that is water from one element can overflow on adjacent elements with or without a channel. Such concept enables graphical displays of hazardous lands (areas where sediment and pollution originates) as well as areas where sediment and pollutants are deposited.

The CREAMS (Chemicals, Runoff, and Erosion from Agricultural Management System) model developed by group of scientists from the U.S. Department of Agriculture[31] consists of three major components: hydrology, erosion/sedimentation and chemistry. This model is relatively simple yet it enables adequate estimation of water, sediment, and pollutant yields from field size agricultural units.

REFERENCES

1. Sanders, W. M. 1976. Non-point source modeling for section 208 planning. In: "Best Management Practices for Non-Point Source Pollution Control," EPA-905/9-76-005, U.S. EPA, Washington, D.C.

2. Beasley, D. B. 1976. Simulation of the environmental impact of land use on water quality. In: "Best Management Practices for Non-Point Source Pollution Control," EPA-905/9-76-005, U.S. EPA, Washington, D.C.
3. Owen, P. T., Dailey, N. S., Johnson, C. A., and Martin, F. M. 1979. An inventory of environmental impact models related to energy technologies. ORNL/EIS-147, Oak Ridge National Laboratory, Oak Ridge, Tennessee.
4. Wanielista, M. P. 1978. "Stormwater Management; Quantity and Quality." Ann Arbor Science, Ann Arbor, Michigan.
5. Anon. 1971. Storm water management model (4 volumes). EPA No. 11 024D0C07/71 to 11 024D0C10/71, U.S. EPA, Washington, D.C.
6. Anon. 1975. Urban storm water runoff-STORM. The Hydrologic Engineering Center, U.S. Army Corps of Engineers, Davis, California.
7. Branstetter, A., Engel, R. L., and Cearlock, D. B. 1973. A mathematical model for optimum design and control of metropolitan wastewater management systems. *Water Res. Bull.*, **9**(6):118–120.
8. Free, M. H., Onstad, C. A., and Holtan, H. N. 1975. ACTMO—An Agricultural chemical transport model. ARS-H-3, Agricultural Research Service, USDA, Washington, D.C.
9. Donigian, A. S., and Crawford, N. H. 1975. Modeling pesticides and nutrients on agricultural lands. EPA-600/2-76/043, U.S. EPA, Washington, D.C.
10. Fulkerson, W., Shulz, W. D., and VanHook, R. I. 1974. Ecology and the analysis of trace contaminants. ORNL-NSF-EATC-6, Oak Ridge National Laboratory, Oak Ridge, Tennessee.
11. Donigian, A. S., and Crawford, N. H. 1976. Nonpoint pollution from the land surface. EPA 600/3-76/083, U.S. EPA, Washington, D.C.
12. Novotny, V., Chin, M., and Tran, H. V. 1979. LANDRUN—An overland flow mathematical model: Users manual, calibration, and use. International Joint Commission, Windsor, Ontario.
13. Anon. 1975. "Hydrocomp Simulation Operation Manual," 4th ed. Hydrocomp, Inc., Palo Alto, California.
14. Heaney, J. P., and Huber, W. C. 1973. Storm water management model: refinements, testing and decision making. Department of Environmental Engineering Science, University of Florida, Gainsville.
15. Meinholz, T. L., Hansen, C. A., and Novotny, V. 1974. An application of the storm water management model. Paper Presented at the National Symposium on Urban Hydrology and Sediment Management, University of Kentucky, Lexington. July 29–31.
16. Holtan, H. N., and Lopez, N. C. 1971. USDAHL-70 Model of watershed hydrology. USDA-ARS Tech. Bull. No. 1435, USDA, Washington, D.C.
17. Crawford, N. H., and Linsley, R. K. 1966. Digital simulation in hydrology: The Stanford Waterford Model IV. Tech. Rep. No. 39, Department of Civil Engineering Stanford University, Palo Alto, California.
18. Linsley, R. K. 1976. Rainfall-runoff models. In: A. K. Biswas (ed.). "System Approach to Water Management." McGraw-Hill, New York.
19. Donigian, A. S., and Crawford, N. H. 1977. Simulation of nutrient loadings

in surface runoff with the NPS model. EPA-600/3-77/065, U.S. EPA, Washington, D.C.

20. Crawford, N. H., and Donigian, A. S. 1974. Pesticide transport and runoff model for agricultural lands. EPA-660/2-74/013, U.S. EPA, Washington, D.C.
21. Beyerlein, D. C., and Donigian, A. S. 1979. Modeling soil and water conservation practices. In: R. C. Loehr et al. (eds.). "Best Management Practices for Agriculture and Silviculture." Ann Arbor Science, Ann Arbor, Michigan.
22. Donigian, A. S., Beyerlein, D. C., Davis, H. H. and Crawford, N. H. 1977. Agricultural runoff management (ARM) Model version II: refinements and testing. EPA 600/3-77/098, U.S. EPA, Washington, D.C.
23. Ritter, W. F., and Jensen, P. A. 1979. Water quality modeling in the Delaware coastal plain. In: R. C. Loehr et al. (eds.). "Best Management Practices for Agriculture and Silviculture." Ann Arbor Science, Ann Arbor, Michigan.
24. Sharp, B. M. H., and Berkowitz, S. J. 1979. Economic, institutional and water quality considerations in the analysis of sediment control alternatives: A case study. In: R. C. Loehr et al. (eds.). "Best Management Practices for Agriculture and Silviculture." Ann Arbor Science, Ann Arbor, Michigan.
25. Novotny, V., Tran, H., Simsiman, G. V., and Chesters, G. 1978. Mathematical modeling of land runoff contaminated by phosphorus. *J. WPCF*, **50**(1):101–112.
26. Munro, J. R., Jr., Luxmore, R. S., Begovich, C. L., Dixon, K. R., Watson, A. P., Patterson, M. R., and Jackson, D. R. 1976. Application of the Unified Transport Model to the movement of Pb, Cd, Zn, Cu, and S through the Crooked Creek Watershed. ORNL/NSF/EATC-28, Oak Ridge National Laboratory, Oak Ridge, Tennessee.
27. Eschenroeder, A. 1979. Multimedia modeling of the fate of environmental chemicals. Paper Presented at the American Society for Testing and Materials, 4th Symposium on Aquatic Toxicology, Chicago, Illinois.
28. Bonazountas, M., and Scow, K. M. 1979. Seasonal solute movement for unsaturated soil zone. Paper Presented at the American Geophysical Union Symposium, San Francisco, California.
29. Perez, A. I., Huber, W. C., Heaney, J. P., and Pyatt, E. E. 1974. A water quality model for a conjunctive surface-groundwater system. EPA-600/5-74/013, U.S. EPA, Washington, D.C.
30. Beasley, D. B., and Huggins, L. F. 1980. ANSWERS–Users Manual. Agr. Eng. Dept., Purdue Univ., West Lafayette, Ind.
31. Kniesel, W. G. 1979. A field-scale model for non-point source pollution evaluation. Proc. Hydrol. Transport Modeling Symp., ASAE Publ. 4-80, ASAE, St. Joseph, Mich.

10

Land use and nonpoint pollution

LAND-USE EFFECTS ON NONPOINT POLLUTION

The term land use often is associated with zoning, which denotes a method used by regional planners to divide an area into districts in which certain activities are permitted.[1] The land-use description or categorization identifies the principal activity taking place in such districts or areal units.

Although zoning or division of areas according to the primary activity originally had little relationship to water quality, it has been realized that water quality loadings from nonpoint sources can—to a certain degree—be correlated with land use and intensity of land-use activities. An example of such a correlation is shown in Fig. 10-1.

The problem of land use and its effect on water quality is associated in the minds of the public with urban and agricultural developments. Spreading urban and uncontrolled shoreline developments can result in deterioration of water quality. Imperviousness of urban areas increases their hydrological activity, and even small rains are capable of washing accumu-

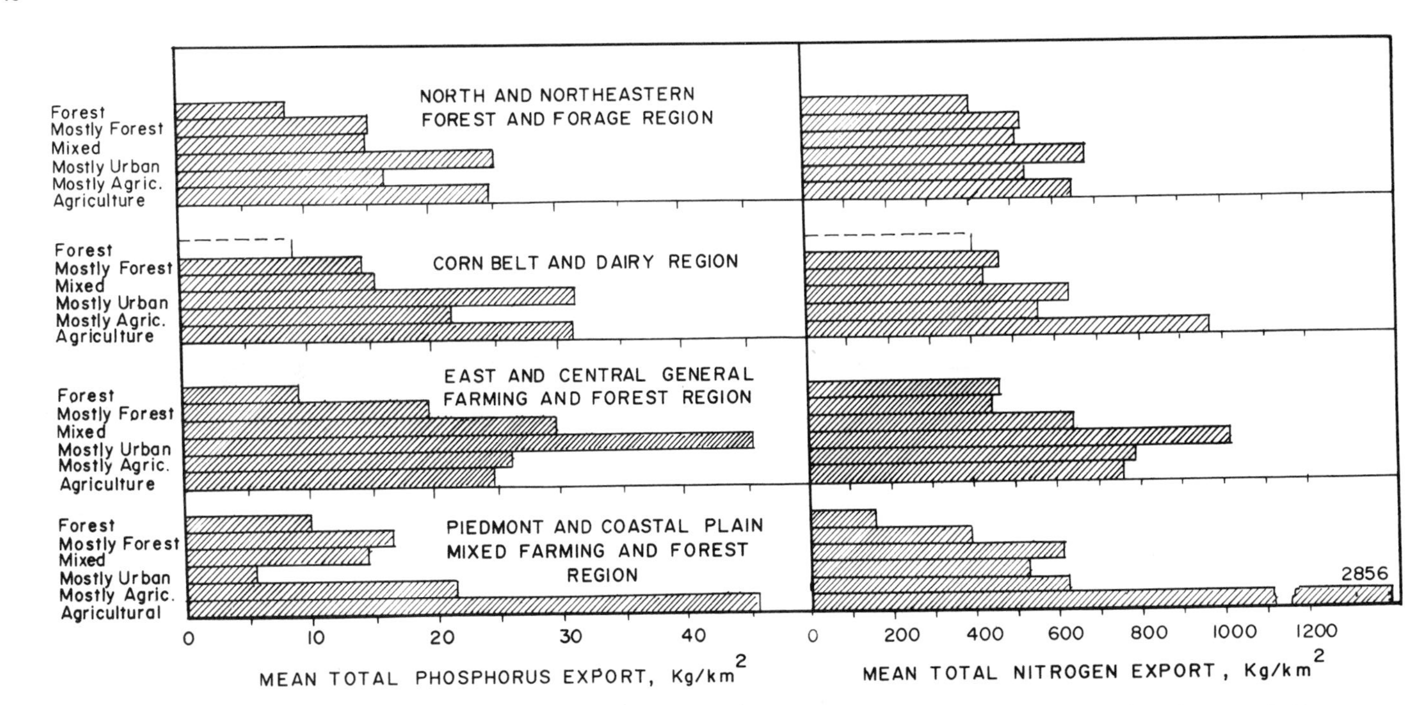

Fig. 10-1. Regional relationship between general land use and annual average stream export of total phosphorus and nitrogen. (From Omernik.[2])

lated pollutants into surface waters. Unsewered residential areas, where septic subsurface disposals of household sewage are predominant, are an additional source of elevated pollutional loadings.

Uncontrolled urban sprawl also may have other adverse effects which have an indirect impact on water quality, for example, loss of valuable farm and forest land, overtaxation and general deterioration of urban centers, overloading of sewage treatment facilities, and curtailment of services to the suburbs.

In rural areas, animal barnyards and feedlots as well as farming on hazardous soils without adequate erosion control also can produce high pollutional loads, especially if overfertilization is prevalent.

Due to the fact that some degree of correlation exists between pollutional loading and land use, zoning has been suggested as a means to control pollution from nonpoint sources.[1]

Pollution from land-use activities must be distinguished from background quality contribution, that is, loading if the area were retained as idle uncultivated land or natural forest.

The pollutional loading potential of land-use activities can be classified into three categories:[3]

1. Land not in need of control, including unmanaged forest land or rangeland.
2. Lands sometimes needing control measures such as agricultural lands in general and croplands in particular.
3. Lands usually requiring control measures; typical examples are some urban areas, mining operations, construction sites, and animal feedlots. These land-use activities generally are considered to be most hazardous.

Land-Use Categories

As noted before, the term land use describes the prevailing activity taking place in an essentially uniform demographic area. The uniformity often is related only to the zoning ordinances of the community and in most cases is not intended as a water pollution control measure. Lands classified into a single land-use category may be quite diversified with regard to topography, soil types, slope, and other important factors; therefore, wide variability in pollutional loadings within a single land-use category should be expected.

Land-use categories are divided into urban and rural types. Although originally only three to five land-use categories were defined for zoning purposes, present land-use inventories often recognize as many as 50 categories and subcategories. Due to the wide variations in pollutional loadings within each land-use category, it is not possible to estimate pollution impact for each detailed land-use category. For pollution studies, land uses are grouped together into more general categories, which bear a certain distinct relationship to generation of pollutants.

The Task Group for determining pollution loadings from nonpoint sources to the Great Lakes recommended the following categories of land uses:[4]

Rural (Nonurban) Lands

1. General agriculture–a broad category that encompasses all rural land uses.
2. Cropland–land used for annual crops, orchards, and vineyards. This category includes row crops such as corn, tobacco, and vegetables, and close-grown crops such as wheat, oats, and other grains.
3. Improved pasture–land used to grow forage and other close-grown crops and managed for pasturing livestock or making hay by techniques such as fertilization, reseeding, and/or overfertilization.
4. Spray irrigation and sewage sludge disposal land–agricultural land used for these purposes.
5. Woodland–land bearing forests, short trees, or brush.
6. Idle land–land not used for active agricultural purposes. It includes open water such as lakes, ponds, and rivers; wetlands such as swamps and marshes; barren land; and perennial grassland not used for pastures.

Nonpoint water quality contributions of the same magnitudes as that from unmanaged woodlands and idle lands can be considered as background water quality contributions.

Urban Lands

1. General urban land–a broad category that encompasses all urban land uses including developed and developing urban zones.
2. Residential land–land used for housing. It includes single and multiple dwelling units in built-up portions of cities, towns, and villages; and areas of urban sprawl such as strip residential developments and subdivisions. It can be divided into low-density, medium-density, and high-density zones. The distribution between low-density, medium-density, and high-density areas may vary reflecting local housing ordinances.
3. Commercial land–land used for commercial purposes including office buildings, shopping centers, warehouses, etc.
4. Industrial land–land used for industrial purposes including manufacturing and extractive mining.
5. Transportation corridors–includes lands used primarily as high-density traffic corridors for vehicular and rail transport.
6. Developing urban land–land that is under development for residential, commercial, transport, or industrial purposes.

Unit Area Loads

Hydrological and Pollutional Characteristics of Land Uses. Land use is a simple term describing the prevailing activity occurring in an area. As such, it bears little relationship to pollution generated from that area. Although the activity per se may produce some pollution directly, many other factors must be considered in assigning unit area loads. If one intends to trace the origin and causes of the pollution, the land-use activity description loses its meaning, and more meaningful factors such as dust and dirt accumulation rates on impervious areas, soil type and slope, vegetative cover, atmospheric deposition, etc., are more closely related to the pollutant loading. A partial list of factors that determine pollutional loadings from areal sources and their relation to land uses are listed as follows:

1. Factors strongly affecting pollution generation and correlated closely with land uses:
   a. Population density.
   b. Atmospheric fallout.
   c. Degree of impervious area usually correlated with population density.
   d. Vegetation cover.
   e. Street litter accumulation rates.
   f. Traffic density.
   g. Curb density and height.
   h. Street cleaning practices.
   i. Pollution conveyance systems.
2. Factors strongly affecting pollution generation but correlated poorly with land uses:
   a. Street surface conditions.
   b. Degree of impervious area directly connected to a channel.
   c. Delivery ratio.
   d. Surface storage.
   e. Organic and nutrient content of soils.
3. Factors strongly affecting pollution generation but unrelated to land uses:
   a. Meteorologic factors.
   b. Soil characteristics and composition.
   c. Permeability.
   d. Slope.
   e. Geographical factors.

From this list of causative factors it can be seen that many are, in part, correlated to land use. Therefore, attempts to relate pollution loadings from diffuse sources to land-use activities are justified. Factors not related to land use such as slope, soil texture and fertility, drainage density, and vegetative cover are less dominant for urban lands, which primarily have impervious surfaces, than for rural lands. Thus, it is often easier to relate pollutant generation to land use for urban settings.

Despite its questionable accuracy, the concept of relating pollution loading to land-use categories has found wide application in areawide pollution abatement efforts and planning. A simple reason explains this situation; the concept provides a simple mechanism and quick answers to pollutant problems of large areas where more complicated efforts would fail because of the enormous amounts of information required. The land-use/pollutant loading concept also is compatible with so-called "overview modeling," whereby unit loadings are combined with information on land use and soil distribution, and other characteristics to yield watershed loadings, or identify areas producing the highest amount of diffuse pollution.

Data can be adjusted by use of a delivery ratio to arrive at results similar to measured pollutant yields at the watershed outlet. Figure 10-2 shows source areas of high pollution yields in the Great Lakes region identified by this concept.

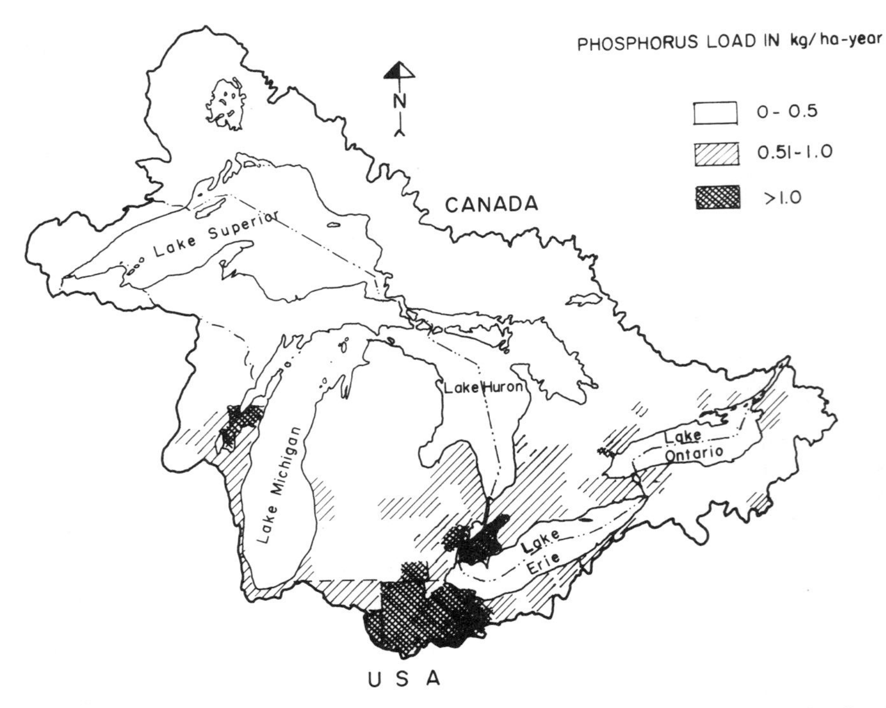

Fig. 10-2. Regions in the Great Lakes basin yielding the largest unit loads of phosphorus.[4]

Definition of Unit Loads. Unit loading is a simple value or function expressing pollution generation per unit area and unit time for each land use. The units usually are expressed in kilograms per hectare-year.

Prior to further discussion, a word of caution should be provided to potential users of this concept. Regardless of the method of estimation, unit loadings are highly inaccurate, often expressing averages from a wide range of measured values. The variations not only are caused by the randomness of the process but include some cyclic and systematic factors such as seasonal variations, geographical factors, meterological and land cover factors, soil characteristics, and reliability of measured data due to insufficient frequency of measurements or analytical and equipment errors. Figure 10-3 shows the variability of measured loadings from six pilot watersheds in the Great Lakes basin.

Those unit loads expressed as a single number for each land use are the least reliable. An example of such values is shown in Table 10-1, but as seen from Fig. 10-3, difference ranging up to several orders of magnitude should be anticipated when comparing such unit loads with measured yields of pollutants.

Improved data may result from coupling unit loads with such other factors as soil type and slope, climate, and traffic density. Examples of such approaches were discussed in previous chapters. A statistical summary of all curb loading data related to climatic factors, land use, and geographical location measured in urban areas of the United States through 1972 was reported.[6] The amount of reported data has more than doubled since the time the study was undertaken.

Other loading functions are based on simplified (annual or seasonal) sediment generation equations such as the Universal Soil Loss Equation (Chapter 5). These concepts are more reliable and suitable for areawide planning when detailed modeling or monitoring approaches are not feasible.

Use of the unit load concept presumes that an adequate inventory of land data is available from maps, aerial and terrestrial surveys, remote sensing, and local information. The unit load concept is applicable—in most cases—to long-term estimates such as average annual loading figures. Application of unit loads on an event basis should be avoided.

Example 10-1: Estimation of Unit Loads

Estimate the approximate unit load of sediment and total phosphorus from a medium-density residential area (40 persons/ha) of 50 ha located on a loam soil with an average slope of 2.4% in the midwestern United States. The area is 85% impervious.

The approximate curb loading of particulates in residential areas can be read from Table 8-1 as 52 g/m-day. On an annual basis the loading is:

$$52(\text{g/m-day}) \times 365\ (\text{days}) = 19\ \text{kg/m-yr}$$

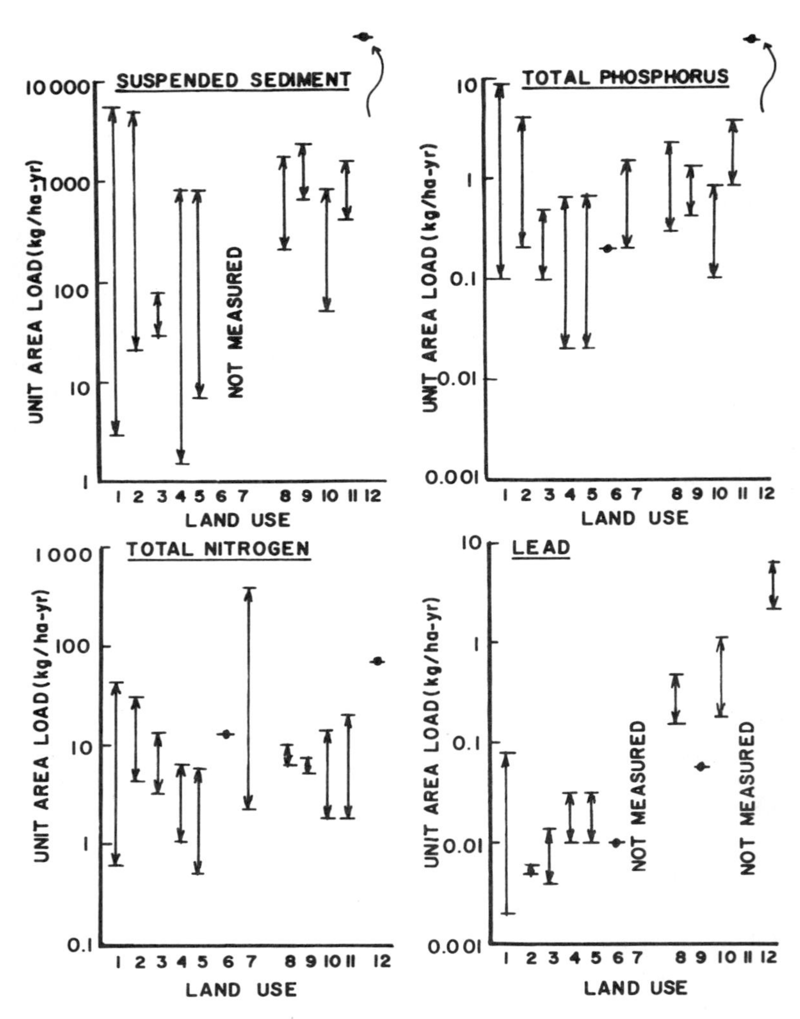

Fig. 10-3. Ranges of unit loads of sediment, total P, total N, and Pb from pilot watershed studies in the Great Lakes region.[4] Land uses: 1, general agriculture; 2, cropland; 3, improved pasture; 4, forested/wooded; 5, idle/perennial; 6, sewage sludge; 7, spray irrigation; 8, general urban; 9, residential; 10, commercial; 11, industrial; 12, developing urban.

The curb density is estimated from Equation 8-1 as:

$$CL = 311.67 - 266.07 \times 0.839^{(2.48\,PD)}$$

$$= 311.67 - 266.07 \times 0.839^{(2.48 \times 40)} = 311.6 \text{ m/ha}$$

TABLE 10-1. Urban Land Unit Area Loadings[a] of Total P and Suspended Solids.[b]

Land Use		Parameter	
		Total P (kg/km²-yr)	Suspended Solids (tonnes/km²-yr)
Areas of combined sewer systems	High industry	1100	72.6
	Medium industry	1000	74.3
	Low industry	900	75.9
Areas of separated sewer systems	High industry	300	66.0
	Medium industry	250	52.3
	Low industry	125	38.5
Unsewered areas		125	38.5
Towns of 1000–10,000 people		250	52.3

[a] For developing urban land an aggravated unit area load applies (for suspended solids, 225 tonnes/km²-yr was used for all land uses; however, for P a simple multiple of the values as they appear above was employed).
[b] Taken from reference 5.

Then sediment loading from impervious surfaces is

$$P_i = 19 \text{ (kg/m-yr)} \times 311.6 \text{ (m/ha)} = 5920 \text{ kg/ha-yr}$$

The above loading represents the upper limit of the loading range since no street cleaning or loss of particulates by wind was assumed.

The solids contribution from the pervious portions of the area can be approximated roughly by the Universal Soil Loss Equation (USLE). From Fig. 5-2, the annual rainfall factor, R, for the analyzed midwestern area is 2.24 × 125 = 280 (tonnes/ha-yr). Using the values estimated in Example 5-1, the annual sediment loss for grassed areas becomes:

$$A = (R)(K)(LS)(C)(P) = 280 \times 0.33 \times 0.47 \times 0.01 \times 1.0$$

$$= 0.434 \text{ tonnes/ha-year} = 434 \text{ kg/ha-yr}$$

Since only 15% of the area is pervious, the sediment contribution from the pervious portion is:

$$P_p = 0.15 \times 434 = 65 \text{ kg/ha-yr}$$

and the total sediment unit load becomes:

$$P = P_i + P_p = 5920 + 65 = 5985 \text{ kg/ha-yr}$$

Phosphate unit loads can be estimated using the potency factors from Table 8-2. Thus, for residential urban areas, the phosphorus content of street refuse is about 900 µg/g of solids. The phosphorus contribution from impervious areas is:

$$Pi_p = 900 \text{ (g/ton)} \times 5.92 \text{ (tons/ha-yr)}$$
$$= 5328 \text{ g/ha-yr} = 5.33 \text{ kg/ha-yr}$$

The contribution from pervious areas can be estimated in a similar way if the soil content of phosphorus and an enrichment ratio are known (see Example 6-3 for a sample computation).

The computed values for suspended solids in this example are higher than average urban area loadings given in Table 10-1. It must be remembered that Table 10-1 contains averages over the entire Great Lakes watershed adjusted to river mouth loadings, while Table 8-1 reflects average daily cumulation rates in the curb storage. In order to make Tables 8-1 and 10-1 compatible, the curb loadings would have to be readjusted to account for the loss by sweeping, wind, and delivery between the source and the watershed outlet.

Nonpoint Calculator. The "nonpoint calculator" was developed by the Midwest Research Institute as a tool to estimate nonpoint pollution loads from different land uses.[7,8,9] It is a computerized package containing "loading functions" developed for use in nonpoint pollution investigations.[7]

The calculator estimates nonpoint pollution loadings from rural areas by the USLE. For urban areas, estimates of sediment loads are based on an assumed loading rate per unit length of curb and an estimate of curb density for the urban area under consideration. No attempt is made to adjust estimated potential erosion to actual river mouth loadings. Loadings of pollutants other than sediment are estimated by multiplication of the sediment loadings by appropriate potency factors.

Output from the calculator delineates areas in a region that have significant nonpoint pollution problems, and estimates the severity of the problems in a semiquantitative relative sense.

Example 10-1 is similar to the methodology employed by the "nonpoint calculator." The nonpoint calculator is an efficient, quick tool for large areas ranging from county size[9] to the entire Great Lakes basin.[8]

COMPARATIVE ASSESSMENT OF POLLUTION IMPACT FROM LAND USES

As stated previously, pollution loads vary depending on land-use activities in the area. Although the literature contains a large amount of data on pollution generated by various lands, most data represent an integrated effect of several land uses and only a few can be considered loadings from homogeneous land uses.

Breakdown of a watershed into smaller homogenous units is necessary for identifying "hazardous" lands, land uses, and land practices, that is, areas requiring control treatments. For example, Black Creek (Indiana) studies[10] indicate that treating 32 ha of the most highly erosive areas rather than the entire 770 ha of an experimental watershed would reduce total sediment load by 40%. The cost of treatment of the 32 ha would be almost $600 compared to $135,000 for the entire watershed.

In the Menomonee River (Wisconsin) watershed, 2.6% of the total area that was under construction at the time of the investigation contributed almost 40% of the sediment loading measured at the river mouth.[11] Similarly, most of the lead originated from transportation corridors and other areas with heavy traffic, while the lead contribution from rural lands was negligible.

On a larger scale, watersheds rich in clay soils such as those found in the Lake Erie basin exhibit higher unit loads for sediments and nutrients.

It must be realized that both land and land-use practices may contribute significant amounts of pollution. Certain streams in the Lake Superior basin, although undeveloped, contribute high suspended solids loads due to the high clay content of soils in the watershed (Red Clay area). On the other hand, sandy soils prevail in the eastern portion of the Lake Michigan basin and, consequently, streams have a high water quality. Clays, because of their high surface area, carry large quantities of pollutants adsorbed on them.

Although unit loadings are expressed mostly on a yearly basis, seasonal variations also are important. As shown in Chapter 5, summer rains in the midwestern United States have the highest erosion potential; however, during summer months the land often is well protected by vegetation and the soils have relatively low moisture contents. Subsequently, most pervious lands have low hydrological activity. The fall is usually drier, resulting in even lower loadings. Generally, midwestern seasons can be categorized into the following periods based on nonpoint pollution potential:

1. Critical runoff period–from snow melt and spring rains.
2. Critical erosion period–from time of spring planting of crops to about 2 months after planting.
3. Noncritical period.

This categorization is applicable to most of the conterminous United States.

In urban areas–where most of the nonpoint pollution originates from impervious surfaces–the seasonal effect is not so profound.

In a situation where a paucity of measured data exists, a computer model may help to arrive at simulated loading values for typical land uses. The hydrological pollution generation model LANDRUN (Chapter 9) was used to generate loadings of pollutants from homogenous 1-km^2 land-use areas during hydrologically different years and seasons. The loadings were analyzed statistically in order to obtain average annual or seasonal values.

The model, calibrated and verified by measurements of water and pollution loads from several diversified, relatively small watersheds, was complemented by measurements from small homogenous land-use areas located in southeastern Wisconsin. Homogeneity was assumed both for land-use activities and surface as well as soil characteristics.

In order to obtain the pollution loading matrix for each land use, the simulation units were located on four different soils, preferably characterizing four basically different hydrological soil groups. Land cover and surface characteristics were made similar to the average features for each particular land use. Unit loadings generated by the model were compared with measured loadings from homogenous pilot land-use areas. The modeling results were synthesized with measured data to provide ranges and characteristic loadings of selected pollutants for typical urban and nonurban land uses.

The four soils selected for simulation were local soils from southeastern Wisconsin, namely:

Boyer sandy loam (hydrologic category A)
Hochheim loam (hydrologic category B)
Ozaukee silt loam (hydrologic category C)
Ashkum silty clay loam (hydrologic category D)

Table 10-2 contains basic information on these soils.

The simulated overland slope for the 1-km^2 units was 4%, which reflects U.S. SCS soil slope category B. This slope was used as a reference for most land-use units with the exception of croplands and woodlands where a reference slope of 9% was used. Since soil loss depends on the size of the area (length of the overland flow) slope correction factors must be applied to other areas of different size and slope. Figures 10-4 and 10-5 present the size correction and slope factor, respectively. These correction factors must be applied to the loading values from pervious surfaces presented in subsequent sections.

Pollution accumulation on impervious areas was computed assuming:

1. a linear accumulation rate with no street cleaning,
2. a decreasing accumulation rate with no street cleaning, and
3. a well-maintained area with weekly street cleaning.

Loadings reflected average midwestern meteorological conditions, resulting in an average annual rainfall energy factor $R = 100$ (the units of the rainfall factor are identical to those in Fig. 5-8). For other meteorological zones, loadings must be adjusted according to the average rainfall energy factor typical of the region.

TABLE 10-2. Soil Data for Simulations.

Soil Type	Hydrologic Group	*A* Horizon Depth (cm)	Clay (%)	Silt (%)	Sand (%)	Median Diameter (mm)	Organic Matter (%)	*A* Permeability (cm/hr)	0.3-bar Moisture (%)	15-bar Moisture (%)	Porosity (%)	USLE *K* Factor	PO_4 Adsorption Unit Factor[a] ($\mu g/g$)	P Content (%)
Bm[b]	A	41	5	15	80	0.415	0.5	40	8	0	30	0.09	243	0.10
Hm[b]	B	20	16	39	45	0.138	2.0	10	20	7	34	0.24	346	0.15
Ou[b]	C	28	20	55	15	0.051	3.0	3.0	30	17	43	0.31	403	0.18
As[b]	D	28	39	56	5	0.021	8.0	0.5	36	24	46	0.15	697	0.31

[a] Approximately equal to the soil adsorption maximum Q^0.
[b] Bm is Boyer sandy loam, Hm is Hochheim loam, Ou is Ozaukee silt loam, and As is Ashkum silty clay loam.

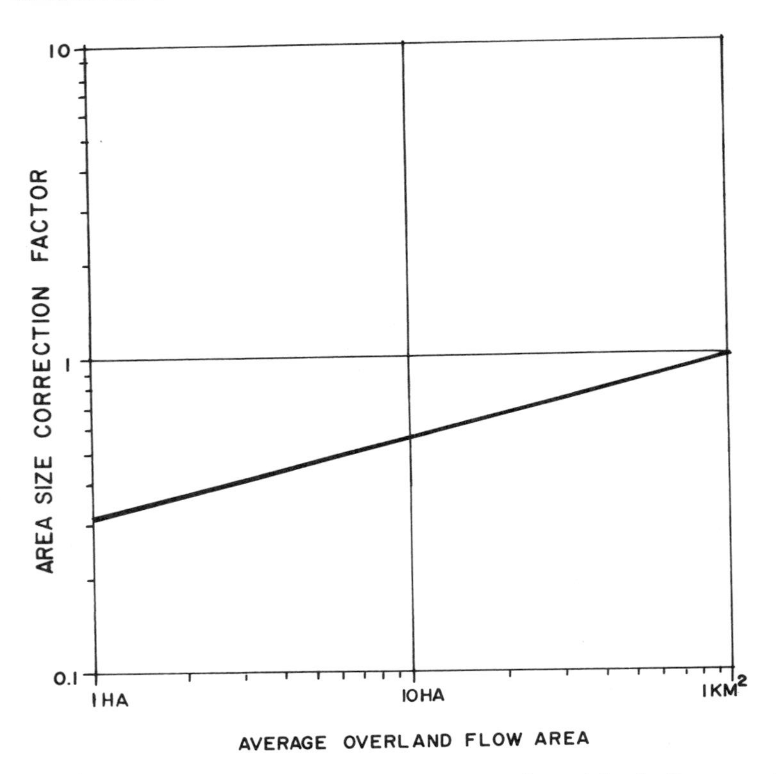

Fig. 10-4. Size correction factor for simulated sediment loads from pervious areas.

Detailed discussions of the comparative simulation methods are presented by Novotny et al.[12,13,14]

Urban Land-Use Unit Loads and Management

Residential Land Use. This term applies to a wide variety of urban sections, ranging from subdivisions with 1 or 2 houses/ha to highly congested urban centers. For planning purposes, residential zones are subdivided according to population density into low-density (1 to 15 people/ha), medium-density (16 to 50 people/ha), and high-density areas (>50 people/ha). Residential lands also are divided according to type of housing into single and multiple family sections. Figure 10-6 shows an example of a medium-density residential zone.

The division of residential areas according to population density may vary according to geographical location. Midwestern suburban and urban zones generally tend to have larger housing lots in single family subdivisions than the more

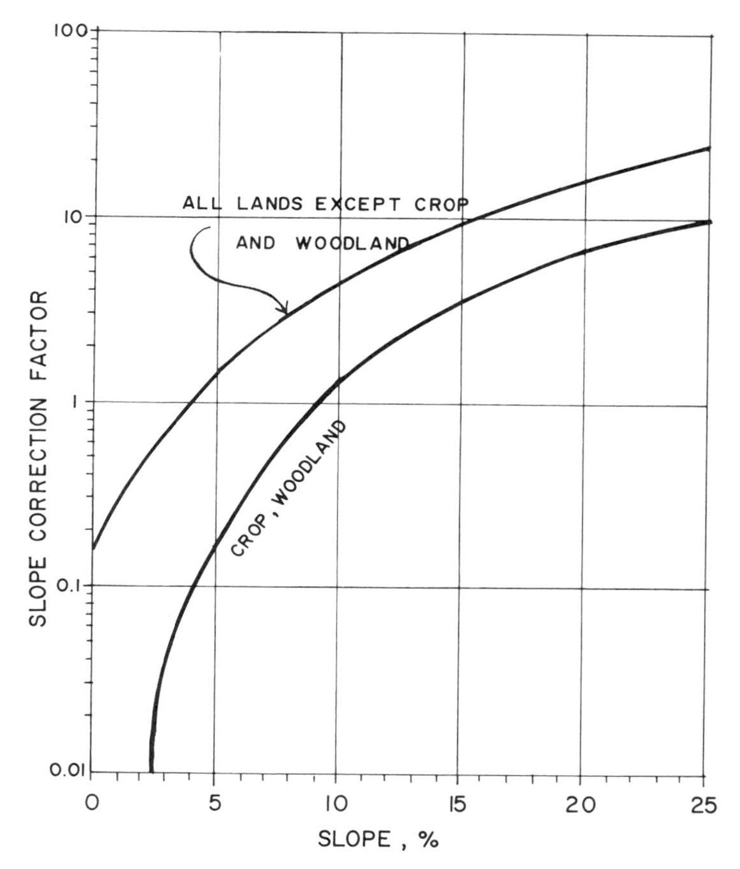

Fig. 10-5. Slope correction factor for simulated sediment loads from pervious areas.

populous areas of the eastern and coastal United States and desert states. Housing densities also are higher in Canada and Europe. Typical housing densities in Ontario, Canada, are 50 people/ha in low-density residential zones, 87 people/ha in medium-density, and 125 people/ha in high-density residential areas.[15]

Factors determining pollution loadings from residential areas include:

1. Degree of imperviousness.
2. Street refuse accumulation and general cleanliness of impervious surfaces.
3. Street sweeping practices.
4. Curb heights.
5. Type of storm-water drainage system.

Fig. 10-6. Aerial view of typical medium-density and commercial land uses. (Photo University of Wisconsin.)

6. Soil type and slope of pervious surfaces.
7. Traffic impact.
8. Atmospheric pollution.

If the residential areas are fully developed (construction sites are included under a separate category), the primary source of pollution varies according to degree of imperviousness of the area. In the low-density residential zones, most pollution originates from erosion of pervious surfaces, especially if located on poor soils and/or high slopes.

In medium-density and high-density residential areas, pollution loads are due mostly to wash-off of accumulated street dust and refuse near the curb. Residential areas with natural storm drainage generally will have lower delivery ratios of pollutants as compared to areas with storm or combined sewers, where for all practical purposes the delivery ratio is close to one.

Urban storm-water runoff is highly variable and its quality often approaches that of treated sewage or may even be worse. Similarly, if loads are expressed in kilograms per hectare per time, a wide range of loading values can be found from the available literature. Loehr[3] reported that COD values from urban areas were in the range of 220 to 310 kg/ha-yr, typical BOD_5 values were 30 to 50 kg/ha-yr, total N loads ranged from 7 to 9 kg/ha-yr, and total P loads were 1.1 to 5.9 kg/ha-yr. These loadings reflect an integration of all urban land uses.

As seen from the bar graphs in Fig. 10-2, the loading figures for residential areas near the Great Lakes were as follows:

	kg/ha-yr
Suspended solids	200-2300
Total phosphorus	0.4-1.3
Total nitrogen	5-7
Lead	0.06 (one number only)

Table 10-3 shows potential sources of nonpoint pollution for urban (residential and commercial) runoff. As mentioned throughout this treatise, evidence indicates that the nonpoint pollution from urban residential areas may be of similar extent as pollution from treated sewage from the same area.

Table 1-4 and 1-5 reported typical concentrations of pollutants found in urban storm water and combined sewer overflow.

It is interesting to note that in addition to other pollutants, high bacterial numbers are often associated with effluents from storm (separate) sewers. Studies on the sources of bacteria found in storm-water runoff from residential and light commercial areas indicated the bacteria were predominantly of nonhuman origin.[16,17] The average numbers of total coliforms found in urban

TABLE 10-3. Sources of Nonpoint Pollution from Urban Residential and Commercial Areas.

Category	Parameters	Potential Sources
Bacterial	Total coliforms, fecal coliforms, fecal streptococci, other pathogens	Animals, birds, soil bacteria, (humans)
Nutrients	Nitrogen, phosphorus	Lawn fertilizers, decomposing organic matter (leaves and grass clippings), urban street refuse, atmospheric deposition
Biodegradable chemicals	BOD, COD, TOC	Leaves, grass clippings, animals, street litter, oil and grease
Organic chemicals	Pesticides, PCBs	Pest and weed control, packaging, leaking transformers, hydraulic and lubricating fluids
Inorganic chemicals	Suspended solids, dissolved solids, toxic metals, chloride	Erosion, dust and dirt on streets, atmospheric deposition, industrial pollution, traffic, deicing salts

storm and combined sewers were $10^6/100$ ml and $10^7/100$ ml, respectively, and those for fecal coliforms were $10^5/100$ ml and $10^6/100$ ml, respectively.[17] In addition to the magnitude of the observed numbers of coliform bacteria, it is important to note the fecal coliform : fecal streptococci ratio. Ratios of greater than 2 : 1 are generally indicative of human contamination by sewage, while ratios less than 1 : 1 indicate pollution primarily derived from warm-blooded animals other than humans. The observed ratios in combined sewers is approximately 4.7, while in separate sewers it is 0.6.[17]

On a distributed annual basis, the typical nonpoint loadings from urban residential and light commercial areas may seem less detrimental to receiving water bodies than corresponding point source pollution by sewage. But it must be realized that these loadings occur during storm events when over a short period of time the "shock" loadings from nonpoint sources may have an impact many times greater than that of untreated sewage from the same area.[17]

Simulated results by LANDRUN for sediment and phosphates (Figs. 10-7 and 10-8) reveal that a trade-off exists between loads from pervious and impervious lands. As the degree of imperviousness of residential areas increases with population density, the source of the pollutant shifts from pervious to impervious surfaces. The loading curves (Figs. 10-7 and 10-8) assume average dust and dirt fallout of 0.5 tonnes/km^2-day, which accounts for about 50% of the total street refuse accumulation rate near the curb. Sweeping efficiency was assumed to be 50% for solids and 22% for phosphates.

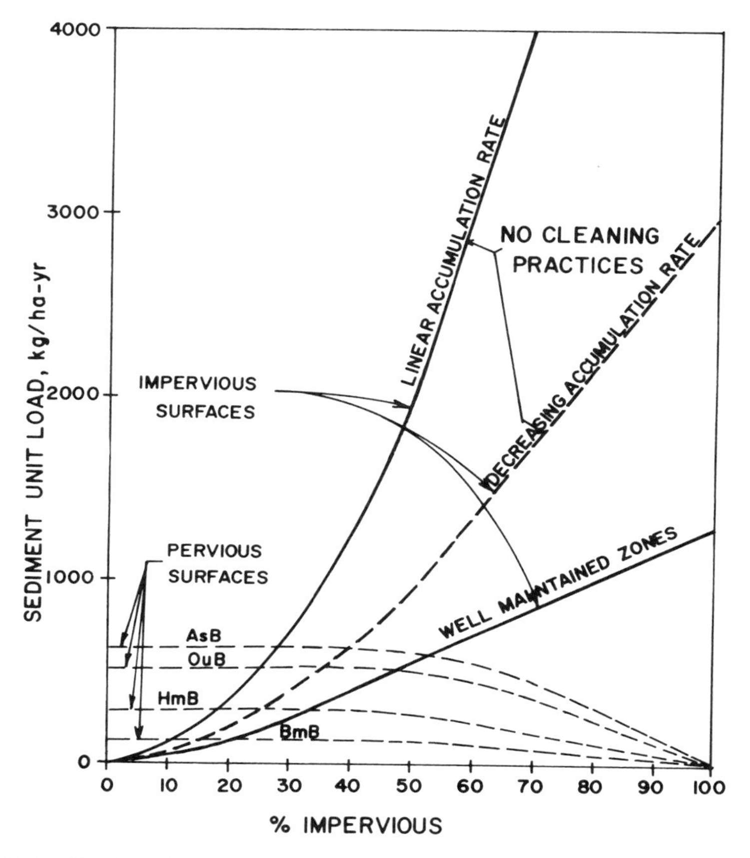

Fig. 10-7. Simulated sediment unit loads from residential and light commercial land uses related to the total imperviousness of the area.

The simulated results show the effect of soils and sweeping practices (streets swept once in 7 days during a dry period) on pollutant loadings from residential areas. The loadings (Figs. 10-7 and 10-8) reflect average Midwestern meteorological conditions, soil slope of 4%, and delivery ratios close to unity, that is, areas with good storm sewer systems. Significantly lower delivery ratios (<10%) can be expected for large, low-density residential zones with natural drainage where grassed storm ditches and surface retention act as effective sediment and pollution traps.

Best Management Practices for Residential Areas. Selection of pollution control practices for residential areas depends on population density and degree of imperviousness of the area. In areas where a large portion of the pollutants origi-

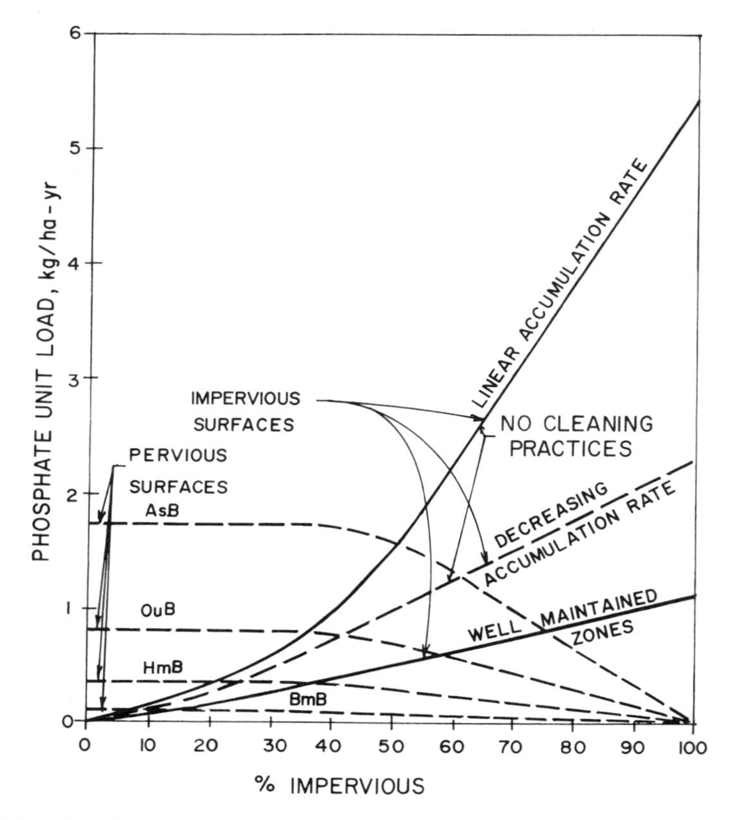

Fig. 10-8. Simulated total phosphate loading from residential and light commercial land uses related to the total imperviousness of the area.

nates from pervious surfaces, soil conservation practices may be the best method to reduce pollutant loads. Besides being aesthetically pleasing, good grass cover plus brush and shrubbery also can reduce pollution loads by more than two orders of magnitude when compared to unprotected bare soils. In addition, natural drainage systems, open grassed ditches, and flood retention basins can reduce delivery ratios of pollutants to a few percent, while areas served by storm or combined sewers will show almost 100% delivery.

Curb loading values can be reduced by lowering curb height if pervious pollution traps are available in the vicinity.

The best management practices for urban runoff control are:[18]

Source control:
 increased infiltration
 retention of runoff

reduction of erosion
reduction of contaminant deposition by controlling sources of atmospheric pollution
street sweeping

Reduction of delivery to surface waters:
change in storm drainage systems
infiltration and sedimentation basins
storage basins to equalize flows
physical, chemical, and biological treatment

In low-density urbanizing zones, the quality of storm-water runoff is handled most efficiently by systems incorporated during development such as zoning, better architectural concepts utilizing more pervious areas in urban zones, increased perviousness and optimal design of storm-water conveyance systems. In high-density developed areas, runoff is handled by good street cleaning practices and one of a series of treatment methods subsequent to collection.

Source control of urban runoff-related pollution, which reduces on-site pollutant generation or prevents pollutants from leaving the small drainage areas (hazardous lands) where a disturbance occurs, is less expensive and more effective than remedial measures designed to treat pollutants after they have left the site and moved downstream. Collection system control and attempts to reduce delivery are more expensive than on-site source control but less costly than treatment at the outfall. Treatment of urban runoff may be feasible only for highly developed congested areas where source control and collection control are not possible and/or for reducing pollution from combined sewer overflow.

Other remedial measures that have been proposed for controlling pollution by urban runoff include increasing the fraction of pervious areas in urban settings and reducing the proportions of impervious areas directly connected to surface drainage systems, installing pervious pavements, and disconnecting roof drains from storm sewers. These measures are ineffective if the area is located on poor soils since the conveyance of surface runoff on pervious areas could create more erosion and pollution wash-off from these soils.

In general, low-density, well-maintained urban areas with natural surface runoff drainage systems generate pollutant loadings that are of the same order of magnitude as background loadings from nonagricultural rural lands. On the other hand, high-density urban areas with poor maintenance must be considered highly hazardous and require control methodologies. However, areas served by on-site septic systems may add significant amounts of nitrate N to interflow and base flow components. Since subsurface septic systems are located below the rooting depths of soils and only on soils of relatively high permeability and low organic matter content, very little N is immobilized and retained by the soils. The N contribution together with other mobile pollutants from septic systems should be added to interflow and base flow quality and considered a form of nonpoint pollution. Streams draining low-density residential areas served by septic systems generally have higher N contents.

Example 10-2: Pollutant Loadings from Residential Areas

Using Figs. 10-7 and 10-8 estimate unit load of sediment for a watershed containing mostly medium-density residential housing zones. The watershed size is 1.5 km^2, the prevailing soil is a clay loam (hydrologic category D), and the average slope is 2%. The watershed is 35% impervious with natural storm drainage (open grassed roadway ditches and gutters).

From Fig. 10-7 the sediment loading for 35% impervious residential lands is:

From impervious surfaces	500 kg/ha-yr–assuming no cleaning and decreasing accumulation rate
Pervious surfaces	700 kg/ha-yr–for hydrologic category D

Size correction factor for an area of 1.5 km^2 from Fig. 10-4 (applies to pervious areas only) is $\xi_A = 1.05$. Slope correction factor from Fig. 10-5 (applies to pervious areas only) is $\xi_S = 0.5$. Sediment delivery ratio assuming natural drainage and 1.5 km^2 size from Fig. 5-14 (applies to the entire watershed)–DR = 0.3.

The estimated average annual unit sediment load is:

$$A = (700 \times 1.05 \times 0.5 + 500) \times 0.3 = 185 \text{ kg/ha-yr}$$

A similar subdivision with storm sewer drainage systems would have a sediment unit area load of 620 kg/ha-yr (DR $\simeq$ 1).

For geographical locations other than the midwestern United States (the corresponding rainfall energy factor R $\simeq$ 100), the loading from pervious surfaces would have to be corrected by an appropriate rainfall energy factor for the area of interest divided by 100.

Commercial Land Use. This category covers a broad scale of land-use activities that include shopping centers (top portion of Fig. 10-6), warehouse storage areas, parking areas, congested downtown commercial zones, and governmental buildings. Due to the broad definition and number of activities taking place in the areas zoned as commercial, the pollution load range is quite wide.

The degree of imperviousness of commercial areas is usually medium to high. Other than buildings themselves, most of the available land is occupied by parking lots and traffic access roadways. The remaining pervious lands are often covered by lawns and shrubbery.

Although higher traffic densities and higher intensities of commercial activities result in elevated pollutant accumulation, frequent sweeping and cleaning of commercial areas is common, and lower values of sediment loadings from commercial areas are typical. Figures 10-7 and 10-8 can be used to obtain a rough estimate once the degree of imperviousness of the area and the cleaning practices are known.

The PLUARG studies measured pollution unit loading from commercial lands in the following ranges:

	kg/ha-yr
Suspended sediment	50-830
Total phosphorus	0.1-0.4
Total nitrogen	1.9-11
Lead	0.17-1.1

The foregoing unit loads also include loadings from transportation zones which PLUARG included under commercial land.

Curb loading values and potency factors for commercial areas were given in Chapter 8.

Best management practices for commercial areas are similar to those for residential medium-density to high-density zones.

Industrial Land Use. Industrial land use is divided generally into two categories: (1) manufacturing, and (2) extractive industrial activities (mining).

Manufacturing. This category ranges from light manufacturing with relatively low pollution impact to heavy industries such as steel mills, foundries, ore smelting, and cement manufacturing. These activities are potential major pollution hazards to surface waters from point and nonpoint sources.

It is difficult to assess pollution from industrial manufacturing operations. The primary sources of nonpoint pollution in many industrial areas are air pollution and subsequent atmospheric deposition caused by industrial operations. Although a major reduction in air pollution has been achieved by implementing advanced stack effluent and air pollution control techniques (including scrubbers, electrostatic precipitators, and others), elevated aerosol and particulate levels are still common near many industrial zones. In addition, disposal sites for waste materials such as fly ash lagoons are of environmental concern and represent a potential source of groundwater contamination.

Table 8-1 shows curb loadings of particulates in industrial areas as about 120 kg/km-day; a value higher than that for residential areas. Also the pollutant strength factors reported in Table 8-2 often are higher than corresponding factors for residential zones.

Best management practices often must be focused on air pollution and its removal and on general cleanup measures. In many cases, contamination of surface deposits in industrial areas is of such magnitude that the entire surface runoff must be collected and treated in a plant.

Extractive Land Use–Mining. Mine drainage water represents a special danger to receiving waters, and coal mining is considered a hazardous and highly disturbing land use (Fig. 10-9).

Two different kinds of mining methods are practiced, namely, underground or deep mining, and surface mining (strip or open-pit mining). The principal pollution problems resulting from mining activities are erosion and acid mine drainage.

Due to strip mining activities, bare land surfaces are exposed resulting in high erosion yields. The erosion yields can range up to 100,000 tonnes/km^2-yr. In addition to high sediment yields, the particles themselves may carry high concentrations of associated or adsorbed metals, ore or coal residues, and chemicals associated with the mined materials.

The mining activity least damaging to surface waters is mining for sand and gravel. The mining activities that are the most damaging include oil fields, pumpage of brine waters from deep mines, piling of mine spoils (gob piles), discharge of acid drainage water, strip mining, drift of particulates from open min-

Fig. 10-9. Land disturbance by strip mining. (Photo University of Wisconsin.)

ing areas, and the traffic of heavy machinery and trucks commonly associated with mining activities.

Acidity of coal mine drainage waters[19-23] is of special concern. As water flows through a mine or mine spoils it comes into contact with sulfur-bearing minerals, primarily pyrites and marcasites (chemical composition FeS_2). When pyrite is exposed to air and water, it is oxidized to ferrous sulfate ($FeSO_4$) and sulfuric acid (H_2SO_4). The reaction proceeds as follows.[20,23]

$$2FeS_2 + 7O_2 + 2H_2O \rightleftarrows 2FeSO_4 + 2H_2SO_4$$

Flowing water leaches away the $FeSO_4$ and H_2SO_4 (sulfuric acid); the leachate is called acid mine water, a phenomenon similar to acid rain.

The $FeSO_4$ can further oxidize to ferric sulfate ($Fe_2(SO_4)_3$) which hydrolyzes to form insoluble ferric hydroxide ($Fe(OH)_3$) and H_2SO_4.

$$4FeSO_4 + 2H_2SO_4 + O_2 \leftrightarrows 2Fe_2(SO_4)_3 + 2H_2O$$

$$2Fe(SO_4)_3 + 6H_2O \rightleftarrows \underset{\text{precipitate}}{\underbrace{2Fe(OH)_3}\downarrow} + 3H_2SO_4$$

In addition to the acid and iron salts, various other constituents may be found in mine drainage such as sulfates of aluminum, calcium, magnesium, potassium, and sodium.[20]

Acid mine drainage has lethal and sublethal effects on the biota of receiving streams including bacteria.[21] The streams affected by acid mine drainage exhibit characteristic color.[22]

Control of pollution by mining must include erosion control practices limiting loss of particulates from stripped lands or gob (mine refuse) piles, and control of acid mine drainage water.

Three factors are responsible for the formation of acid drainage, namely water, air, and contact time of water with mined minerals. Thus, control measures that have been recommended are aimed at limiting the effects of these factors. Control measures include:[20,22]

- Reducing entry of waters into mines through diversion of surface runoff and/or by sealing.
- Minimizing contact time of water in the mine with acid-forming minerals.
- Equalizing the flow of water from the mines over a 24-hr period as opposed to the common practice of intermittent pumping of "slugs" of acid water.
- Employing adequate mine-closure procedures immediately following termination of mining activities.
- Showing greater care in disposal of gob and other mine refuse materials.
- Flooding of abandoned surface mines.
- Treating acid mine drainage by neutralizing chemicals, e.g., by liming.

- Diluting acid mine discharge in streams by low flow augmentation.
- Land reclaiming and protecting strip mines and mine refuse piles.

Transportation. Traffic corridors and their impervious surfaces (roads) or partially pervious surfaces (railroads and secondary roads) are another source of pollutants in urban and interurban land use.

Lead has been identified as the primary pollutant attributed to road traffic; however, other pollutants—hydrocarbons, phosphates, asbestos, and particulate matter—are correlated closely with traffic density, road conditions, and abrasion.

The effect of traffic on curb loadings of pollutants was discussed in Chapter 8. Management practices should be focused on reducing pollutant emission from vehicles and on road maintenance.

Park and Recreation. This land-use category involves land in urban areas that is most likely to be least hazardous, requiring little or no control measures. These lands include municipal parks and picnic areas, golf courses, waterfronts and beaches, and urban forests. The surfaces usually are well protected against erosion. Impervious areas commonly represent only a small fraction of the land, and often a large portion of the impervious surfaces is not directly connected to a drainage system.

A potential source of nonpoint pollution is overfertilization or excessive use of herbicides on parks and golf courses.

Hydrologic activity of these areas is low and, in general, pollution loads are of similar magnitude to those from idle lands or forests.

Construction Sites. In zoning terminology, these lands are called "developing urban zones." Construction sites, in general, produce the highest amount of pollutants, ranging up to 50,000 tonnes/km^2-yr of sediment particles with corresponding high amounts of other pollutants.[24]

The principal cause of high pollution loads arises from stripping topsoils and exposing bare soils with no protection (Figs. 10-10 and 10-11). Furthermore, compaction of soils by construction machinery reduces permeability and surface storage of soils and increases hydrologic activity. Hence, construction sites are areas of highest pollution potential and require the application of control methodologies.

Factors affecting sediment and other pollutant loss from construction sites are:[25] slope, proximity of the site to a stream channel, existence of buffer zones of natural vegetation, erodibility of the soils, erosion control practices on the site, meterological factors, use of heavy machinery, and length of time the soils are exposed and unprotected.

Reported sediment loads from construction sites range from 12 to 500 tonnes/ha-yr (1200 to 50,000 tonnes/km^2-yr).[4,24-26] Yorke and Herb[25] statistically

Fig. 10-10. Land disturbance and nonpoint pollution from construction sites. (Photo authors.)

analyzed sediment yields from construction sites in Washington, D.C., and reported the following equation:

$$\log S_y = 1.21 + 0.143S - 0.01C \qquad (10\text{-}1)$$

with a standard error of estimate of 0.22 log units, where:

S_y = sediment yield (tonnes/ha-yr)
S = average slope (%)
C = total construction area with adequate sediment control (%)

The equation has only local applicability since the effect of meteorological factors and soil type (soils were fairly uniform for all investigated construction sites) was not evaluated.

Simulated loadings for construction sites for soil slope category B for a 1-km^2 area and midwestern United States meteorological conditions (rainfall energy factor $R \simeq 100$) yielded the following data:

Fig. 10-11. View of heavily sediment-laden discharge from a construction site. (Photo University of Wisconsin.)

Soil Type	Sediment (tonnes/ha-yr)	Phosphorus (kg/ha-yr)
Boyer sandy loam	11.0	10.9
Hochheim loam	27.5	41.3
Ozaukee silt loam	43.7	78.7
Ashkum silty clay loam	55.6	172.3

Figures 10-4 and 10-5 are used to correct for watershed size and slope, that is, other than 1 km^2 and 4%. To correct for other meteorological conditions, divide the rainfall factor from Fig. 5-8 by 100 and apply to the unit loading.

Example 10-3: Estimation of Sediment Loss

Estimate sediment loss from an uncontrolled construction zone located in Washington, D.C. The size of the site is 30 ha, average slope is 6.4%, and the soil is a silt loam. About 10% of the area receives adequate sediment control treatment. Use Equation 10-1 and simulated unit loads to arrive at an average annual load.

Sediment load from Equation 10-1:

$$\log S_y = 1.21 + 0.143S - 0.01C$$

$$= 1.21 + 0.143 \times 6.4 - 0.01 \times 10 = 2.0252$$

The sediment unit load is:

$$S_y = 106 \text{ tonnes/ha-yr or } 3180 \text{ tonnes/yr for } 30 \text{ ha}$$

The unit load based on the simulation results for silt loam soils

$$S'_y = 43.7 \text{ tonnes/ha-yr}$$

From Fig. 10-4 the correction factor for a 30-ha area is $\xi_A = 0.7$. The slope correction factor from Fig. 10-5 is $\xi_S = 1.9$. The rainfall energy factor for Washington, D.C., from Fig. 5-8 is $R \simeq 200$.

Then the sediment load becomes (assuming that 90% of the area contributes the sediment)

$$S_y = 0.9 \frac{200}{100} \xi_A \xi_S S'_y = 0.9 \frac{200}{100} \times 0.7 \times 1.9 \times 43.7$$

$$= 105 \text{ tonnes/ha-yr or } 3150 \text{ tonnes/hr}$$

To estimate loads of pollutants other than sediment, appropriate pollutant strength factors typical for A (topsoil) and B (subsoil) horizons of soils in the area must be employed along with corresponding enrichment ratios for the pollutants and the soil.

Control Practices. Control practices for construction sites can be divided into two categories:[26] those that require little cost for implementation but require timing and coordination with construction activities; and those that require some financial resources but can be implemented at some time during construction.

No-cost practices include:

- Storing excavated basement and foundation soils at a reasonable distance from roadway curbs, thereby increasing the distance that eroded soil must travel to reach the drainage system (usually a storm sewer). The effect of increasing the distance of overland flow on delivery ratio can be estimated from Equation 5-13.
- Using only one route (preferably the future driveway) to approach the

building site with trucks and heavy construction equipment. The approach road should be covered by gravel.

- Roughgrading the construction site as soon as possible after excavation, eliminating soil mounds that are easily eroded.
- Removing excess soil from sites as soon as possible after backfilling, reducing sediment loss from surplus fill.

Other practices include:

- Covering an area behind the curb or drainage ditch (approximately 10 m wide) with such protective materials as mulch, filter fabric, or netting. This measure reduces raindrop impact and serves as a sediment trap.
- Covering sides and backyards (if needed) with mulch or protective covering to limit erosion.
- Stabilizing the soil surface by seeding and mulching or sodding as soon as possible.
- Cleaning streets frequently in the construction site zone and removing the sediment from the curbs.
- Installing roof downspout extenders that aid in dispersing rainwater in a diffuse manner, thereby reducing runoff intensity from rooftops.

The effect of some of these practices can be expressed by the *C* factor of the USLE (Table 5-4). For example, covering the soil surface by woodchips reduces the *C* factor from 1.0 to 0.08, thus reducing soil loss by 92%. Some practices that can affect soil loss from construction sites (raking, heavy machinery) are listed in Table 5-6.

All construction sites are hazardous zones requiring control. Frequently, more than 50% of the pollution load from an urban watershed is caused by a small portion of the basin area under development.

Nonurban Unit Loads and Management

Agricultural Cropland. This category covers general agricultural lands. Row crops and close-grown crops are included; however, other lands—orchards, idle farmland—also may be counted as cropland.

Many factors affect pollutant emission from farm croplands. Pollutants arise from surface runoff (erosion of topsoil particles and irrigation return flow), interflow (mostly tile drainage and leachate of excess irrigation), and groundwater base flow. Often the reduction of one component of pollution results in an increase in other components. Figure 10-12 shows a typical example of nonpoint pollution from farm runoff.

Soils of some farms have poor drainage characteristics, and tile drainage systems often are installed to improve the soil moisture characteristics and crop

Fig. 10-12. Pollution by cropland runoff. (Photo University of Wisconsin.)

yields. Drainage tile water contains soluble salts, nitrate N, and organic materials leached from the soils.

In areas where rainfall during the growing season is sparse or where crops require large amounts of water (e.g., some vegetables), artificial irrigation must supplement deficiencies. The amount of water for irrigation must exceed the evaporation requirements of crops in order to prevent salt buildup. Irrigation water from surface sources has some salinity, while soil water evaporated to the atmosphere leaves its salt content in the soil. Therefore, irrigation return flows and tile drainage water have elevated salinity which should be considered as nonpoint pollution.

The characteristics of tile drainage water and irrigation return flows were summarized by Loehr,[3] and ranges of N contents were reported in Chap. 7. Loading ranges for cropland drainage water are:[3]

Drainage	Nitrogen (kg/ha-yr)	Phosphorus (kg/ha-yr)
Irrigation return flow	3–30	1–4
Subsurface tile drainage	5–20	3–10

However, erosion and soil loss by surface runoff is considered a predominant source of pollution from croplands. The disturbing activity associated with tillage substantially increases the erosion potential of croplands. Conversely, increased hydrologic surface storage and permeability of tilled fields reduce hydrologic activity, which sometimes balances the increased erosion potential.

Of the nutrient loss (N and P), over 90% is associated with soil loss.[27] Nutrient losses usually represent only a small fraction of the applied fertilizer and often are economically insignificant.[27] Nevertheless, their pollution impact almost always exceeds the standards accepted for preventing accelerated eutrophication of surface waters.

Dudley and Karr[28] noted that bacterial contamination from agricultural drainage in the Black Creek Watershed, Indiana, exceeded accepted standards for recreation. Organic pollutants and fecal contaminants originate from manure application, unconfined livestock, and septic tank drainage fields. Applying manure to frozen fields is a practice that is most damaging to surface water bodies. PLUARG[4] attributed the following loading ranges to surface runoff from croplands:

Pollutant	Range (kg/ha-yr)
Suspended sediment	30-5,100
Total phosphorus	0.2-4.0
Total nitrogen	4.3-31

Load variations amounting to several orders of magnitude are common for croplands. The variability is caused by several factors, most of which have been discussed previously. The factors include: slope, soil erodibility and texture, drainage characteristics, vegetational cover, tillage and planting practices (up and down the slope or contouring), and meteorological factors.

Simulated loads by LANDRUN for 1-km^2 fields, soil slope category C (9%), and midwestern meterological conditions ($R \simeq 100$) were as follows:

Soil Type	Sediment (kg/ha-season)			Phosphate (kg/ha-season)		
	Spring	Summer	Fall	Spring	Summer	Fall
Boyer sandy loam	2,240	560	150	2.22	0.56	0.15
Hochheim loam	4,400	1,280	296	6.6	1.92	0.44
Ozaukee silt loam	13,600	2,400	578	24.4	4.31	0.94
Ashkum silty clay loam	15,000	5,340	800	46.1	16.85	2.50

These loading values are based on the assumptions that the USLE cover factor is $C = 0.8$ in spring and $C = 0.08$ in summer and fall and that the soil conservation practice factor is $P = 1.0$. The factor to correct for field sizes other than 1 km^2

(100 ha) is read from Fig. 10-4, and slope correction factor for slopes other than 9% can be obtained from Fig. 10-5. The slope correction factor indicates that soil loss caused by rain and runoff from flat agricultural lands is minimal, probably about 100 kg/ha-yr. This low erosion loss is due to the high surface water storage of tilled flat fields.

Example 10-4: Estimation of the Pollution Potential of Agricultural Lands

Estimate soil and phosphorus loss from a field growing corn in southeastern Wisconsin. The field size is 500 ha and average slope is 5.5%. The field was plowed along the contour. The hydrologic category of the soil is B.

The annual loading for a 1-km² unit with 9% slope and hydrologic category B gives a soil loss of:

$$A_s' = 4400 + 1280 + 296 = 5976 \text{ kg/ha-yr}$$

and phosphorus loss of:

$$A_p' = 6.60 + 1.92 + 0.44 = 8.96 \text{ kg/ha-yr}$$

The rainfall correction factor R for southereastern Wisconsin is 125. Area size correction ξ_A (from Fig. 10-4) is 0.8. Slope correction ξ_S (from Fig. 10-5) is 0.11. Soil conservation factor P for contouring (from Table 5-5) is 0.5. Then soil loss becomes:

$$A_s = \frac{125}{100} \times 0.8 \times 0.11 \times 0.5 \times 5976 = 329 \text{ kg/ha-yr}$$

and phosphorus loss is:

$$A_P = \frac{125}{100} \times 0.8 \times 0.11 \times 0.5 \times 8.96 = 0.49 \text{ kg/ha-yr}$$

If the field was irrigated or had tile drainage, an additional increment of dissolved phosphorus would be added.

Management Practices. A wide range of management practices are available to reduce pollution loads from agricultural lands and are categorized as:[29,30]

1. Soil and water conservation practices, including contour plowing, terracing, cover crops, grassed waterways, and runoff diversions.
2. Crop management practices, including conservation tillage, selection of best time for plowing, chisel plowing, minimum or no tillage, monocropping, or alternate meadows.

3. Nutrient management practices, including fertilizer formulation, application rate, application technique, and timing of application.
4. Pesticide management practices, including application methodology, timing and rate of application, use of degradable compounds, and selection of pest-resistant crops.

See Chapter 11 for greater details on the management of agricultural lands.

Pasture and Rangeland: This category accounts for the largest proportion of total land use in the United States[31] and includes about 40% of all nonfederal land. Range and pasture land is used directly for livestock production. The grazing practices include continuous and seasonal or rotational grazing.

As seen in Fig. 10-3, unit loads of most pollutants from pasture and rangeland are at least an order of magnitude less than loads from croplands. Lead is an apparent exception although little information is available to substantiate a trend. Generally, pastures are considered nonhazardous land uses requiring little or no pollution control. Renovation practices on pastures—including mechanical and chemical methods—improve grass quality and density, and reduce soil loss. Converting hazardous agricultural lands to pasture may be a possible control strategy.

Timmons and Holt[32] studied organic pollution and nutrient losses from native prairies in west-central Minnesota. Average annual total COD losses ranged from 2.26 to 40.32 kg/ha, annual total N losses ranged from 0.11 to 1.71 kg/ha, and annual total P losses ranged from 0.01 to 0.25 kg/ha. It should be noted that N and P loads contributed annually by precipitation are significantly higher than the nutrient losses from native prairies. Thus, native prairies are actually nutrient sinks. When cattle are allowed close to a watercourse, pasture may become a limited pollution hazard (Fig. 10-13). To prevent such pollution, fences should be installed to confine the cattle. Runoff from pastures with large numbers of unconfined cattle should be intercepted by grass or other vegetational strips.

Simulated loadings for pastures in soil slope category (4%) ranged from 100 to 500 kg/ha-yr of sediment and <0.1 to 1.7 kg/ha-yr of P. These values are below the soil loss limits calling for control treatment.

Woodlands. Undisturbed forest or woodland represents the best protection of lands from sediment and pollutant losses. Woodlands and forests have low hydrologic activity due to high surface water storage in leaves (interception), ground, mulch, and terrain roughness. Furthermore, forest soils frequently have improved permeability. Even lowland forests with a high groundwater table absorb large amounts of precipitation and actively retain and retard runoff. In addition, tree canopy and ground cover as well as the increased organic con-

Fig. 10-13. Pastures can become a source of organic and bacterial pollution if cattle are allowed in close proximity to watercourses. (Photo University of Wisconsin.)

tent of forest soils significantly reduce erosion losses. Surface runoff from forested areas is often almost nonexistent.

However, streams draining lowland forests may have elevated organic and nutrient levels caused by leaching from soils by interflow and base flow. Despite this effect, woodlands are the determinants of background pollution levels against which other land uses are judged.

Reported and simulated sediment loadings from forested lands are <100 kg/ha-yr.

Uncontrolled logging operations—clearcutting—often disturb the forest's resistance to erosion (Fig. 10-14). Observations and records indicate that almost all sediment reaching waterways from forest lands originates from construction of logging roads[33,34] and from clearcuts.[34] Chief sources are roads that disrupt or infringe upon natural drainage channels.

Two immediate means are available to reduce erosion and solve the drainage problem of logging roads.[33] One is to install open culverts, water-drop structures, paving, and diversion trenches to carry water safely away to protected pervious areas. This practice may be quite expensive considering the required length of logging roads.

Fig. 10-14. Appreciable amounts of sediment and pollution can reach surface waters from logging roads and clearcutting. (Photo University of Wisconsin.)

Another management practice is to reduce runoff from road surfaces by diversion of runoff from upslope drainage and/or by cross-road culverts and drainage tiles, and to have sufficient natural downslope distance below the road to dissipate the runoff and eroded materials before they reach the surface water drainage system.

Feedlots. Feedlots and barnyards can be the most hazardous land uses in rural areas. With the advent of improved feeding methods and handling of ensiled materials, cattle are no longer put out to pasture but are held in relatively small areas. An obvious potential problem exists in feedlots, barnyards, and exercise areas where herds are confined on a year-round or seasonal basis (Fig. 10-15).

Commonly, feedlots themselves (beef, dairy, swine, sheep, and poultry lots) are classified as point rather than nonpoint sources of pollution.[31] Indirectly, however, the solid waste disposal requirements and methods of management cause this subcategory to have major nonpoint source environmental implications.

The majoriy of feedlot wastes reaching surface waters are transported by

Fig. 10-15. Typical view of unsanitary barnyard or feedlot operation. (Photo USDA Soil Conservation Service.)

surface runoff. The quantity of runoff and the hydrologic activity of feedlot areas depend on the degree of perviousness of the lot and the permeability of the soils, the antecedent moisture conditions, the number of cattle on the lot, surface storage characteristics, and rainfall intensity.[35] The topsoil portion of feedlots contains large quantities of compacted, decomposing manure residues, which reduces permeability. In fact, permeability and infiltration rates in many feedlots are only remotely related to the native soil.

The surface storage characteristics of cattle feedlots range from 0.56 to 0.9 cm.[35] A hard crust of soil extending several centimeters below the manure surface resists manure penetration unless disturbed.

The high organic matter content of the surface crust protects against erosion, and consequently sediment yields from feedlots are somewhat lower than expected.[35,36] Nevertheless, barnyard and feedlot runoff has extremely high BOD_5 concentrations (1000 to 12,000 mg/liter); COD (2400 to 38,000 mg/liter; 6 to 800 mg/liter of organic N;[3,35] and 4 to 15 mg/liter of P). Runoff from barnyards is turbid and represents a high-nutrient, short-term (shock) loading to receiving waters. The high level of pollutants that feedlot effluents generate is reflected in the following ranges of unit loads:[3]

	Unit Area Load (kg/ha-yr)	
	Average	Range
Suspended sediment	14,000	7,300-27,000
TKN	510	46-3,110
Total P	77	10-620

Waste deposited on feedlot surfaces is readily biodegradable and its decay rate depends on temperature and moisture content. Increased moisture dissolves more of the waste constituents and increases the soluble components of feedlot runoff. Since biodegradation rates are proportional to temperature, wastes are decomposed more slowly during the winter.

The primary source of pollution is animal waste stored on feedlot surfaces and in the surface several centimeters of soil. Only a small portion of the stored wastes are mobilized during storm or snow-melt events. In semiarid regions of the southwestern United States, cattle wastes become dehydrated until it rains. In more humid climates, the waste decomposes and organic matter is solubilized[3] when moisture and temperature cnditions permit accelerated microbial activity. Runoff from melting snow in humid regions contains concentrations and loadings that are about 10 times as high as runoff during warmer conditions.[35]

Feedlot runoff quality does not depend on the amount of manure on the lot.[36] Once the feedlot surface is covered, manure depth is not an important parameter of water quality. Parameters affecting quality of feedlot runoff include rainfall intensity, antecedent water content of the manure pack, and type of feedlot surface.[37] The nutrient characteristics of feedlot manure are shown in Table 10-4.

Management Practices for Barnyard and Feedlot Runoff. Control of barnyard effluents may depend on simple management techniques even though BOD and nutrient concentrations are high. Runoff control systems for feedlots are

TABLE 10-4. Pollution Characteristics of Manure from Domestic Animals.[38]

	Average Weight (kg/animal)	Wet Manure (g/kg/day)	NH_4-N (g/kg/day)	Total N (g/kg/day)	P_2O_5 (g/kg/day)	Total N (kg/animal/yr)	P (kg/animal/yr)
Chickens	2	62	0.26	0.74	0.60	0.5	0.2
Ducks	2	–	–	8.0	0.8	5.8	0.35
Swine	125	74	0.24	0.51	0.43	23.0	8.0
Dairy cattle	450	84	–	0.23	0.34	38.0	25.0
Beef cattle	450	66	0.11	0.32	0.18	53.0	13.0
Sheep	50	72	–	0.60	0.25	11.0	2.0

governed by two basic principles.[36] First, all clean water originating outside the feedlot should be diverted so that it does not come in contact with feedlot pollutants. Second, the water originating inside the feedlot should be disposed of in a way that minimizes its pollution potential (Fig. 10-16). To meet these two objectives, four components should be considered in the design of runoff control systems: clear water diversion; runoff collection; runoff containment; and controlled disposal.

Clear water diversion systems include terraces, which direct water from upland watersheds away from the feedlot or barnyard, and gutters with downspouts on buildings, which prevent roof drainage from entering the area. When a natural waterway (creek) crosses a feedlot, it may be necessary to pass it through a culvert or relocate the animals.

Collecting runoff with a system of curbs, gutters, or terraces prevents it from directly entering defined watercourses and concentrates the polluted runoff for treatment before disposal.

Runoff may be contained for short periods of time in settling channels or basins, vegetative filters, or infiltration areas. Holding ponds provide long-term storage. Water contained in retention basins can be treated by biological oxidation,[35] can infiltrate the soil, or can be disposed of through irrigation systems.

Vegetative filters and infiltration areas combine the containment and disposal aspects of runoff control. Runoff flowing over grass before reaching a retention basin or watercourse is reduced in volume and pollution content by soil percolation and the filtration capability of the grass.

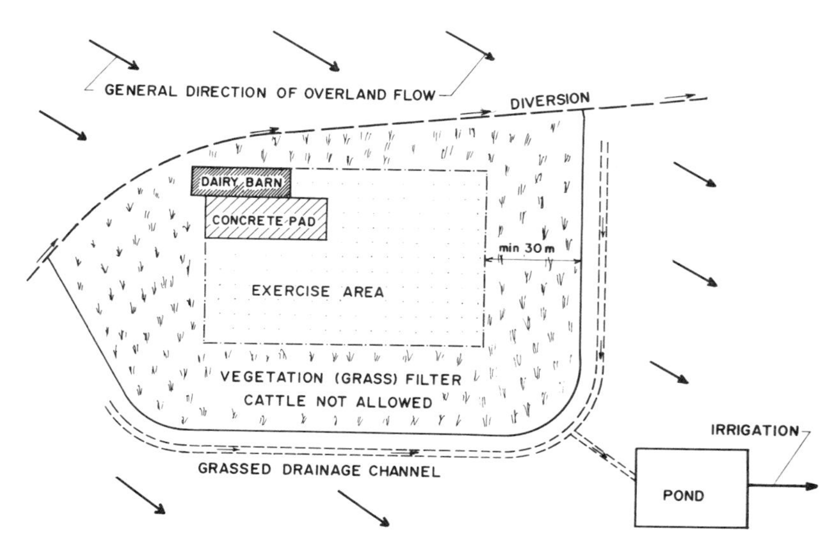

Fig. 10-16. Example of best management practices for containment of feedlot runoff.

The *DR* of barnyard and feedlot pollutants—the ratio of pollutants delivered to a watercourse to the amount generated at the source—decreases rapidly with the overland distance that the runoff must travel. At a certain critical distance, *DR* drops to almost zero. The critical distance has been estimated[36,39,40] to be 30 to 120 m, depending on soil characteristics, grass type, and density of cover.

EFFECT OF HYDROLOGIC MODIFICATIONS

Hydrologic modifications are human activities that result in nonpoint pollution and directly or indirectly affect, or have affected, natural stream flow and the associated water regime detrimentally. Almost any use or modification of land by humans results in a hydrologic change in the water regime and consequently increases the potential for nonpoint pollution. Construction activities reduce permeability and surface storage; incorporation of impervious areas into a watershed increases its hydrologic activity; drainage and irrigation systems in agricultural zones disrupt the groundwater regime; etc. These causes and effects of nonpoint pollution have been discussed previously (Chapters 3, 5, and 8).

However, other activities—mainly in-stream modifications—exist that disrupt natural flow and result in pollution loadings similar to those from diffuse areal sources. These activities include dredging, channelization, dam and impoundment construction and operation, and other in-water or close-to-the-stream activities.

Dredging

Dredging has an impact on water quality primarily because it resuspends or redissolves pollutants on the channel bottom. This resuspension or redissolution occurs when material is lifted from the water or otherwise transported, and when it is disposed of, either by dumping back into the water or on land.

Most current dredging activities in the United States are for maintaining existing navigational waterways. Approximately 220×10^6 m^3 of bottom materials are dredged annually from rivers and estuaries for navigational purposes, and an additional 60×10^6 m^3 are dredged in construction of new navigation channels. These totals include 8×10^6 m^3 dredged annually from 115 lakes. Over 35,000 km of waterways have been modified for commercial navigation, and each year approximately 30,000 km of waterways and 1000 harbors are dredged to maintain waterborne commerce in the United States.[41,42] Figure 10-17 shows the mechanisms of disposal of the dredged materials.

The nature and magnitude of pollution loadings introduced into surface waters by dredging depends on the character of the dredged materials. A few decades ago, dredging evidenced no apparent environmental pollution problems. However, in recent years sediments accumulated in harbors and channels have

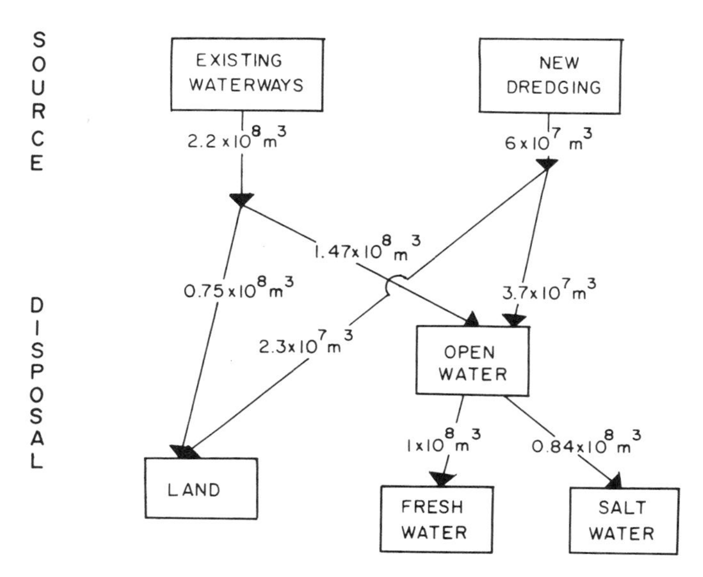

Fig. 10-17. Dredge spoil disposal flow chart.[41,42]

become increasingly polluted as a result of point and nonpoint pollution, and as a consequence, dredged materials may range in quality from relatively clean sand and gravel to organic muck and sludge of natural or man-made origin. Examples of concentrations of pollutants in bottom samples are given in Table 10-5.

TABLE 10-5. Chemical Comparison of Slightly and Heavily Polluted Stream Bottom Samples.[43]

Parameter	Lightly Polluted		Heavily Polluted	
	Mean	Range	Mean	Range
Total volatile solids (%)	2.9	0.7–5.0	19.6	10.2–49.3
Chemical oxygen demand (g/kg)	21	3–48	177	39–395
Kjehldal nitrogen (g/kg)	0.55	0.01–1.31	2.64	0.58–6.80
Total phosphorus (g/kg)	0.58	0.24–0.95	1.06	0.59–2.55
Grease and oil (g/kg)	0.56	0.11–1.31	7.15	1.38–32.1
Initial oxygen demand (g/kg)	0.50	0.08–1.24	2.07	0.28–4.65
Oxygen uptake (g/kg)	–	–	–	–
Sulfides[a] (g/kg)	0.14	0.03–0.51	1.70	0.10–3.77

[a]Values are conservative because of the preservation method used.

Channelization

Channel modifications are implemented primarily for flood control, erosion control, navigation, and drainage. Seven different types of modifications serve as potential sources of pollution.[44]

1. Clearing of debris and opening of blockages to restore the former hydraulic capacity of a stream. These operations have the least nonpoint source pollution consequences.
2. Channel excavations that enlarge and restore an existing channel, or that provide a new channel in its place. Heavy nonpoint pollution can result.
3. Channel realignment to eliminate meanders that have developed in the natural streambed. Heavy nonpoint pollution can result.
4. Construction of floodways to relieve the streambed of excessive flows of storm water. The floodways are normally dry, and if stabilized properly by vegetation, minimal pollution can result when flood flows subsequently enter the floodway.
5. Construction of flood retardation basins for the temporary storage of excess flows. These structures can be in-stream or off-stream, with the latter having low pollution potential during construction.
6. Construction of debris and sediment retention basins to hold back pollution during periods of high water. The amount of nonpoint pollution is influenced by the amount of side disturbance.
7. Construction of drainage ditches or deepening of existing ditches. In-channel vegetation, such as grassed waterways, limits nonpoint source pollution.

The major pollutant from channel modifications is sediment. Depending on the amount of organic matter present in the sediment (tree residuals, vegetation, etc.) products of organic matter decomposition also may be present.

In-water Construction

This category includes:

1. Pile driving or placement of piers in the bottom of a waterway.
2. Placement of bulbheads (vertical walls) in or adjacent to a water body.
3. Placement of structural sections (such as pipes or tunnels) on the bottom of a water body.

These water quality effects are similar to those of dredging.

Dam Constructions and Operations

An artificial dam or impoundment is a structural feature located on a stream that–in most cases–irreversibly changes its flow and quality regime. The most profound changes occur during construction and immediately after completion of the dam during the first flooding.

The effect of construction activities is similar to dredging and other in-stream and off-stream construction activities. Heavy nonpoint pollution can result if proper control measures are not implemented. Stream bypasses of the construction area plus control of runoff and drainage water from the construction sites are the most important control techniques.

Other activities cause nonpoint pollution during construction and postconstruction periods. Cutting trees and removing vegetation in the inundated area may improve water quality to some extent in the future lake, but it may be a significant if not enormous source of pollution due to logging operations, exposure of bare lands to storm action, and temporary transportation of pollutants on unpaved roads.

Irreversible water quality changes produced by reservoirs are caused by stratification, peak operation for hydropower production, accelerated eutrophication, density currents, and deposition of sediment in the reservoir. The effect of these phenomena on water quality were discussed elsewhere.[45]

REFERENCES

1. Lienesch, W. C. 1977. Legal and institutional approaches to water quality management, planning and implementation. U.S. EPA, Washington, D.C.
2. Omernik, J. M. 1976. The influence of land use on stream nutrient levels. EPA-600-76-014. U.S. EPA, Corvallis, Oregon.
3. Loehr, R. C. 1974. Characteristics and comparative magnitude of nonpoint sources. *J. WPCF*, **46**:1849–1872.
4. Chesters, G., Robinson, J., Strefel, R., Ostry, R., Bahr, T., Cootl, D. R., and Whitt, D. M. 1978. Pilot watershed studies. Summary report. International Joint Commission, Windsor, Ontario.
5. Johnson, M. G., Comeau, J. C., Heidtke, T. M., Sonzogni, W. C., and Stahlbaum, B. W. 1978. Management information base and overview modeling. International Joint Commission, Windsor, Ontario.
6. Bradford, W. C. 1977. Urban stormwater pollutant loadings: A statistical summary through 1972. *J. WPCF*, **49**:613–622.
7. McElroy, A. D., Chiu, S. Y., Nebgen, J. W., Aleti, A., and Bennett, F. W. 1976. Loading functions for assessment of water pollution from nonpoint sources. EPA 600/2-76-151, U.S. EPA, Washington, D.C.
8. Drynan, W. R., and Davis, M. J. 1978. Application of the Universal Soil Loss Equation to the estimation of nonpoint sources of pollutant loadings to the Great Lakes. International Joint Commission, Windsor, Ontario.

9. Davis, M. J., and Nebgen, J. W. 1979. Estimation of agricultural nonpoint loads to the Waukarusa River Basin using the Nonpoint Calculator. In: R. C. Loehr et al. (eds.). "Best Management Practices for Agricultural and Silviculture." Ann Arbor Science, Ann Arbor, Michigan.
10. Groszyk, W. S. 1979. Nonpoint pollution control strategy. In: R. C. Loehr et al. (eds.). "Best Management Practices for Agriculture and Silviculture." Ann Arbor Science, Ann Arbor, Michigan.
11. Konrad, J. G., Chesters, G., and Bauer, K. W. 1978. Menomonee River Pilot Watershed Study–Summary pilot watershed report. International Joint Commission, Windsor, Ontario.
12. Novotny, V., and Goodrich-Mahoney, J. 1978. Comparative assessment of pollution loadings from nonpoint sources in urban land use. *Prog. Water Tech.*, **10**:776–785.
13. Novotny, V., Basliger, D., Cherkauer, D. S., Simsiman, G., Chesters, G., Bannerman, R., and Konrad, J., 1979. Simulation of pollutant loadings and runoff quality. Vol. 5. Menomonee River Pilot Watershed Study, International Joint Commission, Windsor, Ontario.
14. Novotny, V., and Bannerman, R. 1979. Model enhanced unit loading concept for estimating pollution from nonpoint sources. Proceedings of the Symposium on Hydrologic Transport Modeling, New Orleans, Louisiana, Dec. 10–11. ASAE, St. Josephs, Minnesota.
15. Marshalek, J. 1978. Pollution due to urban runoff: Unit loads and abatement measures. International Joint Commission, Windsor, Ontario.
16. Quereshi, A. A., and Dutka, B. J. 1979. Microbial studies on the quality of urban stormwater runoff in southern Ontario, Canada. *Water Res.*, **13**:977–985.
17. Schenk, J. E., and Sherger, D. A. 1974. The effect of residential and commercial-industrial land usage on water quality. PLUARG Task A1, International Joint Commission, Windsor, Ontario.
18. Oberts, G. L. 1977. Water quality effects of potential urban best management practices: A literature review. Tech. Bull. 97, Dept. of Nat. Res., Madison, Wisconsin.
19. Anon. 1968. Stream pollution by coal mine drainage: Upper Ohio River Basin, Ohio River Basin Project, FWPCA, U.S. Dept. of Int., Washington, D.C.
20. Yeasted, J. C., and Shane, R. 1976. pH profiles in a river system with multiple acid loads. *J. WPCF*, **48**:91–106.
21. Hackney, C. R., and Bissonnette, G. K. 1978. Recovery indicator bacteria in acid mine streams. *J. WPCF*, **50**:775–780.
22. Cleary, E. J. 1967. "The ORSANCO Story–Water Quality Management in the Ohio Valley under an Interstate Compact." Johns Hopkins Press, Baltimore, Maryland.
23. Rogowski, A. S., Pionke, H. B., and Broynan, J. G. 1977. Modeling the impact of strip mining and reclamation process on quality and quantity of water in mined areas: A review. *J. Env. Quality*, **6**:237–244.

24. Ports, M. A. 1975. Urban sediment control design criteria and procedures. Paper No. 75-2567, Winter Meeting ASAE, Chicago, Illinois.
25. Yorke, T. H., and Herb, W. J. 1976. Urban area sediment yield—of construction-site conditions and sediment control methods. Proc. of the Third Fed. Inter-Agency Sedimentation Conf., Denver, Colorado.
26. Hagman, B. B., and Konrad, J. G. 1979. Method for controlling erosion and sedimentation from residential construction activities. Paper presented at 3rd Water Research Conference, Wisconsin Section AWRA, Oshkosh, Wisconsin.
27. Alberts, E. E., Schuman, G. E., and Burwell, R. E. 1978. Seasonal runoff losses of nitrogen and phosphorus from Missouri Valley Loess watersheds. *J. Env. Quality*, **7**:203–208.
28. Dudley, R. R., and Karr, J. R. 1979. Concentrations and sources of fecal and organic pollution in an agricultural watershed, *Water Res. Bull.*, **15**:911–923.
29. Bailey, G. W., and Waddel, T. E. 1979. Best Management Practices for Agriculture and Silviculture: An integrated overview. In: R. C. Loehr et al. (eds.). "Best Management Practices for Agriculture and Silviculture." Ann Arbor Science, Ann Arbor, Michigan.
30. Mueller, D. H., Wendt, R. C., Daniel, T. C., and Madison, F. W. 1979. Effect of selected conservation tillage practices on the quality of runoff water. Paper presented at 3rd Annual Conference, Wisconsin Section AWRA, Oshkosh, Wisconsin.
31. Unger, S. G. 1979. Environmental implications in agriculture and silviculture. In: R. C. Loehr et al. (eds.). "Best Management Practices for Agriculture." Ann Arbor Science, Ann Arbor, Michigan.
32. Timmons, D. R., and Holt, R. F. 1977. Nutrient losses in surface runoff from a native prairie. *J. Env. Quality*, **6**:369–373.
33. Haupt, H. F. 1959. Road and slope characteristics affecting sediment movement from logging roads. *J. Forestry*, **61**:329–332.
34. Beschta, R. L. 1978. Long-term patterns of sediment prediction following road construction and logging in Oregon coast range. *Water Resources Res.*, **14**:1011–1016.
35. Loehr, R. C. 1972. Agricultural runoff characteristics and control. *J. Sanitary Eng. Div., ASCE*, **98**:909–925.
36. Madison, F. W., Peterson, J. D., and Bubenzer, G. D. 1979. Control methodology for barnyard effluents. Paper presented at 3rd Annual Water Research Conference, Wisconsin Section AWRA, Oshkosh, Wisconsin.
37. Miner, J. R., Lipper, R. I., Fina, L. R., and Funk, J. W. 1966. Cattle feedlot runoff—its nature and variations. *J. WPCF*, **38**:1582–1591.
38. Porcella, D. B., et al. 1874. Comprehensive management of phosphorus water pollution. EPA 600/5-74-010, U.S. EPA, Washington, D.C.
39. Draper, D. W., Robinson, J. B., and Cooke, D. R. 1979. Estimation and management of the contribution by manure from livestock in the Ontario Great Lakes basin to the phosphorus loading of the Great Lakes. In: R. C.

Loehr et al. (eds.). "Best Management Practices for Agriculture and Silviculture." Ann Arbor Science, Ann Arbor, Michigan.

40. Moore, I. C., Madison, F. W., and Schneider, R. R. 1979. Estimating phosphorus loading from livestock wastes: Some Wisconsin results. In: R. C. Loehr et al. (eds.). "Best Management Practices for Agriculture and Silviculture." Ann Arbor Science, Ann Arbor, Michigan.
41. Bhutani, J., Holberger, R., Spewak, P., Jacobsen, W. E., and Truett, J. D. 1975. Impact of hydrologic modifications on water quality. EPA-600/2-75-007, U.S. EPA, Washington, D.C.
42. Boyd, M. B., et al. 1972. Disposal of dredge spoil. U.S. Army Corps of Eng. Waterways Experimental Station, Vicksburg, Mississippi.
43. O'Neal, G., and Sceva, J. 1971. The effect of dredging on water quality in the Northwest. U.S. EPA Office of Water Programs, Reg. X., Seattle, Washington.
44. Whalen, N. A. 1977. Nonpoint source control guidance–hydrologic modifications. Tech. Guidance Memorandum TECG-29, U.S. EPA Water Plann. Div., Nonpoint Sources Branch, Washington, D.C.
45. Krenkel, P. A., and Novotny, V. 1980. "Water Quality Management." Academic Press, New York.

11

Management practices of nonpoint pollution control

INTRODUCTION

Nonpoint pollution control measures are as diversified as the sources of pollution themselves, and no uniform technology can be proposed to control diffuse sources. The traditional means of control—collection and treatment—which works well for most point sources, would be prohibitively expensive for nonpoint pollution control and could be used only in rare cases when all other controls at the source fail.

The "Best Management Practices," introduced in Chapter 1, represent a practice or combination of practices, determined by a state (or designated planning agency) to be the most effective. It is a practicable means of preventing or reducing the amount of pollution generated by nonpoint sources to a level compatible with water quality goals.

The "Best Management Practices" (BMPs)[1] should:

1. Manage pollution generated by nonpoint sources.
2. Achieve water quality compatible with water quality goals.

3. Be "most effective in preventing or reducing the amount of pollution generated."
4. Be practicable.

The application and selection of the BMPs should be based on:

1. Type of land-use activity.
2. Physical conditions in the watershed.
3. Pollutants to be controlled.
4. Site-specific conditions.

Since several land-use activities defined by planners have similar character and pollution potential, BMPs for some land uses may be similar. For example, strip mining, construction sites, and land under development are characterized by the same disturbing activity and require similar control techniques. Also, park and recreation zones, pastures, idle lands, and even some low-density residential areas could be considered under one category in terms of types of control measures that could be applied.

It is not possible to control all pollutants and/or all lands. Practices must be assigned—in a priority order—to the most hazardous lands, that is, lands generating the largest pollutant loads to surface waters. Furthermore, the pollutants to be removed or controlled should be those that are of greatest concern in preserving the quality of surface water and that violate water quality standards.

There is a known rule that applies to point and nonpoint sources of pollution, stating that the difficulty and cost of pollutant control is inversely related to the extent of pollutant dilution by the runoff. Highly concentrated pollutant loads require less expenditures per unit of pollutant removed than pollutants occurring in low concentrations. Thus, control costs increase with the distance the runoff travels from the source to the point of treatment. This introduces the basic philosophy behind the selection of "Best Management Practices," which states that the materials that belong on the land should be kept there. Once the materials begin to move with runoff or soil water, their control is more tedious and expensive.

In general, control measures can be divided into:

1. Source control of hazardous lands and land uses.
2. Collection control and reduction of pollutant delivery to receiving waters.
3. Treatment of runoff.

Control methods also can be grouped into structural and nonstructural. Structural methods include sedimentation basins, seepage beds, treatment facilities, etc., while nonstructural methods rely on better management of agricultural fields, rotation of crops, seeding and sodding, air pollution control, etc.

In this chapter, the methods available in each group of control measures are reviewed with respect to their selection, application, effectiveness, and design.

SOURCE CONTROL MEASURES

The management practices designed to keep potential pollutants at the source and/or to prevent them from leaving the source area are preferred. What may be considered as a pollutant to surface waters may be valuable topsoil and nutrients for growing crops or urban lawns.

Source control can be achieved through:

1. Source ordinances that prohibit certain land-use activities on hazardous lands.
2. Erosion control practices that prevent loss of valuable topsoil.
3. Infiltration enhancement and surface water storage of land that reduce the quantity of runoff and erosion.
4. Reduction of contaminant levels in soils and on impervious surfaces.

Zoning

Zoning has been the principal means of land-use control in America since 1926.[2] Zoning was devised originally with the idea that conflicting land uses in urban areas should be separated, and only casual interest was expressed in water quality and environmental preservation. However, more recently the legitimacy of restricting land use for environmental reasons has been recognized.

In addition to zoning, other similar land-use control measures include: critical area protection, environmental impact review, property acquisition, and taxation and charges.

Zoning and land-use control are applicable to areas that are in the process of development. Zoning and architectural control can be effective in controlling nonpoint pollution. As shown in previous chapters, pollution potential from one land-use activity to another differs by orders of magnitude and prohibiting or limiting certain land-use activities on hazardous lands may provide a solution.

One extreme of the preventive approach to nonpoint pollution management is represented by zoning regulations designed to maintain land in an undeveloped form. Despite public pressure in favor of such regulations, attempts to prohibit development—short of purchasing open space for public uses—gives rise to serious constitutional questions.[2]

A practical alternative to extreme restrictions on future development in site planning is based on ecological considerations.[2] The essence of this approach is not to prohibit development but to direct it to diminish environmental damage.

Zoning methods that can be used to control pollution include:[2]

1. Large lot zoning whereby minimum lot size requirements are imposed.
2. Zoning for protection of open space, which can be used for limiting the extent of impervious areas.
3. Subdivision control and performance-oriented zoning techniques.

Larger building lots may be required to provide sufficient sorption area for septic tank effluents, to reduce runoff from residential areas, and to control erosion in developments on hazardous lands.

Protection of open space is applicable primarily to floodplain management. Generally, development restrictions under a floodplain program (National Flood Insurance Act of 1968-Pl 90-443) are similar to those desirable for water quality protection. For example, flood retention basins can enhance sedimentation of soil particles in runoff, and restriction of development in floodplains and creation of open spaces along stream banks may serve as effective pollutant traps and buffers.

Floodplain regulations provide a valuable tool for integrating water quality control with other programs.

In many states, local governments have the authority to regulate subdivisions and to require certain performance standards from developers. Soil tests commonly are required before septic tank disposal systems can be installed. Many local governments require incorporation of open spaces and, legally, no obstacles exist to adding additional requirements to the subdivision approval process.

Erosion Control Practices

The goal of erosion control practices is to keep erosion rates and subsequent soil and pollutant losses within acceptable limits compatible with receiving water quality standards and requirements for preservation of some lands. The process of erosion has been discussed in detail in Chapter 5. Erosion control practices can be divided into two categories, namely, those that minimize raindrop impact by protecting soils due to surface stabilization or vegetative cover and those that weaken the erosive forces of runoff by reducing slope, diverting runoff or otherwise reducing runoff velocity by channelization, and/or reducing the magnitude of rainfall excess on the land.[3]

Vegetative Cover. The most elemental practice in protecting soil surfaces against erosion is to provide vegetative cover. Vegetative cover not only dissipates rainfall energy but also improves soil permeability characteristics and increases interception storage.

The water quality effects of vegetation are quite marked (Table 5-3). Tem-

Fig. 11-1. Sod laid on prepared and fertilized soil provides instant protection against erosion. (Photo USDA Soil Conservation Service.)

porary fast-growing grasses can reduce erosion by an order of magnitude, and sodding by two orders of magnitude, which is reflected by the decrease of the vegetative factor (C) in the Universal Soil Loss Equation (USLE).

Revegetation, either by sodding (Fig. 11-1), by hydroseeding (Fig. 11-2), by manual seeding, or by shrubbery planting, should begin as soon as possible after disturbance occurs, keeping in mind that such climatic variables as rain (or lack of it), wind, and temperature may inhibit germination, or may totally remove the desired vegetation from the site.

A fescue-bluegrass mixture, because of its rapid germination and fast growth, is a very effective vegetational cover on properly prepared and fertilized, gentle to moderate slopes. Vegetation is most effective when it is young, sturdy, and resilient, and of least benefit in winter and early spring.[4]

Vegetation Strips. Vegetation strips and filters are effective as sediment and pollution traps for surface runoff. Kao et al.[5] concluded that grass filter strips provide excellent trapping efficiency especially for construction sites while Moore et al.[6] recommended grass strips for control of barnyard and feedlot runoff. Grass strips are also used as borders to control soil and pollutant loss from agricultural fields.

The factors affecting the pollution trapping efficiency of grass filters are

Fig. 11-2. Hydroseeding. (Photo Soil Conservation Service.)

increased infiltration and detained particulate matter and runoff by increased hydraulic resistance of grasses. The hydraulic resistance of grasses is greater by an order of magnitude than that of ungrassed areas (Table 3-11), thus slowing the flow and reducing the sediment carrying capacity.

Grassed areas also are effective traps for pollutants associated with sediments such as pesticides.[7] A minimum of 85% sediment removal can be achieved with a 2.5-m grass strip during shallow (nonsubmerged) overland flow.[5,8] This efficiency is increased when alternating grass and bare land strips are used.

The distance at which close to 100% removal is achieved by shallow flow over a grass filter is called the *critical distance*.[6] Draper et al.[9] concluded that the critical distance generally falls between 30 and 120 m. Wilson[10]–in an empirical study on Bermuda grass–found that maximum percentages of sand, silt, and clay were removed at about 3, 15, and 122 m distances along the test strip.

Experimental research at the University of Kentucky[8,11] yielded an empirical relationship for particle trapping efficiency of grass filters. It was found to be a function of two dimensionless numbers, namely, the flow turbulent Reynolds number, Re_t, and particle fall number, N_f (a determination of the probability of how many times a particle can reach bottom during the flow period). These two variables were defined as:

$$Re_t = \frac{v_s R_s}{\nu} \tag{11-1}$$

and

$$N_f = \frac{L_T w}{v_s D_f} \tag{11-2}$$

where:

v_s = flow velocity through the grass media
L_T = overland flow length
R_s = spacing parameter defined as

$$R_s = \frac{sD_f}{2D_f + s} \tag{11-3}$$

s = spacing of grass blades
D_f = depth of flow
w = settling velocity of the particle
ν = kinematic viscosity

The relationship of the removal efficiency to Re_t and N_f is shown in Fig. 11-3. From the figure it can be seen that a removal efficiency approaching 100% can be achieved when:

$$\chi = \left(\frac{v_s R_s}{\nu}\right)^{0.82} \left(\frac{L_T w}{v_s D_f}\right)^{-0.91} \simeq 5 \tag{11-4}$$

Example 11-1: Grass Strip Design

What should be the minimum width of a grass strip dividing a construction site from the nearest drainage channel, that would reduce clay loads to receiving waters by 90% during a storm that yielded an average surface runoff rate of q = 0.003 m^3/sec-meter width. Slope of the strip, S, is 5%.

Assume:

Grass blades spacing s = 3 mm
Clay particle size d = 0.002 mm
and its settling velocity (from Equation 5-27) w = 0.001 cm/sec
Manning roughness factor for grass (from Table 2-8) n = 0.35

The flow depth can be estimated from Manning's equation:

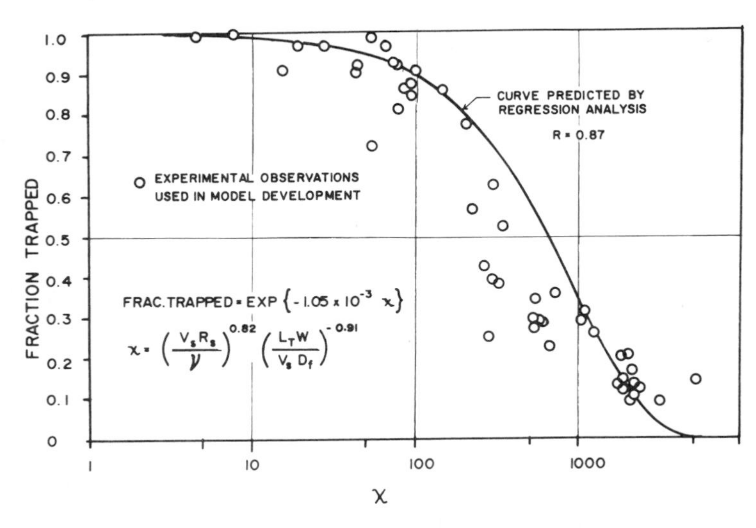

Fig. 11-3. Relationship of particle removal efficiency of grass filters to Re_t and N_f. (From reference 8.)

$$v_s = \frac{1}{n} D_f^{2/3} S^{1/2} = \frac{q}{D_f} \tag{11-5}$$

and assuming shallow flow in a wide channel:

$$D_f = \frac{(qn)^{0.6}}{S^{0.3}} = \frac{(0.003 \times 0.35)^{0.6}}{0.05^{0.3}} = 0.0365 \text{ m} \tag{11-6}$$

Assume that the grass height is greater than the flow depth (nonsubmerged flow conditions). Then the flow velocity becomes:

$$v_s = \frac{q}{D_f} = \frac{0.003}{0.0365} = 0.082 \text{ m/sec} \tag{11-7}$$

The spacing parameter, R_s, is

$$R_s = \frac{sD_f}{2D_f + s} = \frac{0.003 \times 0.0365}{2 \times 0.0365 + 0.003} = 0.00144 \text{ m} \tag{11-8}$$

Ninety percent removal occurs when the removal parameter χ from Fig. 11-3 is $\leqslant 100$. Hence

$$\chi = \left(\frac{v_s R_s}{\nu}\right)^{0.82} \left(\frac{L_m w}{v_s D_f}\right)^{-0.91} \leqslant 100 \tag{11-9}$$

from where the critical distance

$$L_m = \frac{\left(\frac{v_s R_s}{\nu}\right)^{0.9}}{100^{1.1}} \left(\frac{v_s D_f}{w}\right)$$

$$= \frac{\left(\frac{0.82 \times 0.00144}{10^{-6}}\right)^{0.9}}{158.47} \times \frac{0.082 \times 0.0365}{0.00001} = 138.4 \text{ m} \qquad (11\text{-}10)$$

Example 11-2: Efficiency of Grass Strips

For the critical distance, L_m, computed in Example 10-1, estimate removal efficiencies of silt and sand fractions.

For silt, assume the particle size d = 0.02 mm and the settling velocity v_m = 0.03 cm/sec. Then the removal factor χ is

$$\chi = \left(\frac{0.082 \times 0.00144}{10^{-6}}\right)^{0.82} \left(\frac{138.4 \times 0.03}{0.082 \times 0.0365}\right)^{-0.91} = 2.46$$

Since χ for silt is <5, more than 99% of silt will be removed. The same is true for sand fractions, which would yield even lower values of χ.

The estimate of grass filtering efficiency is probably on the conservative side since it does not include the effect of improved permeability on the grassed strip or the effects of increased surface storage.

Chemical Stabilization of Soils. This is a temporary measure employed on bare soils until permanent vegetation cover is established or other erosion control methods are accomplished. The chemical emulsions that have been recommended for erosion control include polyvinyl acetate emulsions; vinyl acrylic copolymer emulsions; and copolymer emulsions of methacrylates and acrylates. Hydrated lime and cement also is used for stabilization of clayey soils.

Other Surface Covers. Covering an exposed area with any of a number of available mulches generally increases surface roughness, protects the surface against rainfall impact, and, subsequently, reduces erosion. The effect of land cover materials is reflected in a reduction in the land cover factor (C) of the USLE (Table 5-5). Straw mulch (small grain) has proven effective on 12% slopes at application rates of 5 tonnes/ha and on a 15% slope at an application rate of 10 tonnes/ha. These application rates should reduce erosion by 75 to 80%.

Figure 11-4 shows an example of straw mulch application protected by plastic and paper fibers. Straw mulching loses its effectiveness on steep slopes because of rill formation and its tendency to be washed away by overland flow. Alternate materials include woodchips, crushed stone, and blankets and mats from textile materials (Fig. 11-5).

Fig. 11-4. Mulch net applied over straw mulch, seed, and fertilizer. The net is of paper fiber with three plastic strands that are pinned down over the straw with wire "hair pins." (Photo USDA Soil Conservation Service.)

The most economical, effective, and practical method is hydromulching, the application of a slurry mixture of seed, mulch, fertilizer, and lime.

Soil Conservation on Agricultural Lands. A list of agricultural erosion control practices is given in Table 11-1.

No-till Planting. No-till planting is considered a most effective erosion control practice applicable to agricultral lands (Fig. 11-6). Planting is accomplished by placing seeds in the soil without tillage and retaining previous plant residues. The previous planting can be killed and weed control accomplished by the use of chemical herbicides. The plant and root residues provide the necessary surface protection. The magnitudes of the cover factor (C) for no-till practices are given in Table 11-2. No-till planting in chemically killed sod can reduce soil loss to less than 5% as compared to conventional planting and plowing practices.

Fig. 11-5. Control of erosion using burlap netting. (Photo Soil Conservation Service.)

However, a greater use of herbicides is required and sometimes lower yields are experienced on some soils. Conversely, an investigation at Ohio State University[13] showed that reduced or no-till practices on well-drained soils may produce greater crop yields than under conventional plowing techniques.

On poorly drained soils under reduced tillage, crop yields frequently are less than those under conventional tillage practices. Therefore, drainage of the soil largely determines the economical success of reduced-till practices.

Conservation Tillage. This practice replaces conventional plowing with some form of noninversion tillage that retains some of the residue mulch on the surface. A chisel, field cultivator, or disk can be used for tilling.

The plowing can be performed either in strips or in row zones with interrow zones or larger areas left untilled; strips should be plowed across the slope (Fig. 11-7). Crop yields seem to be unaffected by conservation tillage practices and they have beneficial effects on the C factor (Table 11-2).

TABLE 11-1. Principal Types of Cropland Erosion Control Practices and their Highlights.[3]

	Erosion Control Practice	Benefits and Impact
E1	No-till plant in prior-crop residues	Most effective in dormant grass or small grain; highly effective in crop residues; minimizes spring sediment surges and provides year-round control; reduces man, machine, and fuel requirements; delays soil warming and drying; requires more pesticides and nitrogen; limits fertilizer- and pesticide-placement options; some climatic and soil restrictions.
E2	Conservation tillage	Includes a variety of no-plow systems that retain some of the residues on the surface; more widely adaptable but somewhat less effective than E1; advantages and disadvantages generally same as E1 but to lesser degree.
E3	Sod-based rotations	Good meadows lose virtually no soil and reduce erosion from succeeding crops; total soil loss greatlv reduced but losses unequally distributed over rotation cycle; aid in control of some diseases and pests; more fertilizer-placement options; less realized income from hay years; greater potential transport of water-soluble P; some climatic restrictions.
E4	Meadowless rotations	Aid in disease and pest control; may provide more continuous soil protection than one-crop systems; much less effective than E3.
E5	Winter cover crops	Reduce winter erosion where corn stover has been removed and after low-residue crops; provide good base for slot-planting next crop; usually no advantage over heavy cover of chopped stalks or straw; may reduce leaching of nitrate; water use by winter cover may reduce yield of cash crop.
E6	Improved soil fertility	Can substantially reduce erosion hazards as well as increase crop yields.
E7	Timing of field operations	Fall plowing facilitates more timely planting in wet springs, but it greatly increases winter and early spring erosion hazards; optimum timing of spring operations can reduce erosion and increase yields.
E8	Plow-plant systems	Rough, cloddy surface increases infiltration and reduces erosion; much less effective than E1 and E2 when long rain periods occur; seedling stands may be poor when moisture conditions are less than optimum. Mulch effect is lost by plowing.

E9	Contouring	Can reduce average soil loss by 50% on moderate slopes, but less on steep slopes; loses effectiveness if rows break over; must be supported by terraces on long slopes; soil, climatic, and topographic limitations; not compatible with use of large farming equipment on many topographies. Does not affect fertilizer and pesticide rates.
E10	Graded rows	Similar to contouring but less susceptible to row breakovers.
E11	Contour strip cropping	Rowcrop and hay in alternate 50- to 100-ft strips reduce soil loss to about 50% of that with the same rotation contoured only; fall seeded grain in lieu of meadow about half as effective; alternating corn and spring grain not effective; area must be suitable for across-slope farming and establishment of rotation meadows; favorable and unfavorable features similar to E3 and E9.
E12	Terraces	Support contouring and agronomic practices by reducing effective slope length and runoff concentration; reduce erosion and conserve soil moisture; facilitate more intensive cropping; conventional gradient terraces often incompatible with use of large equipment, but new designs have alleviated this problem; substantial initial cost and some maintenance costs.
E13	Grassed outlets	Facilitate drainage of graded rows and terrace channels with minimal erosion; involve establishment and maintenance costs and may interfere with use of large implements.
E14	Ridge planting	Earlier warming and drying of row zone; reduces erosion by concentrating runoff flow in mulch-covered furrows; most effective when rows are across slope.
E15	Contour listing	Minimizes row breakover; can reduce annual soil loss by 50%; loses effectiveness with postemergence corn cultivation; disadvantages same as E9.
E16	Change in land use	Sometimes the only solution. Well managed permanent grass or woodland effective where other control practices are inadequate; lost acreage can be compensated for by more intensive use of less erodible land.
E17	Other practices	Contour furrows, diversions, subsurface drainage; land forming, closer row spacing, etc.

Fig. 11-6. No-till planter planting corn in rye and corn stalk mulch. (Photo USDA Soil Conservation Service.)

TABLE 11-2. Effect of No-Till Practices on the Magnitude of the Cover Factor, *C*, in the USLE[a]

Tilling Practice	*C* Factor Production Level High	Moderate
No-till		
Corn after sod[b]	0.017	0.053
Corn after corn	0.18	0.18
Soybeans after corn	0.18	0.22
Grain	0.11	0.18
Corn after small grain[b]	0.062	0.14
Conservation tillage[c]		
Corn-chisel plowing	0.19	0.26
Corn-fall chisel-spring disk	0.24	0.30
Corn-strip till-row zones	0.16	0.24

[a]Adapted from Stewart et al.[3]
[b]Removed by chemical treatment.
[c]Residue not removed.

Fig. 11-7. Strip-cropping of contour-plowed alternating strips. (Photo University of Wisconsin, Department of Agricultural Journalism.)

The extent of farming using no-tillage or conservation tillage practices is increasing rapidly. From about 11 million ha in 1972, the field area has increased to 30 million ha in 1979, and it is expected that by the year 2000 about 81% of the farm area of the United States will use conservation tillage. An important reason for this trend of increasing acreage under conservation tillage practices is not only the effect on improved surface water quality but also decreased use of fuel.

Mueller et al.[14] investigated the effect of tillage practices on surface runoff and sediment, and phosphorus loadings to surface waters. The tillage operations included conventional, moldboard plowing followed by disking; chisel plowing; and no-tillage operations. The effect of a 45 tonnes/ha application of organic dairy manure prior to tillage on sediment and P losses also was evaluated for each tillage method.

Due to lower infiltration and surface storage, significantly greater amounts of runoff occurred at no-till sites than from chisel-plowed or conventionally tilled sites. No significant reduction in pollution loads were achieved from the no-till sites, in spite of reduced crop yields and increased use of herbicides to control weeds. Conventional-till practices yielded somewhat elevated loadings as compared to chisel-plowed soils, but the increase was not significant.

Manure application generally lowered sediment and P loses due to an increase in the stability of organically bound soil aggregates.

Sod-based Rotations. This system has been used for many years to reduce erosion from the conventional plow-based systems in regions adapted to rotation meadows. Soil loss from a good-quality grass and legume meadow is negligible, and plowing the sod leaves residual effects that improve infiltration and reduce erodibility. Sod most frequently is planted in 2- to 4-yr rotations.

The economy of the farm is reduced by the substitution of hay for major crops for 1 yr during the rotation cycle. However, legumes in the meadow mixture can help restore the nitrogen balance of the soils through fixation of atmospheric nitrogen.

Meadowless rotations employed by farmers to restore fertility to their soils is a far less successful erosion control measure than sod-based systems.

Winter Cover Crops. Shredded stalks of corn or sorghum can be left on fields after harvest until the spring planting period. This type of protection is more satisfactory than that provided by late-seeded small-grain winter cover on plowed grounds. If the cover crop is chemically killed and left in place for no-till planting of a row crop, it provides excellent control during the crucial erosion period.

Improved Soil Fertility. The use of fertilizers to increase crop yield has the additional advantage of substantially reducing soil erosion because of the improved crop cover, increased water use by plants, better infiltration, and availability of more residues for soil protection. The general decrease in the magnitude of factor C in the USLE is attributable to the yield increase as indicated in Table 5-4.

Timing of Field Operations. The timing of field operations can influence erosion losses. For example, early-seeded wheat protects against erosion during the spring thaw and snow-melt. A delay in spring plowing usually increases erosion potential because the period of greatest erosion may coincide with rains of greatest rainfall erosivity.

Plow-plant Systems. In plow-plant systems, the field is turn-plowed but secondary tillage is minimized. The plant residues are covered, but the surface remains rough after planting, thereby increasing surface storage and reducing rainfall excess and the velocity of surface runoff.

Contour Plowing. Contour plowing is a widely recommended supporting practice. Plowing and crop rows follow field contours across the slope. Contouring provides excellent erosion control for moderate rainstorms. The effect of contouring on the conservation (erosion control) factor (P) of the USLE is reported in Table 5-5. Contouring is not very effective on high slopes and during extreme storms when rainfall exceeds the available surface storage and infiltration.

Strip-cropping. Strip-cropping alternates contour-plowed strips of row crops and close-grown crops (Fig. 11-7).

Terraces. Terraces reduce the slope effect on erosion rates by dividing the field into segments with lesser or even near-horizontal slopes. This practice slows down runoff and prevents the formation of gullies. The excess water is conveyed from the terraces to grassed outlets or is removed by subsurface drains (Fig. 11-8).

The effect of terracing can be expressed by corresponding reductions in the slope-length factor (LS) of the USLE.

Fig. 11-8. Parallel terraces. (Photo USDA Soil Conservation Service.)

Other practices listed in Table 11-1, including grassed outlets (which act as sediment pollution traps), ridge planting, contour listing, change in land use, contour furrows, and their effect on soil loss are discussed by Stewart et al.[3]

Example 11-3: Effect of Conservation Practices

A corn-growing field on 10% slope, plowed up and down, had estimated erosion losses of soil, organic matter, and P of 20 tonnes/ha/yr, 1.0 tonnes/ha/yr, and 30 kg/ha/yr. Estimate the effect of: (1) no-till planting, (2) sod-based rotation, and (3) terracing on the erosion losses.

Most of the loss (80%) occurred during the critical erosion period (i.e., a period of 2 months after spring plowing). The overland flow length (downslope) is 200 m.

For the reference practice of conventional plowing up and down the slope, the average cover factor during the critical period is $C = 0.8$ and afterwards $C = 0.15$ (see Table 5-3 for appropriate magnitudes). The erosion control factor $P = 1.0$. In the no-till practice, crop residues and root systems from the previous harvest remain on the soil as a protective cover. This can reduce significantly erosion loss during the critical period. If the preceding crop was small grains, the C factor would approximate 0.1. Hence, the effect of this practice on the soil and pollutant loss is:

$$\text{Percent loss reduction} = \left[1 - \frac{\text{weighted } C \text{ with no-till}}{\text{weighted } C \text{ without no-till}}\right] 100$$

$$= \left[1 - \frac{0.8 \times 0.1 + 0.2 \times 0.1}{0.8 \times 0.8 + 0.2 \times 0.15}\right] 100 = 87.8\% \tag{11-11}$$

If all other variables remain the same (a crude and often unrealistic estimate) the soil loss with no-till practice would be:

$$A_s = (1 - 0.879) \times 20.00 = 2.44 \text{ tonnes/ha/yr}$$

and similarly, organic matter and P losses become:

$$A_{OM} = (1 - 0.878) \times 1.00 = 0.122 \text{ tonnes/ha/yr}$$

and

$$A_P = (1 - 0.878) \times 35 = 4.27 \text{ kg/ha/yr}$$

With sod-based rotations, sod is planted once during the rotation cycle. For the purpose of this example, assume that the rotation cycle is 2 yr and the sod is plowed during the spring season by conventional plowing methods. The C factor during the off-year with sod is 0.01. In the following year when corn is planted, the sod residues may reduce C to a value of 0.6. Average soil loss reduction over the 2-yr cycle is:

Percent loss reduction

$$= \left[\frac{1 - \dfrac{0.01}{0.8 \times 0.8 + 0.2 \times 0.15} + 1 - \dfrac{0.8 \times 0.6 + 0.2 \times 0.15}{0.8 \times 0.8 + 0.2 \times 0.15}}{2} \right] \times 100$$

$$= 61.1\%$$

If slopes are reduced by terracing, the slope-length factor (LS) is decreased. If terraces are built 1 m high on a 10% slope, the overland flow length is reduced to 10 m. From Fig. 5-12, the original LS factor for 200-m overland flow distance and 10% slope is 3.7. Having terraces 10 m wide with an average slope of 1% reduces the LS factor to 0.1. Hence the soil and pollution loss reduction becomes:

$$\text{Percent loss reduction} = \left[1 - \frac{(LS) \text{ factor after terracing}}{(LS) \text{ factor before terracing}} \right] \times 100$$

$$= \left[1 - \frac{0.1}{3.5} \right] 100 = 97.1 \qquad (11\text{-}12)$$

Although the cover factor or slope-length factor seems to be predominant for the situations discussed, other factors should be included in a more exact analysis. For example, reduction of the slope by terracing increases surface storage (Fig. 3-3); no-till practices, on the other hand, reduce surface storage and permeability. For more complex analysis, a mathematical model is the best approach to evaluate the combined effects of several factors. Examples of these approaches were discussed in Chapter 9.

Hydrological Modifications by Increased Infiltration and Surface Storage. Reducing the occurrence and frequency of surface runoff by hydrological means may be a feasible technique to control pollution from urban areas.

Porous-pavement utilization on parking lots and roadways reduces the hydrological activity of these otherwise impervious surfaces, and coupled with good cleaning practices, may substantially reduce pollutant loadings. One investigation concluded[15] that among the benefits of this method were: relief from combined sewer overflows, possible augmentation of municipal water supplies, improved traffic safety, preservation of vegetation, relief from flush flooding, and aesthetic improvements. The water quality benefits of porous pavements include: dissipation of runoff energy and associated suspended sediment loss, infiltration of soluble pollutants and some fine materials, recharge of groundwater, and elimination of most hydraulic collection systems. Applications of porous pavements have proven effective in Europe (Germany, Switzerland, and Denmark) and in the United States (University of Delaware; Woodlands, Texas).

The use of porous pavement is effective only in combination with periodic cleaning. Otherwise, clogging may reduce its usefulness. Accumulated pollutants can be washed off easily during a storm that exceeds the infiltration capacity of the pavement. Wherever soils with poor drainage characteristics underlay the porous pavement, proper subdrains must be installed.

Caution must be exercised in utilizing porous pavements in cold climates. Although freeze-thaw cycle tests have been conducted and found successful, the pavement freezes and buckles if subsurface drainage is inadequate or if severe subsoil freezing is frequent.

The costs for porous paved areas are equal to or less than those for conventional paving because the requirements for storm conveyance systems are substantially reduced.

A second effective method of inducing on-site infiltration is the use of small infiltration basins or ditches in conjunction with a drainage system. Such infiltration basins have been used successfully for controlling pollution from municipal parking lots. However, the most frequent application on a larger scale is retention and infiltration control of storm-water or combined sewer overflows.

An additional method of cutting down pollution loadings caused primarily by atmospheric fallout is disconnecting roof drains from storm sewers and allowing them to drain over a pervious surface.

Increased surface storage has similar effects in containing and retaining surface runoff. The best means of increasing surface storage is use of vegetative cover, which combines the ameliorating effects of reduction of the hydrologic activity and provides protective cover to the underlying soils. Use of other means—such as mechanical raking—should only be temporary, and the overall effect is questionable.

The infiltration capacity of clayey soils can be improved by liming.

Reduction of Contamination Levels in Soils and on Impervious Surfaces. Controls aimed at reduction of contamination levels are diversified and depend on the source and type of pollutant. The basic sources of nonpoint pollution (discussed earlier) include soil and its components, atmospheric fallout and washout of air pollutants, litter in urban areas, excessive use of fertilizers and persistent pesticides in agricultural areas, and street salting.

Since a large portion of pollutants deposited on surfaces originates from airborne particles, control of air pollutant levels eventually results in a reduction of contamination. However, it should be realized that a portion or all atmospheric fallout may not originate from the drainage basin, and often, in the absence of a distinct nearby industrial source of pollution, detection of the origin of atmospheric particles is difficult.

In urban areas, street litter deposition rates are reduced by litter and control programs that involve educational programs for citizens, litter control ordinances, and the availability of numerous, well-maintained litter collection areas and systems.

Nonpoint pollution by pesticides can be reduced by proper selection of materials. Such persistent organic chemicals as DDT remain within the environmental systems for decades, and the reduction in their quantities in one part of the system (e.g., soil) will increase them somewhere else (e.g., aquatic ecosystem).

The general use of less persistent pesticides results in a reduction of loading levels and in the best management practice (see Table 11-3 for other recommended management practices).

Removal of Contaminants from Surfaces. The removal of contaminants from surfaces before they are incorporated into runoff is a common management practice for impervious sections of high-density urban developed areas (see Chapter 8). Methods that can be included are street sweeping, leaf and grass clipping removal, and general cleanup of urban areas.

As indicated in Chapter 8, the quantity and nature of materials on urban impervious surfaces is variable and depends on the length of time since the last rain or street flushing together with a number of other factors.

The efficiency of conventional mechanical street sweepers is about 50%, but it decreases rapidly as average particle size decreases. Several investigators believe[4] that carefully planned, frequent sweeps with better equipment—specifically brooms that do not redistribute the materials—can reduce the loading of pollutants to urban waters but alone cannot reduce pollutants to an acceptable level. In spite of the relatively low efficiencies of street sweepers, active street cleaning programs can markedly reduce pollutant loads.

Grass clippings that accumulate on streets in low-density residential areas may cause a problem. Some conventional lawn mowers are equipped for collection of clippings that can be disposed of properly. The importance of this practice in improving water quality is questionable since additional fertilizer often is applied to replace that removed in the clippings.

A significant input of organic compounds and nutrients reaches surface waters from leaves deposited on street surfaces. A rigorous program of leaf removal in the fall should markedly decrease this effect. The burning of leaves in gutters or on street surfaces is unacceptable because most of the nutrients remain in the ash and are transported by subsequent rains.

TABLE 11-3. Practices for the Control of Pesticide Losses from Agricultural Applications.[a]

Control Practices	Impact
	Broad Applicability
Using alternative pesticides	Applicable to all field crops; can lower aquatic residue levels; can hinder development of target species resistance.
Optimizing pesticide placement with respect to loss	Applicable where effectiveness is maintained; may involve moderate cost.
Using crop rotation	Universally applicable; can reduce pesticide loss significantly; some indirect cost if less profitable crop is planted.
Using resistant crop varieties	Applicable to a number of crops; can sometimes eliminate need for insecticide and fungicide use; only slight usefulness for weed control.
Optimizing crop planting time	Applicable to many crops; can reduce need for pesticides; moderate cost possibly involved.
Optimizing pesticide formulation	Some commercially available alternatives; can reduce necessary rates of pesticide application.
Using mechanical control methods	Applicable to weed control; will reduce need for chemicals substantially; not economically favorable.
Reducing excessive treatment	Applicable to insect control; refined predictive techniques required.
Optimizing time of day for pesticide application	Universally applicable; can reduce necessary rates of pesticide application.
	Limited Applicability
Optimizing date of pesticide application	Applicable only when pest control is not adversely affected; little or no cost involved.
Using integrated control programs	Effective pest control with reduction in amount of pesticide used; program development difficult.
Using biological control methods	Very successful in a few cases; can reduce insecticide and herbicide use appreciably.
Using lower pesticide application rates	Can be used only where authorized; some monetary savings.
Managing aerial applications	Can reduce contamination of nontarget areas.
Planting between rows in minimum tillage	Applicable only to row crops in nonplows based tillage; may reduce amounts of pesticides necessary.

[a]After Reference 3.

COLLECTION CONTROL AND REDUCTION OF DELIVERY

The second group of management practices for nonpoint pollution control involves methods and structures for removing pollutants from runoff after they leave the source area. The pollutants are diluted by the runoff and move in lower concentrations than were present at the source, and in this form they are generally more difficult to control. Available control methods included in this category of control measures can be divided into the following groups:

- Removal of pollutants by sedimentation (e.g., reduction of stream bank erosion and installation of sediment traps).
- Removal of pollutants by increased infiltration from the collection systems.

Fig. 11-9. Concrete-lined diversion carries excess runoff from slopes above to gutter and drop structure in the street below. (Photo USDA Soil Conservation Service.)

- Removal of contaminants from collection systems (e.g., dredging; combined and storm sewer overflow control).

Most of the methods require engineering structures and design and are more expensive than source control measures.

For best results, polluted waters should be segregated as much as possible from the rest of the runoff. Segregation can be achieved by diversion dikes, culverts, or other drainageways excavated or constructed to divert upslope runoff from a source of nonpoint pollution (disturbed area). Diversion structures can be temporary or permanent and should not cause in-channel erosion. The diversion structures should allow a shallow, nonconcentrated flow (Fig. 11-9).

If the diversion structure is used for collecting runoff from a disturbed area, the outlet from the diversion channel should be conveyed to a sediment trap or sedimentation basin.

Prevention of In-Stream Erosion in Collection Systems

If unprotected banks of natural and man-made channels are severely damaged by stream flow (Figs. 11-10 and 11-11), channel erosion can become a significant

Fig. 11-10. Erosion of unprotected channels in construction zones. (Photo University of Wisconsin, Department of Agricultural Journalism.)

Fig. 11-11. Loss of unprotected topsoil and channel erosion following a heavy rain in a housing development. (Photo USDA Soil Conservation Service.)

source of sediment and pollution. To minimize or prevent this occurrence, collection systems must be designed in such a way that the shear stress caused by the flow on the stream bottom and banks is less than the resistance of the channel to erosion.

Several engineering means are available to prevent channel erosion. The shear stress caused by flow is proportional to flow velocity, channel slope, and depth. Thus, collection systems should be designed to keep the velocity of flow or slope of the channel below certain critical values representing the resistance of the channel to erosion.

The resistance of unprotected earth-dug channels against erosion is very low. Therefore, these channels should always be protected. Materials used for channel protection include grass (sod), stone, wire mesh with stone, matting (Fig. 11-12), vegetation, concrete lining, or concrete pipes.

Grassed Waterways. Grassed waterways are probably the cheapest but most effective means of conveying water (Figs. 11-13 and 11-14). If waterways are

Fig. 11-12. Laying jute matting to give additional protection in a channel where grass cover has failed during excessive flush runoff periods. (Photo USDA Soil Conservation Service.)

designed properly, in-channel erosion should be minimal, and the grass lining may even serve as a sediment trap.

Grassed waterways are natural or man-made channels of parabolic or trapezoidal cross section (Fig. 11-15) and are used to control and convey sediment-laden concentrated runoff flows and protect waterways from erosion. Grassed waterways are not as effective for sediment control as vegetative buffers due to the more concentrated flow conditions and higher velocities. Although some sediment may be trapped, the effect on removal of small particles is minimal.

The basic design criteria for grassed waterways are given in Table 11-4 and 11-5 and in Fig. 11-15. Table 11-4 shows maximum flow velocities that would cause minimum erosion for different types of vegetation linings. Table 11-5 classifies vegetative linings into retardance groups, which basically express the hydraulic roughness of the vegetal cover.

For design velocities less than 1 m/sec, seeding and mulching are needed to establish the vegetation. For design velocities over 1 m/sec, the waterway should be stabilized with sod, with the seeding protected by jute or excelsior matting,

Fig. 11-13. Diversion grassed waterways in residential zones. (Photo USDA Soil Conservation Service.)

or with seeding and mulching together with diversion of water until the vegetation is established.[16]

Channels can be designed using the common Mannings formula:

$$Q = A \frac{1}{n} R^{2/3} S^{1/2} = vA \qquad (11\text{-}13)$$

where:

Q = flow rate in the channel (m^3/sec)
A = flow cross-sectional area (m^2)
n = Mannings roughness coefficient for the channel
R = hydraulic radius defined as cross-sectional area divided by the wetted perimeter (m)
S = slope of the energy grade line (under steady state it equals the bottom slope of the channel)
v = mean velocity (m/sec)

Fig. 11-14. Grassed waterways in agricultural zones. (Photo USDA Soil Conservation Service.)

Trapezoidal

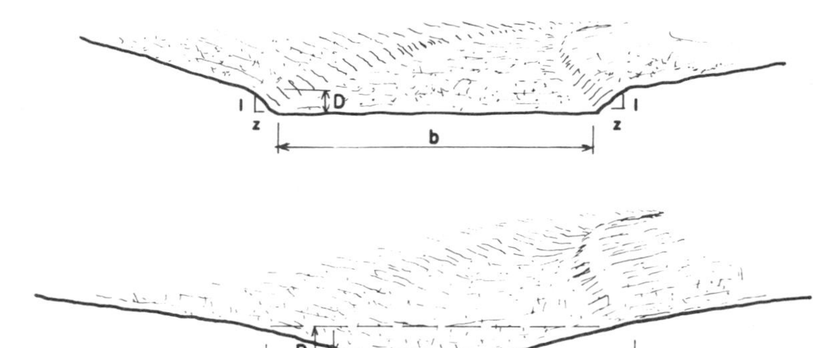

Parabolical

Fig. 11-15. Design sketch for grassed waterways. (From reference 16.)

TABLE 11-4. Maximum Permissible Design Velocities—Grassed Waterway Stabilization.[a]

Cover	Range of Channel Gradient (%)	Permissible Velocity (m/sec)
Vegetative[b]		
1. Tufcote, Midland, and Coastal Bermuda grass[c]	0–5.0	1.8
	5.1–10.0	1.5
	over 10	1.2
2. Reed canary grass, Kentucky 31 tall fescue, Kentucky bluegrass	0–5.0	1.5
	5.1–10.0	1.2
	over 10.0	0.9
3. Red fescue	0–5.0	0.75
4. Annuals[d]		
Small grain (rye, oats, barley millet)	0–50	0.75
Ryegrass		
Vegetative with stone center for base flow	As determined for the vegetative portion from above	

[a]After Ree and Palmer.[17]
[b]To be used only below stabilized or protected areas.
[c]Common Bermuda grass is a restricted noxious weed in Maryland.
[d]Annuals—use only as temporary protection until permanent vegetation is established.

TABLE 11-5. Classification of Vegetative Cover in Waterways Based on Degree of Flow Retardance by the Vegetation.[a]

Cover	Stand	Condition and Height	Retardance
Reed canary grass	Excellent	Tall (average 1 m)	A
Kentucky 31 tall fescue	Excellent	Tall (agerage 1 m)	
Tufcote, Midland, and Coastal Bermuda grass	Good	Tall (average 30 cm)	B
Reed canary grass	Good	Mowed (average 30–40 cm)	
Kentucky 31 tall fescue	Good	Unmowed (average 50 cm)	
Red fescue	Good	Unmowed (average 40 cm)	
Kentucky bluegrass	Good	Unmowed (average 40 cm)	
Redtop	Good	Average	
Kentucky bluegrass	Good	Headed (15–30 cm)	C
Red fescue	Good	Headed (15–30 cm)	
Tufcote, Midland, and Coastal Bermuda grass	Good	Mowed (average 15 cm)	
Redtop	Good	Headed (40–60 cm)	
Tufcote, Midland, and Coastal Bermuda grass	Good	Mowed (6 cm)	D
Red fescue	Good	Mowed (6 cm)	
Kentucky bluegrass	Good	Mowed (5–12 cm)	

[a]After Ree and Palmer.[17]

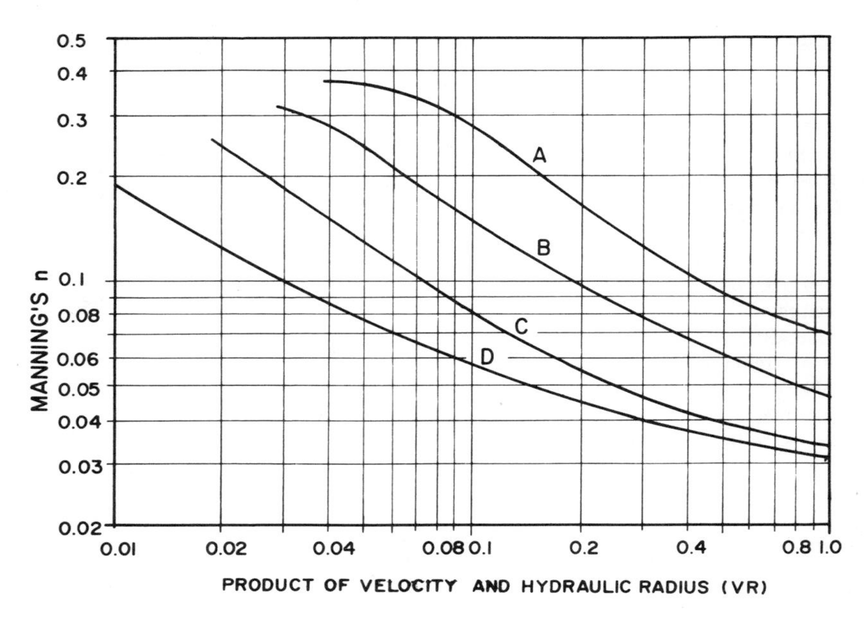

Fig. 11-16. Manning's n for grassed waterways related to velocity v (m/sec), hydraulic radius R (m), and vegetal retardance. (After reference 17.)

The roughness coefficient for grassed waterways depends on flow conditions and vegetative cover. For the four retardance categories defined in Table 11-5, the Manning roughness coefficient can be related to the product of velocity and hydraulic radius (Fig. 11-16).

For shallow flow conditions (width of flow is more than 12 times greater than depth), the hydraulic radius is approximated by depth of flow.

Example 11-4: Channel Design

Determine the nonerosive velocity and dimensions of a trapezoidal waterway given that runoff $Q = 2\ m^3/sec$, slope of channel is 5%, bank slope is 3:1 (Z in Equation 11-15 is 3), and the vegetative cover is a good stand of headed-out Kentucky bluegrass (C retardance curve-see Table 11-5).

From Table 11-4, the permissible velocity $v_{max} = 1.5\ m/sec$. The solution is a trial-and-error process since the Manning roughness coefficient depends on the product of velocity and hydraulic radius.

First determine approximate cross-sectional area:

$$A = \frac{Q}{v_{\max}} = \frac{2}{1.5} = 1.33\ \text{m}^2 \qquad (11\text{-}14)$$

Then try depth of flow $H = 0.3$ m

$$A = BH + ZH^2 \tag{11-15}$$

where B is bottom width, B is computed as:

$$1.33 = B \times 0.3 + 3 \times 0.3^2 \text{ or } B = 3.53 \text{ m}$$

Wetted perimeters (P) is then calculated

$$P = B + 2H\sqrt{Z^2 + 1} = 3.53 + 2 \times 0.3\sqrt{3^2 + 1} = 5.43 \text{ m} \tag{11-16}$$

and the hydraulic radius R is:

$$R = \frac{A}{P} = \frac{1.33}{5.43} = 0.24 \text{ m} \tag{11-17}$$

Estimate the Manning's n coefficient using the product of velocity and hydraulic radius

$$vR = 1.5 \times 0.24 = 0.33 \text{ m}^2/\text{sec}$$

From Fig. 11-16, for a value of $vR = 0.33$ m²/sec and C-retardance the Manning's factor $n = 0.043$.

Next compute v and compare with the initial assumption.

$$v = \frac{1}{n} R^{2/3} S^{1/2} = \frac{1}{0.043} \times 0.24^{2/3} \times 0.05^{1/2} = 2.03 \text{ m/sec} \tag{11-18}$$

which is not acceptable since $v > v_{max}$. Therefore try $H = 0.2$ m.

$$B = \frac{A - ZH^2}{H} = \frac{1.33 - 3 \times 0.2^2}{0.20} = 6.05 \text{ m}$$

$$P = 6.05 + 2 \times 0.20\sqrt{3^2 + 1} = 7.31 \text{ m}$$

$$R = \frac{A}{P} = \frac{1.33}{7.31} = 0.18 \text{ m}$$

$$vR = 1.5 \times 0.18 = 0.27 \text{ m}^2/\text{sec}$$

From Fig. 11-16, $n = 0.048$ and

$$v = \frac{1}{n} R^{2/3} S^{1/2} = \frac{1}{0.048} \times 0.18^{2/3} \times 0.05^{1/2} = 1.5 \text{ m/sec} \tag{11-19}$$

which is acceptable since the computed v agrees with the initial estimate.

Therefore, for $Q = 2.0$ m³/sec, design a grassed waterway about 6.1 m wide at the bottom, which will result in a flow depth of roughly 0.2 m.

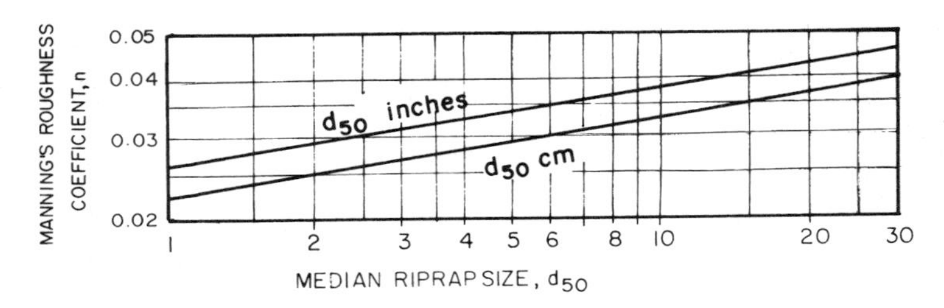

Fig. 11-17. Manning's n for riprap surfaces. (Redrawn from reference 16.)

Riprap. Riprap is a layer of loose rock or aggregate placed over an erodible soil surface. Riprap is used where soil conditions, water turbulence and velocity, expected vegetative cover, and groundwater conditions are such that soil may erode under design flow conditions. Riprap may be used at such places as storm drain outlets, channel banks and bottoms, roadside ditches, and drop structures. The roughness coefficient, n, used for determining flow on the constructed riprap surface, is given in Fig. 11-17.

Riprap is composed of a well-graded mixture of stones down to sizes of about 25 mm, such that 50% of the mixture by weight shall be larger than d_{50} size obtained from Fig. 11-17. The diameter of the largest stones should not exceed 1.5 times the d_{50} size.[16]

Sediment Traps. These are small temporary structures used at various points within, and at the periphery of, disturbed areas to detain runoff for a short period of time and trap heavier sediment particles.[16] As expected, their effect on retaining fine soil particles and pollutants associated with them is minimal.

Various types of sediment traps include sandbags and straw bales (Fig. 11-18), stone or prefabricated check dams, grade reduction drops (Fig. 11-19), log and pole structures, excavated ditches, and small pits. The volume of a sediment trap used to control erosion from surface mining areas is approximately 125 m^3/ha of drained disturbed area.[16]

Sediment should be removed and the trap restored to its original dimensions when the accumulated sediment reaches 50% of the design depth of the trap.

Sediment Detention Basins. Flood flow retardation basins have been in use for some time. However, recently severe problems associated with particulate pollutants have led to construction of small control reservoirs designed specifically to trap sediments. Such basins can be located at outlets from disturbed areas such as construction sites[18] and mining areas,[16] downstream from combined and storm sewer overflows, or at some point to prevent sediment-laden flow from reaching the receiving body of water.

A sediment basin is constructed on a waterway by a dam, which impounds

Fig. 11-18. Temporary sediment traps. Sod is laid immediately after excavation to prevent excessive channel erosion. (Photo authors.)

runoff (Fig. 11-20). The sediment control basin must be sized to accomplish two functions; namely, it must effectively remove a certain percentage of the suspended sediment, and it must provide sufficient storage capacity for the sediment removed.

A considerable amount of research has been devoted to sizing sedimentation basins according to influent particle size distribution. The simplest design methods are based on so-called "overflow rate," defined as a flow rate in the reservoir divided by the surface area of the reservoir. A particle is removed if

$$OR = \frac{Q}{A_s} \leqslant w \tag{11-20}$$

where:

OR = overflow rate (m/day)
Q = flow rate (m^3/day)
A_s = surface area of the reservoir (m^2)
w = particle settling velocity (m/day)

Fig. 11-19. Aluminum drop structure for erosion control. (Photo USDA Soil Conservation Service.)

Fig. 11-20. Sediment detention basin. (Photo USDA Soil Conservation Service.)

The particle settling velocity, w, is estimated from Equation 5-32.

Under the "ideal settling basin" assumption, all particles having settling velocities greater than *OR* will be trapped in the reservoir. The ideal settling basin has been defined as a basin in which settling takes place in the same manner as in a quiescent settling container of the same depth.[19]

The depth of the reservoir should be at least twice the sediment volume accumulated between the cleaning periods.

Example 11-5: Sedimentation Basin Design

Design a settling basin that would remove sediment with particle sizes up to 0.05 mm at a flow rate of $Q = 1\ m^3/sec$.

From the Stokes law (Equation 5-32), the critical settling velocity for a particle diameter $D = 0.05$ mm and specific gravity of the particle $\rho_s = 2.5$ is ($\nu = 10^{-6}\ m^2/sec$, $g = 9.81\ m/sec^2$)

$$w = \frac{gD^2}{18}(\rho_s - 1) = \frac{9.81 \times 0.00005^2}{18 \times 10^{-6}}(2.5 - 1)$$

$$= 0.003206 \text{ m/sec} \qquad (11\text{-}21)$$

The basin size can be calculated from Equation 11-5:

$$A_s = \frac{Q}{w} = \frac{1}{0.003406} = 293.6 \text{ m}^2$$

To account for nonideal settling, increase the computed area by 20%.[16] *Hence, the surface area of the basin should be*

$$A_{\text{adjusted}} = 1.2A_s = 1.2 \times 293.6 = 352.3 \text{ m}^2 \qquad (11\text{-}22)$$

The design of the settling basin based on overflow rate is simple and has been in use for some time for the design of sedimentation basins in sewage and wastewater treatment plants and for control of sediment problems from strip mines.[16] It may be difficult to use for the design of settling basins for stormwater when discharge is not uniform and the surface area of the reservoir varies considerably during the storm event.

A better method of sizing sedimentation basins is based on the investigations of Camp,[19] Brune,[20] Churchill,[21] and Chen.[22] Figure 11-21 shows trap efficiency curves for a settling reservoir related to the reciprocal of the overflow rate.[22] Note that overflow rate is based on outflow discharge, which is commonly less variable than inflow rate. The plot also indicates that the earlier models[20,21] overestimated trap efficiency for coarse materials (both methods are based on overall volumetric removal rates of sediments in reservoirs and not on particle size fraction removal).

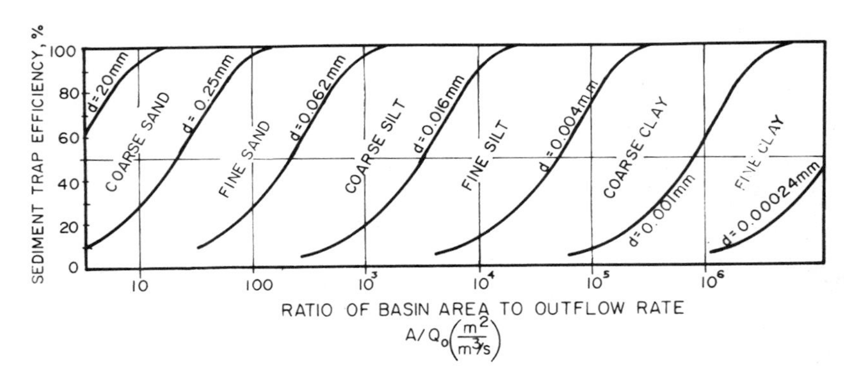

Fig. 11-21. Trap efficiency of sediment detention basins related to overflow rate. (From reference 22.)

Example 11-6: Efficiency of Sedimentation Basins

Estimate trap efficiency of a sedimentation reservoir with surface area of A_s = 352 m^2 and average outflow rate $Q = 1$ m^3/sec, using Fig. 11-21. The particle size of the sediment D is 0.05 mm.

From Fig. 11-21

$$\frac{A_s}{Q} = \frac{352}{1} = 352$$

and then for particle size D = 0.05 mm, the trap efficiency is roughly 55%. Complete removal is achieved for particle sizes >0.1 mm.

It can be seen that the "ideal settling basin" concept overestimates removal rates.

For nonuniform particle size distribution, trap efficiency must be estimated for each particle size fraction separately and integrated to obtain effluent particle size distribution and concentrations.

The entrapment of clay-sized particles usually is obtained by chemical flocculation or by use of basins with large dimensions. It is important to locate the sedimentation basins near the source of the sediment where clay particles move mostly in aggregates and are not separated by turbulent conditions of flow.

Several other models have been developed in recent years. Of note is a model by Ward et al.,[18] which is a computer model based on a plug flow concept. The model estimates effluent sediment concentrations and particle size distribution. It requires inflow hydrograph, inflow sediment graph, sediment particle size distribution, detention basin stage-area relationship, and detention basin stage-discharge relationship as inputs. Based on this information, the model routes the water and sediment through the basin.

Comparison of predicted results with measured data indicate that the model accurately predicts the sedimentation process in detention basins.

Methods Based on Increased Infiltration and Evaporation

Several benefits suggest that infiltration should be considered an effective means of controlling nonpoint pollution. Increased infiltration improves recharge into groundwater aquifers, infiltration in the form of irrigation improves agricultural yields and may provide nutrients to crops. Furthermore, infiltration may be 100% effective for the removal of fine soil particles and other particulates and even for dissolved solids.

The most popular method of increasing infiltration is through trenches or ponds. Trenches usually are shallow excavations with a very permeable material such as gravel or sand as bottom material, laid over a permeable substratum thereby allowing runoff water to percolate freely through the groundwater systems.[4,23] These systems may require a high degree of maintenance due to the tendency of the coarse bottom material to clog with fine particles, a problem that can be solved, in part, by routing the water over grass or vegetation filters prior to allowing its entrance into the infiltration area. Some flow equalization may be necessary; otherwise, land requirements may be excessive.

Infiltration ponds also are used commonly and are an attractive means of disposing of storm runoff; they also must have a highly permeable bottom. Infiltration can be enhanced and clogging minimized in four ways:[24]

1. The basin floor is at two levels, allowing presettling of fines in the deeper part.
2. Retention basins are regulated with overflow discharging to the recharge area.
3. Diffusion wells are punched through the basin bottom to provide additional recharge area.
4. The bottom is scarified or broken up, or a thin layer of it is removed periodically.

Fresno, California, has established recharge basins to collect storm water and induce infiltration for subsequent use as a groundwater recharge.[25] The basins were excavated up to 8 m deep and were planted with Bermuda grass so that they could serve as recreation areas. Nightingale[25] studied the basins to determine the mobility of selected toxic metals and to document toxicity potentials relative to the groundwater drinking supply. The results of this study show that Pb, Zn, and Cu have migrated to some degree up to a depth of 60 cm into the underlying soils, but they were effectively filtered out in the upper 5 cm, reaching background levels at 15 to 30 cm.

Other methods utilized for inducing infiltration include dry wells, wet and dry ponds, evaporation ponds, and special fill impoundments.

The use of wetlands for storage and treatment is becoming a promising means of controlling water pollution and flooding caused by urbanization. Marshes and swamps act as giant sponges, soaking storm water, removing a significant part of pollutants and nutrients, and releasing water gradually. For that reason, protection of existing wetlands is a very effective way of controlling nonpoint pollution.

Projects that use wetlands to control nonpont pollution and flood control from urbanizing areas have been undertaken in Boston, Massachusetts and Waysota, Minnesota.

Removal of Pollutants from Collection Systems

A portion of the pollutants moving in collection systems may become entrapped, settle out, or otherwise accumulate in the collection system during periods of slow flow velocities or they may resuspend during a storm or high flow event. These phenomena are typical for sewers and natural drainage systems. Large sediment deposits in estuaries and harbor areas are a continuous problem for navigation; faulty sediment deposits in sewers cause so-called "first-flush" pollutant loading during a storm even when pollutants in overflows from the sewer system reach high levels.

Dredging. Dredging (Fig. 11-22) has been used for a long time to remove excessive sediment deposits from waterways in zones of low flow velocities. Many navigation routes would cease to exist without dredging. Dredging also has been suggested as a remedial measure to control nonpoint pollution.

In spite of the fact that dredging removes potential pollutants from surface waters, it must be considered more as a source of pollution rather than as a remedial measure. When benthic materials are scooped or sucked up from the streambed by the dredging equipment, some of the material is resuspended or redissolved by the turbulence. If the material is dumped into water at another location, more resuspension can occur. If it is placed on land, part of the material may be lost in surface runoff, or may leach into groundwater.

Regardless of the adverse effects of dredging on water quality, this method may be one of the few control means left to deal with "after the fact" nonpoint pollution, that is, when large sediment and pollution deposits have occurred in receiving water bodies. All precautions should be taken to minimize the adverse effects of dredging.

Control of Sewer Systems. Sewer systems are either combined or separate. Combined sewers are typical for older urban areas and, basically, were brought about by the connection of sewage to existing storm drainage systems during the last century. In the combined systems, sewage and storm water are conveyed to

Fig. 11-22. Dredging is a cosmetic "after the fact" measure that may create significant local water pollution problems. (Photo authors.)

the disposal area by the same hydraulic conduit. To avoid large conduit sizes that would necessitate larger drainage areas, part of the combination of storm water and sewage is relieved to adjacent water bodies. The overflow occurs when the ratio of storm-water flow to that of sewage exceeds a certain allowable dilution ratio, which is between 4:1 and 8:1.

In the separate systems, sewage and storm water are conveyed by separate systems and, supposedly, do not mix. Separate systems are typical for newer sections of urban zones.

Water discharging from storm sewers of the separate systems and that portion of flow in combined sewers that overflows into receiving waters can be considered nonpoint pollution.

During a storm event, high flows have a tendency to scour contaminants that have settled previously in the collection system during the preceding low-flow and low-velocity period. These resuspended contaminants become a part of the "first flush," a more polluted portion of the storm-water overflow or discharge, containing materials from sewers and catch basins. The first-flush problem is multiplied in combined sewers, where the initial part of the overflow also contains untreated sewage stored in the sewers. Reported first-flush concentrations from combined sewer overflows were two to four times higher than those in the extended overflow and were substantially higher than average typical concentrations of untreated dry-weather sewage.[16]

Several references discuss and describe in detail the available control techniques for storm-water discharge from storm and combined sewer systems in urban areas.[27-29] Control efforts are aimed at containment of the first-flush volume. In this case a storage basin or the storage capacity of sewers is utilized in which the excess storm discharge or combined sewer overflow volume is contained and stored during wet periods and released—after some treatment—gradually into receiving waters (Fig. 11-23) or after the storm into a centralized treatment facility.

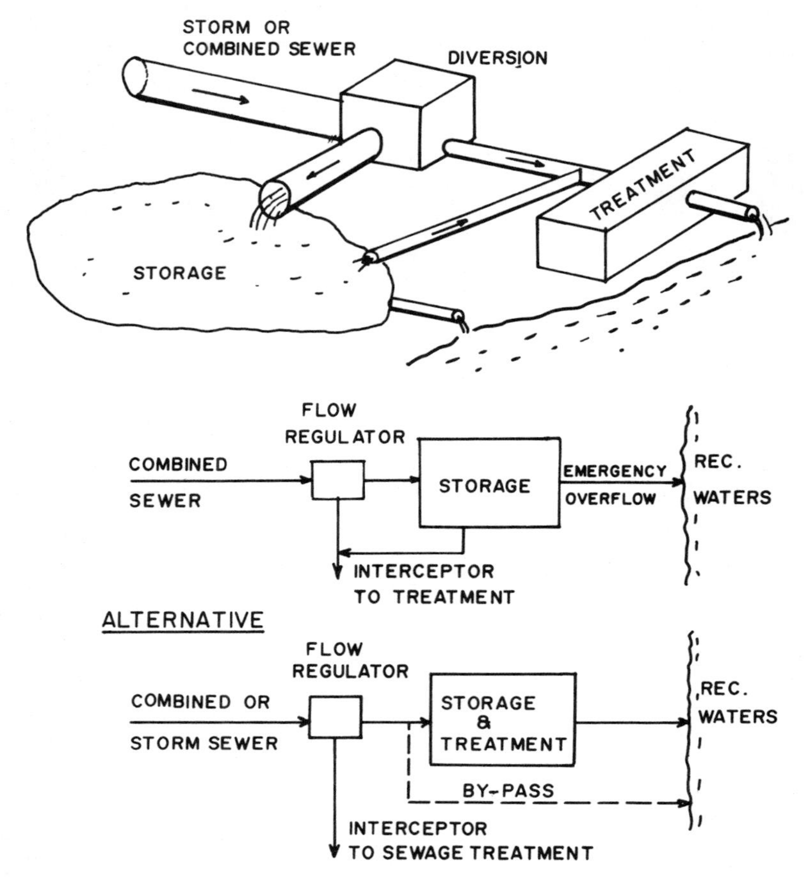

Fig. 11-23. Schema of off-line storage systems to control pollution from storm or combined sewer systems.

Fig. 11-24. Storm-water retention pond in Wauwatosa, Wisconsin. Note large sediment deposits in the middle of the basin that must be periodically dredged. (Photo authors.)

Storage Facilities. Storage facilities that have been employed for this purpose can be either in-line or off-line and include ponds or surface basins, underground tunnels, excess sewer storage, and underwater flexible or collapsible tanks.

Perhaps the most attractive off-line storage methods use ponds (Fig. 11-24), if land is available. Ponds can have other uses besides the storage, including infiltration and groundwater recharge, sedimentation, recreation, general aesthetics, or water for irrigation and fire protection. Vitale and Sprey[30] reported that an 85% decrease in BOD can be realized by capturing the first 8 to 25 mm of runoff. A very successful program has been conducted in Chippewa Falls, Wisconsin, where in 1969–70, 94% of the total combined sewer overflow was withheld from the receiving stream, stored, and subsequently pumped to a treatment plant.[4,31]

Storage of storm water also can be accomplished by using underground geologic formations and man-made tunnels. The system that has been developed for Chicago, Illinois, consists of 190 km of underground deep tunnels intercepting 640 overflow points in an approximately 1000-km^2 drainage area.[32] Such systems may be prohibitively expensive if designed for the sole purpose of storm-water pollution control (cost of the Chicago deep tunnels is in billions of dollars.)

Utilization of existing in-line storage for a drainage conduit has been accomplished effectively in Seattle, Washington, in that city's Computer Augmented Treatment and Disposal (CATAD) system.[33] The system is computer directed to maximize storage volume available in trunk and interceptor sewers. Other in-line storage systems have been built in Minneapolis-St. Paul, Minnesota, and Detroit, Michigan.[27]

The sizing of storage basins can best be accomplished by continuous simulation models since optimizing two variables is involved in the process. These variables are storage volume and the rate at which the basin is gradually emptied during the subsequent dry period. The latter variable determines the capacity and cost of treatment. From the models discussed in Chapter 9, STORM and SWMM have the capability of computing storage capabilities.

A statistical occurrence of a storm-water bypass or emergency overflow must be considered and incorporated into the design. Different storage volumes will be computed if bypass is acceptable once in 10 yr or once in 100 yr.

A simplified method of determining the storage capacity has been suggested as[27]

$$S = 10.0CiAt_1 - 3600Qt_2 \qquad (11\text{-}23)$$

where:

S = storage volume (m^3)
C = runoff coefficient
i = rainfall intensity (mm/hr)
A = service area (ha)
t_1 = rainfall duration (hr)
Q = treatment capacity required to treat the captured volume (m^3/sec)
t_2 = runoff duration (hr)

The first term in the equation is total runoff to be stored, the second term is the amount treated during the storm.

Catch Basins. Catch basins are installed at the point where storm water enters the sewer systems, and may be a significant source of pollution. Catch basins remove large particles and organic debris from the runoff, but they are ineffective in removing fine materials including most of the organic matter. The material removed by the catch basins decomposes over a period of time in the standing pools of water, and unless cleaned out, it is flushed into sewers during a subsequent storm. Average concentrations of pollutants in the catch basin liquid are similar to those for untreated sewage.[27] Thus, catch basins may contribute to high pollution loads to the first flush of a storm. To alleviate the problem, catch basins must be cleaned periodically or their usefulness is greatly impaired.

Other Methods. Other methods of collection control include control of deposition of pollutants in combined sewers during a dry period. Most of the methods are based on maintaining flow velocities above the critical range at which organic particles settle out. This can be accomplished by increasing sewer slope and/or choice of sewer pipes during design and construction. In the postconstruction period, hydraulic roughness and deposition of particles can be controlled by frequent cleaning and maintenance of sewers, flushing of sewers, sewer lining, and polymer addition.

Flushing increases the volume of water and scour forces resuspend the settled particles. Flushing as a pollution control measure is feasible only for combined systems, and the flush flow : dry-weather (sewage) flow ratio should not exceed the critical range at which overflow occurs.

Polymer-gelled slurry injections into sewage have resulted in significant reductions in hydraulic friction and subsequent temporary increases of in-line capacities. The reader is referred to references 27 and 28 for detailed discussions of inline sewer control and maintenance methods. The purpose of the in-line methods is to prevent deposition or to remove deposited solids from highly polluted sewage and convey them to the treatment facility so they do not become part of the first flush of overflow. These methods are applicable and make sense only for combined sewer overflows.

TREATMENT

In areas that are highly developed and have existing combined or separate sewer systems, a large-scale structural treatment method may be the only feasible way to eliminate pollutants from storm-water discharges. Almost all treatment facilities are attached to a sewer system; however, the authors know of at least one instance where a large treatment plant treats the entire flow from a river that has become heavily polluted by point and nonpoint sources from an industrial district (River Emsher in the Ruhr area of German Federal Republic).

Treatment is primarily required for combined sewer overflows; however, pollution from storm sewers may be of such proportions that in many cases (e.g., storm-water discharges into a lake) they also require treatment.

Many treatment alternatives are available for the control of storm-water pollution.[4,27,28] Treatment units are located mostly before the storm water reaches the receiving water body. However, several in-stream or in-lake treatment and remedial techniques have been used routinely to alleviate the temporary effects of nonpoint pollution as it is in cases of excessive nutrient inputs to lakes and reservoirs.

Since storm-water discharges and overflows are highly irregular, some form of in-line or off-line storage must be included to reduce the cost of the treatment facility. As stated before, there is a trade-off between storage-equalization volume and the treatment rate.

TABLE 11-6. Efficiencies of Various Storm-Water Treatment Processes.[4,28]

Process	Efficiency (%)				
	Susp. Solids	BOD_5	COD	Total P	Total Kjeldahl N
Physical-chemical					
Sedimentation					
without chemicals	20–60	30	34	20	38
with chemicals	68	68	45		
Swirl	40–60	25–60	50–90		
Screening					
microstrainers	50–95	10–50	35	20	30
drum screens	30–55	10–40	25	10	17
rotary screens	20–35	1–30	15	12	10
static screens	5–25	0–20	13	10	8
Dissolved air flotation	45–85	30–89	55	55	35
High-rate filtration[a]	50–80	20–55	40	50	21
Magnetic separation[b]	92–98	90–98	75		
Biological[c]					
Contact stabilization	75–95	70–90		50	50
Bio-disks	40–80	40–80	33		
Trickling filters	65–85	65–85			
Oxidation ponds	20–57	10–17		22–40	57
Aerated lagoons	92	91			
Facultative lagoons	50	50–90			

[a]With chemical addition.
[b]Only bench scale.
[c]Applicable only to combined sewer overflows.

The storage volume can be designed to intercept either first flush or the entire storm-water discharge with a certain probability that a bypass or emergency overflow will not occur for a specified period of time, or in other words, that a bypass will occur only at a specified (or greater) recurrence interval.

There are three types of treatment used for treating water and wastewaters, namely, physical, chemical, and biological. Many treatment systems use a combination of two or three types to achieve the best quality of storm-water discharge. The treatment efficiencies of some unit processes are given in Table 11-6.

Physical and Physical-Chemical Treatment Alternatives

Physical and physical-chemical treatment alternatives that have been demonstrated on either pilot or prototype scale to be applicable for treatment of storm-water runoff or combined sewer overflows include:[27] sedimentation with or without chemical clarification, swirl separation, screening, dissolved air flotation, high-rate filtration, and magnetic separation.

Plain Sedimentation. Sedimentation is used primarily and is effective for removal of coarse particulates up to silt size. Many storage facilities described previously double also as sedimentation and flood control basins. To remove particles finer than silt size, sedimentation must be accompanied by flocculation and coagulation of particles into aggregates or larger flocs by the addition of chemicals.

Swirl and Helical Concentrator/Regulator. Swirl and helical concentrator/regulators achieve flow and quality control of streams laden with particulate materials. The swirl unit has no moving parts but instead regulates flow by a central circulator-weir-spillway while simultaneously treating flow by swirl action and solid-liquid separation. Swirls are used mostly for control of combined sewer overflows. During high flow (overflow), the skimmed floatables from the top of the unit and the heavy solids from the bottom are diverted to an interceptor sewer for subsequent treatment while the high-volume supernatant is discharged.

Screening. Screening is used mainly as a pretreatment step to eliminate larger particles and debris. However, with the recent advent and development of microstrainers, screening is used as a primary treatment step. Several types of screening processes are available, including static screens, drum screens, rotary screens, and microstrainers. Microstrainers are effective even for very small (clay sized) particles.[28]

Dissolved Air Flotation. This method seems to be very effective for removal of particles with low specific mass such as oil and grease and light organic materials. Solids are removed from the flow by small bubbles of air that are released in the reaction tank after depressurization and rise to the surface carrying the solids.

High-rate Filtration. The use of multimedia filters (a combination of sand and anthracite) has been considered on a pilot scale as a possible treatment step. So far, only one city (Cleveland, Ohio) utilizes filtration for treatment of combined sewer overflows.[4,28]

Magnetic Separation. This is a new process that utilizes seeding of flow with magnetic iron oxide (magnetite) and adding of coagulants and polyelectrolytes to form a floc amenable to removal in a magnetic gradient. Removal was found more effective for fine particles because the magnetic force on the small particles

Fig. 11-25. Treatment plant for combined sewer overflows in Racine, Wisconsin. The treatment processes include screening and dissolved air flotation. (Photo authors.)

is many times greater than the gravitational force.[28] As of the 1970s, the process had not gone beyond the experimental and pilot scale stage.

Chemical additives often are used to enhance separation of particles from the liquid. Sedimentation and separation of fine particles is greatly enhanced by adding chemical coagulants such as lime, alumn, ferric chloride, and various polyelectrolytes. During the process of chemical clarification–which consists of chemical additions–flocculation, and particle separation (clarification or filtration), the flocs are formed from finer particles mainly by reduction of surface charge and/or by adsorption on surfaces provided by the added chemicals. Chemical addition is also effective for removal of P, toxic metals, and some organic colloids.

Often even the most simple treatment systems utilize a combination of several unit processes. Screening and dissolved air flotation have been used in Racine (Fig. 11-25) and Milwaukee, Wisconsin; swirl separation, sedimentation, and screening have been installed in Lancaster, Pennsylvania; screening and filtration with addition of chemicals are used in Cleveland, Ohio.

Biological Processes

Biological processes are essential for the removal of colloidal and dissolved organic materials. However, many other pollutants can be removed by biosorption, decay, and other processes taking place in the biological treatment unit.

Unfortunately, in many cases this highly effective treatment method can not be used successfully for the treatment of storm-water overflows for the following reasons:

1. Biological treatment processes are effective for flows where organics are concentrated while storm-water discharges exhibit lower BOD concentrations.
2. Biological treatment processes do not perform satisfactorily when waters of highly irregular flow and quality are treated.
3. It is difficult and expensive to keep the biomass alive between events.
4. Biomass is very sensitive to toxic materials present in urban storm-water runoff.

The processes used for the treatment primarily of combined sewer overflows include contact stabilization, high-rate trickling filters, rotating biological contactors (biodisks), and treatment lagoons. Best results are achieved if the storm-water or combined sewer overflow treatment unit (wet-weather treatment) is adjacent to a sewage treatment facility that provides biomass during the storm event.

Contact Stabilization. This method is a modified activated sludge process in which water is in contact with the biomass for a relatively short time. During the contact period the biomass adsorbs the organic pollutants, which are oxidized and stabilized in a separate tank. The biomass is usually provided from a nearby sewage treatment plant.

Trickling Filters. Filters treat water by a biological slime growing on the filter media. Water is distributed uniformly over the filter surface and allowed to pass through the filter. The filters must be supplied continuously with water-containing organic materials that, between the events or when the storage of storm water is exhausted, must be obtained from elsewhere.

Rotating Biological Contactors. These units operate on the same principle as biological trickling filters except that the biological growth is supported by rotating, large-diameter, closely spaced disks that are submerged partially and rotate at a relatively low speed. As with trickling filters, flow must be maintained continuously.

Contact stabilization, trickling filters, and biological contactors require secondary clarification to remove or return the solids from the effluent.

Lagoons. Lagoons are more promising for the treatment of storm water due to lesser difficulty of maintaing the biota alive. In fact, many storage ponds can be, and often are, used as biological lagoons.

The lagoons can be categorized according to type and amount of aeration they receive. Lagoons relying mostly on atmospheric aeration and photosynthesis for their oxygen supply are called *oxidation ponds*. Aerated lagoons rely on mechanical floating or mounted aerators for their oxygen supply. If the aeration is incomplete—that is, when a portion of the lagoon volume is anaerobic—the term *facultative lagoon* is used to describe the process.

Oxidation ponds and aerated lagoons are relatively shallow. Lagoons require some posttreatment to remove generated solids and algae (from oxidation ponds). Typical posttreatment methods are settling ponds, clarifiers, screening, or sand filtration.

Most of the storage ponds are actually oxidation ponds. However, these types of treatment have relatively low (20 to 30%) treatment efficiencies. Lagoons have been installed as a treatment process for combined sewer overflows in Springfield, Shelbyville, Mount Clements, and East Chicago (Illinois).

Disinfection. Adding chlorine to storm water does not only control bacterial viral pollution but also oxidizes organics and some reduced components such as iron, manganese and ammonia present in water. Since these processes must be accomplished first before chlorine acts as a disinfectant, up to 20 to 40 mg of Cl_2/liter may be required for effective bacteria kill. Even higher dosages of chlorine may be necessary if it is to be used to control viruses. Disinfection is always used in combination with other treatment unit processes.

Treatment of runoff and, hence, nonpoint pollution is very expensive and may be economically feasible only for concentrated flows such as combined sewer overflows, feedlot runoff, and runoff from some industrial operations. In all other nonpoint pollution problems, source control seems to be the most effective approach.

Since storm-water pollution events are random, the designing of treatment-storage facilities should be based on long-term simulations and statistical evaluation.[34,35]

REFERENCES

1. Anon. Nov., 1976. Guidelines for state and areawide water quality management program development. U.S. EPA, Washington, D.C., pp. 7-1 to 7-24.
2. Lienesch, W. C. 1977. Legal and institutional approaches to water quality management planning and implementation. U.S. EPA, Washington, D.C.
3. Stewart, B. A., Woolhiser, D. A., Wischmeier, W. H., Caro, J. H., and Frere, M. H. 1975. Control of water pollution from cropland, Volume I. U.S. EPA/600/2-75/026a, Washington, D.C.
4. Oberts, G. L. 1977. Water quality effects of potential urban best management practices. Tech. Bull. No. 97, Wisconsin Department of Natural Resources, Madison, Wisconsin.

5. Kao, D. T. Y., Barfield, B. J., and Lyons, A. E. 1975. On site sediment filtration using grass strips. Proceedings of the National Symposium on Urban Hydrology and Sediment Control, University of Kentucky, Lexington, July 28–31.
6. Moore, I. C., Madison, F. W., and Schneider, R. R. 1979. Estimating phosphorus loadings from livestock wastes: some Wisconsin results. In: R. C. Loehr (ed.). "Best Management Practices in Agriculture and Silviculture." Ann Arbor Science, Ann Arbor, Michigan, pp. 175–192.
7. Asmussen, L. E., White, A. W., Jr., Hauser, E. W., and Shendon, J. M. 1977. Reduction of 2,4-D load in surface runoff down a grassed waterway. *J. Env. Quality*, **6**(2):109–162.
8. Barfield, B. J., Kao, D. T. Y., and Toller, E. W. 1975. Analysis of the sediment filtering action of grasses media. Res. pap. No. 90, University of Kentucky Water Research Institute, Lexington.
9. Draper, D. W., Robinson, J. B., and Coote, D. R. 1979. Estimation and management of contribution by manure from livestock in the Ontario Great Lakes basin to the phosphorus loading of the Great Lakes. In: R. C. Loehr (ed.). "Best Management Practices in Agriculture and Silviculture." Ann Arbor Science, Ann Arbor, Michigan, pp. 159–174.
10. Wilson, L. G. 1967. Sediment removal from flood water. *Trans. ASAE*, **10**(1):35–37.
11. Toller, E. W., Barfield, B. J., Haan, C. T., and Kao, D. T. Y. 1977. Suspended sediment filtration capacity of simulated vegetation. *Trans. ASAE*, **19**(5):678–682.
12. Morison, W. R., and Simmons, L. R. 1977. Chemical and vegetative stabilization of soils. REC-EOC-76-13, Engineering and Research Center, Bureau of Reclamation, Denver, Colorado.
13. Forster, D. L. 1978. Economic impact of changing tillage practices in the Lake Erie Basin. A report to U.S. Corps of Engineers, Department of Agricultural Economics and Rural Sociology, The Ohio State University, Columbus, Ohio (August).
14. Mueller, D. H., Wendt, R. C., Daniel, T. C., and Madison, F. W. 1979. Effect of selected conservation tillage practices on the quality of runoff water. Paper presented at the III Annual Water Research Conference, Wisconson Section AWRA, Oshkosh, Wisconsin (February).
15. Thelen, E., Grover, W. C., Hoiberg, A. J., and Haigh, T. I. 1972. Investigation of porous pavements for urban runoff control. U.S. EPA 11034 DUY 03/72, Washington, D.C.
16. Anon. 1976. Erosion and sediment control, surface mining in Eastern U.S.A. U.S. EPA/625/3-76/006, Washington, D.C.
17. Ree, W. O., and Palmer, V. J. 1949. Flow of water in channels protected by vegetative lining. U.S. Soil Conservative Bull. No. 967.
18. Ward, A. J., Haan, C. T., and Barfield, B. J. 1977. Simulation of sediment detection basins. Res. Rep. No. 103, Water Resources Research Institute, University of Kentucky, Lexington.
19. Camp, T. R. 1945. Sedimentation and the design of settling tanks. *Trans. ASCE*, **71**:445–486.

20. Brune, G. M. 1953. Trap efficiency of reservoirs. *Trans. AGU.* **34**(3):407–418.
21. Churchill, M. A. 1948. Discussion of analysis and use of reservoir sedimentation data, by Gottschalk, L. C., Proceedings Federal Interagency Sedimentation Conference, Washington, D.C.
22. Chen, C. 1975. Design of sediment retention basins. Proc. Nat. Sym. on Urban Hydrl. and Sedim. Control, University of Kentucky, Lexington. July 28–31.
23. Stem, G. L. 1975. An evaluation of practical experience in storm water management. Proc. Nat. Sym. on Urban Hydrl. and Sedim. Control, University of Kentucky, Lexingon. July 28–31.
24. Aronson, D. A., and Seaburn, G. E. 1974. Appraisal of the operating efficiency of recharge basins on Long Island, N.Y., in 1969. U.S. Geol. Survey Water Supply Pap., 2001-D.
25. Nightingale, H. I. 1975. Lead, zinc, and copper in soils of urban storm-runoff detention basins. *J. AWWA*, **67**(8):443–446.
26. Rex Chainbelt, Inc. 1972. Screen-flotation treatment of combined sewer systems. U.S. EPA-11020 FDC01/72, Washington, D.C.
27. Lager, J. A., and Smith, W. G. 1974. Urban storm water management and technology. U.S. EPA/670/2-74/040, Washington, D.C.
28. Lager, J. A., Smith, W. G., Lynard, W. G., Finn, R. M., and Finnemore, E. J. 1977. Urban storm water management and technology: update and user's guide. U.S. EPA/600/8-77/015, Washington, D.C.
29. Wanielista, M. P. 1978. "Storm Water Management: Quantity and Quality." Ann Arbor Science, Ann Arbor, Michigan.
30. Vitale, A. M., and Sprey, P. M. 1974. Total urban water pollution loads: the impact of storm water. U.S. Council on Environmental Quality.
31. Anon. 1973. Storage and treatment of combined sewer overflows. U.S. EPA/430/9-73/016, Washington, D.C.
32. Cook County Flood Control Coordinating Committee. 1972. Development of flood control and pollution control plan for the Chicagoland area. Summary of technical reports. Metropolitan Sanitation District, Greater Chicago, Illinois.
33. Leiser, C. R. 1974. Computer management of a combined sewer system. Municipal Metropolitan Seattle, U.S. EPA/670/2-74/022, Washington, D.C.
34. Small, M. J., and DiToro, D. M. 1979. Stormwater treatment systems. *J. Env. Eng. Div., ASCE*, **105**:557–569.
35. DiToro, D. M., and Small, M. J. 1979. Stormwater interception and storage. *J. Env. Eng. Div., ASCE*, **105**:43–54.

12

Planning for nonpoint pollution control

INTRODUCTION

Water resources planning is not a new process. Plans have been developed in the past for many water-related projects, on both large and small scales throughout the United States and the world. Until recently, watershed planning efforts traditionally employed simple means to achieve simple or narrow goals, with the empahsis on those goals that could be directly measured in monetary terms. Such past planning activities included irrigation, flood control, soil and water conservation, energy, and navigation.

After the passage of the Public Law 92-500, water quality planning became a prominent part of environmental development and protection activities. Sections 208 and 303 of the act[1] provide directives to states to set up management programs for implementation of water quality goals specified by the Act. All states have subsequently developed a continuing planning process within the boundaries of the State Water Quality Management Plan that is aimed at: (1) determining effluent limitations needed to meet the water quality goals, and (2) development of area-wide water quality management programs.

In the traditional approach to water pollution control that still persists in local planning, the main objective is to clean up raw wastewater and comply with the effluent standards imposed by a responsible regulation agency. Due to the high cost involved in pollution control and often nonmonetary benefits associated with improved water quality, the planning process is more a series of conflicts between diversified groups rather than a smooth continuous process. The priorities in such a process may frequently be related to the "visibility" of the pollution input and its consequences. Nonvisible and often unrecorded nonpoint pollution has thus an obvious handicap when pollution control priorities are considered. Nevertheless, it has been realized that cleanup limited to point sources only may not lead to the desired water quality goals. Thus, identification, control, and management of nonpoint pollution must be included in the area-wide pollution abatement programs set up by the Clean Water Act.

Planning Unit

Planning related to water pollution could conceivably be carried out within the boundaries of various geographical units ranging from a farm to a state, including areas defined by governmental jurisdiction, economic linkages, or watershed boundaries. None of these are perfect, but there are many advantages in selecting the watershed as a planning unit as compared to other alternatives (e.g., a county). As noted throughout this book, nonpoint pollution is a hydrological problem where rainfall-runoff simulation and analyses are of paramount importance, and the integrated pollution abatement efforts should result in an acceptable water quality in the receiving water bodies.

Urban drainage often involves problems that cross watershed boundaries but strong watershed implications are involved if sewerage systems discharge into the surface water systems. Groundwater divides and groundwater flow do not often coincide with surface water system. Therefore, if the planning process also deals with groundwater problems, both intrawatershed and interwatershed considerations must be included. Changes in land-use and transportation requirements are ordinarily not controlled by watershed boundaries, but can have major effects on watershed problems. The land use and transportation may significantly affect the hydrologic and hydraulic characteristics of the watershed and significantly affect the pollution loadings.

Finally, the related physical problems of a watershed tend to create strong community interest among the residents of a watershed. Many citizen river or lake restoration committees and action groups have been formed and have assisted in planning activities. The existence of community interest, around which citizen participation in the planning process is organized, is one of the most important factors contributing to the success of such a process. It may therefore be concluded that the watershed is a logical area unit to be selected for resource planning purposes, provided that the relationships existing between the watershed and the surrounding region are recognized.

The foregoing discussion implies that the term watershed may have two meanings. Defined in a strictly physical sense, a watershed is simply a geographic area of overland drainage contributing surface runoff to the flow of a particular stream or watercourse at a given point. Under this definition, the terms watershed and drainage basin are synonymous. To include the planning concept, the definition of the term watershed may be expanded to include the phrase: whose natural and man-made features are so interrelated and mutually interdependent as to create a significant community interest among its residents.

WATER QUALITY PLANNING PROCESS

General Description of the Planning Process

A water quality planning process scheme does not deviate significantly from the classical definition of planning. The planning process generally involves four related steps:[2]

1. Identifying the objectives of design.
2. Translating these objectives into design criteria.
3. Using the criteria to devise plans for the development of specific alternatives that fulfill the criteria in the highest degree.
4. Evaluating the consequences of the plans that have been developed.

In the most advanced countries, the water quality objectives are set nationally or statewide by legislative action and by a responsible regulatory agency. In the United States, the general water quality objectives were mandated by the Water Pollution Control Act. These objectives are translated state by state into water quality standards and effluent limitations. Thus, the designated planning agencies are required to include these objectives and standards in their plans, to focus on the development of the alternatives and optimum solutions that meet these objectives, and to evaluate their consequences (steps 3 and 4).

The national guidelines[3] for the area-wide water pollution control planning activities recommend that the planning process proceed along the following lines:

1. Identify the problems in meeting the water quality goals established by the state and federal water pollution control agencies.
2. Identify all constraints and priorities pertaining to planning.
3. Identify all possible solutions to the problem.

4. Develop alternative plans to meet the statutory requirements.
5. Analyze the alternative plans for technologic and economic feasibility.
6. Select an area-wide plan.
7. Seek approval for the plan.
8. Periodically update the plan.

According to the U.S. EPA guidelines, the State Water Quality Planning activities have two components: (1) technical, and (2) management segments.

The technical planning portion of the planning process involves identifying the priority water quality problems of the area, recognizing any constraints in dealing with the problem, and developing alternatives to achieve the water quality goals. The purpose of technical planning is to develop a coordinate water quality management strategy to meet water quality goals. The strategy may be a combination of

1. Municipal wastewater treatment systems.
2. Industrial pretreatment or treatment.
3. Residual waste management.
4. Urban storm-water management.
5. Nonpoint source management.

Management planning, which concerns the selection of a management agency or agencies and the development of appropriate institutional arrangements for plan implementation, should be conducted concurrently and in coordination with technical planning. Management planning should identify water quality management problems, analyze the capability of existing agencies, and make arrangements to cary out the regulatory and management requirements of Section 208 of the Water Pollution Act. Institutional problems, lack of authority, or lack of financial capacity for meeting Section 208 requirements should also be identified.

The planning activities specified by the Water Pollution Control Act, or Clean Water Act, should be classified as "pollution control planning." Overall water quality planning, as shown in Fig. 12-1, should include such problems as water resources development, groundwater aquifer protection, and also such water quality control means as in-stream and reservoir aeration, low flow augmentation, controlled effluent storage during critical flow conditions and release during higher flows, interbasin and intrabasin transport of water or wastewater, and others.

Many water resources projects are required to include in their planning process elaboration and evaluation of the Environmental Impact Statement. This statement is required by the National Environmental Policy Act passed by

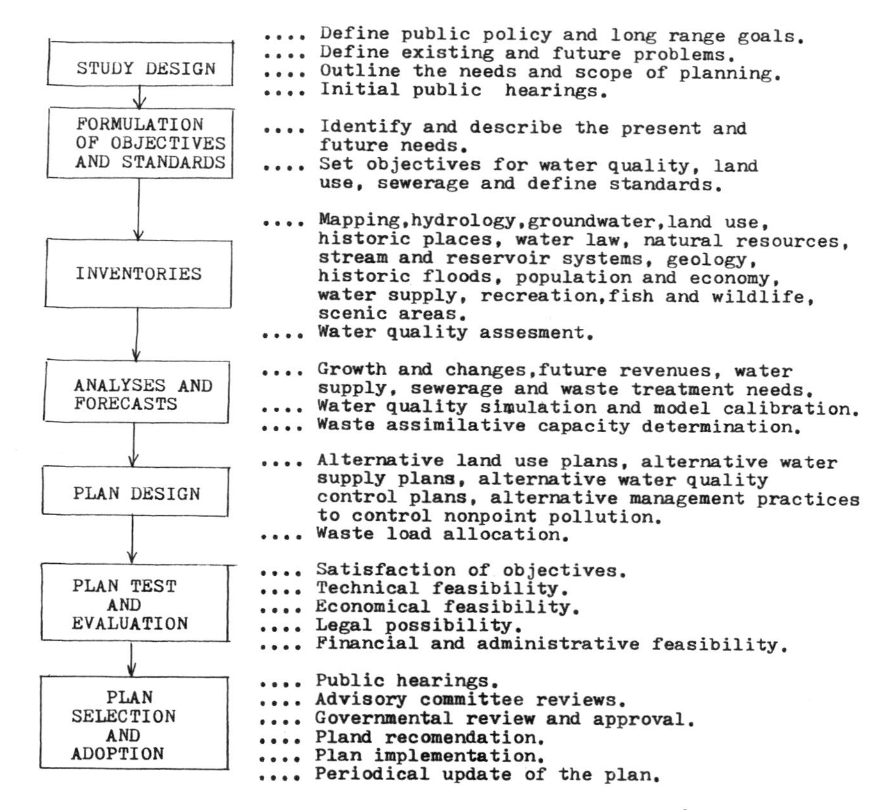

Fig. 12-1. Water quality planning process.[4]

Congress in 1969.[5] The preparation of the Environmental Impact Statement is mandatory with regard to planning, construction, and operating phases of any federally supported project. The Environmental Impact Statement should include at least:

1. A description of the proposed action, including information and technical data necessary to allow adequate assessment.
2. The probable impact of the proposed action on the environment.
3. Any adverse environmental effects that cannot be avoided.
4. Alternatives to the proposed action, their environmental impact, and the reasons they were discarded as a primary choice.
5. The relationship between local short-term uses of the environment and the maintenance and enhancement of long-term production.

6. Any reversible and irreversible commitments of resources that would be involved should the planned action be implemented.

At various stages of the planning process, public participation is necessary. Adverse public reaction may sometimes slow down or even cease the planning of some economically and technically feasible projects that may not be acceptable due to their water quality and environmental impact or lack of public acceptance.

Planning Objectives

The prime objective of planning is often stated as maximization of national welfare.[2] This definition is one nobody would object to, but in a particular case of water quality planning, it may be difficult to translate it into the design or planning criteria. Concepts such as maximization of national income or per capita production may not work for water quality problems. Since water quality planning is always a part of the overall water resources development, the planning must recognize the multipurpose use of water and reconcile competitive uses.

The basic objective in the formulation of the water quality plans is to provide the best use, or combination of uses, of water and related economic and land resources to meet all forseeable short-term and long-term needs. Water quality in a watershed is a multipurpose system, and it is subjected to various constraints. In order to find the best alternative of uses, the objectives must be optimized. In translating verbal objectives (e.g., water quality suitable for recreation) into design criteria, the planner or the regulatory agency must choose with care and discrimination the objectives that will be maximized and those that will be treated as constraints. For example, a farmer would treat profits from his production of agricultural products as an objective and maintenance of water quality as a constraint. A planning agency would, on the other hand, consider water quality as the primary objective and maintenance of a viable farming economy as a constraint.

The general water quality objectives can be divided into the following three categories, and all three will be considered in the water quality planning process:[6]

1. Development. The impact of water quality on development, as well as the adverse effects of economic and urban developments on water quality, have been recognized for many years. Thus planning activities should lead to controlled water use without destruction of water quality by over-use and uncontrolled depletion and degradation of water resources. But it must be realized that national economic development, and development of each region within the country, is essential to the maintenance of national strength and the achieve-

ment of satisfactory levels of living. Water quality and water resources development with related land-use development and control is essential for:

- Adequate supplies of surface and groundwater of suitable quality for domestic, municipal, industrial and agricultural uses.
- Land stabilization measures to prevent soil loss and associated pollution loadings and to protect land and beaches for beneficial purposes.
- Outdoor recreational and fish and wildlife opportunities where these can be provided or enhanced by development works.
- Water navigation, which provides transportation services to the nation's economy.
- Soil and groundwater protection to provide an adequate supply of moisture for growing crops and use of groundwater.
- Other related uses, which include use of water for hydroelectric power production, flood control and flood zone management, and watershed protection.

2. Protection and conservation are becoming the most important objective in developed and industrialized countries. The rehabilitation and protection of quality of water resources to insure availability for their best use should be considered in all plans. Besides protection of water resources for human use, considerations also should be given to maintenance and conservation of areas of natural beauty, wild areas of rivers and lakes, beaches, green spaces, and water balance and quality in forests and parks. Recreation and fish and wildlife protection are the primary objectives for water quality planning according to PL 92-500.

3. The well-being of people should be the overriding determinant in considering the best use of water and related land resources. Care should be taken to avoid resource use and development for the benefit of a few or the disadvantage of many. In particular, policy requirements and guides established by Congress and aimed at assuring that the use of natural resources, including water resources, safeguard the interests of all our people should be observed.

The water quality goals specified by the Water Pollution Control Act have been discussed and presented in Chapters 1 and 2. These goals basically provide for surface water quality suitable for recreation and fish and wildlife protection by 1983, and provide advanced treatment technology to all point sources and best management practices to nonpoint pollution sources.

Planning Criteria

The planning process and its final product are usually evaluated according to three basic criteria:[3]

1. Cost Effectiveness. The Environmental Protection Agency has defined cost effectiveness analysis as a systematic comparison of alternatives that reliably meet given goals and objectives to identify the solution that minimizes total cost to society over time. The total cost of the proposed actions to society that should be minimized include:

- resources cost
- economic cost
- social cost
- environmental cost

In general, cost-effectiveness analysis must include a comparison of the total cost associated with the planned actions and benefits that result if the plan is realized.

2. Implementation Feasibility. The feasibility of the implementation must be evaluated based on the availability of:

- adequate legal authority
- adequate financial capacity
- practicability
- managerial capacity
- public accountability

3. Public Acceptance. The success of planning public water quality control projects depends on its acceptance by affected units of government and the general population. Therefore, the acceptability of the plan to the public and elected officials in the planning area should also be regarded as a basic planning criterion.

Assessment of the Present Situation

Data and Inventories. Reliable basic planning and engineering data collected on a uniform watershed-wide basis are essential to the formulation of design criteria and workable development plans. The crucial need for factual information in the planning process should be evident, since no sound and realistic forecasts can be made or alternative courses of action selected without knowledge of the historical and current state of the system being planned. The realistic formulation of comprehensive watershed development plans related to water quality requires that factual data must be developed on the quantity and quality of surface and groundwater, precipitation, hydraulic and self-purification characteristics of stream channel systems, wastewater sources, water uses, soil characteristics, land uses, economic activity, population, recreation facilities, fish and wildlife habitat, unique natural areas, historic sites, water supply and sewerage systems and other public utilities, and water law.

Not only are the collected data necessary as background and fundamental information for forecasting, but they also represent a so-called zero action alternative serving for comparative purposes when the best alternative solution is sought.

The data base will include historical data, present monitoring, and surveys designed to fulfill the information gaps. The success and objectivity of the planning report then depends on:

1. How well the collected data reflect the real situation.
2. How well the data were interpreted to assess the present situation and future needs.
3. How the alternatives leading to the water quality objectives are evaluated with respect to technical and economical feasibility and resources capability.
4. How much the public perception of water quality needs agrees with the report objectives and goals.

The data required to assess specific problems can be obtained through a four-step approach:[7,8]

1. Identification of the types of data required to assess each problem.
2. Collection and analysis of the pertinent existing data.
3. If the data appear insufficient, adequate assessment of the problem by the design of a reconnaissance study. This study will determine:
   a. The major factors that actually affect the surface waters quality problem, point source inputs, and nonpoint loadings.
   b. The anticipated range of values for each factor.
   c. The amount of data needed to analyze each cause-effect effectively.
4. Design and conduction of field and laboratory data-collection programs at an intensity appropriate to adequately assess the problem.

The actual amount of data necessary to properly assess the water quality problem depends on various factors and will obviously differ for each study. The amount of data necessary to assess the problem depends on the following factors:

1. Number and complexity of specific water quality problems and objectives (e.g., dissolved oxygen, nutrient load, algae nuisance, reservoir or lake sediment control).
2. Planning level (local, area-wide).
3. Planning horizon (short-term forecast, long-range planning).

4. Techniques and manpower available to collect, evaluate, and analyze the data.
5. The factors involved in the planning process.
6. Modeling level of the system.

Critical Period on Which Water Quality Projections Are Based. Water quality is a highly variable phenomenon. In the classical approach to water quality planning, most of the assumptions and proposed actions are based on a so-called critical dry-weather period, with a defined duration and with a probability of occurrence once in x number of years. A typical example of such a planning period is the 7 days duration–10 years expectancy low flow characteristic that is used by many states for determining the waste assimilative capacity of receiving water bodies.

Such an approach is not applicable to water quality planning that includes contributions from nonpoint sources. By definition and its physical nature, the 7 days–10 years expectancy flow can occur only after a prolonged period of severe drought—an assumption that a priori excludes any significant surface runoff contributions, which are a primary source of nonpoint pollution.

The worst water quality may not even occur during conditions of extremely low flow. In many urban centers and areas receiving significant nonpoint pollution, the lowest D.O. concentrations, and the worst water quality conditions in general, are observed when surface runoff from a large storm enters the receiving water body after a prolonged period of low flow. The sudden change of flow plus resuspension of deposited organics usually produce the "worst case scenario."

From the foregoing discussion it follows that it is very difficult to define a critical period for planning analyses of nonpoint pollution effects on surface water quality. The best method of analysis involves long-term flow and quality simulations or a consequent analysis of meteorological and hydrological time series.

It is also quite likely that the period of surveys and data collection will not reflect the design critical rain or low-flow characteristics. A mathematical model calibrated and verified by the collected data is then used to project the water quality conditions into the design conditions.

Analyses and Use of Models. The data collected during the first stages of the planning process, although analyzed and summarized, will not provide enough background information to allow immediate selection of the optimal solution to the problem. The collected data, however, should provide enough information for adequate evaluation of the future trends of most important inputs affecting water quality (land-use development, economic and population growth, industrial development and potential) and also past and future trends in surface water quality if no action to control water quality is taken. The data base should also provide information on ranges of water quality system parameters such as statistical evaluation of flow and meteorological data typical for the region of interest, geology and hydrogeology, topography, and demography.

To find a solution to a particular water quality problem, planners need a model of the system that would reflect the major interrelationships between the inputs and outputs or cause-effect relationship of the water quality system. The role of physical models in water quality planning is limited mostly to hydraulic and waste treatability studies, but mathematical and sometimes socio-economical models have found wide application and have been used extensively, especially for nonpoint pollution problems. The level of models may vary, and in water quality planning, the models can be categorized basically into four levels:

1. Simple unit loading-complete mixed systems. These systems relate the pollution generation to units of land areas under a given use or to loading per capita and industrial production units. These pollution loadings are then divided by the river flow to obtain the pollution concentration.

2. Simple statistical regression models. The pollution loadings and concentration in receiving waters sometimes can be correlated to various causative factors which include population, geology, land-use distribution within the watershed, meteorological factors, and flow, among others.

3. Deterministic process-oriented models. In many cases it is possible to describe the processes participating in pollution generation and transport mathematically using analytical or empirical or semiempirical formulas or submodels. These subunits can then be combined into a larger pollution transport model. Steady-state models assume that there is no change of the input parameters over the period of simulation, while more complex dynamic models enable simulation of the system under dynamic input conditions. The structure and types of nonpoint loading models have been discussed in Chapter 9.

4. Stochastic process models. In many cases both system inputs and system parameters are statistical quantities with certain probabilities of occurrence. In some instances, statistical stochastic models can be designed that would provide the transformation of statistical input variables and system parameters into a range of occurrence of the outputs.

5. Socio-economic–water quality models may apparently represent an ideal tool for planners. These models would relate the water quality response to various socio-economical inputs. However, it must be understood that while the state-of-the-art of pollution generation and transport models has been brought to an adequate modeling level and accuracy, socio-economic–water quality models still require large amounts of background information for development

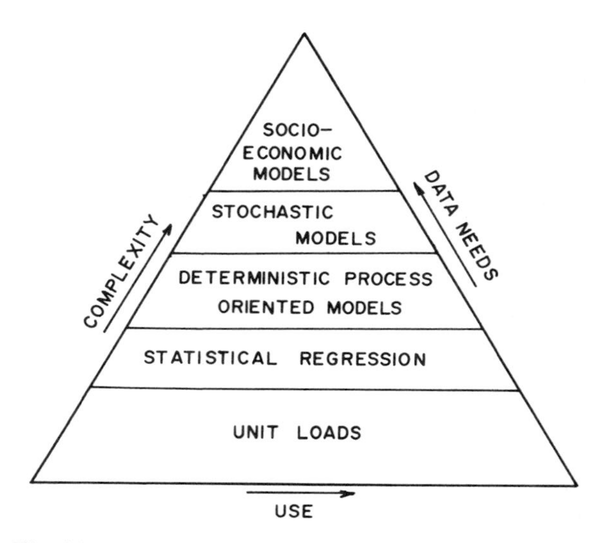

Fig. 12-2. Complexity, data needs, and use of models.

and implementation. These models are more simplistic in detail and description of subprocesses and more complex in scope.

Each model requires a different amount and character of data for its development. Unit loadings, statistical, and socio-economical models are usually developed from past historical data. These data should be collected (or selected) randomly in such a way that each independent variable would have about the same weight. The accuracy of the model is then directly related to the amount of data used for its development. Figure 12-2 depicts the complexity, data needs, and use of various kinds of models.

In concept, mathematical modeling of water quality management and control systems—especially those dealing with nonpoint pollution—provides a powerful tool to resource planners and managers, pollution control officers, and governmental decision-makers. But it should be understood that a model is just a tool and not an answer. Two extremes have occasionally evolved with the use of models in water quality planning. The reviewers and decision-makers have sometimes distrusted complex models and tended to accept the "rule-of-thumb" concepts due to lack of knowledge on model structure, general misinformation, and even failure of some models to adequately represent the system. On the other hand, some planners tend to accept uncritically the results of complex models that were designed without proper and thorough calibration and verification or without proper understanding of the particular system involved (e.g., models applicable to small homogenous watersheds may fail if applied to large basins). Both extremes may have serious consequences. "Rule-of-thumb" and low-level concepts may represent a cheap and easy-to-understand method of planning, but generally they require large safety factors. The second extreme limits the planners only to the selections incorporated in the model, and sometimes due to lack of verification of the model and its results, it may lead to erroneous conclusions.

Statistical regression models should not be used to model conditions outside of the ranges of data used for their design.

Future Population and Economic Developments. The term planning implies regulating or controlling future conditions and evaluating the future impact of the proposed actions. Planning is not forecasting, it is an intentional scientific and engineering process of proposing one or a series of actions that will undoubtedly modify the present situation. This concept of planning, especially in the water quality control area, is in apparent conflict with the classical concept of free-market regulated development. But, theoretically, the planning process should provide optimum conditions for future developments of all branches of the economy.

An essential part of all planning efforts is a forecast of future population and economic growth. This may seem a futile and impossible task especially to those who try to invest in the stock market. Engineering forecasting is a simple evaluation of trends and their projection in the future. It is not applicable to short-range economic forecasts such as market or local economy growth and developments. Engineering forecasting provides actually a kind of safety factor by which capacities are increased to accommodate future growths.

Simple forecasting is based on past trends and developments. The forecasts that do not include the future feedback of the proposed actions can be considered "no action" or "zero effect planning." If, as a result of the planning process, a project or a series of projects is to be realized in the future, its feedback on the forecasted population and economy levels must be evaluated.

As an example, consider an action that would regulate and control nonpoint pollution in an urbanizing area. If unregulated trends were to be used for the forecast of the future levels, the population and, therefore, the pollution would increase linearly or exponentially, resulting in an urban sprawl. If the delegated planning agency proposed zoning and strict pollution control ordinances, the urban sprawl could be significantly slowed down and both population and pollution could be reduced.

Forecasting of future conditions must be undertaken for almost every project. This is not limited to public pollution control works but also applies to the private sector. It is a very risky business. Overestimation and high safety factors will lead to oversizing and unnecessary loss of capital; underestimation and low safety factors may lead to future failures and breakdowns of the pollution control systems.

The pollution and economic forecast is an essential but probably the least reliable part of the planning process. The forecast accuracy obviously decreases with the forecast period. From the data published by Berthouex and Polkowski,[9] the average forecasting error in percent equals the length of the forecasting period in years. The errors are defined as the deviation of the forecasted population from the actual population related to the magnitude of the population. Hence, if the forecasted population growth for a city is from 800,000 to 1,000,000 and the actual growth is to 900,000, the error of the estimate is 100 (1,000,000 - 900,000)/900,000 = 11%.

The forecast error and its sign are affected by the economical conditions of the period at which the forecasts were made. One can expect that the forecasts made in the period of 1930 to 1940 were pessimistic, while those made in the postwar boom were optimistic, but generally it is impossible to know whether the forecast will be optimistic or pessimistic. Unpredictable events in areas such as food production, energy, and public health may upset all past trends, and usually these events are excluded from the forecast assumptions.

The forecasts for large areas are also more reliable than those for small communities. A local change of ordinances, discovery or exhaustion of a local economic resource can often upset all past population trends. This is especially true for areas relying on tourism and mining.

The change of population can occur only in three ways: (1) increase by birth, (2) decrease by deaths, and (3) change by migration.

The methods of population forecasting are graphical and mathematical. An excellent review of the population forecasting methods was presented by McJunkin.[10] In any case, the population forecast must start with a graphical visual plot of past census years versus time. The most simple method utilizes engineering judgment in extrapolating the population curve in the future. Since most engineers hesitate to use any other extrapolation but a straight line, there are several graphical plotting methods that force the data to fit a straight line.

The arithmetic progression method utilizes a plain coordinate paper and the planner either can extend the population curve from the last census according to most recent trends or try to fit a line through the last census and the rest of the data using a visual or a least-squares method. The arithmetic progression method can be mathematically expressed as

$$\frac{dP}{dt} = K_a \tag{12-1}$$

which integrates to

$$P_t = P_0 + K_a(t - t_0) \tag{12-2}$$

where:

P_t = the forecasted population
P_0 = the last census
t = the future time level
t_0 = the time of the last census
K_a = the slope of the population line

The geometric progression method utilizes a semilogarithmic paper. This method indicates that there is an increasing population growth rate that mathematically can be expressed as

$$\frac{dP}{dt} = K_g P \tag{12-3}$$

which integrates to

$$P_t = P_0 e^{K_g(t-t_0)} = P_0 10^{K_g(t-t_0)/2.301} \tag{12-4}$$

and can be plotted as log P vs t.

Other methods include a logistic S-shape curve, geometrical comparison with other large cities, correlation to other factors, ratio comparison with larger cities and others. It is interesting to note that by serveral comparisons the simple arithmetic progression curve provides results with errors of a lesser magnitude than most of the other methods.[9]

Selection of Optimal Solution

The calibrated and verified model (the term model is applied here to all levels of modeling, including both simple formulas and large complex computer models) is used to generate response of water quality to various levels of pollution loadings generated by future developments and nonpoint and point source abatement measures. The first step of the planning process is usually to predict future levels of pollution if no action to control pollution is taken and base it on predicted trends of future population, economy growth, and land-use development. These pollution levels represent then a "zero action" or "no cost" alternative.

A number of alternative solutions may be available to achieve the water quality goals. The planners must use their engineering and scientific creativity and broad knowledge of the problem to immediately eliminate those that are not feasible or realistic and create or develop alternatives that at first approach would be attractive, technically and economically possible, and acceptable to public and responsible governmental agencies.

These alternatives will result in changed inputs to water quality systems and, thus, also corresponding predicted water quality levels. Those alternatives that will not meet the design criteria based on the water quality objectives should be rejected. Those that will meet the criteria represent feasible policy or feasible solution matrix. Traditionally the optimal alternative would be selected on a "minimum cost and/or maximal benefit (profit)" basis using one of the available optimization procedures. Typical optimization procedures applicable to water resources and water quality planning include Lagrangean polynomial, linear and nonlinear programing, and integer programing.

Linear programing, which is probably the most common optimization procedure, can be summarized as follows:[11]

$$\text{maximize or minimize } Z = c_1x_1 + c_2x_2 + \cdots + c_nx_n \tag{12-5}$$

subject to

$$\begin{aligned} a_{11}x_1 + a_{12}x_2 + \cdots + a_{1n}x_n &\leqslant b_1 \\ a_{21}x_1 + a_{22}x_2 + \cdots + a_{2n}x_n &\leqslant b_2 \\ \vdots \qquad \vdots \qquad\qquad \vdots \quad &\quad \vdots \\ a_{m1}x_1 + a_{m2}x_2 + \cdots + a_{mn}x_n &\leqslant b_m \end{aligned}$$

for all $x_i \geqslant 0$ and $x_i \leqslant x_{imax}$ where:

x_i = decision variables
Z = production function
a_{ji}, b_i, c_j = constants

The constants b_j represent the constraints, both physical and economic, imposed on the system. These constraints may include the water quality standards, equipment and space limitations as well as financial resources available for realization of the proposed action. The symbol Z usually represents the overall cost of the best management practices and treatment for various units of the system; c_i represents the unit cost of the best management practices or treatment per unit of pollutant removed.

The a_{ij} are coefficients transforming a unit of a decision variable into the output that is subject to a constraint, and b_j is then the appropriate constraint (e.g, a water quality standard) imposed on the system.

The linear programing problems can be easily solved by digital computers. Many computer program packages are available for the solution of optimization problems.

SELECTION OF THE BEST ALTERNATIVE

Economic Impact of Water Quality and Pollution Abatement

For water quality as for other desirable goods and services, the best alternative in a planning process is the one that would maximize the difference between the occurring social benefits and its associated social cost.[11] The best alternative presented in the simplest example would most likely be the one that yields the maximum profit (benefits minus cost).

This approach may not be feasible and directly applicable to water quality problems. Although the cost associated with water quality control can usually be quantified, the evaluation of benefits attributed to water quality improvements is difficult. Many times, the benefits cannot be expressed in monetary terms or even be given a value.

The economic value of water cannot be directly measured using traditional approaches because it does not have a direct market value and users usually do not have a choice as to where the water will be delivered and at what price. Under normal circumstances, the price of water is determined by the cost of alternatives necessary to make it suitable for intended beneficial uses. Only in times and places of shortage does a disproportion between the supply and demand cause an increase in the price of water, the purpose being to limit demand. For example, the price of water rose significantly during the 1976-1977 drought period in some areas of California.

Deteriorating and unacceptable water quality has the same impact as water shortage. Again, the economic value of quality cannot be defined as a free-market value based on demand and supply. It is difficult to imagine that a water user would have the option to buy water at different levels of quality and the demand for a given water quality would then determine its value. It must be concluded, fortunately or unfortunately, that the interactions of market forces do not perform satisfactorily in the case of water quality control. At present, the upstream polluter is not directly affected by the deterioration of water quality caused by his discharge unless his intake of water is located downstream from the discharge. This is especially true for nonpoint pollution sources where a farmer or a developer does not even anticipate the effects of his pollution since he often relies on other water sources.

Economists refer to water pollution as a "technological external diseconomics."[12] This may be defined as a situation where a particular action, for example, wastewater discharge, produces economic changes such as higher cost or less valuable products and the cost is transferred from the decision unit—polluter—to a managerially independent unit—downstream user. The transfer is by a technical or physical linkage (e.g, by a watercourse) between the production processes and not by a market transaction.

Benefit-Cost Analysis

In the absence of market forces to determine the value of water quality, the value must be determined by institutional and/or legislative means. The value of water quality determined in this manner is based on an overall comparison of alternatives with and without pollution or adverse water quality. Knowing what the overall economic impact of "good" water conditions would be as compared to "bad" or existing water quality conditions, the delegated authorities and/or public must evaluate the benefits, both monetarily and socially, associated with improved water quality conditions. However, the decision as to what level of water quality is acceptable is not based only on benefits.

The economic impact of water quality on pollution depends on such factors as water availability, water use, and the public perception of water quality. Even minimal contamination may have a devastating impact on a water supply source for a community. On the other hand, water used for navigation can tolerate much higher levels of pollution with only marginal economic impact. Similarly, biodegradable organic pollution can have a significant effect on fish and shellfish production but only a marginal or even beneficial effect on water used for irrigation.

Furthermore, based on this concept, the increase of benefits with higher levels of quality is at a reduced rate while the rate of cost increase is higher (Fig. 12-3). The assumption of decreasing benefits implies that as water quality continues to improve, the public's perceived benefit from further increments of improvements decreases. For example, changing the quality of a water body from septic to aerobic conditions may require relatively low capital expenditures but demonstrate dramatic economic improvement by establishing a viable fishery. Obviously, because of this low cost and easily demonstrable benefit, the eco-

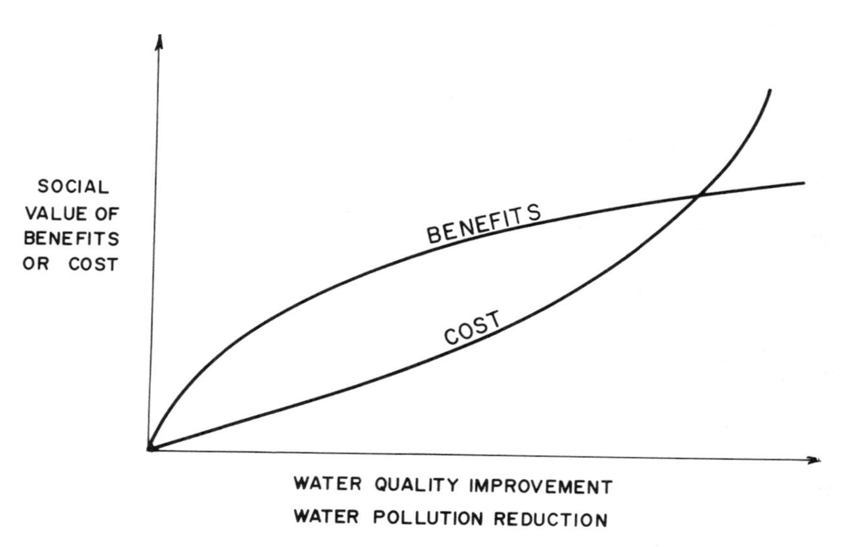

Fig. 12-3. Typical benefit and cost function.[11]

nomic efficiency of this action is high. On the other hand, control of widespread chemicals or reduction of nutrient levels to those required to control eutrophication, require expensive control measures but their apparent efficiency may be low.

The economic value of water quality improvement also depends on the economic state of the area being considered. Inasmuch as the cost of pollution control is high, it is obvious that basic needs such as food and housing will have a higher priority in developing nations than in highly developed ones. Thus, the value of water quality enhancement will increase with an increase in the standard of living. When a society is able to spend more time in the pursuit of recreation and nature's wonders, the economic value of improving water quality and its concomitant benefits increase dramatically. In addition, the advanced society will be more aware of water quality and the potential adverse public health effects of poor water quality. Water quality enhancement in advanced societies may have the following economic impacts:[13]

1. Increased income by providing jobs associated with an improved recreation industry.
2. Reduced cost of treatment for various uses of water.
3. Increased value of property located near the water.
4. Expanded economic development of the region.
5. Decreased damages to health and property associated with adverse water quality conditions.
6. Increased income to commercial and sport fisheries due to increased productivity.

Obviously, society must dictate the level of water quality that will yield desirable economic impact and the level of pollution that is acceptable. The economic value of good water quality can usually be correlated with the economic standard of the society and, to a lesser degree, with such factors as population density, water availability, and natural beauty of the region.

Definition and Classification of Benefits. Benefits are generally defined as increases or gains, net of associated or induced cost, in the value of goods and sevices that result from conditions with the project as compared with conditions without the project. Benefits include tangibles and intangibles and may be also characterized as primary and secondary.

Tangible benefits are those that can be expressed in monetary terms based on or derived from actual or simulated market prices for the product or services, or, in the absence of such measures of benefits, the cost of the alternative means that would most likely be utilized to provide equivalent products and services.

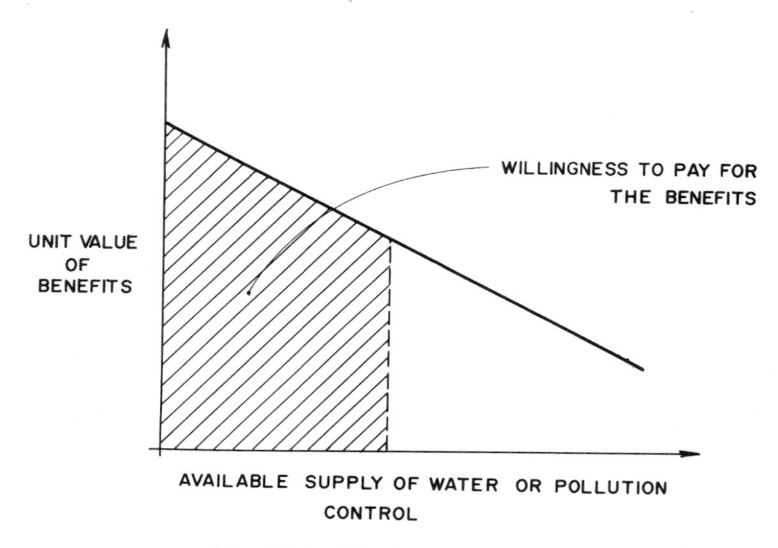

Fig. 12-4. The demand curve.

The latter should also include the cost of interest, taxes, insurance, and other cost elements that would actually be incurred by such alternative means.

Intangible benefits are those benefits that, although recognized as having real value in satisfying human needs or desires, are not fully measurable in monetary terms.

The price of drainage and irrigation water, increased commercial fishing, increased agricultural yields due to reduced soil loss, or savings on treatment that result from improved water quality are examples of tangible water pollution control benefits, while aesthetic improvement is a typical intangible benefit.

Primary benefits include the value of goods and services directly resulting from the project, less the cost associated with the realization of the benefits.

Secondary benefits cover the increase in the value of goods and services that indirectly result from the project under conditions expected with the project as compared to those without the project. This increase will be net of any nonproject costs that need to be incurred to realize these benefits. Such benefits usually include increased economic and commercial activity connected to recreation possibilities and improved aesthetics, better transportation, and other water quality related economic gains.

The nature of the benefits, including those derived from improved water quality, generally decreases with their available supply (Fig. 12-4). The market price of fish will go down if there is an abundance of fish. Similary, but not identically, the first 50% sediment or BOD load reduction to a stream may bring more benefits than the removal of the remaining load.

Types of Primary Benefits Related to Water Pollution Control. These benefits include:

1. Water quality control benefits—the net contribution and improvements of public health and safety, recreational opportunities, enjoyment of water for all purposes that are related to improved water quality. The net benefits are usually evaluated in terms of cost associated with avoidance of adverse effects in the absence of the proposed water quality control measures. Such cost can be expressed in damages to existing structures, increased cost of treating and pumping water from larger distances, impairment of health and welfare, damage to fish and wildlife, siltation by sediment, impairment of recreational opportunities, decreased farm production, and others. These benefits can be both tangible and intangible.

2. Water supply benefits can be measured as the cost water consumers will be willing to pay for improved water quality. The consumer group includes domestic, municipal, industrial, and agricultural users. In the absence of adequate information on the willingness to pay for water, the benefits can be approximated as a value of water paid by domestic, municipal, industrial, and agricultural users situated in the region.

3. Recreation benefits can be expressed as the net increase in the quantity and quality of boating, swimming, camping, picnicking, and recreation activities. It has been stated that recreation benefits account for about 75% of all water quality benefits.[14] To avoid classification of these benefits as intangible, several methods have been developed to assign a monetary value to the recreation demand. These methods of quantifying the magnitude of a recreation experience are based on the Recreation Day or Visit.[14] The Recreation Day unit can be defined as "a presence of one person on lands or waters, generally recognized as providing outdoor recreation, for continuous, intermittent, or simultaneous periods of time totaling twelve hours."[15] The methods of estimating the monetary value of the Recreation Day range in accuracy and include (a) evaluation of gross expenditures by recreationist, (b) estimation of market value of fish, (c) evaluation of income increase of the population in the area due to the increased recreation, (d) determination of property value increases in the area, (e) estimation of willingness to pay for recreation by interviewing a sample of the population, and (f) estimation of the demand function for a given site.

Figure 12-5 and Table 12-1 show examples of the Recreation Day values estimated for 1975 levels. The value of the Recreation Day unit varies with recreation use, distance of the area from population centers, the availability and attractiveness of existing and potential alternative recreational opportunities, leisure time, recreational habits, and living standards.

The intangible benefits of recreation include aesthetics, preservation of areas

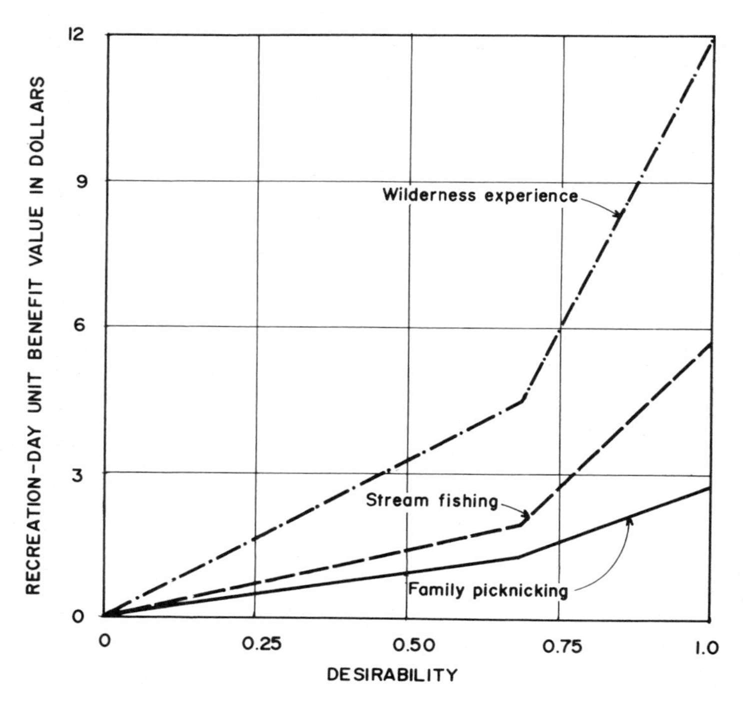

Fig. 12-5. Recreation Unit Benefit Function.[16]

of unique natural beauty, and maintenance of scenic, historical, and scientific interests.

4. Fish and wildlife benefits can be estimated as a value of net increase in commercial aspects of fishing, water game hunting, and resource preservation. Recreational benefits connected with fish and wildlife are enumerated in the previous category.

5. Watershed preservation benefits are related to land stabilization, erosion control, and nutrient lost from nonpoint sources. These benefits can be evaluated from improved farming yields, reduction in the loss of net income, or loss in land values due to erosion of soils and its adverse effect on water quality.

6. Irrigation benefits of pollution can be evaluated from increased farming yields and the value of land which results from reduction in salinity of irrigation water and general improvement of water quality and irrigation return flow.

7. Other water pollution control benefits—since the water pollution control facilities are often a part of multipurpose water resource development, other benefits associated with pollution control measures in a basin may include the following: (a) navigation benefits derived from improved transportation and less damage due to water quality, (b) flood control benefits derived from lesser flood

TABLE 12-1. Recreational Unit Benefits.[16]

	Unit Benefit Values, B (dollars per user day)									
	"Average" Conditions $\overline{B}$					"Uniquely Beautiful" Conditions, B_{max}				
Activity (1)	Low (2)	25% (3)	50% (4)	75% (5)	High (6)	Low (7)	25% (8)	50% (9)	75% (10)	High (11)
Power boating	2.0	3.0	4.0	5.0	15.0	4.0	5.0	6.5	10.0	30.0
Sailing	1.0	2.3	4.7	6.0	20.0	3.0	4.5	8.0	14.0	40.0
Stream fishing	0.5	1.6	2.2	4.0	15.0	1.5	3.3	5.7	7.3	30.0
Lake fishing	0.5	1.8	2.3	3.5	20.0	1.5	4.0	5.3	9.9	30.0
Waterfowl hunting	2.5	3.3	4.8	5.0	20.0	4.0	5.0	7.0	10.0	35.0
Small game hunting	1.5	2.0	2.9	4.0	15.0	3.0	3.5	4.8	7.7	25.0
Deer hunting	2.5	3.8	6.0	10.0	25.0	5.0	7.9	11.0	18.0	50.0
Tent camping	0.5	1.5	2.9	4.5	10.0	2.0	3.8	5.0	8.5	20.0
Recreation vehicle camping	0.5	2.6	3.5	4.8	10.0	1.0	4.6	7.0	9.0	15.0
Family picnics	0.5	0.7	1.0	2.3	7.0	1.0	1.8	2.5	4.5	14.0
Nature observation and photography	0.5	0.8	1.5	2.8	5.0	1.0	3.5	4.6	8.0	15.0
Swimming	0.5	0.8	1.0	2.0	5.0	1.0	1.5	2.0	4.3	10.0
White-water canoeing	2.0	2.7	4.0	5.0	15.0	5.0	5.0	8.0	15.0	30.0
Wilderness experience	1.0	2.3	4.5	8.0	20.0	4.0	6.0	12.0	15.0	40.0
Hiking	0.5	0.8	1.8	2.9	10.0	2.0	2.0	3.8	6.0	20.0
Saltwater sport fishing	2.0	3.0	4.6	8.0	13.8	5.0	6.0	8.8	15.0	23.8
Swimming in ocean and related beach activities	0.5	0.8	1.4	2.0	5.0	1.0	1.8	3.0	4.9	15.0

damage due to water storage in low-flow augmentation reservoirs, (c) electric power benefits derived from electricity produced at some multipurpose reservoirs and impoundments, and (d) property value benefits derived from increased value of land and real estate located near watercourses and increased tax revenues.

Example 12-1: Nonpoint Pollution Benefits

A lake is anticipating water quality problems of accelerated eutrophication caused by excessive nutrient loadings from nonpoint and point sources located in its drainage areas. Define the primary benefits associated with the Best Management Practices implemented in the watershed that would result in reduced nutrient inputs and, subsequently, improved water quality.

Water from the lake is used for water supply and for recreation. There is a potential for a substantial controlled lake development.

A. Water Quality Control Benefits

Cost of weed cutting equipment used to control excessive algal growths during the summer productive period or other associated costs necessary to control the symptoms of eutrophication.

B. Water Supply Benefits

Cost of chemicals and treatment used to reduce taste and odor problems caused by algae.

C. Recreation Benefits

Increased frequency of visits due to improved water quality, resulting in an increase in Recreation Days.

D. Watershed Preservation Benefits

Cost of fertilizer that would be necessary to replace that lost to watercourses.

Improved agricultural yields due to implemented soil conservation practices.

E. Property Value Benefits

Increased value of shoreline property due to improved water quality and derived taxation income.

Increased desirability of the area for residential and recreational development and related increased land value and taxation.

Definition and Classification of Cost. Cost of the Best Management Practices and pollution control facilities can be generally classified into two basic categories—capital and operating costs. The capital or installation cost includes the value of goods and services necessary for the establishment of the project such as initial project construction; land, easement, right-of-ways, and water rights; capital outlays to relocate facilities or prevent damages; and all other expenditures for investigations and surveys, designing, planning, and constructing a project after its authorization.

Operation, maintenance, and replacement (OMR) costs include the value of goods and services needed to operate a constructed project and make repairs and replacement necessary to maintain the project in sound operating condition during its economic life.

The investment for capital cost is spent during a relatively short time period during the initial phase of the project. The operation, maintenance, and replacement expenses are spent usually in yearly installments throughout the lifetime of the project, such as maintenance, plant supplies, chemicals and energy, depreciation, real estate taxes, insurance, etc.

For planning purposes, the cost for both capital and OMR can be expressed in the form of a cost function.[17] The cost function is defined as any relationship that provides an estimate of control cost in terms of the values of more basic variables, called cost determinants or factors. The cost function is obviously a statistical variable that will change year to year as the market value of the technology involved in the pollution control varies. Continuous cost functions

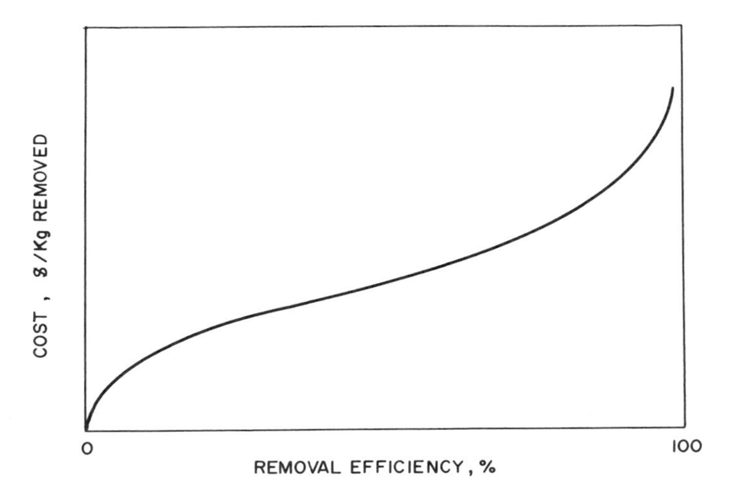

Fig. 12-6. Typical cost function related to percent removal.

are curves or multidimensional surfaces that predict costs over a complete interval of determinant values. Cost functions can be expressed as matrices, graphs, or statistical equations. A simple one-parameter cost function is shown in Fig. 12-6. In mathematical terms the cost function is

$$C = f(X_1, X_2, \ldots X_n) \tag{12-6}$$

where C is the cost based on determinants $X_1, X_2, \ldots X_n$.

The most important determinants of cost function for a given pollutant are capacity (size) and efficiency of reduction. Other characteristics include composition of the waste, temperature, labor wage levels, topography, transportation, and climatic conditions in addition to the above two primary determinants.

Most of the cost functions utilize the following single determinant formula

$$C = aP^b \tag{12-7}$$

where:

C = the cost unit
P = the magnitude of the primary determinant, usually capacity in cubic meters per day or tonnes of pollutants removed per day
a, b = satistical coefficients

Other alternatives to which the above function can be applied are population, storage volume for basins, surface area for reservoirs, or tonnes of sediment restriction or removal.

TABLE 12-2. Control Measures Most Suitable for Reduction and Management of Nonpoint Pollution from Urban Areas.[18]

Control Measures	Potential Pollutant Source Areas							
	Rain	Rooftops	Street Surfaces	Parking Lots	Landscaped Areas	Vacant Land	Construction Sites	Other (Industrial and Solid Waste Runoff)
Street cleaning			X	X				
Leaf removal				X	X			
Alternate deicing methods			X	X				
Control grass types					X	X		
Repair streets			X	X				
Control fertilizer, pesticides, etc.					X	X		
Control use of vacant land						X		X
Control litter			X			X		
Control dog litter			X		X	X		
Control direct discharge of pollutants to storm drains		X						X
Eliminate cross connections with sanitary sewers								X
Clean catch basins			X				X	
Clean storm sewers and drainage channels			X				X	
Prevent roof drainage from directly entering storm sewer		X						
Direct runoff away from contaminated areas							X	X
Retain runoff from contaminated areas							X	X
Regrade disturbed areas						X	X	X
Control erosion at construction sites							X	
Store and treat runoff		X	X	X	X	X	X	X

The aggregate cost of improving surface water quality is then a summation of costs of all individual units involved in the pollution abatement program subject to optimization.

In considering the cost of implementing a plan, it is necessary to distinguish between outlays and economic cost. In some instances the cash outlays (buildings, equipment, structure) adequately represent costs, but sometimes a resource is used with no cash outlay. The cost in such a case is the value of the resource in its best alternative use—its "opportunity cost."[3]

Cost of Urban Best Management Practices. As it should be expected, there is no single Best Management Practice that would optimally control pollution from urban areas. For example, street sweeping practices will control only areas accessible to street sweeping equipment. Furthermore, present street sweeping techniques are relatively ineffective for removing finer fractions of street refuse. One of the most important steps in selecting the BMPs for an urban area is to identify the problems and source areas most responsible for the urban water pollution loads.

Table 12-2 lists the various control measures that have been considered for controlling urban runoff. The contributions of pollutants from individual sources must be expressed as fractions of the total load, and the best combination of practices should be determined, for example, by employing linear programing techniques (Equation 12-5).

Figure 12-7 and Table 12-3 show the cost of removing street pollutants by sweeping. The cleaning cost in San Jose, California, averaged in 1976–1977

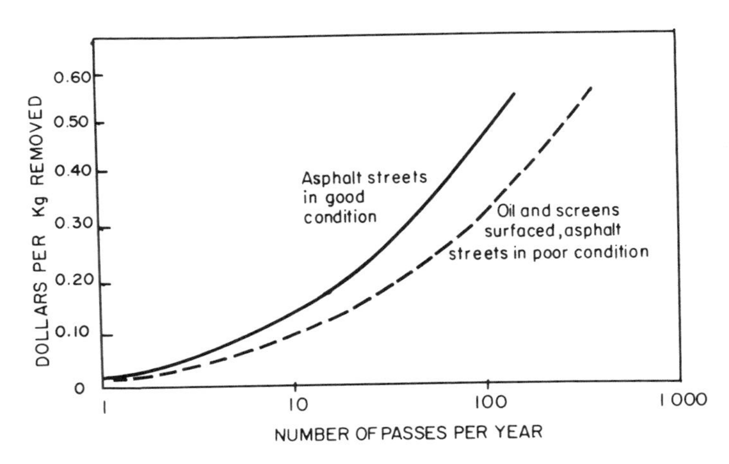

Fig. 12-7. Cost to remove a unit weight of street pollutants as a function of the number of passes per year. Data and the function are for the San Francisco Bay area.[19]

TABLE 12-3. Cost to Remove Various Street Surface Contaminants by Street Cleaning Programs in the San Francisco Bay Area (1977–1978 price levels).[19]

Parameter	Cost ($/kg removed) Minimum	Maximum	Average
Total solids	0.06	0.37	0.11
Suspended solids	0.11	0.73	0.46
COD	0.51	3.08	2.20
BOD_5	1.00	6.40	4.40
PO_4-P	396	3,525	2,026
Total Kjeldahl nitrogen	24	220	139
Pb	31	205	84
Zn	114	639	306
Cr	128	793	529
Cu	62	418	286
Ca	13,436	127,750	74,890

approximately $9/curb-km ($14/curb-mile) and required 0.56 man-hour/curb-km cleaned.

Heaney et al.[20] analyzed a nation-wide cost of wet-weather pollution treatment of combined sewer overflows. The cost function parameters related either to flow rate (m^3/sec) or storage volume (m^3) are given in Table 12-4.

The cost of porous pavement is slightly higher than that of conventional asphalt pavement.[21]

There is now a consensus among the urban environmental planners that preserving natural drainage has beneficial water quality and aesthetical impact on urban residential zones. If natural drainage techniques are developed, the cost of the natural drainage system is approximately $1500/ha less than that of a conventional system.[22]

Cost of Erosion Control from Hazardous Areas. The areas producing the highest sediment and pollutant loadings—primarily construction and developing sites—may have an impact on water quality of the receiving water that may exceed the combined effect of all other lands within the watershed. The control practices to prevent or minimize excessive soil loss should include a combination of the following practices:[22]

1. Reduction of the area and duration of soil exposure.
2. Protection of the soil by mulch and vegetative cover.
3. Reduction of the rate and volume of runoff by increasing of infiltration

TABLE 12-4. Cost Function for Treatment of Urban Runoff and Combined Sewer Overflows.[20]

		Annual Cost[a] ($/year)					
		Amortized Capital $CA = lT^m$ or lS^m		Operation and Maintenance $OM = pT^q$		Total $TC = sT^z$ or sS^z	
Device	Control Alternative	l	m	p	q	s	z
Primary	Swirl concentrator	17,600	0.70	4,400	0.70	22,000	0.70
	Microstrainer	79,100	0.76	19,800	0.76	98,900	0.76
	Dissolved air flot.	113,000	0.84	28,200	0.84	141,000	0.84
	Sedimentation	291,000	0.70	72,800	0.70	364,000	0.70
Secondary	Contact stabilization	280,000	0.85	69,900	0.85	350,000	0.85
	Physical-chemical	466,000	0.85	116,000	0.85	582,000	0.85
Storage	High density (37.0/ha)	–	–	–	–	51,000	1.00
	Low density (12.4/ha)	–	–	–	–	10,200	1.00
	Parking lot	–	–	–	–	10,200	1.00
	Rooftop	–	–	–	–	10,200	1.00

[a]Includes land cost, chlorination, sludge handling, engineering, and contingencies; interest rate $i = 8\%$, amortization period $n = 20$ y. No economics of scale beyond $T = 378.5$ m^3/day.
T = treatment rate (m^3/sec).
S = storage capacity or volume (m^3).

rates and surface storage and by planned diversion of excess runoff from surrounding areas.

4. Diminishing of runoff velocity with planned engineering works.
5. Protection and modification of drainage ways to withstand concentrated runoff from adjacent impervious areas.
6. Trapping of as much sediment as possible in temporary or permanent sediment traps and basins.

These principles can be implemented by a variety of simply constructed facilities. The cost of some alternatives is given in Table 12-5.

Cost of Agricultural Best Management Practices. There is a large variety of potential BMPs that are applicable for controlling nonpoint pollution from rural (agricultural) lands. The BMP options range from conservation tillage practices to temporary sediment traps and permanent drainage ways and retention ponds.

A farmer, in most cases, relies on the advice from SCS specialists who evaluate the magnitude of the soil loss potential and suggest or design the Best Manage-

TABLE 12-5. Cost of Erosion Control Practices for Developing Areas (1974–1975 price levels).[22,23]

Practice	Capital Cost ($/ha)	First-year Maintenance Cost ($/ha)
Surface protection and stabilization		
Seeding: seedbed preparation, seed and application, mulching at 4.5 tonnes/ha		
Temporary seeding by machine	600–800	125-300
Temporary seeding by hand	800–1000	125-300
Hydroseeding by machine	1,950–3,000	125-300
Sodding including bed preparation	5,900–9,000	590–7,200
Mulch, 4.5 tonnes/ha		
By hand	300–350	–
By machine	220–300	–
Structures		
Earth diversion berm	0.37–0.75	3–9
Straw bale sediment trap	1.85–2.70	3–9
Silt basin with earth dam for the watershed area:		
0.8–2 ha	1,480–2,960	1,230–1,850
10–40 ha	2,960–8,650	1,850–2,960
40–80 ha	8,650–12,350	2,960–4,450

ment Practices—usually a combination of available options—that would maximize the farmer's economical return.

In order to reduce the soil loss, the farmer has the following options:[24]

1. Reducing row-crop intensity and switching to close-grown crops.
2. Using higher-residue tillage practices.
3. Using no-till planting practices with chemical killing of weeds and residues.
4. Changing primary tillage from fall to spring (not effective in some geographical regions).
5. Contouring.
6. Strip-cropping.
7. Terracing.
8. Some combination of the above.

Reducing row-crop intensity is very costly due to the loss of revenues from the substitution of more profitable crops by less profitable close-grown crops. The use of some of the conservation tillage practices is the least costly alternative in most cases.[24–26]

Other management practices available to farmers for reduction of nonpoint pollution include runoff diversion, grass filter border areas, grassed waterways, irrigation water management, manure management and restriction of spreading in winter months, drainage installation, livestock exclusion, and several other practices. Generally the cost of BMPs may range from no-cost practices such as some conservation tillage methods to serveral thousands of dollars per hectare of severely erodible lands and lands with a high pollution potential.

The costs associated with these practices can be classified as follows:

1. Loss of income due to reduced yield, changed crop rotation, or substitution of less profitable, densely grown crops.
2. Cost of increased herbicide use to kill residues and control weeds in no-till practices.
3. Increased cost of labor.
4. Cost of construction of sediment control facilities such as sediment retention ponds, runoff diversion, and waterways.
5. Cost of cleaning and disposal of trapped sediment.
6. Cost of machinery.
7. Cost of drainage.
8. Cost of livestock exclusion.
9. Cost of winter manure storage.

Examples of some cost figures are presented in Table 12-6. As noted from the table, the costs of most common practices range from no-cost methods to methods that will reduce the farmer's income by a few percent. Obviously the cost of BMPs increases with the level of the pollution abatement measures and soil loss reduction. At high levels, the cost would be overwhelming. Bear in mind that the natural erosion rates are on the order of 1 ton/acre (2.2 tonnes/ha). It is logical that the sediment reduction should be set somewhere above the natural erosion rates.

For the purpose of finding the most optimal level of the sediment loss reduction, the cost of the Best Management Practices associated with each level of soil reduction can be expressed per ton of reduction. Such representation, whereby cost (or benefit), is expressed per one unit of output (or input) is called marginal cost (benefit). Figure 12-8 shows the marginal cost relation versus soil loss reduction. Alt et al.[29] showed that the marginal cost of soil loss reduction starts to increase dramatically when gross erosion is held to less than 5 tons/acre-yr (11 tonnes/ha-yr).

Efficiency of Best Management Practices. Simply stated, a project becomes efficient if the value of all benefits prevails over the value of the total cost both enumerated over the life time of the project. In other words, if B represents the total benefits and C is total cost, the project becomes efficient or feasible if

$$B - C > 0 \quad \text{or} \quad B/C > 1.0$$

TABLE 12-6. Examples of Cost Figures for Some Agricultural Nonpoint Pollution Management Practices (1977–1978 price levels).

Practice	Annual Cost[a] ($/ha)	Reference
Contour strip-cropping[b]	48	White and Partenheimer[27]
Combination of higher residue tillage and crop rotation		
Soil loss reduction		
from 15 to 5 tons/acre[c]	48	McGrann and Meyer[24]
from 10 to 5 tons/acre[c]	22	
Combination of terracing, chisel plowing, and crop rotation	7.1	Seitz et al.[26]
No-till in Ohio		
permeable soils	No cost	Forster[25]
poorly drained soils	13.3–41.1	
Herbicide application	25	Schneider[28]
Ban on winter spreading of manure		
all lands	10	Schneider[28]
only on slopes >4%	3	
Reducing fertilizer to 50% and herbicide to zero	175	McGrann and Meyer[24]
Diversion ($/m)	2.2	White and Partenheimer[27]
Drainage ($/m)	3.3	White and Partenheimer[27]

[a]Net cost, does not include cost sharing.
[b]Soil loss reduction from 15 to 1.4 tons/acre.
[c]To convert from tons/acre to tonnes/ha, multiply by 2.24.

The first relationship is called the efficiency ranking function, the second is known as the benefit-cost ratio. However, this oversimplified relationship cannot be directly applied to a real-world situation. There are five major reasons complicating the benefit-cost relationship:

1. The value of benefits decreases with the increase in their supply.
2. The major portion of cost (capital cost) is spent during the first phase of the project, while the benefits and operation and maintenance costs are realized year by year throughout the lifetime of the project.
3. The economy of the project can be imagined as an interest loan covering the capital cost and annual operation, maintenance, and replacement cost. The loan is then repaid in yearly installments by the annual benefits. Thus the discount or interest rate and repayment period may become very important but uncertain factors in the benefit-cost analysis.
4. The value of both benefits and cost may vary in the future based on the available supply, technology, general economy, and other factors that generally are very difficult to predict.

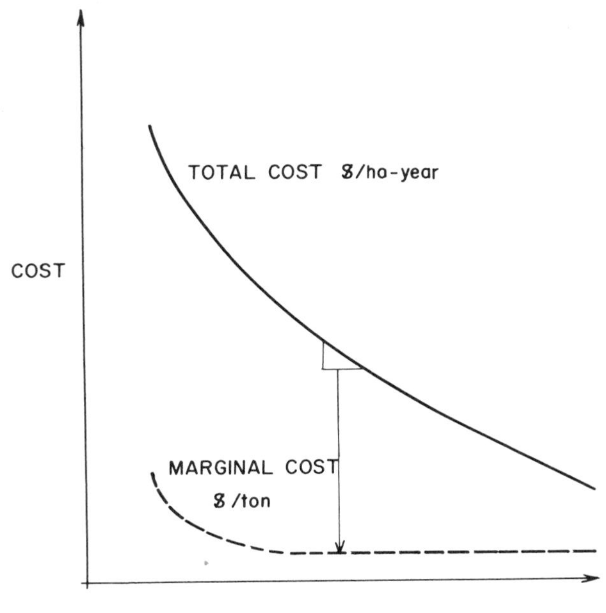

Fig. 12-8. Total and marginal cost of soil loss reduction.

5. Water pollution control measures are usually a part of overall water resource system development with multiple use and multiple benefits.

The Harvard University publication on water resources design and optimization[2] recommends the following efficiency or net benefit ranking function for a multipurpose water resource system development with an economic life of T years.

$$(B - C) = \sum_{t=1}^{T} \frac{E_t - M_t}{(1 + r)^t} - K > 0 \tag{12-8}$$

where:

M_t = operation, maintenance, and replacement cost in year t
K = construction cost
r = the interest rate applied for the analysis of the project

The function E_t represents the benefits of the project at year t. For multiple-purpose projects E_t is a summation of individual benefits related to various uses of the project including benefits associated with the improved water quality and reduced soil loss. If the value of benefits decreases with their increased supply, the benefit function should be replaced by the demand curve expressing the willingness of the user to pay for the benefits (Fig. 12-4). The benefit, E_{tj}, is then represented by the shaded area under the demand curve.

If the benefits and OMR cost are uniform throughout the lifetime of the project, the efficiency ranking function can then be simplified to

$$(B - C) = \sum_{j=1}^{Q} [\theta(E_j - M_j) - K_j] > 0 \tag{12-9}$$

where:

j = subscript denoting an output from the system
E_j = benefit associated with the output in a year
M_j = OMR cost related to the jth output in a year
K_j = capital cost related to the jth output
Q = number of outputs
θ = the discount factor given by

$$\theta = \frac{1 - (1 + r)^{-T}}{r} \tag{12-10}$$

The selection of an optimal alternative can then be based on the maximization of the efficiency ranking function (Equation 12-9) subjected to given constraints of the system. In a multiple-output system, a multiple ranking function $(B - C)$ also must be subjected to the production function

$$f(x_j, y) = 0 \tag{12-11}$$

which states that an increase of benefits for one use will result in reduction of benefits for another use unless the input to the system is also increased.

Future annual benefits and costs multiplied by the factor θ are transformed herein to the time level $t = 0$, and therefore, the term present value of benefits and costs is used. Hence

$$(B - C) = PVB - PVK = \sum_{j=1}^{Q} (\theta E_j) - \left[\sum_{j=1}^{Q} (\theta M_j) + K_j\right] \tag{12-12}$$

where:

PVB = the present value of benefits
PVK = the present value of costs

In Fig. 12-9, the shaded areas represent the respective present values of costs and benefits. The above relationship helps to solve the dilemma planners usually have with the prediction of future benefits and costs. As can be seen in Fig.

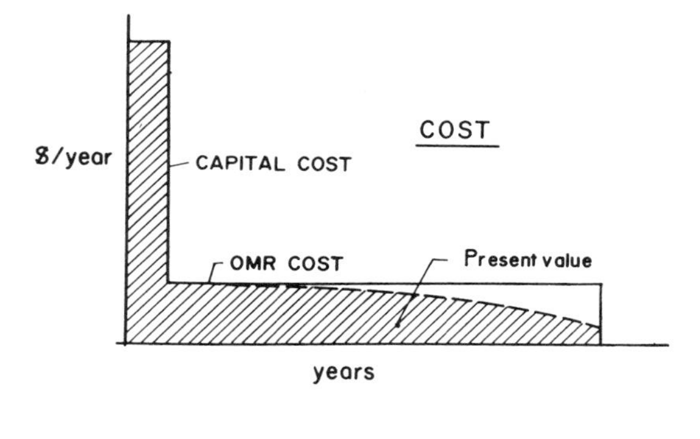

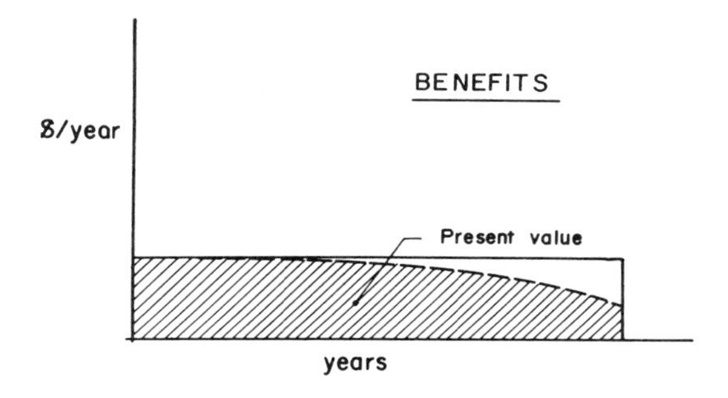

Fig. 12-9. Cost and benefit distribution and their present values.

12-9, distant future benefits and costs are discounted to their present values more than the near future ones; thus, difficulties and errors become less important as they become more pronounced for distant years.

The efficiency ranking function defined by Equation 12-9 requires a knowledge of the demand curve for given benefits. These curves are difficult to estimate since planners usually know the value of benefits at the present level of supply or availability and not for any other future levels. The efficiency of a project should therefore be related to the present knowledge of benefits and costs. Fortunately, an objective criterion of efficiency can be developed from the maximization of the efficiency ranking function. This procedure will lead to the following criterion:

$$\frac{B}{C} = \frac{MVP_j(X) - MM_j(X)}{MK_j(X)} > 1.0 \qquad (12\text{-}13)$$

where:

$MVP(X)$ = marginal present annual benefit j associated with one extra unit input X to the system

$MM(X)$ = the marginal present annual OMR cost associated with one extra unit of input X to the system
$MK(X)$ = the marginal capital (construction) cost.

The B/C ratio states that no project will be efficient if the present value of extra benefits derived from a planned system is less than the increased cost for expanding the magnitude of the system. The benefit cost ratio, B/C, thus represents another efficiency ranking criterion that utilizes the present knowledge of benefits and associated cost.

For actual planning, the B/C ratio should be in excess of 1. No planner or responsible agency would accept a solution whereby benefits would just equal the cost and would not provide a margin of safety for uncertainties in the evaluation of future benefits and costs.

Planning Horizon

The planning horizon is the most distant future time considered in the economic study. Very often the planning horizon is associated with the lifetime or amortization period of the project. The planning horizon depends on the scope of planning, the facilities involved in the system, and the area extent for which the planning is done.

For watershed quality planning, a time of 50 yr or more in the future can be chosen as a planning horizon. However, population and economy forecasts may be highly inaccurate for such distant time levels. Therefore, several planning time levels may be chosen to forecast and solve the system in a near-future (10 to 25 yr) and a distant future (50 to 100 yr) level.

If the planning horizon is related to the lifetime of a particular project, the time span will depend on the type and character of the structures involved and the conditions that will determine the usability of the facilities. The lifetime of a concrete gravity dam may go to a thousand years as can be seen in similar structures built by Egyptians and Romans; however, the reservoir volume will be filled by sediments and debris in a period of 100 yr or even less. Thus the functional capability of the reservoir will be the determinant of the planning horizon.

The effect of the planning horizon on soil erosion control measures was studied by Seitz et al.[26] The authors noted that on an average in corn belt farming areas, all topsoil would be lost in approximately 100 yr if conventional farming practices continued. However, the returns under the conventional practices are higher for the first 25 to 30 yr, after which they fall off drastically to levels much below the levels under a soil loss constraint.

Long planning horizons and low interest rates are required for farmers to apply erosion control practices. The planning horizon should be at least 40 yr before the deferred benefits of erosion control justify the initial cost and lower income in early years. At a 5% discount rate, the planning horizon should be at least 60 yr before an individual would choose to adopt conservation practices without governmental incentives.[26]

A typical planning horizon for the economic analysis of treatment plants is 15 to 30 yr; the period of amortization for sewer systems may be less.

Planners must also realize the time lapse between the preliminary plans and full operation of the planned facility. This time period increases with the magnitude of the project, its B/C ratio, and public and legislative interest. It may take a year or two to build a small treatment plant, but it may take 50 yr to fully realize a complex watershed plan.

Discount Rate. Discounting is a way to account for the opportunity cost of funds invested in a project that could be also used productively in the private sector of the economy or in some other public project. The interest rate to be used in evaluating water-related public projects is prescribed by the Water Resources Council. The rate specified by the council is based on the interest rate on federal securities with maturities over 15 yrs.[3] The council annually determines the rate.

Example 12-2: Urban Nonpoint Pollution Management Practices

A city is faced with a nonpoint pollution problem. A section of the city with an area of 100 ha generates suspended solids loadings of 2000 kg/ha-yr that is discharged into a lake. Ten percent of the loading originates from the local sewage treatment plant effluent and 90% is discharged from storm sewers.

To reduce the nutrient input into the lake, the sediment loading must be reduced by 70%. The city engineer's office has the following options:

1. On-site removal of pollutants from impervious surfaces by sweeping that will affect a maximum of 60% of the total solids load and a minimum of 20%, since sweeping is intended also for aesthetic reasons. The cost of sweeping has been estimated as $0.10/kg of solids removed for the equipment acquisition and $0.10/kg for operation and maintenance of the equipment.

2. Solids removal in a retention basin. The expected removal efficiency is about 85%. The cost of the basin is $0.15/kg for capital cost and $0.08/kg for solids removal and cleaning.

3. Upgrading of the sewage treatment plant for higher solids removal efficiency. The capital cost would be $0.4/kg and OMR cost would be $0.20/kg of solids removed.

Determine the best selection of management practices. Assume an amortization period of 20 yr and an interest rate of 10%.

The decision variables are:

X_1 . . . kg of solids removed by sweeping
X_2 . . . kg of solids removed in the retention basin
X_3 . . . kg of solids removed in the STP

Estimate discount factor for annual cost.

$$\theta = \frac{1 - (1 + r)^{-T}}{r} = \frac{1 - (1 + 0.10)^{-20}}{0.10} = 8.51$$

Estimate the present cost per hectare of the city area:

sweeping cost

$$PV_{SW} = X_1(0.10 \times 8.51 + 0.10) = 0.95X_1$$

sediment retention

$$PV_{SR} = X_2(0.08 \times 8.51 + 0.15) = 0.83X_2$$

STP upgrading

$$PV_{STP} = X_3(0.20 \times 8.51 + 0.4) = 2.1X_3$$

In this case the efficiency ranking function will minimize the total cost of the practices. Then

$$\min Z = PV_{SW} + PV_{SR} + PV_{STP} = 0.95X_1 + 0.83X_2 + 2.1X_3$$

The constraints imposed on the combination of the practices are:

a. *70% solids restriction, or* $2000 - X_1 - X_2 - X_3 \leqslant 600$,
b. *ranges of sweeping efficiency (20 to 60%), or* $400 \leqslant X_1 \leqslant 1200$,
c. *max solids removed in the retention basin (90%), or* $X_2 \leqslant 0.9(1800 - X_1)$,
d. *ranges of the solids removed by the STP* $0 \leqslant X_3 \leqslant 200$.

Formulate the linear programing model.
The ranking function is

$$\min Z = 0.95X_1 + 0.83X_2 + 2.1X_3$$

subjected to

a. $X_1 + X_2 + X_3 \geqslant 1400$
b_1. $X_1 \geqslant 400$
b_2. $X_1 \leqslant 1200$
c. $X_2 + 0.9X_1 \leqslant 1620$
d_1. $X_3 \geqslant 0$
d_2. $X_3 \leqslant 200$

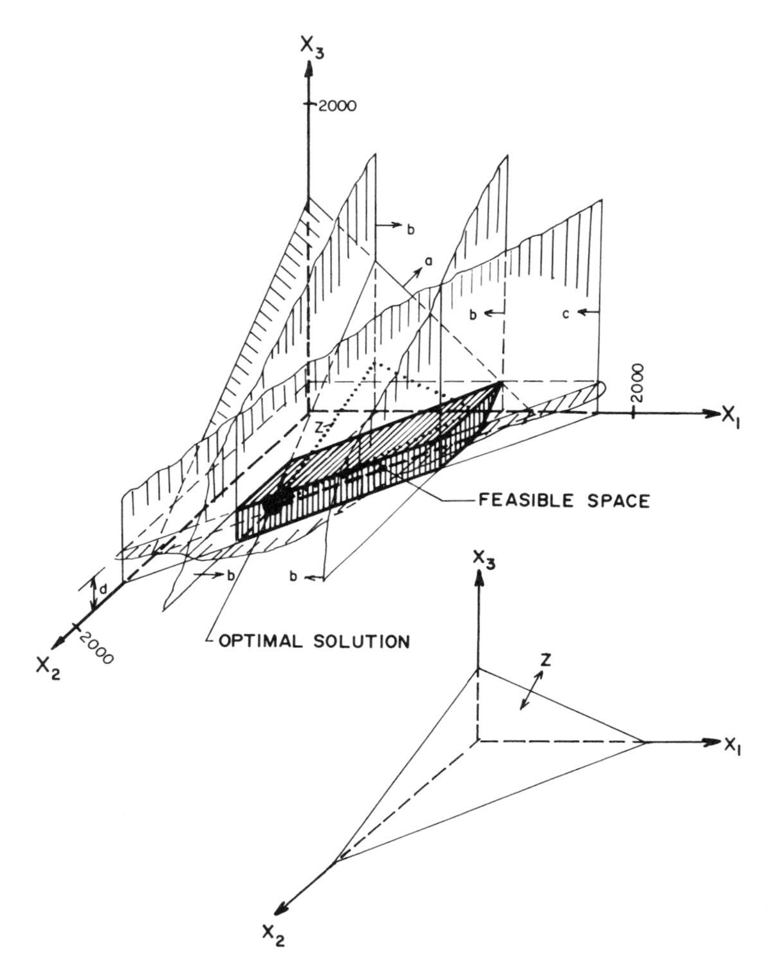

Fig. 12-10. Feasible space and solution for Example 12-2.

This problem can be easily solved using a standard computer routine for linear programing. Since the number of the decision variables is only three, the problem can be plotted and solved graphically as shown in Fig. 12-10.

The optimum solution can be located where a Z plane with the lowest value touches the feasible prism constructed from constraints. The optimal solution is

$$X_1 = 400 \text{ kg/yr}$$
$$X_2 = 1000 \text{ kg/yr}$$
$$X_3 = 0$$

and the present value of the cost for this solution is

$$Z = 400 \times 0.95 + 1000 \times 0.83 = \$1210/\text{ha}$$

Total present value of cost: $Z \times$ area = $1210 \times 100 = \$121{,}000$.

In this example the removal of solids by a silting basin is the most effective, as can be seen from the marginal cost of $0.83/kg as compared to $0.95/kg for sweeping and $2.10 for upgrading the sewage treatment plant.

Example 12-3: Benefit-Cost Analysis

The benefits associated with the solids removal from Example 12-2 can be enumerated as follows:

1. Recreation benefits: The water quality improvement that resulted from the implemented practices will result in an approximate increase of 2,000 Recreation Day units at a value of $5.00 per unit.

2. Property value benefits: The property value of real estate that will be affected by water quality improvement is estimated as $4 million with a local property tax of $25 per $1000 of the value. A 20% property value increase is a reasonable estimate based on property values at nearby water bodies.

Estimate the efficiency ranking function $(B - C)$ of water quality improvement.

Recreation benefit

$$B_1 = 5 \times 2000 = \$10{,}000/\text{yr}$$

Tax increase benefit

$$B_2 = 0.2 \times 4{,}000{,}000 \times 25/1000 = \$20{,}000/\text{yr}$$

Total benefits $B_T = \$30{,}000/yr$.

Cost estimates

$$\text{Capital cost} = (0.10 \times X_1 + 0.15 \times X_2)\text{area}$$
$$= (0.10 \times 400 + 0.15 \times 1000)100 = \$19{,}000$$

Annual OMR cost

$$(0.10 \times X_1 + 0.08 \times X_2)100 = \$12{,}000$$

Discount factor for $T = 20$ yr and $r = 10\%$

$$\theta = 8.51$$

The net benefit ranking function becomes

$$(B - C) = PVB - PVK = 30{,}000 \times 8.51 - (8.51 \times 12{,}000 + 19{,}000)$$
$$= \$134{,}180$$

and the benefit/cost ratio is

$$\frac{B}{C} = \frac{8.51 \times 30{,}000}{8.51 \times 12{,}000 + 19{,}000} = 2.11$$

Example 12-4: Selection of Agricultural BMPs

A farmer growing primarily corn using conventional tillage is losing approximately 30 tonnes of topsoil per hectare of tilled land. The soil lost by erosion also represents nitrogen loss of about 200 kg/ha-yr and phosphorus loss of about 100 kg/ha-yr that must be replaced by a fertilizer to sustain the soil fertility. The cost of the fertilizer is about $1.0/kg of N. The present yield of corn is about 350 bushels per hectare at a price of $3/bushel. Estimate the marginal cost to the farmer to reduce the soil loss to 15 tonnes/ha and 6 tonnes/ha.

For the simplicity of the example, assume that the SCS specialist advising the farmer has the following options:

1. Chisel plowing that reduces the soil and nutrient losses by about 60% (see Table 11-2) at an annual cost of $40/ha.

2. Crop rotation and substituting close-grown crop (alfalfa) that reduces the soil loss by 90%. The yield of alfalfa is about 10 tonnes/ha at a price of $40/ton. Assume that the alfalfa has approximately the same nitrogen requirement as corn.

This simple optimization problem can again be solved by the linear programing concept. The objective is the maximiation of the farmer's income subjected to the soil loss restriction. The decision variables are:

X_1—fraction of the area under conventional tillage with corn.
X_2—fraction of the area under conservation tillage with corn.
X_3—fraction of the area growing alfalfa.

Income under conventional tillage per hectare

corn yield:	350 bushels × $3/bushel	$1050
nutrient loss:	200 kg N/kg × $1/kg	$ 200
	Net income	$ 850

Income under conservation tillage per hectare

corn yield:	$1050
nutrient loss: 0.4 × 200 kg N/ha × $1.0/kg	$ 80
cost of conservation practices:	$ 40
Net income	$ 930

Income for alfalfa

hay income: 10 tonnes/ha × $40/ton	$400
nutrient loss: 0.1 × 200 kg N/ha × $1.0/kg	$ 20
Net income	$380

Formulate the net benefit ranking function

$$Z = 850X_1 + 930X_2 + 380X_3$$

Subject to the soil loss reduction

$$30X_1 + 12X_2 + 3X_3 \leqslant b$$

where b is the desired soil loss level.

By investigating the objective function, one can note that implementation of the conservation tillage actually improves the farmer's income due to the reduced cost of the fertilizer. Hence, let us assume that the farmer will first convert all of his production to conservation tillage. This will reduce the annual soil loss to 12 tonnes/ha. For further soil loss reduction, up to 3 tonnes/ha, the problem can be easily solved by substituting alfalfa on a portion of his farm and rotating the crops. Then for the 6 tonnes/ha reduction or

$$12X_2 + 3X_3 \leqslant 6$$

The above equation can be solved knowing that

$$X_2 + X_3 = 1.0$$

Then

$$X_2 = \tfrac{1}{3}$$

and

$$X_3 = \tfrac{2}{3}$$

Under the 6 tonnes/ha reduction, the farmer can grow corn on one-third of his farm and substitute alfalfa on two-thirds of the area. This will give him an income of

$$Z = 930X_2 + 380X_3 = 930 \times \tfrac{1}{3} + 380 \times \tfrac{2}{3} = \$563/\text{ha}$$

or the 6 tonnes restriction will cost the farmer

$$C = 930 - 563 = \$367/\text{ha}$$

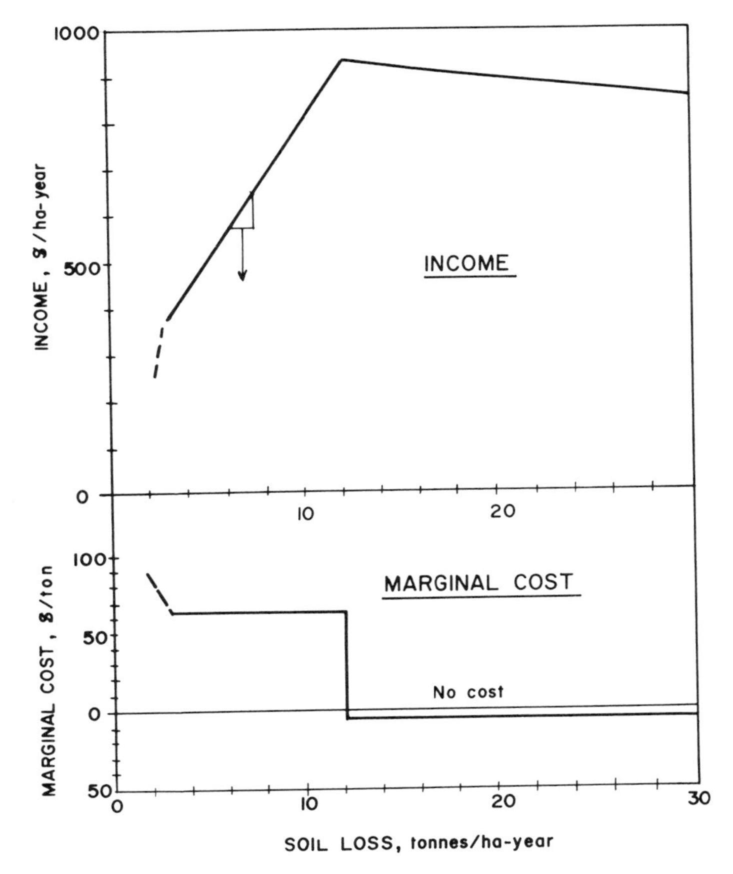

Fig. 12-11. Income function and marginal cost of BMPs in Example 12-4.

This social cost should be returned to the farmer in the form of a subsidy.

The marginal cost of the nonpoint pollution control, which in this case is the cost of one extra tonne of soil loss reduction, can be obtained from the tangents to the cost curve as shown in Fig. 12-11.

STRATEGY OF NONPOINT POLLUTION CONTROL

Economic Mechanics and Incentives to Control Nonpoint Pollution

As mentioned previously, the market forces do not perform satisfactorily in controlling water pollution, especially that related to nonpoint emissions of pollutants. An owner of a car has no economic incentive or pressure to buy unleaded gasoline; similarly, there is no economic linkage between a farmer responsible for nutrient loadings and downstream lake users suffering from the symptoms of eutrophication. Nonpoint pollution is also often "invisible." For example, users of a lake often blame a small industry or municipality discharging small amounts of treated organic wastes for the eutrophication problems instead of the nutrient input from nonpoint sources. Very often firms responsible for nonpoint pollution (farmers, developers) argue that the benefits associated with their production activity (food production, development) outweigh the damages to surface waters. Unlike the pollution from point sources, which is controlled by the National Pollution Discharge Elimination System permits, pollution from nonpoint sources is difficult to regulate.

A host of alternative strategies exist to restrict soil and pollutant input from nonpoint sources.[26] The principal strategies include:

1. Application of emission standards restricting soil and pollutant loss.
2. Taxes on soil and pollutant loss.
3. Subsidies for reduced soil and pollutant loss.
4. Regulation of production processes and inputs to allow only those that reduce soil and pollutant loss from hazardous lands.
5. Subsidies for inputs that reduce soil and pollutant loss.
6. Taxes on inputs that encourage soil loss and produce pollution.

Emission standards, along with subsidies, are more likely to be used in urban areas, where regulation is more acceptable. In agricultural areas, collaboration of farmers with the SCS is more or less voluntary and relies on: (1) education and technical assistance, (2) financing and funding (cost sharing), (3) incentives, and (4) monitoring and enforcement.[30]

From the farmer's or developer's viewpoint, the subsidy is the preferred economic mechanism, regulation is the next preferred, and tax is the least preferred. For the taxpayer, the subsidy is the least preferred and tax is the most preferred. The downstream user of water is indifferent to what method is used to reduce pollution, unless there is an economic linkage between the user and the polluter. The net economic impact to farmers of subsidies, restrictions, or taxes is approximately the same.[31] As stated previously, without subsidies or taxes or some other enforcement or incentive mechanism, the reduction of sediment and pollutant emissions would be attractive to neither farmers nor developers.

Subsidies for the implementation of BMPs can be provided from federal, state, or local tax income. In the period of 1973 to 1983, the federal government provided subsidies for pollution abatement facilities and BMPs amounting up to 75% of the cost, provided that they were included in the water quality plan under Section 208 of the PL 92-500.

Public Perception of Water Quality

The public is sensitive to water quality and reacts to adverse water quality changes. Water quality affects property values near surface water bodies, recreation activities, and the related economy. For example, the price of shoreline property on some lakes located in southeastern Wisconsin may range from $100/m of the shoreline at some low-quality and eutrophic lakes to over $1000/m at a high-quality oligotrophic lake. Weed cutting in eutrophicated lakes increases taxation of lake communities, which increases public pressure on planners and responsible agencies to remove the water quality nuisance. The public groups directly involved and interested in water quality include recreationists and recreation businesses, marinas, fishermen and boat owners. Environmentalists, who worry about potential destruction of natural areas, are one of the most vocal groups concerned with water quality problems.

Many communities have established citizen groups for the protection or restoration of local surface waters. Water quality planners must locate and contact these groups and involve them in the overall planning process.

Public Participation in Planning. Public participation in the water quality planning process is not only advisable, it is mandatory.[1,3]

The public should have an input into all phases of the planning process (Fig. 12-12). Citizen opinion should be sought on the following issues:[3]

1. The identification of water quality problems and priorities to resolve them.
2. The relative importance of water quality goals in relation to other community goals.
3. The role that water quality management can or should play in achieving community goals.
4. The use of land-use controls to protect water quality.
5. Compatibility of water pollution control approaches (municipal and industrial source control for point and nonpoint sources) with other community goals.
6. Public reaction to management alternatives to implement the plan.
7. Impact assessment.

A formally constituted Citizen's Advisory Committee (CAC) has the potential of providing the best communication between the planners and citizen groups.

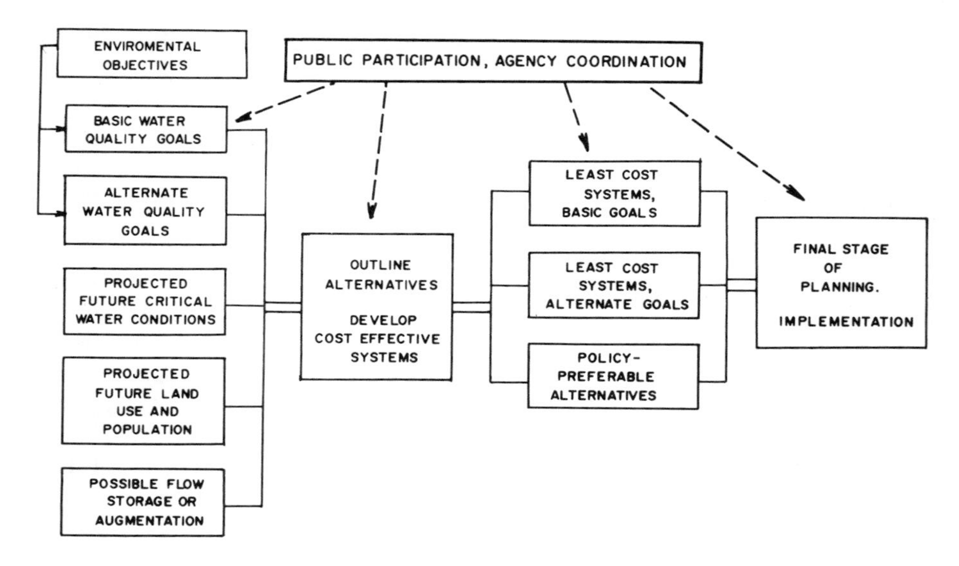

Fig. 12-12. Public participation in water quality planning.[33]

EPA's public participation regulations[32] contain the following guidelines for the formation of a CAC:

1. The CAC membership should be comprised of substantially equal representation from four interests: private citizens; public interest groups; governmental officials; and economic interests.
2. A notice of intent to form an advisory committee should be circulated widely in the community so that those wishing to serve on such a committee may have an opportunity to do so.
3. The task of the CAC should be clearly defined, including its role relative to other public participiation mechanisms. In this way, the CAC can maintain its contacts and credibility outside the CAC.
4. The CAC work should be supported by adequate staff and budget.

The CAC should be appointed in the early stages of the process. To be fully effective, the committee must have direct input into all major decisions affecting the plan.

Public Hearings. Public hearings that must be set up during several phases of the planning process are a reactive and consultative mechanism. These hearings are meant to give individuals and organizations a formal opportunity to express their opinions on the plan and issues prior to a decision on the alternatives being made.

Other Forms of Public Participation. Many communities have some type of planning and/or zoning board that is responsible for the development and approval of pollution abatement programs and serve also as a link between the local community and the planning agency.

Communication of planners with the public also utilizes mass media—passively or actively—town meetings, referenda, conferences and seminars, and surveys.

REFERENCES

1. Anon. 1972. Federal Water Pollution Control Act Amendments. P.L. 92-500, U.S. Congress, Washington, D.C.
2. Maas, A., et al. 1970. "Design of Water Resources Systems." Harvard University Press, Cambridge, Massachusetts.
3. Anon. 1976. Guidelines for state and areawide Water Quality Management Program Development. U.S. EPA, Washington, D.C.
4. Anon. 1976. A comprehensive plan for the Menomonee River Watershed. Planning Rep. No. 26, Southeast Wisconsin Regional Planning Commission, Waukesha, Wisconsin.
5. Anon. 1969. National Environmental Policy Act. P.L. 91-190. U.S. Congress, Washington, D.C.
6. Anon. 1974. Proposed principles and standards for planning water and related land resources. Water Resources Council, Federal Register, 38, No. 174.
7. Rickert, D. A., Hines, W. G., and McKenzie, S. W. 1975. Methods and data requirements for river-quality assessment. Water Res. Bull., **11**:1013–1039.
8. Hines, W. G., Rickert, D. A., McKenzie, S. W., and Bennett, K. P. 1975. Formulations and use of practical models for river quality assessment. *J. WPCF*, **47**:2357–2370.
9. Berthouex, P. M., and Polkowski, L. B. 1970. Design capacities to accommodate forecast uncertainties. *J. San. Eng. Div., ASCE*, **96**:1183–1210.
10. McJunkin, F. E. 1964. Population forecasting by sanitary engineers. *J. San. Eng. Div., ASCE*, **90**:31–58.
11. Schneider, R. R. 1978. Planning diffuse pollution control: An analytical framework. *Water Res. Bull.*, **14**:322–336.
12. Kneese, A. V., and Bower, B. T. 1971. Managing water quality: Economics, technology, institutions. John Hopkins Press, Baltimore, Maryland.
13. Krenkel, P. A., and Novotny, V. 1980. "Water Quality Management." Academic Press, New York.
14. Binkley, C. S., and Haneman, W. M. 1978. The recreation benefits of water quality improvement: Analysis of day trips in an urban setting. EPA-600/5-78-010, U.S. EPA, Washington, D.C.

15. Hawkins, D. E., and Tindall, B. S. 1966. "Recreation and Park Yearbook 1966." Recreation and Park Association, Washington, D.C.
16. York, D. W., Canan, L. W., and Dysart, B. C. 1976. Modeling water related recreation. *J. Water Res. Plan. Man. Div., ASCE*, **102**:409–413.
17. Tihansky, D. P. 1974. Historical development of water pollution control cost function. *J. WPCF*, **46**:813–833.
18. Pitt, R. 1978. Sources, effects, and control of urban runoff. Paper presented at the 51st Annual WPCF Conference, Anaheim, California.
19. Pitt, R. 1979. Demonstration of nonpoint pollution abatement through improved street cleaning practices. EPA-600/2-79-161, U.S. EPA, Cincinnati, Ohio.
20. Heaney, J. P., Huber, W. C., Field, R., and Sulivan, R. H. 1979. Nationwide cost of wet-weather pollution control. *J. WPCF*, **51**:2043–2053.
21. Anon. 1972. Investigation of porous pavement for urban runoff control. Franklin Research Institute, Rep. No. 11034DUY03/72, U.S. EPA, Washington, D.C.
22. Lager, J. A., Smith, W. G., Lynard, W. G., Finn, R. M., and Finnemore, E. J. 1977. Urban stormwater management and technology: Update and user's guide. EPA-600/8-77-014, U.S. EPA, Edison, New Jersey.
23. Task Committee on the Effects of Urbanization on Low Flow, Total Runoff, Infiltration, and Groundwater Recharge of the Committee on the Surface-Water Hydrology of the Hydraulics Division, 1975. Aspects of hydrologic effects of urbanization. *J. Hydraulics Div., ASCE*, **101**:44–468.
24. McGrann, J. M., and Meyer J. 1979. Farm-level economic evaluation of erosion control and reduced chemical use in Iowa. In: R. C. Loehr et al. (eds.). "Best Management Practices for Agriculture and Silviculture." Ann Arbor Science, Ann Arbor, Michigan.
25. Forster, D. L. 1978. Economic impact of changing tillage practices in the Lake Erie basin. Tech. rep. Department of Agricultural Economics and Rural Sociology, Ohio State University, Columbus, Ohio.
26. Seitz, W. D., Osteen, C., and Nelson, M. C. 1979. Economic impact of policies to control erosion and sedimentation in Illionois and other corn-belt states. In: R. C. Loehr et al. (eds.). "Best Management Practices for Agriculture and Silviculture." Ann Arbor Science, Ann Arbor, Michigan.
27. White, G. B., and Partenheimer, E. J. 1979. The economic implications of erosion and sedimentation control plans for selected Pennsylvania dairy farms. In: R. C. Loehr et al. (eds.). "Best Management Practices for Agriculture and Silviculture." Ann Arbor Science, Ann Arbor, Michigan.
28. Schneider, R. R. 1976. Diffuse agricultural pollution: The economic analysis of alternative control. Ph.D. Thesis, Department of Agricultural Economics, University of Wisconsin, Madison.
29. Alt, K. F., Miranowski, J. A., and Heady, E. O. 1979. Social cost and effectiveness of alternative nonpoint pollution control practices. In: R. C. Loehr et al. (eds.). "Best Management Practices for Agriculture and Silviculture." Ann Arbor Science, Ann Arbor, Michigan.

30. Rice, J. M. 1979. Management and financing of agricultural BMPs. In: R. C. Loehr et al. (eds.). "Best Management Practices for Agriculture and Silviculture." Ann Arbor Science, Ann Arbor, Michigan.

31. Forster, D. L., and Becker, G. S. 1977. Economic and land management analysis in Honey Creek watershed. Ohio Agr. Res. and Dev. Center, Ohio State University, Columbus.

32. Rastatter, C. L. 1979. Municipal Wastewater Management Public Involvement Activities Guide, EPA-430/9-79-005, U.S. EPA, Washington, D.C.

33. Whipple, W., Jr. 1976. Strategy of water pollution control. *Water Res. Bull.*, **12**:1019–1028.

Appendix 1

CONVERSION FACTORS
U.S. Customary to S.I. (Metric)

U.S. Customary Units	Multiply by to obtain	S.I. Symbol	S.I. Name
acre	0.405	ha	hectare[a]
acre-ft	1 233.5	m^3	cubic meter[b]
acre-in	102.79	m^3	
foot (ft)	0.3048	m	meter[c]
ft/s (fps)	0.3048	m/s	meter per second
ft^2 (sqft)	0.0920	m^2	square meter[d]
ft^3 (cuft)	0.0283	m^3	cubic meter
ft^3/s (cfs)	0.0283	m^3/s	cubic meters per second
°F	0.555 (°F-32)	°C	degrees Celsius
gallon-U.S. (gal)	3.785	l	liter
gal/acre	9.353	l/ha	liter per hectare
gal/ft^2	40.743	l/m^2	liters per square meter
gal/ft^2	0.0407	m^3/m^2 = m	meter
gal/ft^2-day	4.72×10^{-7}	m/s	meter per second
gal/ft-day	1.438×10^{-7}	m^2/s	square meter per second
hp	0.746	kW	kilowatts

CONVERSION FACTORS (*Continued*)

U.S. Customary Units	Multiply by to obtain	S.I. Symbol	S.I. Name
inch (in)	2.54	cm	centimeter
in.^{2}	6.452	cm^2	square centimeter
in.^{3}	16.39	cm^3	cubic centimeters
in.^{3}	0.0164	l	liter
pound (lb)	0.454	kg	kilograms[e]
lb	454	g	grams
lb/acre	1.121	kg/ha	kilograms per hectare
lb/ft^3	16 042	g/m^3	grams per cubic meter[f]
lb/in.^{2} (psi)	0.0703	kg/cm^2	kilograms per square cm[g]
lb/Mgal	0.120	mg/l	miligrams per liter
lb/mi	0.282	g/m	grams per meter
mile (mi)	1.609	km	kilometer
mile/hour (mph)	0.447	m/s	meter per second
million gallons (Mgal)	3785	m^3	cubic meters
million gallons per day (Mgd)	0.0438	m^3/s	cubic meters per second
mi^2 (sqmi)	2.59	km^2	square kilometer
parts per billion (ppb) in water	1.0	μg/l	micrograms per liter
parts per million (ppm) in water	1.0	mg/l	miligrams per liter
ton (short)	0.907	t	tonne
tons/acre	2240	kg/ha	kilograms per hectare
tons/sqmi	3.503	kg/ha	kilograms per hectare
yard (yd)	0.914	m	meter
yd^3	0.765	m^3	cubic meter

[a] 1 ha = 10 000 m^2 = 0.01 km^2
[b] 1 m^3 = 1 000 l
[c] 1 m = 100 cm = 0.001 km = 1 000 mm
[d] 1 m^2 = 10 000 cm^2 = 10^{-6} km^2

[e] 1 kg = 1 000 g = 0.001 tonne
1 g = 1000 mg = 10^6 μg = 10^9 ng
[f] 1 g/m^3 = 1 mg/l
[g] 1 kg/cm^2 = 0.968 atm = 0.981 bars

Appendix 2

Glossary

GLOSSARY

Acid mine waste. One of the principal pollutants arising from mining operations. Acid water forms when water contacts exposed mine wastes and sulfur-containing ores.

Acid rain. Precipitation with $pH < 5.6$ formed by contact of atmospheric moisture with sulfur and nitrogen oxides in aerosols.

Aerated lagoon. A natural or artificial wastewater treatment lagoon (generally from 1 to 4 m deep) in which mechanical or diffused-air aeration is used to supplement the oxygen supply. It can be used for treatment of combined sewer overflows.

Area-wide water quality management plan. The plan required by P.L. 92-500 for managing water quality. It must consider both point and nonpoint source pollution on an area-wide basis.

Best management practices (BMPs). Nonstructural and low-structural measures that are determined to be the most effective, practical means of preventing or reducing pollution inputs from nonpoint sources in order to achieve water quality goals.

Biochemical oxygen demand (BOD). The quantity of dissolved oxygen used by microorganisms in the biochemical oxidation of organic matter and oxidizable inorganic matter by aerobic biological action. Generally refers to standard 5-day BOD test.

Biological treatment processes. Means of treatment in which bacterial biochemical action is intensified to stabilize, oxidize, and nitrify the unstable organic matter present. Trickling filters, contact stabilization, and lagoons are examples that have been used for treatment of combined sewer overflows.

Buffer strip or zone. Strips of grass or other erosion-resistant vegetation between a waterway and an area of more intensive land use.

Check dam. A small dam constructed in a gully or other small watercourse to decrease the stream flow velocity, minimize channel erosion, and promote deposition of sediment.

Chisel plowing. Seedbed preparation by plowing without complete inversion of the soil, leaving a protective cover of crop residues on the surface for erosion control.

Chute spillway. Overfall structure that allows water to drop rapidly, without causing erosion, through an open lined channel.

Combined sewer. A sewer receiving both intercepted surface runoff and municipal sewage.

Combined sewer overflow (CSO). Flow from a combined sewer in excess of the sewer capacity that is discharged into a receiving water or diverted to a storage facility for subsequent treatment and disposal.

Contour farming. Conducting field operations, such as plowing, planting, cultivating, and harvesting, across the slope.

Contour strip-cropping. Farming operations performed on the contour with crops planted in narrow strips, alternating between row crops and close-growing forage crops.

Conventional tillage. Standard method of preparing a seedbed by completely inverting the soil and incorporating all residues with a moldboard plow. The field is gone over more than once in order to prepare a smooth, fine surface.

Crop rotation. A planned sequence of crops planted in a regular succession on the same area of land, as opposed to continuous culture of one crop.

Detention. The slowing, dampening, or attenuating of flows—either entering the sewer system or surface drainage—by temporily holding the water on a surface area, in a storage basin, or within the sewer.

Disinfection. The killing of the larger portion of microorganisms in or on a substance with the probability that all pathogenic bacteria are killed by the agent used.

Diversion. Individually designed conveyances across a hillside. They may be used to protect bottomland from hillside runoff, divert water from areal sources of pollution, or protect structures from runoff.

Double-disking. See Reduced tillage.

Drainage density. The amount of watercourses per unit area.

Erosion. The wearing away of the land surface by running water, wind, ice, or other geological agents.

Eutrophication. The process of enrichment of water bodies by nutrients. Eutrophication of a lake normally contributes to its slow evolution into a bog or marsh and ultimately to dry land. Eutrophication may be accelerated by human activities and thereby speed up this aging process.

First flush. The condition, often occurring in storm sewer discharges and combined sewer overflows, in which a disproportionately high pollutional load is carried in the first portion of the discharge or overflow.

Grassed waterway. A natural or constructed waterway (usually broad and shallow covered with erosion-resistant grasses) used to conduct surface runoff.

Gully. A channel caused by the concentrated flow of water over unprotected erodible land. Gullies are deep enough to interfere with, and not be obliterated by, normal tillage operations.

Heavy metals. See Toxic metals.

Hydrologic cycle. The circuit of water movement from the atmosphere to the earth and back to the atmosphere through various processes. These processes include: precipitation, interception, runoff formation, infiltration, groundwater movement, storage, evaporation, and transpiration.

Hydroseeding. Dissemination of seed under pressure, in a water medium. Mulch, lime, and fertilizer can be incorporated into the spraying mixture.

Infiltration. The gradual downward movement of water from the surface into the subsoil.

Intercepted surface runoff. That portion of surface runoff that enters a sewer, either storm or combined, directly through catch basins, inlets, etc.

Interception storage. The volume of water intercepted from precipitation by surface vegetation that is released back to the atmosphere by evaporation.

Interceptor. A sewer that receives dry-weather flow from a number of transverse combined sewers and additional predetermined quantities of intercepted surface runoff and conveys such waters to a point for treatment.

Mulch. (1) Any material such as straw, sawdust, leaves, plastic film, loose soil, etc., that is spread upon the surface of the soil to protect the soil and plant roots from the effects of raindrops, soil crusting, freezing, evaporation, etc. (2) To apply mulch to the soil surface.

Municipal sewage. Sewage from a community, which may be composed of domestic sewage, industrial wastes, or both.

Nonpoint source (NPS) pollution. Pollution caused by sediment, nutrients, and organic and toxic substances originating from land-use activities and/or from the atmosphere, which are carried to lakes and streams by runoff. Nonpoint source pollution occurs when the rate at which these materials entering water bodies exceeds natural levels.

No-tillage. A method of planting crops that involves no seedbed preparation other than opening the soil for the purpose of placing the seed at an intended depth. This usually involves opening a small slit or punching a hole into the soil. There is usually no cultivation during crop production. Chemical weed control is normally used. Also referred to as "zero tillage" or "no-till."

Nutrients. Elements or substances, such as nitrogen or phosphorus, that are necessary for plant growth. Large amounts of these substances reaching water bodies can become a nuisance by promoting excessive aquatic algal growth.

Organic chemicals. Manufactured organic chemicals used for controlling weeds and pests. Examples include DDT and other pesticides, and PCBs. They are generally considered toxic, and many of these chemicals may be also carcinogenic.

Organic materials. Carbon-containing substances in plant and animal matter. High concentrations of organic materials are often found in municipal and industrial wastewater and in surface runoff. Also referred to as biodegradable organics.

Pesticides. Chemical agents used to control specific plants or animals; e.g., herbicides and insecticides.

Physical-chemical treatment. A means of treatment in which the removal of pollutants is brought about primarily by sedimentation with or without an addition of chemicals that enhance flocculation and coagulation. The treatment processes generally include primary treatment, chemical clarification, filtration, carbon adsorption, and disinfection.

Point source pollution. This type of water pollution results from the discharges into receiving waters from sewers and other identifiable "points." Common point sources of pollution are discharges from industries and municipal sewage treatment plants.

Polychlorinated biphenyls (PCBs). A group of man-made industrial chemicals known to be highly toxic to humans and other organisms. In the past they were used primarily as coolant insulation fluids in electrical transformers and capacitors, and as additives in paper and plastic materials. They are persistent in the environment and have caused considerable environmental damage.

Pretreatment. The removal of materials such as coarse solids, grit, grease, and scum from sewage flows prior to physical, biological, or physical-chemical treatment processes to improve treatability. Pretreatment may include screening, grit removal, skimming, preaeration, and flocculation.

Reduced tillage. Any tillage system that involves less soil disturbance and retains more plant residue on the surface than conventional tillage. Sometimes called "conservation" or "minimum" tillage. Common practices include plow planting, double-disking, chisel plowing, and strip tillage.

Retention. The prevention of runoff from entering the sewer system by storing it on a surface area or in a storage basin.

Riprap. A facing layer or protective mound of stones placed to prevent erosion or sloughing of a structure or embankment.

Runoff. The portion of rainfall, melted snow, or irrigation water that flows across the surface or through underground zones and eventually runs into streams. The runoff has three components–surface runoff, interflow, and groundwater flow. Runoff may pick up pollutants from the air or the land and carry them to receiving waters.

SCS. Soil Conservation Service.

Sediment. Transported and deposited particles derived from rocks, soil, or biological materials.

Sedimentation. Removal, transport, and deposition of detached sediment particles by flowing water or wind.

Septic systems. These are used to hold and dispose of domestic sewage when a sewer line is not available to carry the wastes to a treatment plant. The wastes are piped to an underground tank directly from a home or a building.

Bacteria in the tank decompose some of the ogranic materials, and sludge settles on the bottom of the tank. The effluent flows from the tank into the underground drains. Most of the pollution is adsorbed by soil in the seepage field; however, some mobile pollutants can penetrate into the groundwater zones. The sludge must be pumped from the tanks at regular intervals.

Sewer systems. A network of underground conduits with free-flowing water surface, with appurtenances, used for collecting and conveying wastewaters from source to discharge. Sewer systems can be separate, where sanitary sewage and storm water are conveyed in separate sewer networks, or combined, where sewage and storm water are carried by the same sewer. When the dilution of sewage by storm water in combined sewers reaches a certain ratio, excess flow is discharged via combined sewer overflows.

Soil texture. The relative size of soil particles; the fineness or coarseness of the soil. More specifically, texture is determined by the relative proportions of sand, silt, and clay.

Storm sewer. A sewer that carries only surface runoff, street wash, and snowmelt from the land. Storm sewers are completely separated from those that carry domestic and industrial and commercial wastewater.

Strip-cropping. The crop-growing practice that requires different types of tillage, such as corn and alfalfa, in alternate strips.

Strip tillage. Tillage operation for seedbed preparation that is limited to a strip not to exceed one-third of the distance between rows; the area between is left untilled with a protective cover of crop residues on the surface for erosion control.

Surface roughening. Any practice that provides for rougher, more permeable surfaces that will slow runoff velocity and reduce erosion potential. Porous pavements, grassed and gravel driveways are examples.

Surface runoff. Precipitation excess that is not retained on the vegetation or surface depressions and is not lost by infiltration, and thereby is collected on the surface and runs off.

Terraces. An embankment, or combination of an embankment and channel, constructed across a slope to control erosion by reducing the slope and by diverting or storing surface runoff instead of permitting it to flow uninterrupted down the slope.

Toxic metals. Metals present in industrial, municipal, and urban runoff, including lead, copper, cadmium, zinc, mercury, nickel, and chromium, in quantities that are harmful to humans or aquatic biota.

Toxic substances. Any substances present in water, wastewater, or runoff that may kill fish or other aquatic life or could be harmful to public health. The substance may exhibit chronic toxicity or buildup in the food chain (biomagnification), or it may show acute toxicity and result in immediate death. Ammonia, acids, cyanides, phenols, toxic metals, and chlorinated hydrocarbons, among others, are examples of toxic substances.

Urban runoff. Surface runoff from an urban drainage area that reaches a stream or other body of water or a sewer.

Watershed. The area of land that contributes runoff of a given point in a drainage system.

Index